INVERSION METHODS
IN ATMOSPHERIC
REMOTE SOUNDING

Academic Press Rapid Manuscript Reproduction

INVERSION METHODS IN ATMOSPHERIC REMOTE SOUNDING

Edited by

Adarsh Deepak

Department of Physics and Geophysical Sciences
Old Dominion University, Norfolk, Virginia
and
Institute for Atmospheric Optics and
Remote Sensing
Hampton, Virginia

ACADEMIC PRESS
New York   San Francisco   London   1977
A Subsidiary of Harcourt Brace Jovanovich, Publishers

COPYRIGHT © 1977, BY ACADEMIC PRESS, INC.
ALL RIGHTS RESERVED.
NO PART OF THIS PUBLICATION MAY BE REPRODUCED OR
TRANSMITTED IN ANY FORM OR BY ANY MEANS, ELECTRONIC
OR MECHANICAL, INCLUDING PHOTOCOPY, RECORDING, OR ANY
INFORMATION STORAGE AND RETRIEVAL SYSTEM, WITHOUT
PERMISSION IN WRITING FROM THE PUBLISHER.

ACADEMIC PRESS, INC.
111 Fifth Avenue, New York, New York 10003

United Kingdom Edition published by
ACADEMIC PRESS, INC. (LONDON) LTD.
24/28 Oval Road, London NW1

Library of Congress Cataloging in Publication Data

International Interactive Workshop on Inversion Methods
   in Atmospheric Remove Sounding, 1st, Williamsburg,
   Va., 1976.
   Inversion methods in atmospheric remote sounding.

   1. Atmosphere, Upper—Remote sensing—Congresses.
2. Inverse problems (Differential equations)—Congresses.
3. Matrix inversion—Congresses.   I. Deepak,
Adarsh.   II. Title.

QC878.5.I56 1976   551.5'14'028   77-14939
ISBN 0-12-208450-0

PRINTED IN THE UNITED STATES OF AMERICA

CONTENTS

Participants   *ix*
Preface   *xiii*
Frontispiece   *xv*

RADIATIVE TRANSFER
Session 1—W. Irvine, Chairman

Hybrid Methods Are Helpful   1
    H. C. van de Hulst

Review of Radiative Transfer Methods in Scattering Atmospheres   21
    Jacqueline Lenoble

MATHEMATICAL THEORY OF INVERSION METHODS
Session 2—L. Kaplan, Chairman

Some Aspects of the Inversion Problem in Remote Sensing   41
    Sean S. Twomey

Generalization of the Relaxation Method for the Inverse Solution
of Nonlinear and Linear Transfer Equations   67
    Moustafa T. Chahine

Session 3—M. Chahine, Chairman

Statistical Principles of Inversion Theory   117
    Clive D. Rodgers

Inverse Solution of the Pseudoscalar Transfer Equation
through Nonlinear Matrix Inversion   139
    Jean I. F. King

INVERSION METHODS IN THERMAL, GASEOUS, AND AEROSOL ATMOSPHERES
Session 4—C. Mateer, Chairman

Backus-Gilbert Theory and Its Application to Retrieval of Ozone and Temperature Profiles   155
Barney J. Conrath

Inversion of Infrared Limb Emission Measurements for Temperature and Trace Gas Concentrations   195
John C. Gille and Paul L. Bailey

Session 5—A. E. S. Green, Chairman

Inversion of Scattered Radiance Horizon Profiles for Gaseous Concentrations and Aerosol Parameters   217
Harvey L. Malchow and Cynthia K. Whitney

Inversion of Solar Aureole Measurements for Determining Aerosol Characteristics   265
Adarsh Deepak

Analytic Model Approach to the Inversion of Scattering Data   297
Alex E. S. Green and Kenneth F. Klenk

Open Discussion—I   323

Session 6—S. Twomey, Chairman

Comparison of Linear Inversion Methods by Examination of the Duality Between Iterative and Inverse Matrix Methods   325
Henry E. Fleming

Inversion of Passive Microwave Remote Sensing Data from Satellites   361
David H. Staelin

Session 7—D. Staelin, Chairman

Application of Statistical Inversion to Ground-Based Microwave Remote Sensing of Temperature and Water Vapor Profiles   395
E. R. Westwater and M. T. Decker

Inversion Methods in Temperature and Aerosol Remote Sounding:
Their Commonality and Differences, and Some Unexplored
Approaches   429
   Alain L. Fymat

Session 8—M. P. McCormick, Chairman

Application of Modified Twomey Techniques to Invert Lidar
Angular Scatter and Solar Extinction Data for Determining Aerosol
Size Distributions   469
   Benjamin M. Herman

The Inversion of Stratospheric Aerosol and Ozone Vertical
Profiles from Spacecraft Solar Extinction Measurements   505
   William P. Chu

Inversion of Solar Extinction Data from the Apollo-Soyuz Test
Project Stratospheric Aerosol Measurement (ASTP/SAM) Experiment   529
   Theodore J. Pepin

Session 9—H. C. van de Hulst, Chairman

Effective Aerosol Optical Parameters from Polarimeter Measurements   555
   Jacob G. Kuriyan

Experience with the Inversion of Nimbus 4 BUV Measurements
to Retrieve the Ozone Profile   577
   Carlton L. Mateer

Temperature Sensing: The Direct Road to Information   599
   Lewis D. Kaplan

Open Discussion—II   613

Index   *617*

PARTICIPANTS

Numbers in parentheses indicate pages on which authors' contributions begin.

J. Abele, Forschungsinstitut für Optik, D74 Tübingen 1, Schloss Kressbach, West Germany

Richard R. Adams, MS 234, NASA-Langley Research Center, Hampton, Virginia 23665

Paul L. Bailey (195), National Center for Atmospheric Research, P.O. Box 3000, Boulder, Colorado 80307

Bruce R. Barkstrom, George Washington University, MS 169, NASA-Langley Research Center, Hampton, Virginia 23665

Jacob Becher, Department of Physics and Geophysical Sciences, Old Dominion University, P.O. Box 6173, Norfolk, Virginia 23508

Sherwin M. Beck, MS 401A, NASA-Langley Research Center, Hampton, Virginia 23665

Michael Box, Old Dominion University, MS 234, NASA-Langley Research Center, Hampton, Virginia 23665

Kurt Bullrich, Institut für Meteorologie, Johannes Gutenberg University, D-6500 Mainz, West Germany

Todd A. Cerni, University of Wyoming, P.O. Box 3095, University Station, Laramie, Wyoming 82071

Moustafa T. Chahine (67), JPL, California Institute of Technology, 4800 Oak Grove Drive, MS 183-301, Pasadena, California 91103

William P. Chu (505), MS 475, NASA-Langley Research Center, Hampton, Virginia 23665

Mark Clayson, Department of Astro-Geophysics, University of Colorado, Boulder, Colorado 80303

Barney J. Conrath (155), Code 622, Goddard Space Flight Center, Greenbelt, Maryland 20771

M. T. Decker (395), NOAA/Wave Propagation Laboratory, R45Y4, Boulder, Colorado 80302

Adarsh Deepak (265), Old Dominion University, Norfolk, and Institute for Atmospheric Optics and Remote Sensing (IFAORS), MS 234, NASA-Langley Research Center, Hampton, Virginia 23665

Diran Deirmendjian, The Rand Corporation, 1700 Main Street, Santa Monica, California 90406

John J. DeLuisi, NOAA/ERL-ARL-GMCC, RF 3292, Boulder, Colorado 80302

S. Roland Drayson, University of Michigan, Department of Atmospheric and Oceanic Science, Ann Arbor, Michigan 48109

PARTICIPANTS

Henry E. Fleming (325), NOAA/NESS, FB #4, Room 0262, Washington, D.C. 20233

Alexander H. Fluellen, Atlanta University Center, Box 320, Clark College, Atlanta, Georgia 30314

Robert S. Fraser, NASA, Code 911, Goddard Space Flight Center, Greenbelt, Maryland 20771

Alain L. Fymat (429), JPL, California Institute of Technology, 4800 Oak Grove Drive, MS 183B-365, Pasadena, California 91103

John C. Gille (195), National Center for Atmospheric Research, P.O. Box 3000, Boulder, Colorado 80307

Aaron Goldman, Department of Physics, University of Denver, Denver, Colorado 80210

Richard B. Gomez, Atmospheric Science Laboratory, DRSEL-BL-MS, White Sands Missile Range, New Mexico 88002

Robert Goulard, The George Washington University, Washington, D.C. 20052

Alex E. S. Green (297), Department of Physics and Astronomy, University of Florida, Gainesville, Florida 32601

F. S. Harris, Jr., Old Dominion University, MS 475, NASA-Langley Research Center, Hampton, Virginia 23665

Benjamin M. Herman (469), University of Arizona, Institute of Atmospheric Physics, Tucson, Arizona 85721

Arlon J. Hunt, Lawrence Berkeley Laboratory, Building 50-205C, Berkeley, California 94720

William M. Irvine, University of Massachusetts, Tower B, Astronomy, Amherst, Massachusetts 01002

Wolfgang Jessen, Forschungsinstitut für Optik, D-7400 Tübingen, Schloss Kressbach, West Germany

Lewis D. Kaplan (599), University of Chicago, Department of Geophysical Sciences, Chicago, Illinois 60637

Jean I. F. King (139), OPI, Air Force Geophysics Laboratory, Hanscom AFB, Massachusetts 01731

Kenneth F. Klenk (297), Department of Physics and Astronomy, University of Florida, Gainesville, Florida 32601

Jacob G. Kuriyan (555), UCLA, Department of Meteorology, 405 Hilgard Avenue, Los Angeles, California 90024

J. Donald Lawrence, Jr., MS 401A, NASA-Langley Research Center, Hampton, Virginia 23665

Jacqueline Lenoble (21), Laboratoire d'Optique Atmos., Univ. de Lille I, B.P. 36, 59650 Villeneuve d'Ascq., France

Julius London, Department of Astro-Geophysics, University of Colorado, Boulder, Colorado 80303

Leonard R. McMaster, MS 234, NASA-Langley Research Center, Hampton, Virginia 23665

PARTICIPANTS

M. Patrick McCormick, MS 475, NASA-Langley Research Center, Hampton, Virginia 23665

Michael McElroy, Pierce Hall, Harvard University, Cambridge, Massachusetts 02138

Harvey L. Malchow (217), C. S. Draper Laboratory, 555 Technology Square, Cambridge, Massachusetts 02139

Carlton L. Mateer (577), Atmospheric Environment Service, 4905 Dufferin Street, Downsview, Ontario M3H 5T4, Canada

Jae H. Park, College of William and Mary, MS 401A, NASA-Langley Research Center, Hampton, Virginia 23665

Edward M. Patterson, National Center for Atmospheric Research, P.O. Box 3000, Boulder, Colorado 80307

William A. Pearce, EG&G, Washington Analytical Service Center, 6801 Kenilworth Avenue, Riverdale, Maryland 20804

Theodore J. Pepin (529), University of Wyoming, Department of Physics and Astronomy, P.O. Box 3095, University Station, Laramie, Wyoming 82071

Walter G. Planet, NOAA/NESS, FB #4, Suitland, Maryland 20233

Heinrich Quenzel, Meteorologisches Institüt der Universität München, 8 München 2. Theresienstrasse 37, West Germany

John A. Reagan, Eng. Bldg. #20, University of Arizona, Department of Electrical Engineering, Tucson, Arizona 85721

Ruth A. Reck, Physics Department, Technology Center, General Motors Research Laboratories, Warren, Michigan 48090

Ellis Remsberg, MS 401A, NASA-Langley Research Center, AESD, Hampton, Virginia 23665

Clive D. Rodgers (117), Oxford University, Clarendon, Lab OX1, 3 PU, England

James M. Russell III, MS 401A, NASA-Langley Research Center, Hampton, Virginia 23665

Philip B. Russell, Stanford Research Institute, 333 Ravenswood Avenue, Menlo Park, California 94025

Eric P. Shettle, AF Geophysics Laboratory, L. G. Hanscom AFB, Bedford, Massachusetts 01731

Walter L. Snow, MS 235A, NASA-Langley Research Center, Hampton, Virginia 23665

David H. Staelin (361), Massachusetts Institute of Technology, Room 26-341, Cambridge, Massachusetts 02139

Larry B. Stotts, Naval Electrical Laboratory Control C-2500, 271 Catalina Boulevard, San Diego, California 92152

Otto N. Strand, NOAA/Wave Propagation Laboratory, Environmental Research Laboratories, R45x7, Boulder, Colorado 80302

Joel Susskind, Goddard Institute for Space Studies, 2880 Broadway, New York, New York 10025

Thomas J. Swissler, Systems and Applied Sciences, MS 475, NASA-Langley Research Center, Hampton, Virginia 23665

PARTICIPANTS

Didier Tanre, Laboratoire d'Optique Atmosphérique, D.U.S.V.A., B.P. 36, Villeneuve D'Ascq., 59650 France

Morris Tepper, Code ERD, NASA Headquarters, Washington, D.C. 20546

Robert W. L. Thomas, EG&G Washington Analytical Service Center, 6801 Kenilworth Avenue, Riverdale, Maryland 20840

Robert E. Turner, Environmental Research Institute of Michigan, P.O. Box 618, Ann Arbor, Michigan 48107

Jerold T. Twitty, Old Dominion University, MS 494, NASA-Langley Research Center, Hampton, Virginia 23665

Sean A. Twomey (41), University of Arizona, Institute of Atmospheric Physics, Tucson, Arizona 85721

Hendrick C. van de Hulst (1), Leiden University, Huygens Laboratory, 78 Wassenaarse Weg, Leiden 240 Netherlands

David Wark, NOAA/NESS, FB #4, Washington, D.C. 20233

Roger L. Weichel, Lawrence Livermore Laboratory 142, P.O. Box 808, University of California, Livermore, California 94550

James A. Weinman, University of Wisconsin, 1225 W. Dayton Street, Madison, Wisconsin 53706

E. R. Westwater (395), NOAA/Wave Propagation Laboratory, R45x4, Boulder, Colorado 80302

Cynthia K. Whitney (217), C. S. Draper Laboratory, 555 Technology Square, Cambridge, Massachusetts 02139

Harold W. Yates, National Environmental Satellite Service, NOAA, Suitland, Maryland 20233

PREFACE

This volume contains the technical proceedings of the First International Interactive Workshop on Inversion Methods in Atmospheric Remote Sounding, held in Williamsburg, Virginia, December 15-17, 1976. Seventy-three invited scientists from seven countries, representing universities, research laboratories, and U.S. Government agencies, participated in the workshop. The purpose of the workshop was to provide an interdisciplinary forum to review and assess the state of the art in inversion methods available for retrieving information about the atmosphere from remotely sensed data.

Twenty-one invited papers covered the mathematical theory of inversion methods as well as the application of these methods to the remote sounding of atmospheric temperature, relative humidity, and gaseous and aerosol constituents. The emphasis was on the assumptions, methodology, resolution, stability, accuracy, and future efforts needed in the various inversion methods. Also included are invited papers on the direct radiative transfer methods and results relevant to the inversion problem. The latter were presented in a special session on radiative transfer methods, held jointly with the Optical Society of America Topical Meeting on Atmospheric Aerosols, which preceded the workshop. One of the major workshop objectives was to enable researchers in different areas of atmospheric remote sounding to compare and optimize the utilization of these inversion procedures in their respective remote sounding techniques. Ample time was allowed for discussions following each paper and in two open discussion sessions. This fulfilled an important objective of the workshop. Discussions presented were recorded and the transcripts postedited. Each discussant edited his/her portion of the statements with the aim of improving its clarity without changing its substance.

Since NASA is involved in developing several remote sensing experiments designed to monitor the atmospheric constituents and properties from aboard space platforms, the editor suggested to M. P. McCormick, Langley Research Center, that organization of an interactive workshop dealing with the mathematical aspects of the inversion methods would greatly benefit all researchers concerned with inversion and radiative transfer methods. He, along with J. D. Lawrence, Jr., Langley Research Center, and M. Tepper, NASA Headquarters, concurred and supported the idea with the result that I undertook the assignment of organizing such a workshop with the goal of making the proceedings of the workshop readily available to the scientific community.

To ensure proper representation of major disciplines involved, a Workshop Program Committee, composed of A. Deepak (Chairman), Old Dominion University; M. P. McCormick (Associate Chairman), NASA Langley Research Center; B. M. Herman, University of Arizona; J. D. Lawrence, Jr., NASA Langley Research Center; and M. Tepper, NASA Headquarters, was set up. The committee was ably assisted in this endeavor by the following program consultants: M. T. Chahine, Jet Propulsion Laboratory; B. J. Conrath, NASA Goddard Space Flight Center; A. L. Fymat, Jet Propulsion Laboratory; J. Russell III, NASA Langley Research Center; and E. Westwater, NOAA/Environmental Research Laboratory.

Dr. Tepper opened the workshop, stressing the significance of inversion problems to NASA and its programs involved in the monitoring of atmospheric environments of the Earth and other planets. The challenges inherent in the inversion problem were perhaps best characterized by his analogy that the problem of the inversion method was like that of unscrambling an egg, wherein one investigates the scrambled egg to determine what it was like originally.

The editor wishes to acknowledge the enthusiastic support and cooperation of the participants, the members of the organizing committee, the program consultants, the session chairmen, and the speakers in making the workshop a very stimulating and valuable experience for everyone. Special thanks are due M. P. McCormick, whose wholehearted cooperation and active support as Associate Chairman assured the success of the workshop. Commendations are due the Science and Technical Information Program Division and especially the Technical Editing Branch, for their cooperation and high quality of workmanship in publishing this volume. Last but not least, it is a pleasure to thank and highly commend the superb job done by Mrs. M. Sue Crotts both in helping with the organization of the workshop and with the excellent quality of typing of the manuscripts.

Behind every successful remote sensing technique is at least one reliable inversion method. I hope this volume will be a lasting contribution to the field of inversion methods.

Workshop Speakers and Chairmen: (Left to right) Front row: M. P. McCormick (Associate Chairman), Langley Research Center; S. Twomey, U. Arizona; L. Kaplan, U. Chicago; M. Chahine, JPL/Cal. Tech; H. van de Hulst, U. Leiden, Netherlands; C. Whitney, C. S. Draper Lab; E. Westwater, NOAA/WPL; D. Staelin, MIT; B. Conrath, Goddard SFC; J. Kuriyan, UCLA; J. Gille, NCAR; W. Chu, Old Dominion U.; A. Deepak (Chairman), Old Dominion U. Second row: J. Lenoble, U. de Lille, France; B. Herman, U. Arizona; A. Fymat, JPL/Cal Tech; J. King, AFGL; A. Green, U. Florida; H. Malchow, C. S. Draper Lab; W. Irvine, U. Massachusetts; H. Fleming, NOAA/NESS; C. Rodgers, U. Oxford, UK; C. Mateer, Atmos. Environ. Serv., Canada; T. Pepin, U. Wyoming.

HYBRID METHODS ARE HELPFUL

H. C. van de Hulst
Leiden University

A basically simple problem like multiple scattering in a plane layer often permits the convenient use of different methods joined together. Sample numerical results to illustrate this point refer to X- and Y-functions, asymptotic fitting, the small-loss approximations, polarization in high orders, and photon path distribution.

I. INTRODUCTION

Methods to solve problems in radiative transfer and in multiple scattering exist in such a wide variety that I shall not attempt another review. In any practical problem, the method must be chosen on the basis of expediency, and this, in turn, depends on many factors, such as: range of variables; desired accuracy of results; occasional or frequent computations needed; cost, available funds; and experience and taste. I emphasize in this paper the fact that in many situations a hybrid approach containing elements from different methods, though not "elegant", is the most practical. A normal rose or fruit tree consists of different varieties skillfully grown together because the desired properties of roots and fruits (or flowers) are not met in a single variety.

Before illustrating this point with a number of examples taken from Ref. 1, I wish to make a general remark. We all have learned to respect the power of mathematics. Solving a problem in mathematical physics often is like going somewhere by train. The mathematics is like the train: we enter at a station called equation and we get

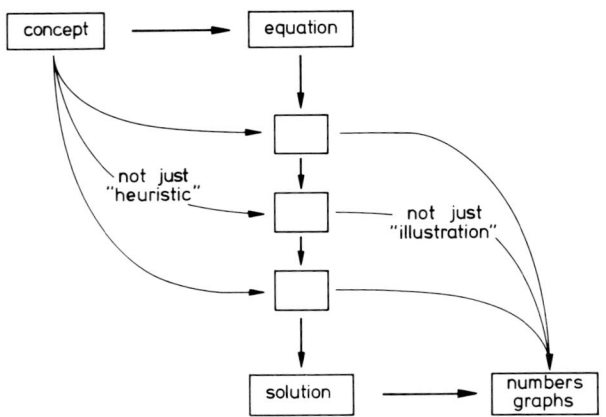

Fig. 1. A schematic diagram illustrating the advisability of accessing from physical concepts to intermediate results, whose interpretation may be as important as that of final numbers.

off at a station called solution (see Fig. 1). Once inside the train, we can relax and look out of the window, for nothing much can go wrong. In contrast, the roads from our home to the station and from the last station to our destination may be time-consuming, uncomfortable, or even hazardous. My point is that in many problems the train is suburban: it stops at certain intermediate stations. Boarding the train at one of those stops may be quite as safe, respectable, and economic as entering at the initial station which in our topic is called equation of transfer. At any rate, it is worth a try to find out which is simplest. Likewise, it often pays to evaluate and physically interpret certain intermediate results which is the same as getting off before the end station.

The philosophical message is that combining physics with mathematics makes a hybrid method anyhow. Therefore, we might as well experiment a little to find the most convenient connection.

II. THE X- AND Y-FUNCTIONS

Consider any given landscape (Fig. 2) with an isotropic light source placed at a point P and no other source of illumination. Looking at this landscape from a distance, say from the direction Q, we see a blob of light in which the source itself (dimmed or not) may still be discernible. We define the *gain* (from P to Q and conversely) as the intensity that reaches Q from source plus illuminated landscape divided by the intensity that would reach Q from the bare source placed at the same distance.

This definition is given in preparation of a discussion of the well-known X- and Y-functions for isotropic scattering. These functions were introduced by Ambartsumian in the early forties and extensively studied by Chandrasekhar about 1945, who defined them as solutions of certain simultaneous nonlinear integral equations.

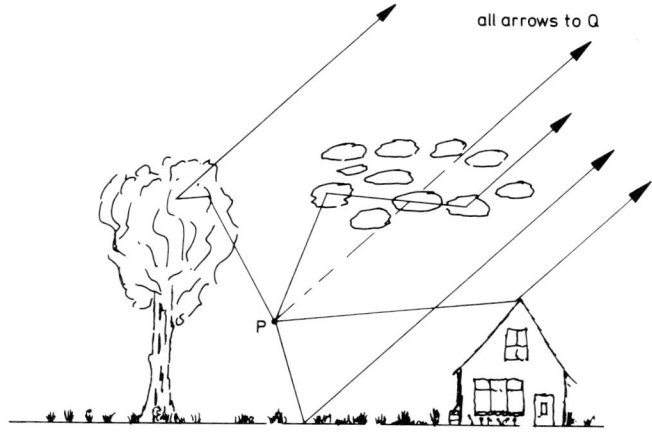

Fig. 2. *A schematic sketch of a "landscape" with an isotropic source at P for illustrating the concept of gain between point P and a direction Q.*

Far simpler, in the gain definition just given, we may take as the "landscape" a homogeneous slab of isotropically scattering particles with optical thickness b, the source P just outside the slab, and the direction Q subtending an angle θ with the normal. The X-function then is the gain with the source seen in front of the slab and the Y-function is the gain with the source (dimmed) seen through the slab. This is all there is to it: no problems of existence or uniqueness if we board the train at this station. Both functions depend on three variables: b, $\mu = \cos \theta$, and a = the albedo for single scattering.

I was quite pleased when in 1947 I rediscovered these definitions, from which Ambartsumian had started, and found that certain properties of these functions can be far more easily derived from these physical definitions. Since that time, I do not hesitate to use the two approaches mixed.

Figure 3 shows a selection of values of these functions. The Y-function usually is less than unity because the blob of scattered and multiply-scattered light does not fully compensate the dimming of the direct source. The X-function always is greater than unity because it includes a term unity arising from the unobstructed light from the source. The X-function for a semi-infinite atmosphere (b = ∞) usually is called the H-function. If b = ∞ and a = 1, all radiation incident on the atmosphere is returned as diffusely reflected light after many scattering events. If the landscape were a mirror reflecting all incident radiation, we would see the source double from any direction, which would mean $X(\mu) = 2$. The diffuse reflection leads to the same average value of 2 for $H(\mu)$, but the distribution with μ is different, ranging from $H(0) = 1.0$ to $H(1) = 2.908$.

As a counter example, in which the physical picture is of little use, I mention the extension of the H-function to arguments outside the domain $\mu = 0$ to 1. Such an extension is needed in a variety of problems. It is then convenient to plot the inverse

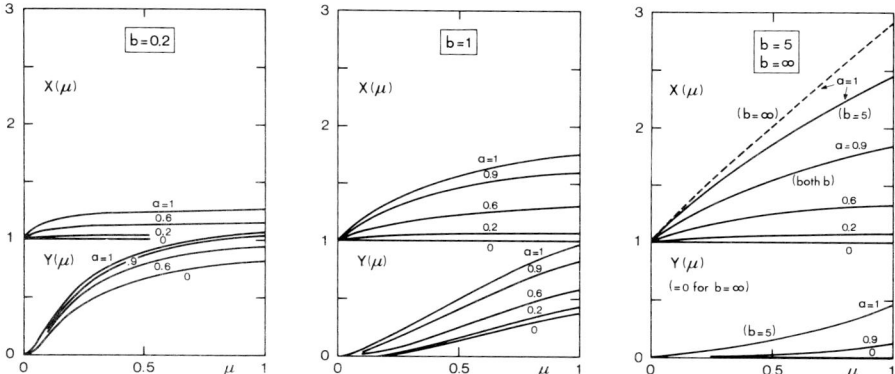

Fig. 3. Reminder of the dependence of the X- and Y-functions for isotropic scattering on the variables b, a, and μ. The X-function for b = ∞ is called H-function.

function $\{H(\mu)\}^{-1}$ against the inverse argument μ^{-1}. Figure 4 gives an example which happens to refer to anisotropic scattering but this makes no essential difference. We see that the graph continues to curve down for $\mu^{-1} < 1$ and reaches 0 at $\mu^{-1} = -k$, where k is the diffusion exponent, i.e., the value for which a self-consistent solution to the transfer equation in an unbounded medium exists in which the dependence on optical depth τ is given by the factor $\exp(\pm k\tau)$. The values and slopes at $\mu^{-1} = 0$ and $\mu^{-1} = -k$ occur in several standard problems.

We have taken these illustrations from isotropic scattering and one example from very simple anisotropic scattering. Phase functions of arbitrary form, or phase matrices with polarization, require a more elaborate set of formulas. Yet, the situation remains basically the same: carefully preparing the access at intermediate stations usually pays off in clarity or in speed of computation.

A final remark on the X- and Y-functions for single-scattering patterns of arbitrary form is that these same functions appear in

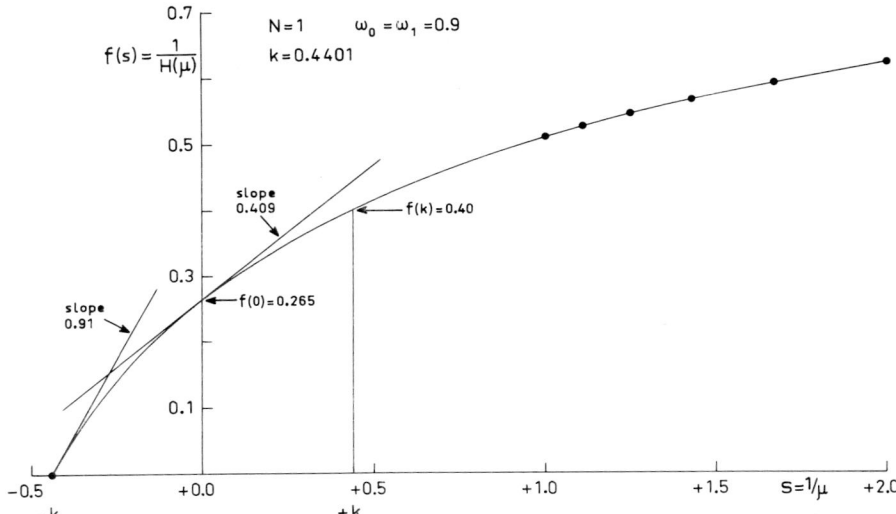

Fig. 4. A plot of $1/H(\mu)$ as a function of $1/\mu$ depicting the best way for visualizing the behavior of the H-function beyond the usual domain $0 \leq \mu \leq 1$ for linearly anisotropic scattering with $\omega_0 = \omega_1 = 0.9$.

conceptually quite different methods. The sketch in Fig. 5, which we shall not explain in detail, shows four important methods of solving radiation transfer problems. In the second method, labeled "invariant embedding," the X- and Y-functions are introduced to describe the effect of a narrow layer added to one or the other side of a slab. The method of singular eigenfunction expansion works from an entirely different concept, in which the complete set of eigenfunctions for the unbounded medium is first established. The proper coefficients of each that match the boundary conditions are then found by applying orthogonality relations and it is in the course of establishing the half-range orthogonality relations that the X- and Y-functions have to be introduced.

HYBRID METHODS ARE HELPFUL

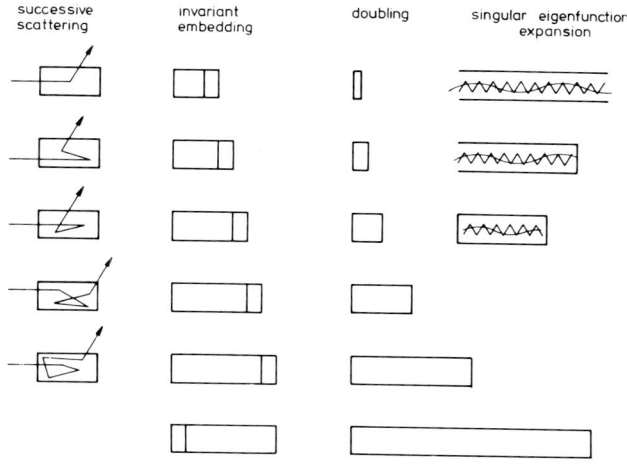

Fig. 5. *Schematic representation of some commonly used methods to solve problems of multiple scattering or radiative transfer.*

III. ASYMPTOTIC FITTING

Before describing this method, I wish to convey by means of Fig. 6 an impression of the range of variables in which the simplest approximations suffice. I refer to the legend for details. The diagram shows that, if the scattering is conservative, or nearly so, a wide range of four decades in the optical thickness b exists, in which we cannot in practice say: the layer is very thin, or it is infinitely thick. This is annoying because the convergence of almost any method is small for large b. This point is illustrated for the successive scattering method in Fig. 7.

The incidence is normal in this example and the scattering is isotropic, so that the source function for first-order scattering J_1 is $1/4 \, e^{-\tau}$ in both examples. In the right-hand side example, the layer thickness is 1. At each successive scattering there are losses at both sides and the net effect is that after some five scattering events, the distribution has become symmetric and drops

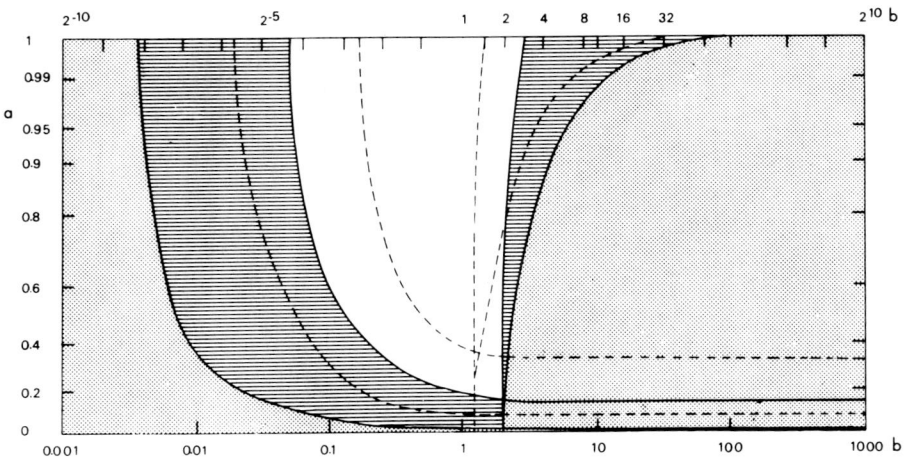

Fig. 6. Representation of the range of the variables b and a in which the simplest approximations suffice. The very simplest approximations are valid with an accuracy better than 1% in the shaded areas in this diagram: left single scattering, right the equations for a semi-infinite atmosphere. The next approximations, shown by hatched areas, are left single plus double scattering and right the thick-layer asymptotic formulae, again to a 1% accuracy. The corresponding 5% limits are shown by dotted curves. Scales are linear in log b and $\sqrt{1-a}$ and the quantity treated is the plane albedo for normal incidence with isotropic scattering.

with every further scattering by a constant factor 0.619. The convergence, then, is as a geometric series with this ratio. However, on the left-hand side example, the thickness is $b \geq 10$. Here, the convergence seems rapid near $\tau = 0$, but is not at all visible yet at $\tau \geq 3$. Eventually, the convergence will be as a geometric series with ratio 0.976 (for $b = 10$), or 0.993 (for $b = 20$), or as a series with terms proportional to $n^{-3/2}$ (for $b = \infty$). In any case, this convergence is too slow to make the method of successive scattering attractive.

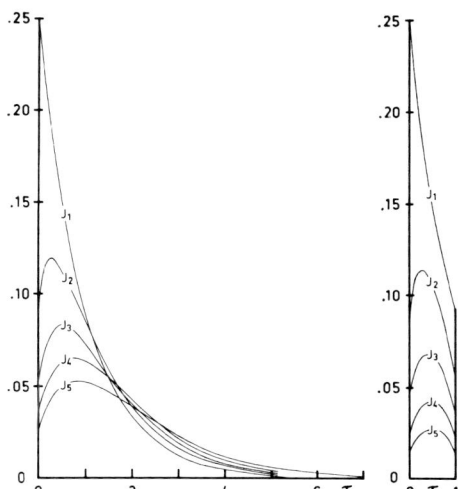

Fig. 7. Demonstration of the convergence of the method of successive scattering by plots of the source function against optical depth τ in successive orders. The example refers to perpendicular incidence on a layer with isotropic scattering. Left: total depth ≥ 10, slow convergence. Right: total depth 1, rapid convergence.

In this domain, we have elaborately used the doubling method. It is clear that by the doubling method we can in ten steps bridge the range of optical thickness b = 1/32, 1/16, 1/8, ..., 8, 16, 32, but that it is silly to try to carry this process to b = ∞, which we can never reach. Instead, we wish to use the known asymptotic forms of the reflection and transmission functions for sufficiently large b. The grafting of the different approach (asymptotic theory) onto the numerical computation (doubling) makes a hybrid computational method and we have called this method asymptotic fitting. It uses the doubling results in exactly the same way as we would use measured data to find the constants in an equation of known form.

The approach to such a known form, prescribed by asymptotic theory is well illustrated by Fig. 8. Again the example is simple, namely isotropic scattering, and the ordinate is the source function. It is seen that not quite at b = 4, but certainly at b = 8 and higher there is a range of τ in which the graph is straight. In this "diffusion domain" the source function goes as exp(-kτ) as in an unbounded medium. The diffusion domain is the range of τ far enough from the top side, where injection takes place

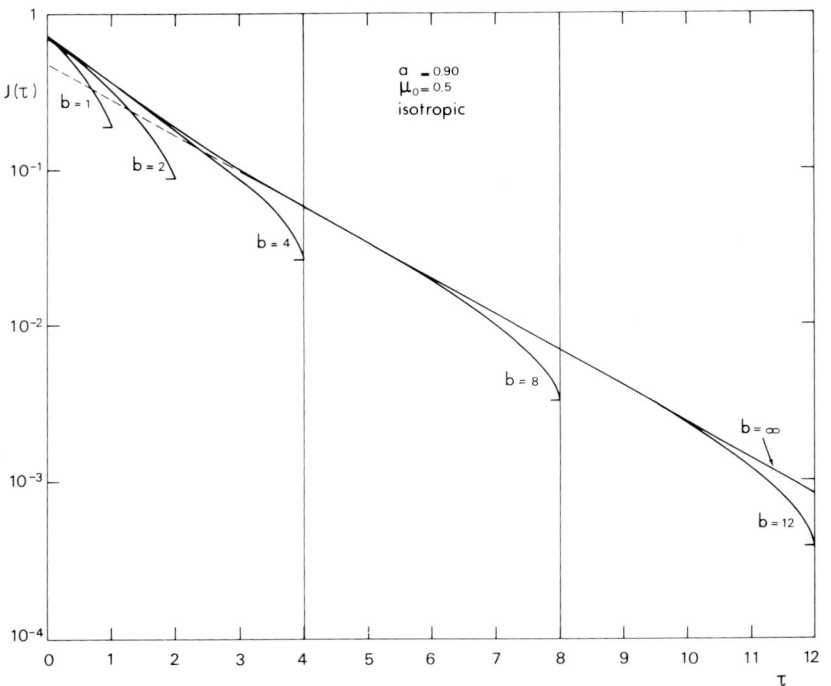

Fig. 8. The source function is plotted against optical depth in six situations which differ only by the assumed thickness b of the layer and in which the single scattering albedo a = 0.90 and the cosine of the angle of incidence μ_o are kept constant. From b = 8 up the curves show a straight middle portion (diffusion domain) and the asymptotic laws are valid.

($\tau = 0$ to 3), and from the bottom side, where the escape of radiation becomes noticeable ($\tau = b - 2$ to b). Using this knowledge and taking the results of three successive steps, say $b = 8$, 16 and 32, yields all desired functions and constants for a semi-infinite medium.

IV. THE CORNER OF SMALL LOSSES

In the problem of reflection and transmission by a homogeneous slab of scattering particles, I like to display the results in a diagram of a, the albedo for single scattering, against b, the optical thickness of the slab. Such a representation is chosen in Fig. 9. This figure may be regarded as a companion to Fig. 6, because it shows in a different way (roughly delineated by the dotted curves) in which area we may say that the result is as for $b = \infty$ and in which area we may say that the result is as for $a = 1$.

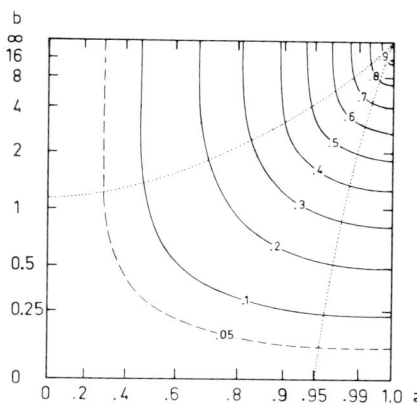

Fig. 9. The full domain of albedo a and layer thickness b is displayed on linear scales of $\sqrt{1 - a}$ and $(1 + b)^{-1}$. The curves show the values of the reflected flux portion (plane albedo) for normal incidence with isotropic scattering.

At one corner, defined by $a = 1$, $b = \infty$, the reflection is complete. Small losses by escape from the bottom of the slab occur if b is not quite ∞. Small losses by absorption in the atmosphere occur if $a < 1$. These losses are not additive and their combined behavior has posed nasty problems in many papers, numerical and analytical alike. Yet, the situation is simple if we refer to asymptotic theory. Both of these losses have the same dependence on angle of incidence and escape because they both occur in deep layers. They add in a way which is universal, except for scale factors. This is shown in Fig. 10 which is an enlarged portion of the top right-hand corner of Fig. 9. The dotted curves, which have known tangents in the corner domain, now indicate the exact loci where the losses by escape from the bottom (i.e., into the black ground) and the losses by imperfect scattering in the atmosphere have a fixed ratio.

V. POLARIZATION IN HIGH-ORDER SCATTERING

The fine visual observations of polarization of the planet Venus at various phase angles published in Lyot's thesis in 1929 have outlasted 40 years before the first tentative interpretation could be replaced by a more definite one. Lyot conjectured that polarization might be completely absent in all but first-order scattering. This guess was not correct. Since 1970, the accuracy of both the measurements and the calculations have been greatly improved. Most authors now agree that the polarization curves of Venus measured at different wavelengths present convincing evidence of concentrated sulfuric acid as the main constituent of the droplets.

Let us return to the basic problem: how much polarization can be present in the light arising from double and multiple scattering? I have worked out an extreme practicing example based on Rayleigh scattering. Radiation is assumed to fall on a semi-infinite atmosphere under $60°$ from the normal and we seek to compute the intensity

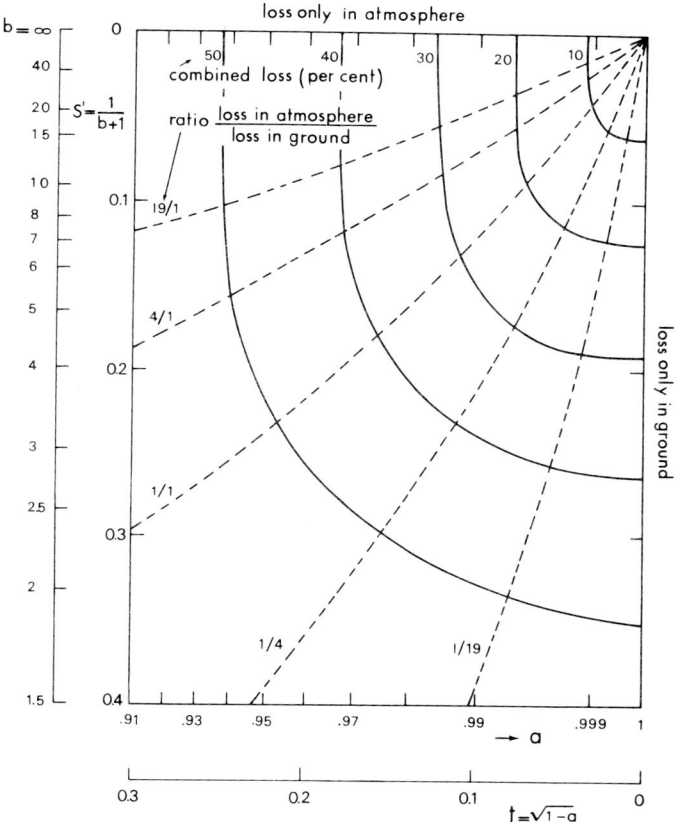

Fig. 10. Enlarged portion of a corner of Fig. 9. Heavy curves give the combined loss = nonreflected fraction of incident flux and dotted curves show the ratio of the two kinds of loss. The curves near the loss-less corner have a universal form independent of the phase function.

and polarization reflected back to the source (as for part of a planet in opposition). The degree of polarization in this example is 0% in the first order, 9% in the second order, and surprisingly goes up even further to 13% and 14% in the third and fourth orders before it settles to 2.2% in very high orders.

If now we add a factor a^n to the nth order and take the sum, we obtain the intensity diffusely reflected back in each polarization with a = single scattering albedo (Fig. 11). The effect just mentioned then shows up as a maximum polarization of 6.1% for a = 0.95. The simplest way to collect these results was to combine data obtained by very different methods.

VI. PHOTON PATH DISTRIBUTIONS

The understanding of many problems, for instance, the formation of planetary absorption lines, is greatly aided by a clear knowledge of the photon path distributions. It is well known that such distributions may be obtained by an inverse Laplace transform from the dependence of the reflection function on a (albedo for single scattering). This knowledge may be put to use in various ways: the inverse Laplace transform may be applied to any form in which the reflection function is known--analytic, numerical, or asymptotic. A systematic exploration of this possibility has made it possible to combine smoothly the results for very low orders, where *ad hoc* calculations are fast, with those of high orders, where an asymptotic approach is more appropriate.

I shall show two examples, both referring to finite layers with isotropic scattering. For very low orders, it is possible to derive the explicit form of the photon path distribution. This is done for n = 1 and n = 2 in Fig. 12 (with b = 1, $\mu = \mu_o = 1$) on the basis of computations made by Irvine. It is much easier to derive only the average path $<\lambda_n>$ and the variation σ_n^2 from the average path. A smooth distribution curve with a convenient form and with the correct average and variation then is

$$p_n(\lambda) = \frac{m}{\Gamma(k)} (m\lambda)^{k-1} e^{-m\lambda}$$

where m and k must be solved from

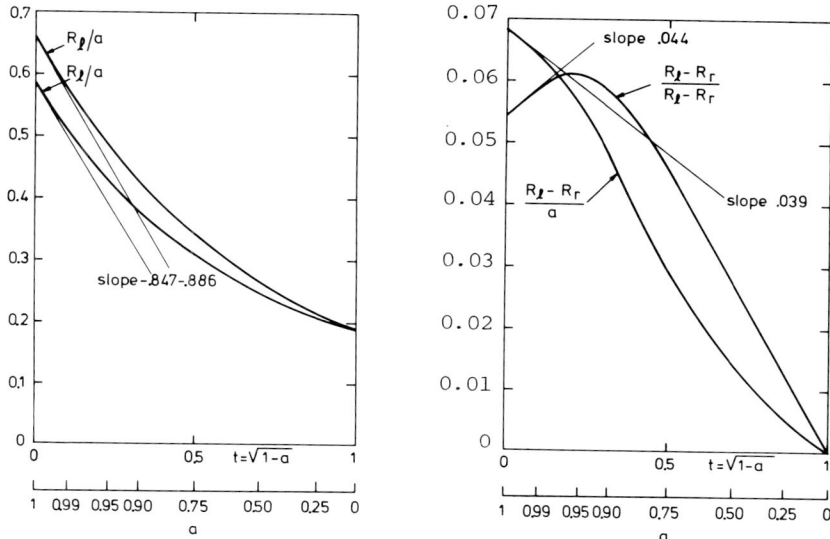

Fig. 11. Construction of the curve of polarization against single scattering albedo a for reflection into the precise backward direction against a semi-infinite Rayleigh atmosphere viewed and illuminated under $60°$ with the normal.

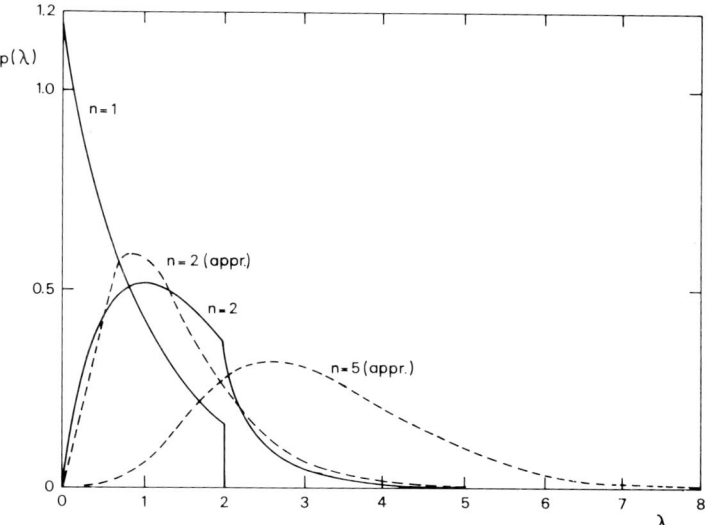

Fig. 12. Average photon path length per scattering event for reflecting against or transmission by a finite isotropically scattering layer with optical thickness b.

$$k = m\langle\lambda_n\rangle = m^2\sigma_n^2$$

and $\Gamma(k)$ is the gamma function.

I have plotted, in Fig. 12, as dotted curves these distributions for $n = 2$ and $n = 5$. The fact that the dotted curve for $n = 2$ comes already close to the exact curve means that at higher n for most practical problems we do not have to worry about the exact curves but can use the approximation with full confidence.

The larger the total number of scattering events, the larger the combined optical path. In an unbounded medium, we simply have an average path length per scattering event equal to 1. In order to show the transition from very thin to very thick layers, I have therefore plotted in Fig. 13 the average path length divided by n. Again, exact results for $n = 1$ and $n = 2$ have been combined with exact asymptotic results (value and slope) near $n = \infty$. Note that toward large values of n, the photon again "forgets" from which side it came so that the curves for transmission and reflection converge toward the same value and become tangent. For very thin layers ($b \leq 0.1$), it is immaterial from the start whether we consider reflection or transmission.

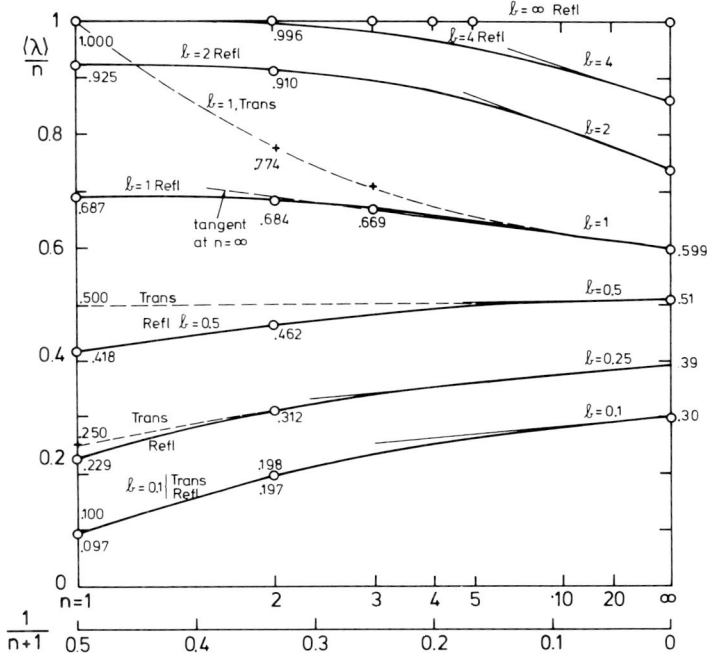

Fig. 13. Path-length distribution of radiation diffusely reflected back in vertical direction after n scattering events from a slab of isotropically scattering particles of optical thickness 1 exposed to vertical illumination. The exact forms, known for small n, rapidly approach an asymptotic theory.

SYMBOLS

a	single scattering albedo
b	optical thickness
H	X-function for a semi-infinite atmosphere ($b = \infty$)
J	source function for first-order scattering
k	$\pm 1/\mu$
$p_n(\lambda)$	photon path distribution function
R_ℓ, R_r	intensity for the polarized components (subscript ℓ is for parallel component; and r is for perpendicular component)

s'	$1/(b + a)$
X	X-function for isotropic scattering
Y	Y-function for isotropic scattering
$\Gamma(k)$	gamma function
θ	zenith angle
λ	wavelength
$<\lambda_n>$	average path
μ	$\cos \theta$
σ_n^2	variation from the average path
τ	optical depth

REFERENCE

1. H. C. van de Hulst, "Multiple Light Scattering in Cloud Layers," Academic Press, New York (to be published in late 1978).

DISCUSSIONS

Green: I wondered if your slide that you did not talk about had something to do with multiple scattering and path lengths?

van de Hulst: Yes, this slide[1] shows the average path length per scattering. It is simplest to think that you have done a Monte Carlo calculation and separate the paths out by the number of scatterings. Those which have two successive scatterings, you divide by two. And those which have five successive scatterings you divide the total path length by five. Then that comes at infinity to a certain value which is not too difficult to calculate. These have been combined in the graph with definite values shown as dots. When the thickness is not very large, in transmission and in reflection, you get approximately the same path lengths. And also, if you go to the very large n, the light doesn't know anymore from which side it came, so the transmission and reflection curves again coincide.

---

[1] See Fig. 13 in this paper.

REVIEW OF RADIATIVE TRANSFER METHODS IN

SCATTERING ATMOSPHERES

Jacqueline Lenoble
Université des Sciences et Techniques de Lille

The problem of radiative transfer in a scattering plane-parallel atmosphere is discussed, considering the exact analytical, the computational and the approximate methods. Some results of numerical comparisons are given. Finally, the difficulties of realistic atmospheric models are emphasized.

I. INTRODUCTION

The problem of radiative transfer in scattering atmospheres can roughly be divided into two cases: (1) when the plane-parallel approximation is acceptable a great deal of methods are now available, and it was worth trying a comparative review of the methods and numerical comparisons between them; and (2) when the atmosphere cannot be assumed to be plane-parallel (twilight, finite clouds), the problem reaches a much higher degree of complexity and there is no more a question of choosing between existing methods, but of seeking the development of new methods.

II. EQUATION OF TRANSFER (Ref. 1)

The general equation of transfer has the matrix form

$$\vec{\Omega}.\vec{\nabla}\underline{I}(\vec{r},\vec{\Omega}) = K(\vec{r}) \{-\underline{I}(\vec{r},\vec{\Omega}) + \frac{\bar{\omega}_o(\vec{r})}{4\pi} \iint \underline{P}(\vec{r};\vec{\Omega},\vec{\Omega}') \underline{I}(\vec{r},\vec{\Omega}') \, d\omega' + (1-\bar{\omega}_o(\vec{r}))\underline{B}(\vec{r})\} \quad (1)$$

where $\underline{I}(\vec{r},\vec{\Omega})$ is the radiance matrix at a point $(\vec{r})$ in the direction $(\vec{\Omega})$. $\underline{P}(r;\Omega,\Omega')$ is the phase matrix; $K(\vec{r})$, the total extinction coefficient (absorption + scattering); and $\bar{\omega}_o(\vec{r})$, the single scattering albedo. $\underline{B}(\vec{r})$ is the source function due to thermal emission; it will be neglected in this paper.

The radiance $I(\vec{r},\vec{\Omega})$ is governed by equation

$$\vec{\Omega}\cdot\vec{\nabla}I(\vec{r},\vec{\Omega}) = K(\vec{r})\left\{-I(\vec{r},\vec{\Omega}) + \frac{\bar{\omega}_o(\vec{r})}{4\pi}\iint\{P_{II}(\vec{r};\vec{\Omega},\vec{\Omega}')\ I(\vec{r},\vec{\Omega}')\right.$$

$$+ P_{IQ}(\vec{r};\vec{\Omega},\vec{\Omega}')\ Q(\vec{r},\vec{\Omega}') + P_{IU}(\vec{r};\vec{\Omega},\vec{\Omega}')\ U(\vec{r},\vec{\Omega}')$$

$$\left. + P_{IV}(\vec{r};\vec{\Omega},\vec{\Omega}')\ V(\vec{r},\vec{\Omega}')\}\ d\omega'\right\} \quad (2)$$

and depends on the three other Stokes' parameters Q, U, V.

Neglecting the polarization, the approximate equation

$$\vec{\Omega}\cdot\vec{\nabla}I(\vec{r},\vec{\Omega}) = K(\vec{r})\left\{-I(\vec{r},\vec{\Omega}) + \frac{\bar{\omega}_o(\vec{r})}{4\pi}\iint p(\vec{r};\vec{\Omega},\vec{\Omega}')\ I(\vec{r},\vec{\Omega}')\ d\omega'\right\} \quad (3)$$

can be used; $p(\vec{r},\vec{\Omega},\vec{\Omega}')$ is the phase function.

Table 1 shows an example of the error done in this approximation; it can be of the order of 10% for molecules and very small particles, but becomes negligible for particles with a Mie parameter α larger than 4 or 5.

III. HOMOGENEOUS PLANE--PARALLEL ATMOSPHERE

A. Generalities

The simplest case is an homogeneous atmosphere limited by two infinite parallel planes and illuminated on its upper boundary by the solar beam. The only position variable is z or better the optical depth $\tau = \int_z^Z K(z)dz$, where Z is the altitude of the upper boundary; the total optical thickness is $\tau_1 = \int_0^Z K(z)dz$. The direction $\vec{\Omega}$ is characterized by the zenith angle $\theta = \text{Arc cos }\mu$ and the azimuth angle ϕ.

REVIEW OF RADIATIVE TRANSFER METHODS

TABLE 1

Error in Percent When Neglecting Polarization in the Computation of I

$$\tau_1 = \infty, \quad \tau = 0, \quad \mu = \mu_0 = -1$$

$\bar{\omega}_o$	Rayleigh	$\alpha = 1$	$\alpha = 2$	$\alpha = 5$
0.99	5.29	5.37	1.19	0.07
0.9	8.40	9.40	3.77	0.27
0.8	9.17	10.86	5.91	0.46
0.6	8.19	10.51	8.25	0.58
0.4	5.89	8.04	7.59	0.48
0.2	3.06	4.40	4.13	0.24

The equation of transfer is in this case

$$\mu \frac{\partial I(\tau;\mu,\phi)}{\partial \tau} = I(\tau;\mu,\phi) - \frac{\bar{\omega}_o}{4\pi} \int_0^{2\pi} \int_{-1}^{+1} p(\mu,\phi;\mu',\phi')$$

$$\times I(\tau;\mu',\phi') \, d\mu' \, d\phi' + \frac{\bar{\omega}_o}{4} p(\mu,\phi;\mu_o,\phi_o) F \, e^{\tau/\mu_o} \quad (4)$$

where πF is the sun irradiance at the upper boundary on a plane perpendicular to the solar direction (μ_o,ϕ_o). Here, I refers to the radiance of the diffuse flux excluding the direct solar beam.

The boundary conditions are

$$\left.\begin{array}{l} I(0;\mu < 0,\phi) = 0, \\ \text{and } I(\tau_1;\mu > 0,\phi) \text{ given by ground reflection.} \end{array}\right\} \quad (5)$$

Integrating Eq. (4) with Eq. (5) we get the integral form of the equation of transfer

$$I^+(\tau;\mu,\phi) = I^+(\tau_1;\mu,\phi) \, e^{-\frac{\tau_1-\tau}{\mu}} + \frac{1}{\mu} \int_\tau^{\tau_1} J(t;\mu,\phi) e^{-\frac{t-\tau}{\mu}} \, dt$$

$$(\mu > 0 \uparrow) \quad (6a)$$

$$I^-(\tau;\mu,\phi) = I^-(0;\mu,\phi)e^{\tau/\mu} - \frac{1}{\mu}\int_0^\tau J(t;\mu,\phi)e^{-\frac{t-\tau}{\mu}} dt$$

$$(\mu < 0 \downarrow) \tag{6b}$$

with

$$J(\tau;\mu,\phi) = \frac{\omega_0}{4\pi}\int_0^{2\pi}\int_{-1}^{+1} p(\mu,\phi;\mu',\phi') I(\tau;\mu',\phi') d\mu' d\phi'$$

$$+ \frac{\bar{\omega}_0}{4} p(\mu,\phi;\mu_0,\phi_0) F e^{\tau/\mu_0}$$

$$= \text{Source function} \tag{7}$$

The transmission (S) and reflection (T) functions are defined by

$$I^+(0;+\mu,\phi) = \frac{1}{4\pi\mu} S(\tau_1;\mu,\phi;\mu_0,\phi_0)\pi F \tag{8a}$$

$$I^-(\tau_1;-\mu,\phi) = \frac{1}{4\pi\mu} T(\tau_1;\mu,\phi;\mu_0,\phi_0)\pi F \tag{8b}$$

They are useful when the main interest is in the outgoing (diffusely transmitted or reflected) radiation.

We have seen that the equation of transfer contains the phase function $p(\theta)$ as kernel. This phase function is either given by a table of numerical experimental values or by a mathematical expression. Between the different possible forms, Eq. (9) gives the expansion in Legendre polynomials:

$$p(\theta) = \sum_{\ell=0}^{L} \beta_\ell P_\ell(\cos\theta) \tag{9}$$

its main advantage is to allow an expansion in azimuth of the radiance (and of all the radiation parameters):

$$I(\tau;\mu,\phi) = \sum_{s=0}^{L} (2 - \delta_{os}) I^s(\tau;\mu) \cos[s(\phi - \phi_0)] \tag{10}$$

using Eq. (10) with

REVIEW OF RADIATIVE TRANSFER METHODS

$$p(\mu,\phi;\mu',\phi') = \sum_{s=0}^{L} (2 - \delta_{os}) \cos[s(\phi - \phi')]$$

$$\times \sum_{\ell=s}^{L} \beta_\ell P_s^\ell(\mu) P_s^\ell(\mu') \quad (11)$$

the equation of transfer with three variables can be split into a system of equations with only two variables

$$\mu \frac{\partial I^s(\tau;\mu)}{\partial \tau} = I^s(\tau;\mu) - \frac{\bar{\omega}_o}{4} \sum_{\ell=s}^{L} \beta_\ell P_s^\ell(\mu) P_s^\ell(\mu_o) F e^{\tau/\mu_o}$$

$$- \frac{\bar{\omega}_o}{2} \int_{-1}^{+1} \sum_{\ell=s}^{L} \beta_\ell P_s^\ell(\mu) P_s^\ell(\mu') I^s(\tau;\mu') d\mu'$$

$$s = 0, 1, \ldots, L \quad (12)$$

I will give here a brief description of the methods for the plane parallel case; it follows a report prepared by a Working Group of the Radiation Commission (Ref. 2).

B. Exact Analytical Methods

By exact analytical methods, we understand the methods leading to a solution for the radiance in terms of mathematical functions, which have finally to be tabulated; therefore, the accuracy of these methods may not be better than the accuracy of more direct numerical methods. Their main interest is in the understanding of the mathematical structure and of the general behavior of the solutions. Their basic drawback is the difficulty to use these methods in the case of a real atmosphere.

Among these methods we will classify the Singular Eigenfunctions (or Case) method, the Wiener-Hopf method, and the reduction to H- or X- and Y- functions which can be founded on the principles of invariance. This reduction is straightforward only for very simple phase functions, such as isotropic or Rayleigh scattering. In these cases, many accurate tables of the H- and X-, Y-functions have been computed and this makes the method a reference for testing other methods in the simplest cases.

C. Computational Methods

By computational methods, we understand methods specifically designed for computers, but they may include some analytical treatment before the numerical procedure, as is the case for the spherical harmonics. The spherical harmonics solution is based on a discrete spectrum of eigenvalues ν and can be seen as an approximation to the exact Case method which uses a continuous spectrum. The discrete ordinates method is indeed very similar to the Spherical Harmonics. In the Monte Carlo method, one photon at a time is followed on its path through the atmosphere and each event is defined by a probability distribution. The Dart method uses a discretization in radiation streams whose arrangement is based on a regular dodecahedron.

The successive scattering is based on an iteration starting from the primary scattering. The Gauss Seidel method uses another possible scheme of iteration with about the same advantages and drawbacks. The matrix operator is based on the interaction principles: the reflection and transmission of the layer (τ_0, τ_2) are expressed in terms of those of the two layers (τ_0, τ_1) and (τ_1, τ_2). The adding method uses the same principles with a different algorithm and it reduces to doubling, which is much faster, for an homogeneous atmosphere. On the other hand, the invariant imbedding uses the addition of infinitely thin layers to obtain differential equations for the reflection and transmission functions.

Finally, in the case of very thick layers, asymptotic relations can be derived; the scattering function of a layer of optical thickness τ_1 is expressed in terms of the same function for a semi-infinite layer; the correcting term contains the solution of the Milne problem for the same atmosphere and decreases as $e^{-2\nu\tau_1}$, where ν is the inverse of the higher eigenvalues which appears in the Case and in the Spherical Harmonics methods; relations exist for the transmitted and the internal radiance.

REVIEW OF RADIATIVE TRANSFER METHODS

Table 2 shows an example of numerical comparisons[1] between the Spherical Harmonics, the Matrix Operator, and the Successive Scattering methods. It gives the intensity at some depths τ and for some directions μ, for the case of a haze layer with $\tau_1 = 1$, $\omega_o = 0.9$ and normal incidence ($\mu_o = -1$). The relative difference Δ between Spherical Harmonics and Matrix Operator

TABLE 2

Intensity Haze L

$\tau_1 = 1$, $\bar{\omega}_o = 0.9$, $\mu_o = -1$

τ	μ	Spherical Harmonics	Matrix Operator	Δ %	Successive Scattering	Δ %
0	1	2.794-2	2.789-2	1.8	2.831-2	12.9
	0.6	3.913-2	3.911-2	0.5	3.935-2	5.6
0.5	1	1.374-2	1.371-2	2.2	1.390-2	11.6
	+ 0.6	2.210-2	2.208-2	0.9	2.218-2	3.6
	- 0.6	1.153-1	1.153-1	0.0	1.157-1	3.5
	- 1	2.240 0	2.244 0	1.8	2.239 0	0.4
1	0.6	2.007-1	2.007-1	0.0	2.019-1	6.0
	- 1	2.967 0	2.972 0	1.7	2.975 0	2.7
Authors		Devaux	Plass Kattawar		Quenzel	

---

[1] All the numerical results presented in this paper have been computed in the framework of a comparison program sponsored by the Radiation Commission and the author is greatly indebted to all contributors.

values is always smaller than two-tenths of a percent. In the Successive Scattering method, the accuracy can be increased by increasing the number of iterations and the computation whose results are given here has been stopped in order to achieve an accuracy of about 1%; it is even better than that.

The computation time is shorter with the Spherical Harmonics method than with the Matrix Operator method, except maybe when a very large number of solar directions are wanted at the same time. For the Successive Scattering method, the computation time is quite competitive in this case, but it increases very fast when $\bar{\omega}_o$ tends to 1 and when the optical thickness increases.

Table 3 shows the accuracy obtained by Monte Carlo and Dart methods for the same case of $\tau = 0.2$ and for various values of μ. The values plotted are the relative differences Δ with the "exact" values obtained by both the Spherical Harmonics and the Matrix Operator methods. The Monte Carlo program has been run by two groups of authors; in both cases, the accuracy is about a few percent, sometimes better than 1%. For the Dart method the error is a little larger, but remains always smaller than 10%. The time is much larger for both these methods than for the semi-analytical methods; but their main interest is in their ability in handling non-plane-parallel cases as we will see later.

Table 4 shows again a comparison of the intensity computed by various methods, now in the case of a thick conservative cloud ($\tau_1 = 64$; $\omega_o = 1$; $\mu_o = -1$). Many methods are unable to treat such a large optical thickness without a prohibitive computation time.

In the Spherical Harmonics method, the computation time is nearly independent of the optical thickness. But the difficulty here is related to the forward peak; Dr. Devaux has approximated the phase function of the cloud by the sum of a Delta-function and an expansion with 36 Legendre polynomials. Therefore, the result is wrong in the forward direction ($\mu = -1$) for small

TABLE 3

Error Δ in Percent   Haze L

$\tau_1 = 1$, $\omega_o = 0.9$, $\mu_o = -1$, $\tau = 0.2$

μ	Monte Carlo	Monte Carlo	Dart
1	+ 0.4	− 4.0	+ 9.4
0.8	+ 1.2	− 3.1	+ 17
0.6	+ 1.8	− 1.5	+ 11
0.4	+ 3.8	− 0.7	− 4.0
0.2	+ 6.2	− 0.5	− 12
0	− 7.3		− 5.2
− 0.2	+ 0.6	+ 3.0	− 4.5
− 0.4	− 1.8	+ 4.7	+ 1.9
− 0.6	− 0.5	+ 2.9	+ 2.0
− 0.8	+ 0.6	+ 6.5	+ 5.6
− 1	− 0.7	+ 3.4	− 4.3
Authors	PLASS − KATTAWAR	MIKHAILOV KUZNETSOV	WHITNEY

optical depths; it might easily be checked that the values found at $\mu = -1$ for $\tau = 6.4$ are smaller than the values expected for primary scattering only. Elsewhere, it is expected that the truncature procedure gives an accuracy better than 1%.

The Monte Carlo method has no problem with the forward peak and the results at $\mu = -1$ and small optical thickness are better than those of the Spherical Harmonics. But the computation time becomes very large for such large optical thickness and only one group of authors has run the Monte Carlo program for the cloud case. Except for $\mu = -1$, the results can be compared with those of the Spherical Harmonics; the accuracy seems of the same order as that for Haze.

Finally, the Asymptotic method finds, in the case of large optical thickness, its own field of application. Of course, it

TABLE 4

Intensity

Cloud C_1, $\tau_1 = 64$, $\omega_o = 1$, $\mu_o = -1$

τ	μ	Spherical Harmonics	Asymptotic	Monte Carlo
0	1	1.042 0	1.03 0	1.05 0
	0.6	8.254-1	8.21-1	7.88-1
	0.2	5.584-1	5.57-1	5.91-1
6.4	1	9.622-1	9.64-1	9.50-1
	0.6	9.876-1	1.01 0	9.59-1
	0.2	1.003 0	1.06 0	1.17 0
	- 0.2	1.000 0	1.11 0	9.09-1
	- 0.6	1.004 0	1.15 0	1.03 0
	- 1	1.840 0	1.20 0	1.15+1
32	1	5.237-1	5.21-1	5.59-1
	0.6	5.697-1	5.68-1	4.49-1
	0.2	6.158-1	6.15-1	5.95-1
	- 0.2	6.618-1	6.63-1	6.78-1
	- 0.6	7.079-1	7.10-1	7.30-1
	- 1	7.54-1	7.57-1	7.45-1
64	- 0.2	8.929-2	9.18-2	7.45-2
	- 0.6	1.458-1	1.50-1	1.25-1
	- 1	1.961-1	2.01-1	1.50-1
Authors		DEVAUX	GERMOGENOVA KONOVALOV	MIKHAILOV KUZNETSOV

does not apply at small optical depth for downward radiation. But elsewhere, its agreement with the Spherical Harmonics is better than 5%.

D. Approximate Methods

By approximate methods, we understand methods which include a very rough approximation of the atmospheric properties or (and) of the transfer problem. They generally give only the flux and not the intensity, but the computation time is reduced by a factor

REVIEW OF RADIATIVE TRANSFER METHODS

Symbols are defined as follows:

- —   *exact*
- ●   *Eddington (Drs. Irvine, Esposito and Shettle)*
- ○   *Eddington + Similarity (Dr. Shettle)*
- ■   *Delta-Eddington (Dr. Wiscombe)*
- ▽   *Double Delta-Eddington (Dr. Bonnel)*
- □   *Two-Stream (Drs. Irvine and Esposito)*
- ▲   *Modified Two-Stream (Drs. Irvine and Esposito)*
- △   *Modified Two-Stream (Drs. Kerschgens, Raschke and Pilz)*
- ⊙   *Exponential Kernel (Dr. Brogniez)*

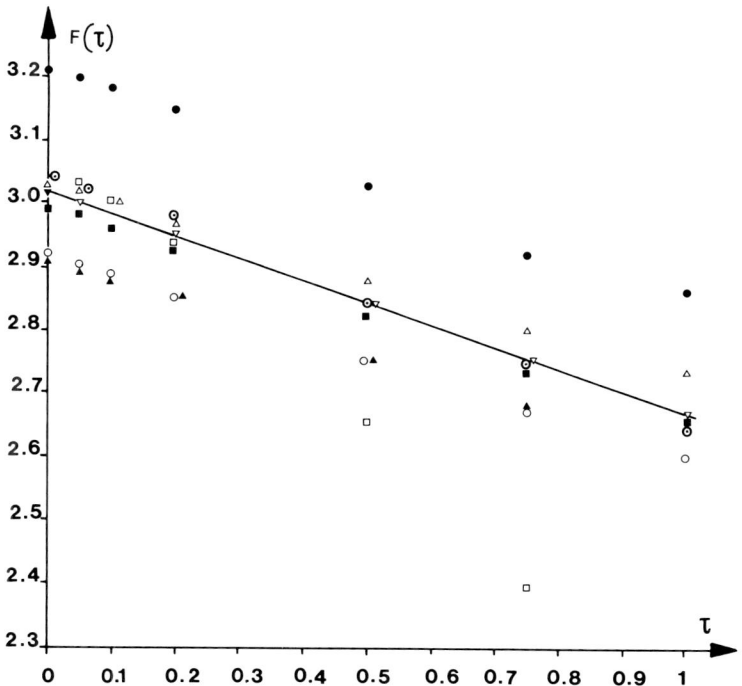

Fig. 1. Haze L, $\tau_1 = 1$, $\bar{\omega}_o = 0.9$, $\mu_o = -1$. (See Ref. 2)

larger than 100 or even the use of a computer can be avoided. Moreover they give simple analytical expressions of the flux. Between them we have classified the similarity relations which reduce the anisotropic problem to an isotropic one, the Eddington, Two-Stream, and Exponential Kernel methods. Various modified versions of Two-Stream and Eddington have been tried with success. Recently, Meador and Weaver[2] have proposed a general theoretical framework to compare these methods. Figure 1 shows the net flux versus τ in the case of a haze layer in normal incidence ($\mu_o = -1$) and for $\bar{\omega}_o = 0.9$. The solid line corresponds to the "exact" values obtained by various methods within an agreement of a few tenth of a percent; the points correspond to various approximate methods.

The worst results are obtained by the Eddington and Standard Two-Stream methods; the accuracy varies from a few to 10%. The best results are given by the Delta-Eddington, the Double Delta-Eddington, and the Exponential Kernel methods, with an accuracy of about 1%. Intermediate performances are achieved by other modifications of Eddington or Two-Stream methods.

The results remain about the same when $\bar{\omega}_o$ tends to 1, but most of the methods are a little less accurate in oblique than in normal incidence, except the Eddington method which becomes better.

If we consider now increasing the optical thickness, we may say that a thick conservative cloud is the worst case for the Standard Two-Stream method (accuracy $\sim$ 24%) and the best for Eddington method (accuracy $\sim$ 1%). But again, always the Exponential Kernel and the Delta- and Double Delta-Eddington achieve an accuracy of about 1%.

---

[2] Personal communication of their work is acknowledged by the author.

REVIEW OF RADIATIVE TRANSFER METHODS

IV. REALISTIC ATMOSPHERES

A. Inhomogeneous Atmospheres

It must be first noted that the case we have called "homogeneous" is actually the case where only the extinction coefficient K is function of the altitude z (see definition of the optical depth τ), with the single scattering albedo and the phase function constant throughout the atmosphere.

A vertically inhomogeneous atmosphere occurs when mixing ratio of scatterers and absorbers varies with height ($\bar{\omega}_o$ function of τ), or when the type of scattering particles varies with height (phase function varying with τ).

The vertical inhomogeneity can be handled by the equation of transfer, Eq. (4), including the variable τ in $\bar{\omega}_o(\tau)$ or/and in $p(\tau;\mu,\phi;\mu'\phi')$. Most of the methods reviewed for the homogeneous case can be applied, perhaps, with slight modifications. At the limit, the vertically inhomogeneous atmosphere can always be approximated by superposition of thin homogeneous layers. Anyway, it must be noted that the computation time may increase rapidly with the inhomogeneity (for example, doubling is replaced by adding) and that in some methods numerical difficulties or instabilities appear (Spherical Harmonics or Discrete Ordinates).

The horizontal inhomogeneity is much more difficult to handle, as it can be imagined from the equation of transfer

$$(1-\mu^2)^{1/2} \cos\phi \, \frac{\partial I(x,y,z;\mu,\phi)}{\partial x} + (1-\mu^2)^{1/2} \sin\phi$$

$$\frac{\partial I(x,y,z;\mu,\phi)}{\partial y} + \mu \frac{\partial I(x,y,z;\mu,\phi)}{\partial z} =$$

$$- K(x,y,z) \{I(x,y,z;\mu,\phi) - J(x,y,z;\mu,\phi)\} \qquad (13)$$

where the three coordinates x, y, z and the three corresponding partial derivatives appear. This case comprises the very important problem of finite clouds. Here, the flexibility of the Monte Carlo Method (Refs. 3, 4, and 5) finds all its advantages, and although a few attempts have been done to develop other methods, for example, invariant imbedding (Ref. 6) or statistical methods (Ref. 7), the only results obtained for finite clouds until now have been obtained by Monte Carlo.

B. Spherical Atmospheres

For some problems, we have to consider the sphericity of the real atmosphere. This is particularly important in remote sensing of the atmosphere from twilight measurements or from limb scanning by satellites.

Using spherical coordinates with reference to the local vertical, the transfer equation has the form

$$\{\mu \frac{\partial}{\partial z} + \frac{(1-\mu^2)}{R+z} \frac{\partial}{\partial \mu} + \frac{(1-\mu^2)^{1/2}(1-\mu_o^2)^{1/2}}{R+z} \cos(\phi - \phi_o) \frac{\partial}{\partial \mu_o}$$

$$+ \frac{\mu_o}{1-\mu_o^2} \sin(\phi - \phi_o) \frac{\partial}{\partial(\phi - \phi_o)} \} I(z,\mu_o,\phi_o;\mu,\phi) =$$

$$- K(z) \{I(z,\mu_o,\phi_o;\mu,\phi) - J(z,\mu_o,\phi_o;\mu,\phi)\} \qquad (14)$$

where R is the earth radius. Here again, the Monte Carlo method (Ref. 8) can be applied and the Dart method (Ref. 9) has been especially studied for this case; but the research in approximate analytical methods seems promising (Ref. 10).

C. Scattering with Gaseous Absorption

All what we have said here concerns monochromatic radiation. A last problem appears when we have a radiative transfer problem in a scattering atmosphere with gaseous absorption. Then, the number of monochromatic problems to be treated becomes prohibitive, even for a single absorption line. Therefore, approximate methods have been sought. One of them consists in introducing the photon path distribution $P(\lambda)$ (Refs. 11 and 12) which is defined by

$$I(\bar{\omega}_\nu) = I(\bar{\omega}_o) \int_0^\infty P(\lambda) \exp\{-\frac{\lambda k_\nu}{\sigma + k_c}\}d\lambda \qquad (15)$$

where the subscript ν refers to the frequency ν and the subscript c to the continuum outside the line; $P(\lambda)$ can be obtained either directly from a Monte Carlo calculation or by the inverse Laplace transform

$$P(\lambda) = L_{r,\lambda}^{-1}\left(\frac{I(\bar{\omega}_\nu)}{I(\bar{\omega}_c)}\right), \quad r = k_\nu/(\sigma + k_c) \qquad (16)$$

When $P(\lambda)$ is once for all obtained, the intensity at any frequency of the spectrum can be obtained from Eq. (15). Moreover, it is possible for some problems to define various mean path lengths.

V. CONCLUSION

In the case of plane-parallel atmospheres, even with vertical inhomogeniety, several methods can be used to obtain the complete radiation field with a good accuracy and a reasonable computer time. If only the flux and the heating rate are sought, fast approximate analytical methods are available.

When more realistic atmospheric models must be considered, the Monte Carlo can be used with a rather good accuracy, but at the price of large computation times, while active research on faster approximate methods is carried on.

SYMBOLS

$\underline{B}$	source function for thermal emission
πF	sun irradiance
I	radiance
$\underline{I}$	radiance matrix
I^+	upward radiance
I^-	downward radiance
J	source function
k	absorption coefficient
$k_c, \bar{\omega}_c$	absorption coefficient and single scattering albedo in the continuum
$k_\nu, \bar{\omega}_\nu$	absorption coefficient and single scattering albedo at frequency ν
$K = k + \sigma$	extinction coefficient
p	phase function
$\underline{P}$	phase matrix
$P_{II}, P_{IQ}, P_{IU}, P_{IV}$	coefficients of the phase matrix
$P(\lambda)$	photon path length distribution
Q, U, V	Stokes' parameters
$\vec{r}$	position variable
R	earth radius
S	diffuse reflection function
T	diffuse transmission function
z	altitude
α	Mie parameter (ratio of circumference to wavelength)
θ	zenith angle
μ	$\cos \theta$
μ_o, ϕ_o	sun direction
σ	scattering coefficient
τ	optical depth
τ_1	total optical thickness
ϕ	azimuth angle

$\bar{\omega}_o = \sigma/K$ single scattering albedo
$\vec{\Omega}$ direction variable

REFERENCES

1. S. Chandrasekhar, "Radiative Transfer," Oxford University Press, Oxford, 1950. [Reprinted by Dover Publications, New York, 1960.]
2. J. Lenoble (Ed.), Standard procedures to compute atmospheric radiative transfer in a scattering atmosphere, *Report of the Radiation Commission*. [Preliminary ed., vols. I and II, 1974.]
3. T. B. McKee and S. K. Cox, Scattering of visible radiation by finite cloud, *J. Atmos. Sci. 31*, 1885 (1974).
4. V. P. Busygin, N. A. Yevstratov, and Ye M. Feigelson, Optical properties of cumulus clouds and radiant fluxes for cumulus cloud cover, *Izv. Atmos. Oceanic Phys. 9*, 1142 (1973).
5. J. F. Appleby and D. J. Van Blerkom, Absorption line studies of reflection from horizontally inhomogeneous layers, *Icarus 24*, 51 (1975).
6. R. Bellman, R. Kalaba, and S. Ueno, Invariant imbedding and diffuse reflection from a two dimensional flat layer, *Icarus 1*, 297 (1963).
7. K. Yu. Niylisk, Cloud characteristics in problems of radiation energetics in the earth's atmosphere, *Izv. Atmos. Oceanic Phys. 8*, 154 (1972).
8. V. S. Antyufeyev and M. A. Nazaraliyev, A new modification of the Monte Carlo method for solution of problems in the theory of light scattering in a spherical atmosphere, *Izv. Atmos. Oceanic Phys. 9*, 463 (1973).
9. C. Whitney, R. E. Var and C. R. Gray, Research into radiative transfer modeling and applications, *Report AFCRL-TR-73-0420*, July 1973.

10. V. V. Sobolev, "Light Scattering in Planetary Atmospheres," Pergamon Press, New York, 1975.
11. W. M. Irvine, The formation of absorption bands and the distribution of photon optical paths in a scattering atmosphere, *Bull. Astron. Inst. Netherlands 17,* 266 (1964).
12. Y. Fouquart, "Contribution à l'Étude des Spectres Réfléchis Par les Atmosphères Planétaires Diffusantes--Application à Vénus," Thesis, Université de Lille, March 1975.

DISCUSSION

Turner: Do you know if anyone has used a method similar to the spherical harmonics method using other orthogonal functions, for example, Gegenbauer or Jacobi polynomials?

Lenoble: I have never heard about that but it would be possible.

Turner: It should be much better than the spherical harmonics method because these functions more nearly represent the anisotropy of the scattering phase function and with fewer terms in the series expansion.

Lenoble: Yes, but that is true for the phase function alone.

Turner: Yes, but one can expand the radiation field function in a series of these more general functions.

Irvine: But when you do the multiple scattering, it is convenient to use spherical harmonics, because you can then expand the intensity in a Fourier series in azimuth and each azimuthal component satisfies an independent equation of transfer.

Turner: But there are addition formulas for these other polynomials although they are considerably more complicated than those for the spherical harmonics.

Irvine: I don't know of any that is being done.

Gal: I would like to answer some of these questions. We, at Lockheed Palo Alto Research Laboratory, are applying the Hartel formulation for multiple scattering by spherical particulates. Hartel (Germany, 1941) reported that multiple scattering may be solved by obtaining solution for the scattering function by successive scattering. The scattering function for each scattering order may be written in terms of a Legendre series expansion. Hartel did not have the mathematical tools to prove his idea. Since then, the Legendre expansion is available for a spherical particle and my colleague, Dr. Chou, solved the radiation transfer equation for a plane parallel geometry. Agreement with currently available solutions, such as Dave's iterative and Monte Carlo methods are excellent. The advantage of our solution is that it requires an order of magnitude less computer time. This work will be published in a few months.

Lenoble: I would like to have it.

Barkstrom: Two comments. The first comment is that in stellar atmospheric work and some recent work which we have done, there is an alternate procedure which breaks up the equation of transfer into a flux conservative finite difference form. The computation time is comparable with that of the doubling method, and it isn't bothered by vertically inhomogeneous atmospheres. Secondly, I am not sure which approximate analytical methods you are referring to in the three-dimensional case. There are a number of diffusion-type approximations that are related to the delta-Eddington approximation. I expect they are going to appear in the literature very shortly.

Lenoble: For the spherical case, I was referring mostly to the work by the Soviet group of Sobolev and Minin and this kind of work. But we are now working on the report for spherical atmospheres and for three-dimentional problems. It is just at its very beginning and I don't have much information on this work.

Barkstrom: These methods are connected with problems, such as finite clouds.

Lenoble: But for spherical problems you mentioned the problem of stellar atmospheres. It is quite different because you don't have the solar illumination so you have really a spherical symmetry.

Barkstrom: It doesn't matter; the procedure is the same.

Unidentified Speaker: What are the publication plans for the report you are preparing for the Radiation Commission?

Lenoble: Well, the first draft of the report has been published. But the final printing is to be done at NCAR. Maybe Professor London can give some more information on the publication.

London: It will be available through the Radiation Commission.

Lenoble: But, presently, it is only the part concerning the plane parallel atmospheres.

SOME ASPECTS OF THE INVERSION PROBLEM

IN REMOTE SENSING

S. Twomey
University of Arizona

A brief discussion of several commonly used methods for inversion--constrained linear inversion, synthesis (Backus-Gilbert) methods and nonlinear iterative techniques for the Chahine type--is given. It is demonstrated that a very close connection exists between Backus-Gilbert solutions and those given by constrained linear inversion.

A number of examples of the application of such methods are presented, showing that resolution is not greatly different for quite different algorithms--a result quite in accord with general theoretical considerations: more "resolution" can be achieved at the expense of introducing greater a priori bias in the procedure.

I. INTRODUCTION

When Kaplan (Ref. 1), in 1959, outlined the possibilities of determining atmospheric structure from infrared radiance measurements, there were initially probably more sceptics than enthusiasts. Nevertheless, the steps toward implementation proceeded faster than most (at least in terrestrial atmospheric physics) and it was only a few years later that a crude (horizontal) temperature sounding experiment was carried out through an open window of the Meteorological Satellite Center in Suitland, Maryland. At that time, inversion methematics tended to be concentrated around the determination of coefficients in arbitrary expansions of the unknown, and results from the first attempted inversion were quite promising but, being restricted to the coefficients of a quadratic,

not capable of coming very close to reality--which contained sharp
"step" in temperature akin to an atmospheric thermal inversion and
could, therefore, be modeled reasonably only if two inflections
could be allowed into the solution. This implied a cubic rather
than a quadratic, and a fairly routine improvement of the inversion
from three unknowns to four was undertaken by the Suitland group.
Disaster soon ensued. The cubic "improvement" indicated negative
absolute temperatures and super-adiabatic lapse rates, a temperature distribution which was physically unacceptable, but which, if
it could exist, would give radiance values experimentally undistinguishable from those given by the much less spectacular temperature structure which actually existed.

These facts are mainly of historical interest and are well
known to the present audience. We still find a great deal of
attention directed to procedures and algorithms for inversion, the
implication being that a more sophisticated numerical technology
can sidestep the obstacles which in the 1960s turned modest quadratic success into cubic nonsense. But the ambiguity of inversions
is fundamental, caused by the kernels, which describe the underlying physical connection between measured and sought functions,
and a successful algorithm can only succeed by making an acceptable
selection from all the possibilities. That selection is arbitrary;
its basis does not lie within the measurement or the integral
relationship connecting sought distribution with the measured
quantities.

The simplest version of the indirect sensing problem:

$$g_i = \int_a^b K_i(x) f(x) dx \qquad (1)$$

or

$$\underset{\sim}{g} = \int_a^b \underset{\sim}{k}(x) f(x) dx$$

describes a kind of convolution of the kernel functions with the
unknown or "sought" function $f(x)$. In real atmospheric physics

SOME ASPECTS OF THE INVERSION PROBLEM

problems, $K_i(x)$ is fundamentally smooth, being in most cases necessarily positive and essentially exponential in character, of the form $\sum_j \xi_j e^{-k_j x}$ or $\int_0^\infty \psi(u) e^{-xk(u)} du$ with ξ_j or $\psi(u)$ non-negative. Smoothness of $K_i(x)$ implies a diminishing sensitivity of g_i to higher frequency components in $f(x)$, and a fundamental, inherent instability and ambiguity in any inference of $f(x)$. This has nothing to do with inversion algorithms, linear or nonlinear, simple or complicated: they do not even have to be brought into the picture to show the fundamental ambiguities. We simply cannot get something for nothing.

Figure 1 shows two curves from an early paper by Wark and Fleming (Ref. 2); it shows kernels dt/d ln p for an Elsasser band with maximum contribution at 100 mb and also for a constant mass absorption coefficient (i.e., simple exponential behavior). (1 bar = 100 kPa.) Note that the width and smoothness of the two are comparable--the more realistic band model is slightly narrower but not markedly so. This implies that a great deal can be learned by looking at simple exponential kernels (which bring us to a LaPlace transform inversion problem) or kernels of the form xe^{-yx} (which have the advantage that they attain maxima at $x = y^{-1}$.

The power spectrum of the kernel xe^{-yx} can easily be calculated; it decreases rapidly with increasing frequency, asymptotically as ω^{-2}.

Figure 2 shows the Fourier transform of a kernel related to the indirect measurement of particle size distribution from light scattering. This illustrates an important and often overlooked point--that a physical kernel may have in its spectrum a blind spot at relatively low frequencies, even though it has not yet begun its ultimate asymptotic decline.

Early inversion procedures were linear in nature and amounted to the inference of coefficients in some expansion of $f(x)$ in the form

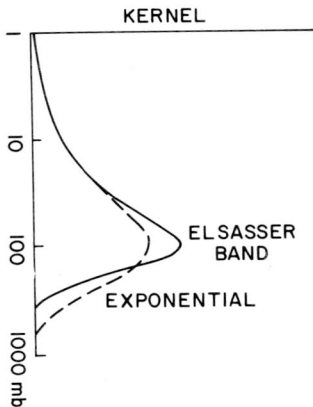

Fig. 1. Kernels for Elsasser band and strictly exponential absorption. (From Ref. 2.)

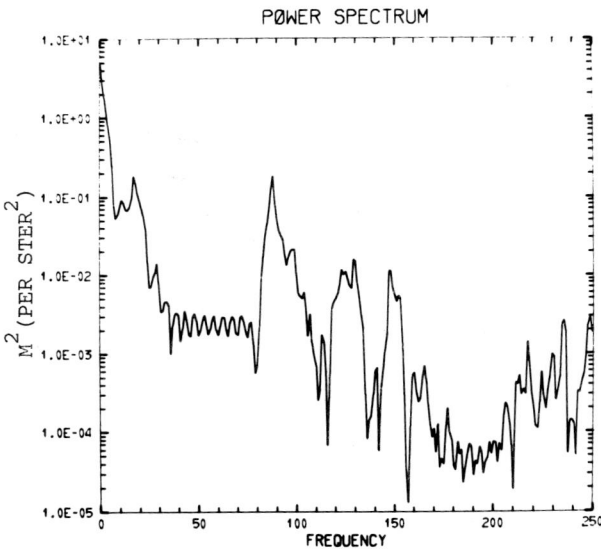

Fig. 2. Power spectrum for Mie scattering kernel.

SOME ASPECTS OF THE INVERSION PROBLEM

$$f(x) = a + bx + cx^2 + \ldots$$

or

$$f(x) = a_o + a_1 \phi_1(x) + a_2 \phi_2(x) + \ldots$$

but soon use of the values $f(x_1)$, $f(x_2)$ ... $f(x_n)$ of $f(x)$ at selected tabular values of x became common practice. There is fundamentally no difference between the two approaches since $f(x_1)$, $f(x_2)$, etc. can be identified with coefficients in an expansion of $f(x)$ in terms of a set of functions $\phi_k(x)$ which are unity at $x = x_k$ and fall linearly to zero at $x = x_{k-1}$ and $x = x_{k+1}$. With further development of inversion algorithms, there tended to be a return to the expansion approach in a somewhat different guise, in which empirical orthogonal functions were used for the functions $\phi_i(x)$. Use of such functions, coupled with constraints resting on the fundamental properties of empirical orthogonal functions (EOF's), led to more "realistic" inversions. Nevertheless, it is important to realize what is going on in such cases: one forms a set of orthogonal functions which are linear combinations of *observed* functions; these latter may be measured with instruments of high resolution and can contain harmonic components which may extend to frequencies to which a given indirect sensing procedure (with finite accuracy, smooth band-limited kernels) is "blind." Furthermore, orthogonality of the functions $\phi_m(x)$, $\phi_n(x)$ does not imply

$$\int K_1(x) [\phi_m(x) - \phi_n(x)] dx \neq 0$$

so two independent EOF's could be undistinguishable to the remote sensing measurement. A stable solution is possible only because, for example, one demands that the expansion coefficients fall off at a rate similar to that found on average in the original population of measured soundings (e.g., a large set of radiosonde profiles). In this way, a greater degree of structure can be retained in a solution but it is in a sense pseudo-resolution--the solution is permitted to "wiggle" in a way which is expected on the basis of past data. This kind of procedure while in many ways eminently reasonable, does nevertheless introduce a strong probability of bias--for example, solutions for unpopulated or oceanic areas (those for which indirect sensing is likely to be most valuable) in essence become pushed toward a behavior pattern representative in some sense of more populated regions where direct measurements are most frequent. Since people and radiosondes are simply not randomly distributed over the globe, the distortions produced through such bias are systematic in nature. On the part of general-circulation and numerical weather prediction researchers, a degree of disenchantment with satellite-based temperature soundings seems to have emerged recently; it is germane to ask how much the aforementioned bias contributes to this disenchantment.

It is also worth pointing out that the atmosphere as a whole is generally close to a balanced condition; one's primary concern in prediction is with departures from that balance. There is no skill needed to predict cloudless conditions with midday temperatures near the 80s for Arizona in winter, or to predict showery conditions with steady trade winds for Hilo, Hawaii. It is the departures from these norms that constitute the prediction problem and if we distort temperature profiles toward the expected norm, it must adversely affect the chances of predicting excursions away from the norm.

II. RÉSUMÉ OF COMMONLY USED METHODS

A. Constrained Linear Inversion

This consists in its most direct application of the conversion of the integral equation to a quadrature form in which the function $f(x)$ is replaced by a vector containing tabular values of $f(x)$ in its elements (the behavior of $f(x)$ between the tabular points being implicitly specified by the quadrature scheme), and the resulting matrix-vector equation is inverted to find the most acceptable vector $\underline{f}$, which satisfies the fundamental relationship to $\underline{g}$ to within a prescribed accuracy. "Most acceptable" is almost always specified numerically in terms of minimization of some quadratic form $\underline{f}^*H\underline{f}$ in $\underline{f}$ which we arbitrarily introduce to gauge "acceptability." In general, the solution is

$$\underline{f}' = (A^*A + \gamma H)^{-1} A^*\underline{g} \qquad (2)$$

We obtain a useful solution provided H is framed to be in some sense a measure of the smoothness of $\underline{f}$. One can, of course, take out any initial guessed or known expected value $\underline{f}_o$ and remove the corresponding term $A\underline{f}_o$ from $\underline{g}$ before inverting. The shape of $\underline{f}_o$ is arbitrary and if it contains wiggles or other high-frequency features these will show up in the solution. But this is artificial in the sense that for many physical kernels these could be filtered from $\underline{f}_o$ without changing $\underline{g}$ perceptibly.

Linear constrained inversion methods are simple extensions of least-squares methods for solving systems of linear equations. Methods, in which f(x) is described by a vector of expansion coefficients, are algebraically equivalent. It is, however, less easy to formulate measures of acceptability in terms of such coefficients if the expansions are polynomials or are made in terms of arbitrary orthogonal functions (e.g., Fourier or Fourier-Bessel expansions, Tchebycheff polynomials, etc.). But if, when empirical orthogonal functions are used, one has *a priori* statistical grounds for asking the coefficients to diminish at a known rate, this provides a valuable method of inversion in situations where a sizable background of data measurements for f(x) is available. In this procedure, one is asking that the solution f'(x) be "like" the population from which the orthogonal function set was constructed.

B. Synthesis Methods

All linear methods of solution produce a solutuion f'(x) which for any value of x consists of a linear combination of the measured g's, i.e.,

$$f'(x_k) = \sum_j b_{kj} g_j \tag{3}$$

The array b_{kj} is often not calculated explicitly, as in the case of constrained linear inversion methods, in which $\|b_{kj}\|$ would be given by $(A^*A + \gamma H)^{-1} A^*$. The Backus-Gilbert (Ref. 3) solution directed its attention explicitly to the array b_{kj} and the constraints are formulated in terms of the b_{kj}. Equation (2) has the consequence that the solution value $f'(x_k)$ can be written

$$f'(x_k) = \int_a^b \left\{ \sum_j b_{kj} K_j(x) \right\} f(x) dx$$

so that $\sum_k b_{kj} K_j(x)$ is a "scanning function" which, of course, would have to be the delta function $\delta(x - x_k)$ if $f'(x_k)$ was to reproduce $f(x_k)$ exactly. That is an evident impossibility and the linear inversion problem can be regarded as a search for realizable scanning functions which approximate the delta function and are as far as possible free from side lobes and other undesirable features which could distort $f'(x)$ and produce in it artificial peaks or troughs. It is sufficient to consider only normalized scanning functions, so that

$$\int_a^b S_k(x) dx = 1$$

Since the coefficients b_k ultimately multiply the errors in g, stability is ensured if the magnitude of $\sum_j b_{kj}^2 = |b_k|^2$ is limited. To complete the Backus-Gilbert procedure, one, therefore, needs to solve a minimization problem which is very similar to that encountered in constrained linear inversion. An array of coefficients $\underset{\sim}{b}_k$ is to be determined which minimizes a spread $\underset{\sim}{b}_k^* S_k \underset{\sim}{b}_k$ measured about the point $x = x_k$, with $\underset{\sim}{b}_k^* \underset{\sim}{b}_k \leq$ Constant and $\underset{\sim}{b}_k^* \underset{\sim}{k}_I$ also constant ($\underset{\sim}{k}_I$ being written here for the array $\int_a^b K_j(x) dx$, $j = 1, 2, 3 \ldots m$). The two constraints (one restricting the magnitude of the coefficients, the other a normalization

SOME ASPECTS OF THE INVERSION PROBLEM

constraint) imply two Lagrangian multipliers, but the solution algebra is again quite straightforward and we obtain

$$\underset{\sim}{b}_k = \beta(S_k + \gamma I)^{-1} \underset{\sim}{k}_I \qquad (4)$$

β is readily calculated explicitly--since $\underset{\sim}{k}_I^* \underset{\sim}{b}_k$ must be unity, β^{-1} must be $\underset{\sim}{k}_I^*(S_k + \gamma I)^{-1}\underset{\sim}{k}_I$. For the particular spread measure $\int (x - x_k)^2 s^2(x) dx$, S_k is $\| \int (x - x_k)^2 K_i(x) K_j(x) dx \|$, and γ plays a role similar to that played by γ in constrained linear inversion, in that as γ decreases in size, error magnification increases and the scanning function becomes narrower, with diminishing γ. Backus-Gilbert solutions involve covariance matrices $C = \| \int K_i(x) K_j(x) dx \|$ and vectors such as $\| \int K_i(x) dx \|$ and $\| K_i(x_K) \|$. To the accuracy of quadrature, these quantities are connected with those occurring in constrained linear inversion through relationships $C = AY$, $\underset{\sim}{k}_i = \| \int K_i(x) dx \| = A\underset{\sim}{e}$, and so on. Y is written for the $m \times n$ tabulation $\| K_i(x_j) \|$ of the kernels. By these relationships, Backus-Gilbert solutions can be transformed into relationships involving the quadrature matrix $\underset{\sim}{A}$ and the measured $\underset{\sim}{g}$, and a very close relationship can be established between solutions obtained by the two methods. One simple such relationship applies to the direct inverse $f' = A^{-1}g$, which can also be obtained by the Gilbert-Backus procedure if the square norm of the difference between the delta function and the scanning function is the basis for optimization; carrying out this optimization (unconstrained) one obtains a solution which the substitution $C = AY$ transforms into the direct inverse $A^{-1}g$.

The trade-off curves provided automatically in Backus-Gilbert inversion shows how much accuracy of measurement is needed to give a certain resolution. As we change γ in Eq. (4), the width of the scanning function and the magnitude of the b-vectors change; the latter magnitude determines how much error magnification can enter into the solution (which is $f_k' = b_k^* g + b_k^* \epsilon$), and so error magnification can be plotted against resolution. Such a plot, for

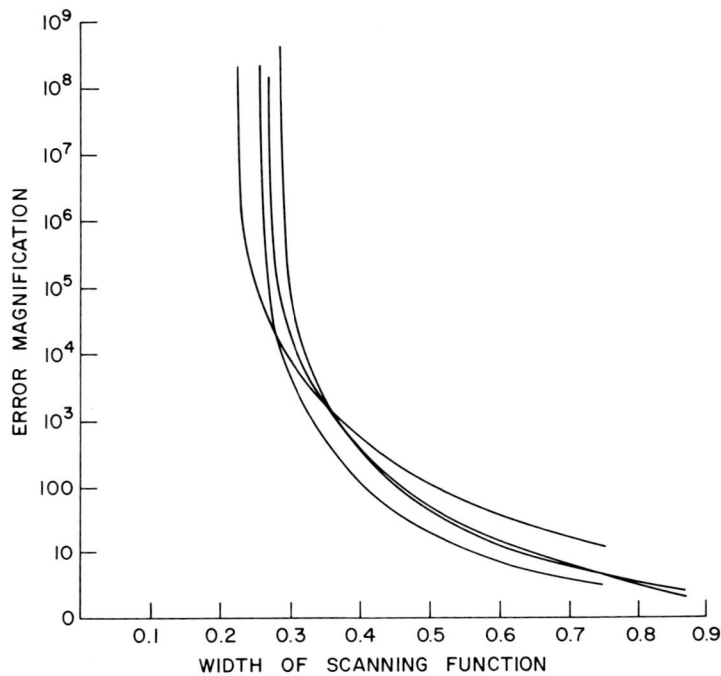

Fig. 3. Trade-off curves by Backus-Gilbert procedure for kernels of the form xe^{-yx}. Several values of y, giving maxima at $x = 0.2, 0.4, 0.6,$ and 0.8, are represented.

kernel xe^{-yx} is shown in Fig. 3. It is apparent that beyond a certain point (one-half width of scanning function ≈ 0.2 times the interval of integration), a huge increase in accuracy is needed to produce even a small increase in resolution. This is *not* an artifact of the Backus-Gilbert method (Ref. 3)--it is a property of the kernels involved and, furthermore, is not significantly influenced by the *number* of kernels used, provided that a reasonable number are employed (7 to 10, say). Similar plots can be made of resolution against number of kernels (for a fixed error magnification) and they show a rapidly diminishing return with further increase in number of kernels.

SOME ASPECTS OF THE INVERSION PROBLEM

C. Statistical Inversion Methods

Quite apart from linear inversion techniques with constraints based on statistics (e.g., EOF expansions), statistics find application even more directly. One sees in inversion linterature a very natural tendency to use comparisons between "measured" data (radiosonde soundings, for instance) and data given by inversion as a basis for judging the relative merits of the various numerical procedures. If we can make enough simultaneous *measurements* of both functions $g(y)$ and $f(x)$, then the relationship between $g(y)$ and $f(x)$ in principle can be "learned" from these measurements, provided only that it is physically reasonable to assume that the $f \leftrightarrow g$ relationship is linear to a good approximation. This is essentially the procedure followed in certain kinds of "statistical" inversions where a matrix ($\underset{\sim}{B}$, say) is inferred from a set of paired distributions $\underset{\sim}{f}$ and $\underset{\sim}{g}$. One is seeking to learn $\underset{\sim}{B}$, where

$$\underset{h \times 1}{\underset{\sim}{f}} = \underset{n \times m}{\underset{\sim}{B}} \underset{m \times 1}{\underset{\sim}{g}} + \underset{\sim}{\varepsilon} \tag{5}$$

and a number of fairly obvious minimization approaches can be followed to minimize $|\varepsilon|^2$ or some other suitable norm. One can, for example, for any specified $x_k = \xi$ (i.e., elemental position within $\underset{\sim}{f}$), solve for the kth row of $\underset{\sim}{B}$, imposing as usual a constraint for stability. Thus, if the kth row of $\underset{\sim}{B}$ is $\underset{\sim}{b}_k$, we have

$$f_k = \sum_j b_{kj} g_j + \varepsilon_k = \underset{\sim}{b}_k{}^* \underset{\sim}{g} \tag{6}$$

for every set of m measurements. If f_{lk} denotes the value of f_k from the lth measurement and $\underset{\sim}{g}_l$ is the corresponding array of g-values, one can write

$$\begin{pmatrix} \longleftarrow & \underset{\sim}{g}_1 & \longrightarrow \\ \longleftarrow & \underset{\sim}{g}_2 & \longrightarrow \\ \longleftarrow & \underset{\sim}{g}_3 & \longrightarrow \\ & \vdots & \\ \longleftarrow & \underset{\sim}{g}_L & \longrightarrow \end{pmatrix} \begin{pmatrix} \uparrow \\ \underset{\sim}{b}_k \\ \downarrow \end{pmatrix} = \begin{pmatrix} f_{1k} \\ f_{2k} \\ \vdots \\ \vdots \\ f_{mk} \end{pmatrix} \quad (7)$$

$$L \times m$$

and solve $\underset{\sim}{b}_k$. For each k (row of $\underset{\sim}{B}$) the procedure can be repeated and thereby a complete matrix $\underset{\sim}{B}$ generated. The constraint $\underset{\sim}{b}_k = $ Constant ensures error magnification, the constraint $\underset{\sim}{b}_k^* \underset{\sim}{b}_k \propto$ (variance of f_k) relates the permitted variation in the solution to the observed variation at the kth level. Rather than computing $\underset{\sim}{B}$ row by row, it can be obtained in toto if the sets of n-dimensioned vectors $\underset{\sim}{f}$ and of m-dimensioned vectors $\underset{\sim}{g}$ are collected into matrices. One version of this procedure considers a set of measured vectors $\underset{\sim}{f}_1, \underset{\sim}{f}_2, \ldots \underset{\sim}{f}_\ell$ and associated measured $\underset{\sim}{g}_1, \underset{\sim}{g}_2, \ldots \underset{\sim}{g}_\ell$ with

$$\left. \begin{aligned} \underset{\sim}{f}_1 &= \underset{\sim}{B} \, \underset{\sim}{g}_1 + \underset{\sim}{\varepsilon} \\ \underset{\sim}{f}_2 &= \underset{\sim}{B} \, \underset{\sim}{g}_2 + \underset{\sim}{\varepsilon} \\ &\vdots \\ \underset{\underset{n \times 1}{}}{\underset{\sim}{f}_\ell} &= \underset{\underset{n \times m}{}}{\underset{\sim}{B}} \underset{\underset{m \times 1}{}}{\underset{\sim}{g}_\ell} + \underset{\underset{n \times 1}{}}{\underset{\sim}{\varepsilon}} \end{aligned} \right\} \quad (8)$$

which gives

$$\underset{\sim}{B} \cong \underset{\sim}{G}^{-1} \underset{\sim}{F}; \quad \left[\underset{\sim}{F} = \overline{g(y_i) f(x_j)}; \; \underset{\sim}{G} = \overline{g(y_i) g(y_j)} \right] \quad (9)$$

$\underset{\sim}{F}$ is essentially the cross-covariance matrix of the $\underset{\sim}{f}$'s and $\underset{\sim}{g}$'s while $\underset{\sim}{G}$ is the covariance matrix of the $\underset{\sim}{g}$'s. Again, constraints must be applied to bring about stable solutions in which the individual elements of $\underset{\sim}{B}$ are not allowed to become large and excessively oscillatory. This can be done in exactly the same way as

SOME ASPECTS OF THE INVERSION PROBLEM

is done in the case of vector equations. It is the rows of $\underset{\sim}{B}$ that need individually to be constrained and this is most easily done if the equation for $\underset{\sim}{b}_r^*$, the rth row of $\underset{\sim}{B}$ can be written:

$$G \underset{\sim}{b}_r = \underset{\sim}{f}_r + \underset{\sim}{\varepsilon}_r \qquad G = \left\| g_i(y_i) \right\| ; \; \underset{\sim}{f}_r = \begin{pmatrix} f_1(x_r) \\ f_2(x_r) \\ \vdots \end{pmatrix} \qquad (10)$$

which can be solved under the constraint $b_r^* b_r$ = constant by

$$\underset{\sim}{b}_r = (G^* G + \gamma I)^{-1} G^* \underset{\sim}{f}_r$$

If more should be known about the error statistics, they can be incorporated into the solution process at the expense of a little additional complexity.

D. Nonlinear Iterative Methods

Linear iterative methods have found application, but they are not fundamentally different from other linear methods. Nonlinear iterative methods, on the other hand, are different and appear to be capable of giving good inversions where more direct methods have difficulty. Such methods were applied to atmospheric sounding by Chahine (Ref. 4) and have been successfully used in cloud- and particle-size distribution problems and elsewhere. They consist of taking a first guess $f_o(x)$ for the unknown $f(x)$ and then modifying it so as to improve the discrepancies between the measured g_i and $g_i' = \int K_i(x) f(x) dx$. In some applications the entire set of g_i' are calculated and the first guess (or subsequent iterate) is updated simultaneously at all tabular x-values; in other applications the first guess and all subsequent iterates are updated as soon as one value for f' has been calculated. There does not appear to be any clear-cut advantage of one procedure over the other.

Adjustment methods in nonlinear iterative algorithms rely on the principle that the change in g_i resulting from an adjustment in $f(x)$ around $x = \xi$ is proportional to the value of $K_i(\xi)$. In some algorithms the value of $f(x)$ is changed only where $K_i(\xi)$ is greatest, but if the tabular intervals are closely spaced, high frequencies may thereby be introduced in proportions that can become excessive. An alternative, which in the writer's experience can be superior, is to make the change at $x = \xi$ proportional to the value of the kernel there. A useful algorithm based on this principle is as follows: given an iterate $f_m(x)$, this is adjusted by comparing g_i and $g_i' = \int_a^b K_i(x) f_m(x) dx$ and computing the new iterate as

$$f_{m+1}(x) = f_m(x) \left[1 + \frac{g_i - g_i'}{g_i'} K_i(x) \right] \tag{11}$$

This makes the greatest proportional change where $K_i(x)$ is greatest and makes no change where $K_i(x)$ is zero.

This and several other similar algorithms have been used on a variety of problems and generally show excellent stability and independence of the first guess (see Fig. 4). With nonnegative kernels and positive first guess, no iterate will become negative anywhere provided none of the g values are zero and max $\{K_i(x)\} < 1$, a condition which can always be ensured by appropriate scaling. One, therefore, has a nonnegativity constraint built into the algorithm and there is little doubt that that is a major influence on the stability of the inversion. Since products of kernels are generated as the iteration proceeds, the bandwidth of admissible frequencies increases rapidly whatever the first guess, but this is a second-order process and in practice high frequency oscillations rarely seem to be a problem in inversions by this method (whether it be applied to computed data or real measurements).

The relative change made in $f(x)$ at every step can be kept small simply by limiting it within the algorithm; to the first

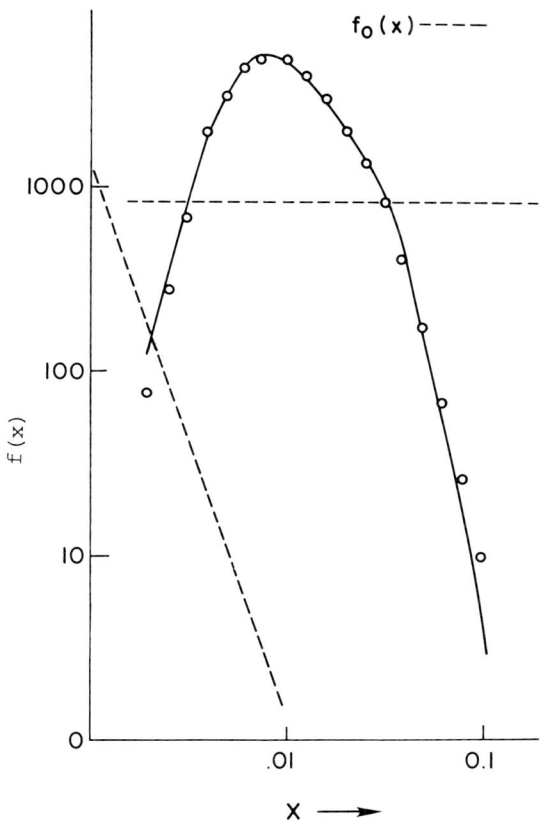

Fig. 4. Iterative nonlinear inversion on a particle sizing problem. Solid line and open circles refer to the two very different first guesses $f_1(x)$ (shown dashed).

order the solution thus produced is a linear combination of the kernel functions. Such a solution can be generated by linear methods also --e.g., we write $f(x) = \sum_i \xi_i K_i(x)$, and solve for $\underset{\sim}{\xi}$. The equation for $\underset{\sim}{\xi}$ is

$$C\underset{\sim}{\xi} = \underset{\sim}{g} + \underset{\sim}{\varepsilon} \tag{12}$$

C being the kernel covariance matrix. The connection between nonlinear iterative solutions, nonlinear iterative solutions with limited adjustments and direct linear methods might usefully be investigated by systematically comparing results given by the three methods for the same problem. The writer is not aware of this having been done at the time of writing.

III. ASSESSMENT OF THE VARIOUS PROCEDURES

A. Comparison of Results on a Standard Problem

There are many ways of doing this and they will not always give the same answer. It is nevertheless informative to take a very simple problem and solve it in a number of ways. This has been done as follows:

$$g_i = \int_0^1 K_i(x) f(x) dx \quad (i = 1, 2, \ldots N; N = 20)$$

$$K_i(x) = x e^{-y_i x} \quad (y_i = 0.1, 0.2, \ldots 1.0, 1.2, 1.4, \ldots,$$
$$2.0, 2.5, 3.0, 3.5, 4, 5, \ldots 10)$$

$$f(x) = 1 - 4(x - 1/2)^2$$

With this integrand, a simple weighted trapezoidal quadrature gave the g_i with an rms error of 0.1% over all y-values. This problem has been solved in a variety of ways and the solutions are shown on the Fig. 5. They are:

a. Constrained linear inversion with minimum constraint H = I (i.e., minimum "power").

b. Backus-Gilbert inversion, error magnification ≈ 100 (Ref. 3).

c. Chahine's original iterative method, only one ordinate adjusted per step (Ref. 4).

d. A modification of the latter, as described above.

e. Constrained linear inversion with end points fixed at their correct values [f(0) = f(1) = 2].

f. Constrained linear inversion with end points fixed at approximately their correct values f(0) = 1.95, f(1) = 2.05.

SOME ASPECTS OF THE INVERSION PROBLEM

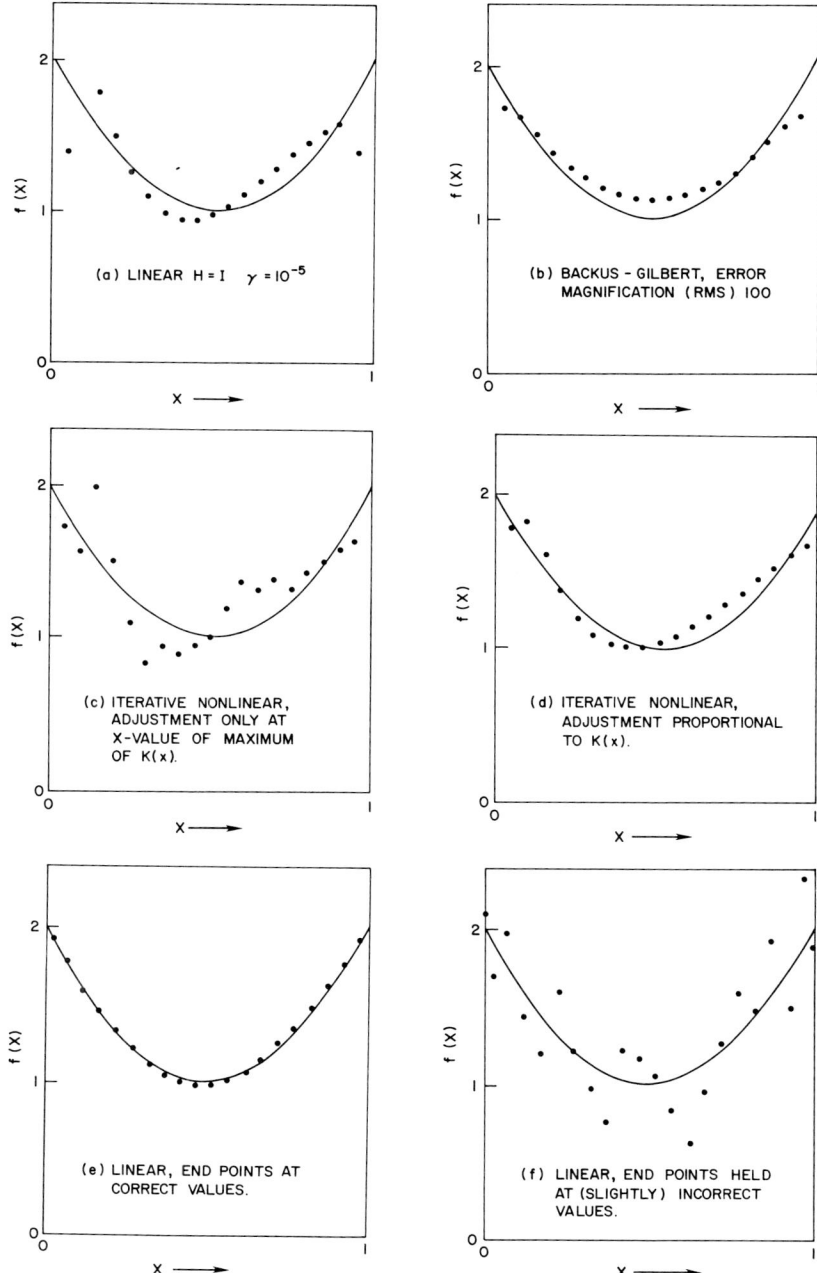

Fig. 5. Inversion by several different inversion methods.

There is no point whatever in trying to grade the methods on the basis of these tests. The results are shown mainly to demonstrate that reasonable inversions can be obtained by each of the methods and that rather dramatic improvements (such as from Figs. 5(a) to 5(e)) can be achieved when we bring in *a priori* knowledge. But, as Fig. 5(f) shows, we must be sure of this *a priori* knowledge; otherwise, more harm than good is done by its incorporation.

B. Resolution

There is a very simple method whereby we can make estimates of the resolution of an inversion procedure in toto. To do this, we simply take the array of g_i values corresponding to the special case $f(x) = \delta(x - x_0)$ and applied the inversion algorithm to that g-vector, obtaining a distribution which we call $r(x_0)$. A perfect inversion procedure would return $\delta(x - x_0)$ when given as input the g-vector $(\underline{k}(x_0),$ i.e., $(K_1(x_0), K_2(x_0), \ldots K_m(x_0))$, and the extent to which any procedure fails to do this is a useful objective gauge of its resolution. (It should be emphasized that "procedure" here encompasses the set of kernels which are used, which play a more important role than the algorithm itself in determining resolution or lack of it.)

Some examples of resolution tests by this method are shown in Fig. 6. We also have included in the figure Backus-Gilbert (Ref. 3) scanning functions for the same set of kernels and it should be noted that for linear inversion methods the result given by the above δ function test is in fact a scanning function, i.e., the solution obtained is a convolution of that function with the original $f(x)$. In the case of nonlinear inversions, that does not hold true, but the significance of the distribution $r(x_0)$ is still clear.

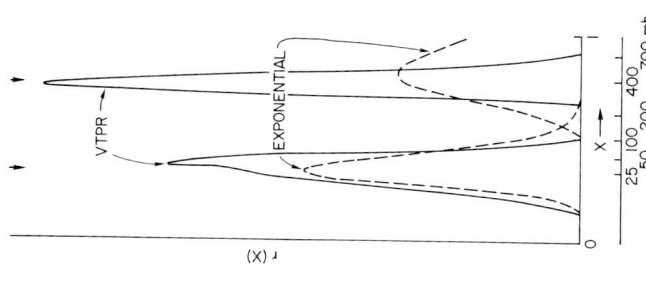

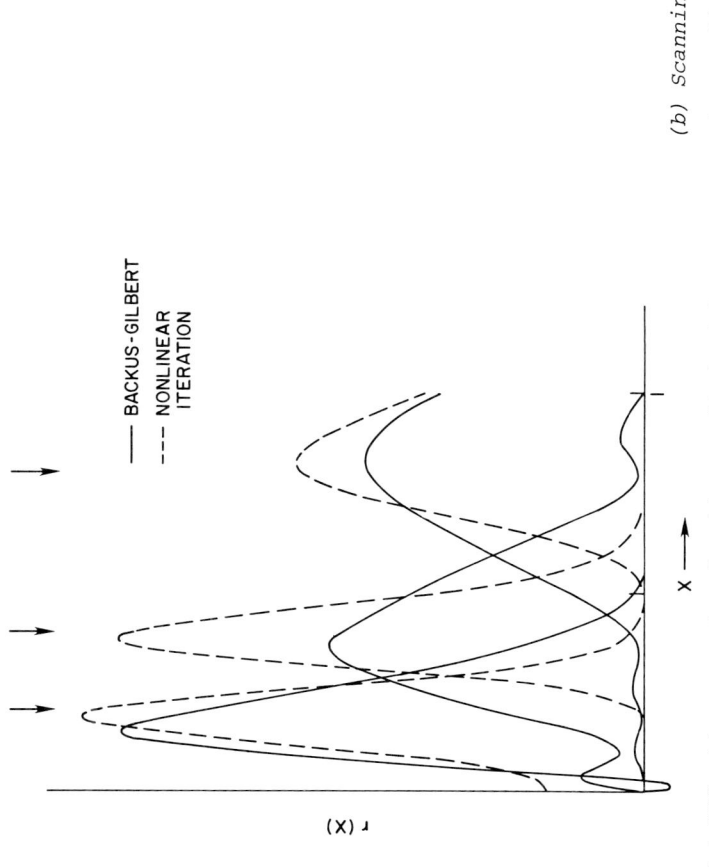

(a) Effective scanning functions, obtained by inverting data which would be given when the unknown $f(x)$ was a delta function. Arrows show positions of the delta functions. Kernels of form xe^{-yx}.

(b) Scanning functions for inversion by iterative nonlinear method of data functions at indicated values of x. Solid curves relate to VPTR kernels, broken curves to exponential transmittance.

Fig. 6. Scanning functions

C. Merits and Disadvantages

Each of the procedures possesses certain merits. The constrained linear inversion is fast, especially when a large number of observed g vectors are to be processed in the same way; it is objective when smoothness constraints are employed, and it allows anything which we may know about f(x) to be built into the constraint. The Backus-Gilbert (Ref. 3) method is aesthetically appealing in that the scanning function can be calculated and graphed. Statistical techniques give more realistic soundings than the other methods. Nonlinear iterative methods avoid the necessity to formulate explicit constraints. They can accommodate a wide range of magnitude in f(x) and they seem generally to be superior with respect to resolution, and tolerance of errors.

IV. CONCLUDING REMARKS

The preceding discussion was not intended to be a step-by-step review of inversion problems and algorithms. Rather, a number of selected points were given attention primarily in order to demonstrate that there is not a great deal of difference fundamentally between the various procedures. The differences lie rather in the basic selection process (without which no stable solution can be obtained). If we decide to select the smoothest solution, we necessarily damp out features such as the tropopause; if we select the solution which is closest to some statistical expectation, such features will appear, but not necessarily in the right place; we may also thereby fail to see rare but real and important excursions. It is misleading to judge the result of an inversion on the basis of "reasonableness." Indeed, in the case of such things as temperature and water vapor soundings, it seems more appropriate that the selection of "most suitable" constraints and filterings be based on what is going to be done with them, not on some subjective judgment which can be greatly influenced by the scaling, etc., chosen to display, in some form or other, "true" and "computed"

SOME ASPECTS OF THE INVERSION PROBLEM

soundings, thickness plots or whatever. Nobody really cares what the temperature is over some point at 9 km (30,000 ft) in mid-Pacific nor for that matter what the 200 mb thickness is there or how much water vapor resides above a particular level. These quantities primarily provide input for analysis and prediction procedures. Many of these procedures are fundamentally nonlinear but may, perhaps, be approximated closely by a linear operation. In a very qualitative and general way, one might look at the prediction of tomorrow's wind field in the Pacific as an operation on today's field and today's satellite radiance data. If today's and tomorrow's fields were accurately known, statistical or other techniques could be utilized to optimize the prediction, virtually ignoring the physics involved. If this is not possible, at least the selection of "best" from the large population of profiles *which are all equally acceptable from the radiance point of view* should surely be based as much as possible on the intended application of the soundings, and users of inversion data should be more actively brought into the selection process.

As was mentioned earlier, *appreciable* improvement on resolution by improving the accuracy of measurement is hardly practicable since a very great improvement in accuracy produces only a very modest improvement in resolution. There does appear to be room for improvement in accuracy so far as the kernels are concerned.

SYMBOLS

$\underset{\sim}{k}(x)$	kernel function
$k_i(x)$	elements of the kernel function
$\underset{\sim}{f}(x)$	"sought" function
$\underset{\sim}{g}$	function representing measured quantities
t	time
p	pressure

REFERENCES

1. L. D. Kaplan, Inference of atmospheric structure from remote radiation measurements, *J. Opt. Soc. Am. 49,* 1004 (1959).

2. D. O. Wark and H. E. Fleming, Indirect measurements of atmospheric temperature profiles from satellites, *Mon. Weather Rev. 94,* 351 (1966).

3. G. Backus and F. Gilbert, Uniqueness in the inversion of inaccurate gross earth data, *Philos. Trans. R. Soc. London, A266,* 123 (1970).

4. M. T. Chahine, Inverse problems in radiative transfer: Determination of atmospheric parameters, *J. Atmos. Sci. 27,* 980 (1970).

DISCUSSIONS

Chahine: You have made a flat statement that this type of integral equation has an infinite number of solutions. Do you say that this is true for any kernel and for any form of the given function, whether it is continuous or discreet?

Twomey: One should, perhaps, modify that to, say, physical kernels, the kernel which one is likely to encounter in a physical measurement. It would certainly not be true for many mathematical kernels, such as delta functions.

Chahine: Not necessarily. Exponential kernels can give you a unique solution sometimes.

Twomey: Well, you can demonstrate in the presence of finite accuracy--you only have to go to a high enough frequency and you get ambiguity.

Chahine: The nonuniqueness can be due to noise in the given function (data) or in the kernel. But it is not necessarily an intrinsic property of this integral equation. I would like to make this point clear.

Twomey: Yes. I would agree. However, I would comment that I think from the point of view of keeping oneself out of trouble, one would be better off to feel that there are an infinity of solutions than one would be to think there is a unique solution.

Chahine: Physically?

Twomey: Yes, from the point of keeping oneself out of trouble. You can't get into trouble mathematically.

King: Don't you think that part of the problem of oversell is related to formatting? Inversion efforts, thus far, have been directed toward mimicking the infinitely resolved temperature profile given by the radiosonde. And this, of course, the inversion is incapable of doing in principle. Isn't it preferable to format the information in terms of, say, the six parameters that one could infer from six radiance observations? These six gross atmospheric structure parameters would, perhaps, be of more use to numerical weather predictors anyway, since their algorithms are not really sensitive to the kind of detail provided by the radiosonde.

Twomey: I agree 100 percent, Jean, yes. I think we should even go a step further than this because (I think I said it in the written version of this paper), I don't think any of us really care what the temperature is 200 millibars above Einewtok or somewhere like that. It is only data that is going into a circulation model or a prediction model and so forth. The first thing those people are going to do is try to apply some of their own numerical processes to that. I mean, for example, if you feed them a superadiabatic lapse rate, they are going to make a convective adjustment to it. So the first thing they are going to do with a nice temperature profile which you have at great trouble and expense fashioned, they are going to change it. So I think you should look at that and incorporate that and see what is best for them. It may be that some of the things inversion people are sweating to try to put into their profiles are not actually needed by the people who are going to use those profiles. I think it is also up to the people who are going to use those profiles to state what they want, not the way they have been stating it. I think when Dave Wark and I talked to people some years ago, they would say, oh, we want a quarter of degree; we want at least as good as the radiosonde; we want better than the radiosonde. But they have a grid in which the whole atmosphere is in about 160 km (100 mi) boxes. So they are only putting themselves on if they say they want radiosondes every 100 meters and with accuracy of 0.3048 (1 ft) resolution, because they are not going to be able to handle it.

Kaplan: Well, if we get a superadiabatic lapse rate, we better not feed it to them to make the convective adjustment, because we are supposed to be sounding the atmosphere. And we better do the adjustment because we know how to do it better than they do. And it isn't the convective adjustment that is required. There are real needs and I think we have to try to answer those needs. And I think probably what is necessary is to see what it is that really is wanted and probably what is wanted in numerical weather prediction is mean temperatures between constant pressure levels or constant height levels. And, I think we ought to see whether that can be produced and the solutions given in those forms. And if it can't, there should be a negotiation going back and forth and trade-offs between the people doing the soundings and the users, because the soundings are not being done for its own sake. I mean, some of us may like the idea of the game, but the real purpose is being able to improve the weather forecast.

Goldman: I was wondering if you could expand your opinion about the accuracy of the constraint against no constraint at all? You have shown in your example that small error in the constraint can give much worse inversion. And I was wondering if you could expand a little bit on this by some actual examples?

Twomey: Well, no, I can't. I have that example there, but one could go on and on, I mean, calculate various examples. I think it is fairly obvious what is going on. If there are real constraints, which there are in many physical problems, of course, especially at the endpoints of your interval, there is likely to be some kind of physical or essential constraint there quite outside your inversion problem. If it is of that nature and you can state accurately what it is, you can only benefit by putting it in. But just exactly how much error you need to be in that constraint before it is doing more harm than good, that would obviously depend on your particular problem. You know the precise numbers that you are working on. It is a simple thing to test, obviously.

GENERALIZATION OF THE RELAXATION METHOD FOR THE INVERSE SOLUTION OF NONLINEAR AND LINEAR TRANSFER EQUATIONS

Moustafa T. Chahine
Jet Propulsion Laboratory
California Institute of Technology

A mapping transformation is derived for the inverse solution of nonlinear and linear integral equations of the types encountered in remote sounding studies. The method is applied to the solution of specific problems for the determination of the thermal and composition structure of planetary atmospheres from a knowledge of their upwelling radiance.

I. INTRODUCTION

The problem of determining an unknown function $g(z)$ from the integral equation

$$I(\nu) = N\, g(z)$$

given the function $I(\nu)$ and the integral operator N, has arisen repeatedly in problems of radiative transfer as in Ref. 1 and in transport theory as in Ref. 2. The mathematical problem here is a difficult one and, in fact, may not always have a solution for an arbitrary function $I(\nu)$. The difficulties are compounded by the facts that $I(\nu)$ is obtained in general from measurements contaminated with noise and that the integral operator itself is an approximation to a real physical process.

In this paper, we treat a specific class of linear and nonlinear integral equations in which *the kernel has the important property of reaching its maximum peak at different values of z for different values of* ν. This mathematical property is very common in transfer and transport problems where the dominant physical process takes place within a narrow segment of the range of integration.

In the following section, we will develop a relaxation method of solution based on the principle of mapping transformations to recover

$$g(z) = N^{-1} I(\nu)$$

and study the stability and accuracy of the solutions in the presence of noise in the given data. The remaining sections of this paper will be devoted to the study of the inverse problems for the determination of atmospheric temperature and composition profiles from remote sounding radiance data.

II. MATHEMATICAL FORMULATION

The formulation of remote sounding problems in radiative transfer leads often to nonlinear integral equations of the form

$$I(\nu) = B[\nu, g(z_o)] C(\nu, z_o) + \int_{z_o}^{\bar{z}} B[\nu, g(z)] K(\nu, z) \, dz \qquad (1)$$

In some cases, however, the resulting integral equations are linear in $g(z)$ and of the form

$$I(\nu) = g(z_o) C(\nu, z_o) + \int_{z_o}^{\bar{z}} g(z) K(\nu, z) \, dz \qquad (2)$$

In Eqs. (1) and (2) $C(\nu, z_o)$ and $K(\nu, z)$ describe specific radiative transfer processes such as absorption, emission, or scattering in the atmosphere. In many problems of remote sounding, this kernel $K(\nu, z)$ reaches its maximum peak at different values of z for different values of ν. $I(\nu)$ is a given function, usually measured at a discrete number of observations $\tilde{I}(\nu_j)$, and $g(z)$ is the distribution of an atmospheric parameter to be determined. Thus, the inverse solution here reduces to finding a function $g(z)$ such that when it is substituted into Eqs. (1) or (2), it will yield values of $I(\nu)$ equal to the corresponding measurements $\tilde{I}(\nu_j)$, with

$$\tilde{I}(\nu_j) - I(\nu_j) = 0$$

for all the given values of ν_j.

In remote sounding problems there is usually no doubt of the physical existence of a solution or perhaps even of its uniqueness. But from a mathematical point of view, the problems of demonstrating existence and ensuring uniqueness of $g(z)$ are of great importance and are directly related to the various simplifying physical and mathematical assumptions made in the derivation of the integral equation and depend also on the information content of the measured data. Therefore, a general treatment of the problems of existence and uniqueness for Eqs. (1) and (2) is difficult unless it is approached from the narrow point of view of the dependence of the solution on the initial guess and the interpolation (or quadrative) method used. However, accurate demonstration of existence and uniqueness may be carried out, in principle, for some problems, depending on the form of the kernel and the degree of discretization of the function $\tilde{I}(\nu_j)$.

A. General Method of Solution

The right-hand side of Eqs. (1) and (2) may be viewed as an integral operation transforming variations of g with respect to z into variations of I with respect to ν as

$$I(\nu) = N g(z). \tag{3}$$

To obtain $g(z)$ we need, therefore, to perform an inverse transformation from the (I,ν) plane into the (g,z) plane with

$$g(z) = N^{-1} I(\nu) \tag{4}$$

We can accomplish this by applying a mapping transformation which maps points on the ν-axis into corresponding points on the z-axis and similarly maps points on the I-axis into points on the g-axis.

1. *The $\nu - z$ Mapping Transformation*

To map the ν-axis into the z-axis, let z_j be a point between z_o and $\bar{z}$ where $K(\nu_j, z)$ reaches its maximum value or

$$\left. \frac{\partial K(\nu_j, z)}{\partial z} \right|_{z = z_j} = 0 \tag{5}$$

From Eqs. (1) or (2) we note that variation in $g(z)$ around z_j should affect the values of $I(\nu_j)$ very strongly while variation in $g(z)$ at values of $z \ll z_j$ and $z \gg z_j$ should not affect $I(\nu_j)$ by the same magnitude. We propose, therefore, to use this property to map this point ν_j into z_j on the z-axis. In general, since $K(\nu_j, z)$ reaches its maximum values at different values of z_j, $j = 1, 2, \ldots J$ for different values of ν_j, $j = 1, 2, \ldots J$. We can use Eq. (5) to derive a relationship between ν and z

$$\nu_j \longleftrightarrow z_j$$

and map different points on the ν-axis into different points on the z-axis.

2. *The $I - g$ Relaxation Transformation*

Mapping of the I-axis into the g-axis is much more difficult and needs to be carried out by iteration. We apply the mean value theorem (or the method of steepest descent) to Eq. (1) and derive a relaxation equation of the form

$$\frac{B[\nu_j, g^{(n+1)}(z_j)]}{B[\nu_j, g^{(n)}(z_j)]} \approx \frac{\tilde{I}(\nu_j)}{I^{(n)}(\nu_j)} \tag{6a}$$

GENERALIZATION OF THE RELAXATION METHOD

Equation (6) is written in an iterative form where $g^{(n)}(z_j)$ is the nth guess of the solution. $I^{(n)}(\nu_j)$ is the corresponding value computed according to Eq. (1). $\tilde{I}(\nu_j)$ is given and $g^{(n+1)}(z_j)$ is the resulting (n + 1) guess. Similarly, the relaxation equation corresponding to Eq. (2) is of the form

$$\frac{g^{(n+1)}(z_j)}{g^{(n)}(z_j)} \approx \frac{\tilde{I}(\nu_j)}{I^{(n)}(\nu_j)} \qquad (6b)$$

Details of this derivation are given by the author in Refs. 3 and 4. Equations (6a,b) transform changes in the ratio $\tilde{I}(\nu_j)/I^{(n)}(\nu_j)$ at different points on the ν-axis into changes in $g^{(n)}(z_j)$ at specific points along the z-axis. The mechanism of this transformation is iterative in which $g^{(n)}(z_j)$ is modified at every step n to yield a new value $g^{(n+1)}(z_j)$.

It is possible to generalize Eqs. (6a,b) formally and write the relaxation transformation as

$$g^{(n+1)}(z_j) = \alpha_j^{(n)} g^{(n)}(z_j) \qquad (7)$$

where $\alpha_j^{(n)}$ are scaling factors computed directly from Eqs. (6a) or (6b).

(It is sometimes necessary to use a weighted form of the scaling factors, as defined in Eq. (10), in which several values of $\tilde{I}(\nu_j)$, $j = j', \ldots j''$, are used to derive each scaling factor $\bar{\alpha}_j^{(n)}$. This approach will be discussed later in this section.)

3. Iterative Steps of the Method of Solution

Assume that a set of measurements $\tilde{I}(\nu_j)$ is given for $j = 1, 2, \ldots J$ and use Eq. (5) to find the corresponding values of z_j, $j = 1, 2, \ldots J$.

(a) Make an initial guess (n = 0) for $g^{(n)}(z)$.

(b) Substitute $g^{(n)}(z)$ into Eq. (1) [or Eq. (2)] and evaluate $I^{(n)}(\nu_j)$ $j = 1, 2, \ldots J$ using an appropriate interpolation (quadrature) formula.

(c) Check the residuals,

$$R^{(n)}(\nu_j) = \frac{\tilde{I}(\nu_j) - I^{(n)}(\nu_j)}{\tilde{I}(\nu_j)} \qquad j = 1, 2, \ldots J$$

If all $R^{(n)}(\nu_j) \to 0$, then $g^{(n)}(z_j)$ is a solution. If not, then obtain a new guess,

$$g^{(n+1)}(z_j) = \alpha_j^{(n)} T^{(n)}(z_j)$$

(d) Go to step (b) and repeat until step (c) is satisfied.

B. Analytical Example

In order to illustrate the relaxation method of solution, we suggest the following analytical example.

Let

$$I(\nu) = \int_{-\infty}^{+\infty} g(z) e^{-(z-\nu)^2} dz$$

and given

$$I(\nu) = e^{-(z_o - \nu)^2}$$

where z_o is a constant, find $g(z)$ by the method of relaxation. (The answer is a delta function, $g(z) = \delta(z_o - z)$, which we will try to determine next.)

1. The $\nu - z$ Mapping Transformation

The kernel $e^{-(z-\nu)^2}$ has a maximum at $\nu = z$ and thus the corresponding mapping transformation from the ν-axis to the z-axis is $\nu = z$.

2. The $I - g$ Relaxation Transformation

The corresponding relaxation equation according to Eq. (6b) is

GENERALIZATION OF THE RELAXATION METHOD

$$g^{(n+1)}(z) = g^{(n)}(z) \left. \frac{\tilde{I}(\nu)}{I^{(n)}(\nu)} \right|_{\nu = z}$$

3. Iterative Solution

In this ideal analytical example we do not need to specify an interpolation formula because we are dealing with a continuous function $\tilde{I}(\nu)$ which maps the entire ν-axis into the entire z-axis.

To start the iteration ($n = 0$) let us assume a constant value for the initial guess; for example

$$g^{(0)}(z) = b$$

and apply the mapping transformations to generate the following iterative solutions. We get

$$g^{(1)}(z) = \frac{1}{\sqrt{\pi}} \exp\left[-(z_o - z)^2\right]$$

$$g^{(2)}(z) = \sqrt{\frac{2}{\pi}} \exp\left[-\frac{3}{2}(z_o - z)^2\right]$$

$$\vdots$$

$$g^{(n+1)}(z) = \sqrt{\frac{1 + C_n}{\pi}} \exp\left[-\left(C_n + \frac{1}{1 + C_n}\right)(z_o - z)^2\right]$$

where the coefficients C_n are obtained from the following recurrence relaxation

$$C_o = 0 \qquad C_n = C_{n-1} + \frac{1}{1 + C_{n-1}}$$

To show that the recovered solution $g^{(n)}(z)$ is a delta function centered at z_o, we can easily prove that, as $n \to \infty$,

$$C_n \to \infty \qquad \int_{-\infty}^{+\infty} g^{(n)}(z) \, dz \to 1$$

and

$$d_n = 2\sqrt{\frac{\ln 2}{C_n}} \to 0$$

where d_n is the width of the function $g^{(n)}(z)$ at half peak as shown in Fig. 1.

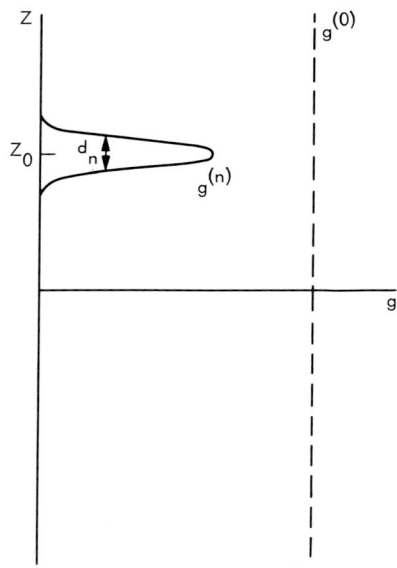

Fig. 1. The analytical illustration.

C. Stability and Convergence of Solutions

The previous analytical example offers an ideal case in which the entire ν-axis maps into the entire z-axis, the number of iterations that can be carried out is infinite, and the given function $\tilde{I}(\nu)$ is exact. In remote sensing problems, however, we are given small (and sometimes incomplete) data sets $I(\nu_j)$, $j = 1, 2 \ldots J$ where the peaks of the kernels are not uniformly distributed and where both $I(\nu_j)$ and $K(\nu_j, z)$ are known with a certain degree of uncertainty. Under these conditions, it becomes sometimes necessary to (i) supplement the data with *a priori* information about the expected result, (ii) use weighted scaling factors $\bar{\alpha}$ instead of α in order to minimize the effects of large measurement noise on the accuracy and stability of solutions, and (iii) establish numerical criteria for determining the optimum number of iterations required. We will examine these aspects of the problem in the remaining part of this section.

D. Properties of the Residuals

Application of the relaxation method to a variety of linear and nonlinear problems has shown that in the absence of noise in the given data, the computed values of the function $I(\nu_j)$ approach the measured data $\tilde{I}(\nu_j)$ after a small number of iterations. Their residual difference $R(\nu_j)$ tends asymptotically to zero (or to the value of the quadrature errors) in a root-mean-square (rms) sense as

$$R^{(n)} = \left\langle \frac{\tilde{I}_j - I_j^{(n)}}{\tilde{I}_j} \right\rangle_{rms} \to 0 \tag{8}$$

But, if we add ε_j noise to the data as

$$I'(\nu_j) = I(\nu_j) + \varepsilon_j$$

we observe that the residuals, in this case, decrease first rapidly and then approach an asymptotic value equal to the noise in the data in an rms sense with

$$R'^{(n)} = \left\langle \frac{\tilde{I}'_j - I_j^{(n)}}{\tilde{I}'_j} \right\rangle_{rms} \to \langle \tilde{\varepsilon}_j \rangle_{rms} \tag{9}$$

(This property will be discussed in more detail in the following section and will be illustrated in Fig. 6.)

Thus, in the presence of random errors in measurements, the residuals do not give a false indication of convergence because they will not tend toward zero. The residuals first decrease and then approach an asymptotic value of the same order of magnitude as the errors in measurements. This property is due to the partial overlapping of the kernels and suggests that the iterative process should be terminated when $R^{(n)}$ approaches its asymptotic value.

The effect of continued iterations beyond this point does not always increase the amount of information extracted from the data, and in the presence of large noise levels in the measurements might lead to oscillations in the recovered solutions. To prevent such oscillations from happening, it is advisable to apply, after a few iterations, say n > 3, certain weights to the scaling coefficients $\alpha_j^{(n)}$ in the form of

$$\bar{\alpha}_j^{(n)} = \frac{\sum_{k=1}^{J} \alpha_k^{(n)} W_j(\nu_k)}{\sum_{k=1}^{J} W_j(\nu_k)} \qquad j = 1, 2 \ldots J \qquad (10)$$

and use $\bar{\alpha}_j^{(n)}$ to generate the new iterations as

$$g^{(n+1)}(z_j) = \bar{\alpha}_j^{(n)} g^{(n)}(z_j) \qquad (11)$$

The use of $\bar{\alpha}_j$ will slow down the rate of convergence considerably but will tend to diminish the effects of random noise in the data on the iterative solution.

Different forms of $W_j(\nu_k)$ can be adopted, such as taking $W_j(\nu_k)$ equal to the fractional value of the kernels $K(\nu, z)$ at z_j for $\nu_1, \nu_2 \ldots \nu_j$. In this case, we write

$$W_j(\nu_k) = K(\nu_k, z_j)/K(\nu_k, z_k) \qquad (12)$$

where $K(\nu_k, z_k)$ is the maximum value of the kernel. Additional details on the convergence properties of the solution can be found in Refs. 5, 6, and 7.

Ultimately, the iteration process should be terminated when the rate of convergence of the solution with respect to itself, defined as

$$\langle \Delta g^{(n)} \rangle_{av} = \frac{1}{J} \sum_{j=1}^{J} \left| g^{(n+1)}(z_j) - g^{(n)}(z_j) \right| \qquad (13)$$

approaches a certain prescribed value.

GENERALIZATION OF THE RELAXATION METHOD

E. Interpolation Methods

The interpolation aspects of the solution are not trivial, particularly in cases where the number of measured data points is deficient. The recovered solution is a function of the number of useful observations available and of the interpolation method used. The selection of a suitable interpolation formula is one of the subjective aspects of this problem. While it is always possible to apply many different interpolation methods, it is obvious that from a finite set of J measurements, it is impossible to recover more than J *independent* parameters.

Application of the relaxation method of solution to a variety of problems has shown that the resulting solutions are independent of the initial guess but depend on the interpolation formula selected to determine the solution $g(z)$ at the intermediate values between z_j ($j = 1, 2, \ldots J$). Linear interpolation methods are recommended in the absence of any information about the expected solutions; however, other interpolation methods can be applied when needed.

In cases when *a priori* knowledge of the shape of the expected solution is given, a perturbation approach is recommended to make use of the available information. This can be accomplished in the iteration process by including in the initial guess all available information about the shape of the expected solution. To preserve this shape in subsequent iterations, we perform the interpolation on the scaling factors $\alpha^{(n)}(z_j)$ and generate scaling factors at all intermediate values of z, $\alpha_{int}^{(n)}(z)$. The complete interpolated solution is then obtained at all values of z as

$$g^{(n+1)}(z) = \alpha_{int}^{(n)}(z) \, g^{(n)}(z) \qquad (14)$$

It is obvious that the same interpolation procedure can be performed also when $\bar{\alpha}_j^{(n)}$ is used. In this process, the final answer is made to depend on the initial guess by preserving the form of the input function $g^{(0)}(z)$ in all the steps of the iterative solution.

In general, however, reliance on *a priori* information about the expected solution should not be considered unless the number of available measurements is insufficient and unless the information content of measurable data is incapable of recovering certain essential features of the solution, such as the location of the tropopause. However, the fundamental mathematical condition that from a set of J measurements it is possible to recover only J' independent parameters such that $J' \leq J$ remains the rule.

In the following sections, we will apply the relaxation method to two remote sounding problems and study the properties of the solution and quality of results specifically for the determination of the thermal and composition structure of planetary atmospheres.

III. THE INVERSE PROBLEM FOR TEMPERATURE PROFILES

In the problem of remote sounding of atmospheric temperature profiles, the following equation occurs

$$I(\nu) = B[\nu, T(z_o)]\tau(\nu, z_o) + \int_{z_o}^{z} B[\nu, T(z)]K(\nu, z)dz \qquad (15)$$

Equation (15) is the integral form of the radiative transfer equation for a plane parallel homogeneous and nonscattering atmosphere in local thermodynamic equilibrium. $I(\nu)$ is the outgoing radiance measured at a vertical distance $\bar{z}$ from the surface z_o of a planet within a narrow solid angle around the local vertical axis, z. B is the Planck function explicitly given as

$$B[\nu, T(z)] = a\nu^3 / (e^{\frac{b\nu}{T(z)}} - 1) \qquad (16)$$

where a and b are two given constants. $\tau(\nu, z)$ is the

GENERALIZATION OF THE RELAXATION METHOD

transmittance of a column of absorbers between levels $\bar{z}$ and z and is defined for monochromatic observations as

$$\tau(\nu, z) = \exp\left[-\int_z^{\bar{z}} \rho_s(z')k(\nu, z')dz'\right] \tag{17}$$

where $k(\nu, z)$ is the absorption coefficient at ν due to all lines, and can be represented for a Lorentz profile by the equation

$$k(\nu, p) = \sum_i \frac{S_i(T)}{\pi} \frac{\alpha'_i(T, p)}{(\nu - \nu_i)^2 + \alpha_i'^2(T, p)} \tag{18}$$

where S_i is the strength, α'_i is the half width of the line and ν_i is the frequency at peak intensity of the line. The density profile of the absorbing gas s is given by $\rho_s(z')$. The kernel $K(\nu, z, \rho_s, \ldots)$ is defined as

$$K(\nu, z) = \frac{\partial \tau(\nu, z)}{\partial z} \tag{19}$$

The pressure $p(z)$ is related to z through the hydrostatic equation

$$dp = -\rho(z) g\, dz \tag{20}$$

where $\rho(z)$ is the density profile of the entire atmosphere.

In practical observations, measurements of $I(\nu)$ are made at a discrete number of frequencies ν_j centered within a finite band $\Delta\nu$ with

$$I(\nu_j) = \int_{\nu''}^{\nu'} \phi(\nu_j, \nu) I(\nu) d\nu \tag{21}$$

where $\phi(\nu_j, \nu)$ is the instrument function.

From a practical point of view, it is advisable (but not necessary for the present method) to substitute Eq. (15) into Eq. (21) and define the transmittance $\tau(\nu_j, z)$ in the interval $\Delta\nu$ as

$$\tau(\nu_j, z) = \int_{\nu''}^{\nu'} \phi(\nu_j, \nu) \tau(\nu_j, z) d\nu \tag{22}$$

Since $B(\nu, T)$ is a smooth function in the interval $\Delta\nu$, we take

$$B(\nu, T) \doteq B(\nu_j, T)$$

and rewrite Eq. (21) as

$$I(\nu_j) = B[\nu_j, T(z_0)]\tau(\nu_j, z_0) + \int_{z_0}^{Z} B[\nu_j, T(z)]K(\nu_j, z)dz \quad (23)$$

In the rest of this paper, we will deal with Eq. (23) whether we refer to the measured radiance as $\tilde{I}(\nu_j)$ or as $I(\nu)$.

From a given set of values of $\tilde{I}(\nu_j)$, $j = 1, 2 \ldots J$, we want to determine the temperature profile $T(z_j)$, assuming that $\rho_s(z)$, $\tau(\nu_j, z)$ and $K(\nu_j, z)$ are known. In this problem, the selection of the set ν_j and the determination of $T(z_j)$ are strongly related and form the basis for the method of inverse solution of Eq. (23).

By selecting a set of frequencies ν_j with varying degrees of atmospheric attenuation such that $\tau(\nu, z_0) \geq \tau(\nu_j, z), \ldots \geq \tau(\nu_j, z)$, we can generate a set of kernels $K(\nu_j, z)$ such that for each value of ν_j the kernel possesses a maximum at a different value of z_j as shown in the illustration (Fig. 2).

A. The Mapping Transformation

Equation (23) is a nonlinear integral equation with fixed limits which may be viewed as a nonlinear transformation from $T(z)$ to $I(\nu)$ as in Eq. (3), namely,

$$\tilde{I}(\nu) = N\, T(z)$$

To obtain $T(z)$, we need to perform an inverse transformation as in Eq. (4), namely

$$T(z) = N^{-1}\, \tilde{I}(\nu)$$

Figure 2 strongly suggests a mapping transformation from the ν axis to the z axis. Since the kernels $K(\nu_j, z)$ are strongly decaying functions, variations of $T(z)$ around z_i will affect the

GENERALIZATION OF THE RELAXATION METHOD

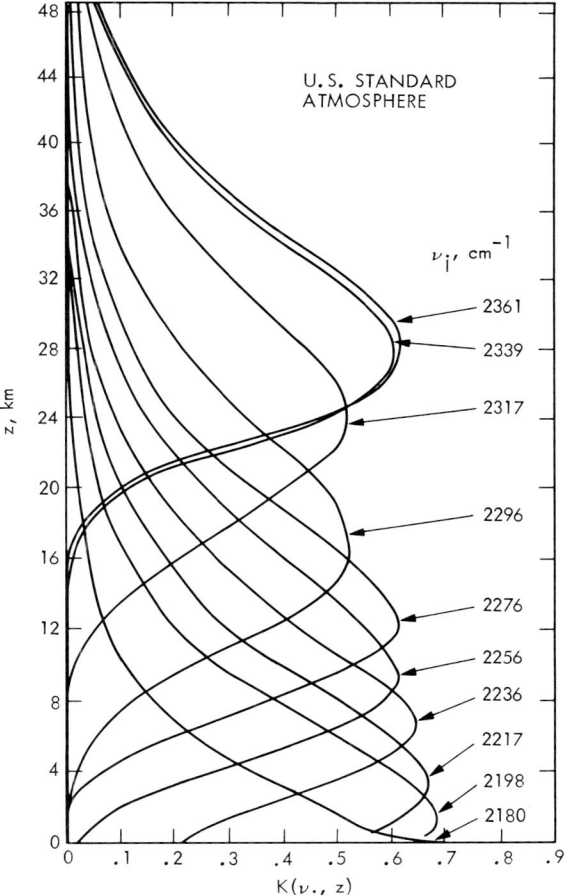

Fig. 2. *The 4.3 μm CO_2 band weighting functions for atmospheric temperature sounding corresponding to $\bar{z} = \infty$.*

values of $I(\nu_j)$ very strongly, while variations of $T(z)$ at values of $z \ll z_j$ and $z \gg z_j$ do not affect $I(\nu_j)$ appreciably. Hence, we propose to map ν_j into z_j where z_j corresponds to the peak value of the kernel $K(\nu_j, z)$. Mathematically, we derive the transformation

$$\nu_j = \nu(z_j) \tag{24}$$

from the solution of the equation

$$\frac{\partial K(\nu_j, z)}{\partial z} = 0 \quad (j = 1, 2, 3 \ldots J) \tag{25}$$

Equation (25) can be used immediately to map the set of J points on the ν axis into a set of J points on the z axis.

But in order to map the I axis into the T axis, we need a relationship between $I(\nu_j)$ and $T(z_j)$. The author (Refs. 3 and 4) has applied the mean value theorem to Eq. (9) and derived the following relaxation equation

$$\frac{B\left[\nu_j, T^{(n+1)}(z_j)\right]}{B\left[\nu_j, T^{(n)}(z_j)\right]} \approx \frac{\tilde{I}(\nu_j)}{I^{(n)}(\nu_j)} \tag{26}$$

Equation (26) relates changes in the outgoing radiance for one frequency ν_j with changes in the Planck function at one level z_j, as illustrated in Fig. 2. Equation (26) is expressed in an iterative form useful for our purposes where $T^{(n)}(z)$ and $T^{(n+1)}(z)$ are two temperature profiles at different orders n of an iterative solution. $I^{(n)}(\nu_j)$ is the radiance computed from Eq. (23) for a given $T^{(n)}(z)$, and $\tilde{I}(\nu)$ is the measured radiance. For additional details regarding the derivation of Eq. (26), see Eqs. (6) to (9) in Ref. 4. We should note here that Eq. (26) can be derived also by applying the method of steepest descent as indicated in Ref. 2 by the author.

B. The Iterative Method of Solution

We proceed to solve Eq. (23) for the determination of T(z) by iteration as follows:

Assume a set of measured radiances, $\tilde{I}(\nu_j)$ is given for j = 1, 2, ... J.

1. Make an initial guess (n = 0) for $T^{(n)}(z)$.
2. Substitute $T^{(n)}(z)$ into Eq. (23) and evaluate the

corresponding $I^{(n)}(\nu_j)$ $j = 1, 2, \ldots J$ using in this case a linear interpolation formula.

3. Check the residuals

$$R^{(n)} = \frac{1}{J} \left[\sum_{j=1}^{J} \left(\frac{\tilde{I}^{(n)}(\nu_j) - I^{(n)}(\nu_j)}{\tilde{I}^{(n)}(\nu_j)} \right)^2 \right]^{1/2} \qquad (27)$$

for each frequency and in an rms sense. If $R^{(n)}$ is small, then $T^{(n)}(z)$ is a solution. If $R^{(n)}$ is not small, go to step 4.

4. Obtain a new guess

$$T^{(n+1)}(z_j) = \alpha_j^{(n)} T^{(n)}(z_j)$$

where the scaling factors are obtained from Eqs. (16) and (26) as

$$\alpha_j^{(n)} = \frac{b\nu_j / T^{(n)}(z_j)}{\ln\{1 - [1 - \exp(b\nu_j / T^{(n)}(z_j))] I^{(n)}(\nu_j) / \tilde{I}^{(n)}(\nu_j)\}} \qquad (28)$$

A second criterion to establish convergence of the iterative solution was applied here. The criterion is obtained by observing the rate of convergence of the temperature profile with respect to itself as

$$<T^{(n)}>_{av} = \frac{1}{J} \sum_{j=1}^{J} \left| T^{(n)}(z_j) - T^{(n-1)}(z_j) \right| \qquad (29)$$

The iteration is terminated when $<\Delta T^{(n)}>_{av}$ is less than some prescribed value, say 0.1 K.

5. Go to step 2 and repeat until step 3 or Eq. (29) is satisfied.

C. Accuracy of the Results

The relaxation method of solution has been applied to invert synthetic radiance data generated by a computer from a set of model temperature profiles in an atmosphere having a constant CO_2 mixing ratio of 462×10^{-6} by mass. The spectral interval selected to illustrate this study corresponds to the set of 4.3-μm frequencies shown in Fig. 2.

The instrumental slit function $\phi(\nu_j, \nu)$ is taken to be triangular, symmetrical with respect to ν_j, and having a base width equal to 60 cm^{-1}. A typical example of the accuracy of the reconstructed temperature values is shown in Fig. 3.

1. Uniqueness of Solution

The typical results of Fig. 3 show two interesting properties which call for comment. First of all, examination of the solutions, obtained by using as an initial guess any isothermal profile or the U.S. Standard Atmosphere temperature profiles, shows that all the reconstructed temperature values reproduced very well the original profile in less than seven iterations and the average absolute error in the reconstructed temperature, $\langle \Delta T \rangle_{av}$, is less than 0.1 K, irrespective of the initial guess. Secondly, when a small perturbation of the order of 1 K was superimposed on the exact profile and the resulting profile used as an initial guess, the solution converged with similar rapidity. The value of $\langle \Delta T \rangle_{av}$ decreased from 1 K to 0.074 K in one iteration only. These results show clearly that the final answers do not depend on the initial guess and that convergence is guaranteed for large, as well as small, perturbation solutions.

We have also observed that the residuals $R^{(n)}(\nu_j)$ of the individual sounding frequencies do not converge simultaneously at the same order n of iteration. The reason for this is because the absorption properties of the atmosphere do not usually allow for, or lead to, the selection of a uniform set of kernels $K(\nu_j, z)$ with equal half-widths and equally spaced peaks z_j. From a numerical point of view, the resulting system of integral equations is poorly discretized, and in case the weighting functions are also broad, it becomes difficult to resolve small details in the profile even in those regions where the

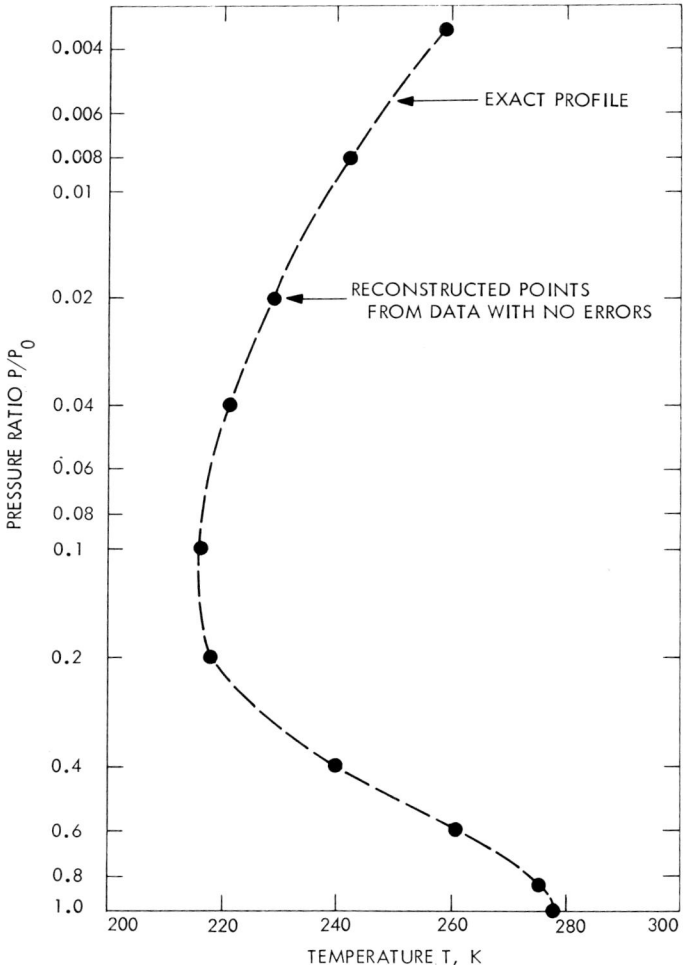

Fig. 3. Comparison between the exact temperature profile and the reconstructed temperature values for synthetic radiance data starting with an isothermal initial guess and using linear interpolations. (From Ref. 4.)

corresponding peaks are narrowly spaced. We note here that, in general, the resolution of small detail in only one region of the profile, to the exclusion of the rest, is not always possible regardless of the accuracy of the measured data.

It is possible, however, to use more than one measurement to recover the solution at one point z_j. This can be done by applying weighted scaling factors as in Eq. (11)

$$T^{(n+1)}(z_j) = \bar{\alpha}_j^{(n)} T^{(n)}(z_j)$$

where each $\bar{\alpha}_i^{(n)}$ is obtained as a weighted average of more than one sounding frequency in a manner similar to Eq. (10).

2. *Stability of Solutions*

The present relaxation method of solution is a discrete numerical process in which convergence is judged according to the extent to which this algorithm suppresses the effects of quadrature, random and systematic errors on the final temperature profiles.

a. *Quadrature errors.* The effect of quadrature errors on the final answer depends on two sources: one of these is computational, resulting from the integrations of Eq. (23) for the evaluation of $I^{(n)}(\nu_j)$; the other is due to interpolations resulting from the inability of a discrete set of points to fit the whole temperature profile exactly even for perfect data. In the results typified by Fig. 3, a modified Simpson's rule was used to evaluate Eq. (23) with a first-order interpolation formula for the intermediate values of temperature. The temperature profile in Fig. 3 is relatively smooth, and the use of a different interpolation formula or a larger number of sounding frequencies for such profiles is not warranted.

b. *Random errors.* The question of the propagation of random errors is a critical one and depends on a number of factors, including the spectral region in which the observations are made.

GENERALIZATION OF THE RELAXATION METHOD

In examining the tolerance of this algorithm to random errors, the effect of their distribution, their maximum values, and their rms values will be taken into consideration.

To obtain a feeling for the stability of the resolution, errors were produced by a uniform random error generator subroutine; they were then added to the exact synthetic data, and inversions were performed for a variety of cases as in the previous section. In the results shown in Fig. 4, a set of 10 random errors having a maximum value of 9.3% and a rms value of 4.8% was superimposed on the exact synthetic data of Fig. 1. The reconstructed temperature values show an excellent tolerance to random errors. The average absolute error $<\Delta T>_{av}$ in the temperature here is 1.5 K.

Since the present inversion scheme is nonlinear, the effects of random errors must be examined for each case separately. The results of some 30 cases studied are summarized in Fig. 5. They show that, in observations made in the 4.3-μm region, a temperature accuracy $<\Delta T>_{av}$ of 1 K can be expected with a 2% rms random error in observations, and an accuracy of 2 to 3 K can be expected with a 5 to 7% rms error in observations.

The effects of random errors in observations vary according to the frequency and can be estimated in principle from Eq. (16). We consider a hypothetical set of sounding frequencies for which the weighting functions form a perfect set of delta functions. The dependence of errors in the temperature solution on random errors can be obtained by differentiation of the blackbody function with respect to temperature which yields

$$\Delta T = \left[\frac{T^2(1 - e^{-b\nu/T})}{b\nu} \right] \frac{\Delta B}{B} \tag{30}$$

Equation (30) shows that a relative error $\Delta B/B$ in measuring the blackbody radiance will be multiplied, by the expression between brackets, when the radiance is inverted to temperature. This ideal multiplication factor is a function of frequency. A comparison of

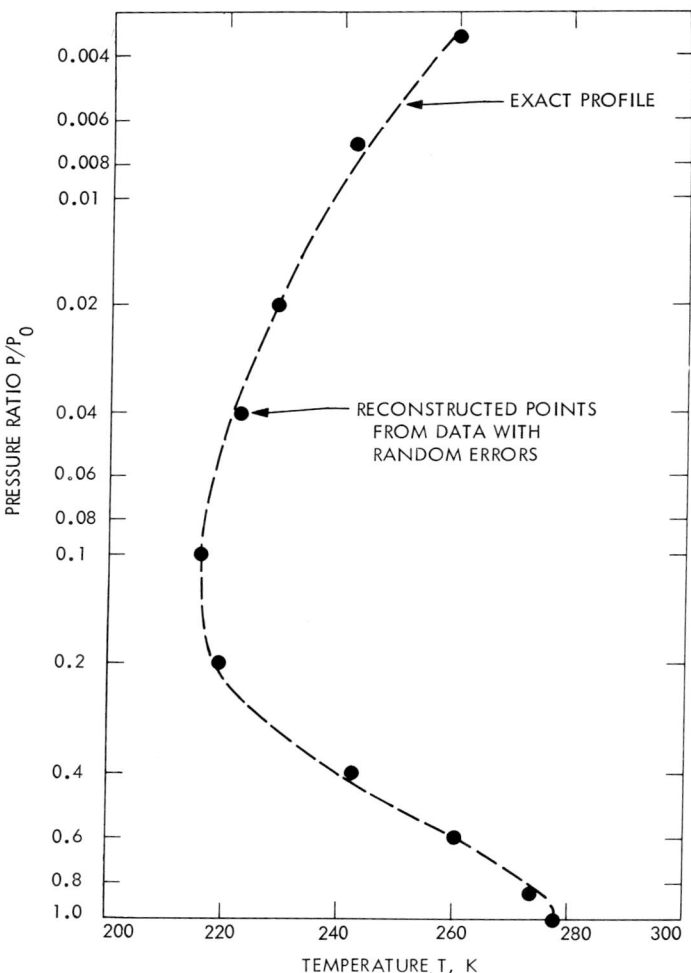

Fig. 4. Same as in Fig. 3 except for synthetic data with random noise. (From Ref. 4.)

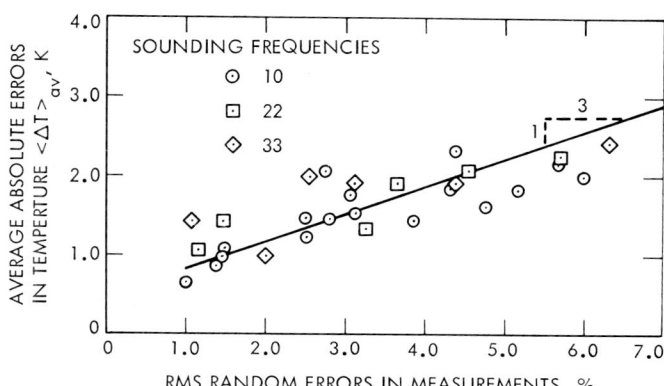

Fig. 5. Effect of random noise in the data on the accuracy of the recovered temperature profile. (From Ref. 4.)

this factor with the slope of the least-squares-fit line in Fig. 5 turns out to be very satisfactory.

Perhaps more significant is the effect of random errors in measurements on the behavior of the residuals $R_j^{(n)}$. In the case of Fig. 3 for zero random errors in observations, we recall that the solution converged after a small number of iterations. The variations of the corresponding rms value of the residuals, $<R^{(n)}>_{rms}$, with respect to the order of iteration, is shown as the lowest curve in Fig. 6. By contrast, the uppermost curve corresponds to the case of Fig. 4 with an rms random error of 4.8%.

A closer examination of the various results shown in Fig. 6 reveals that the residuals tend toward different asymptotic values

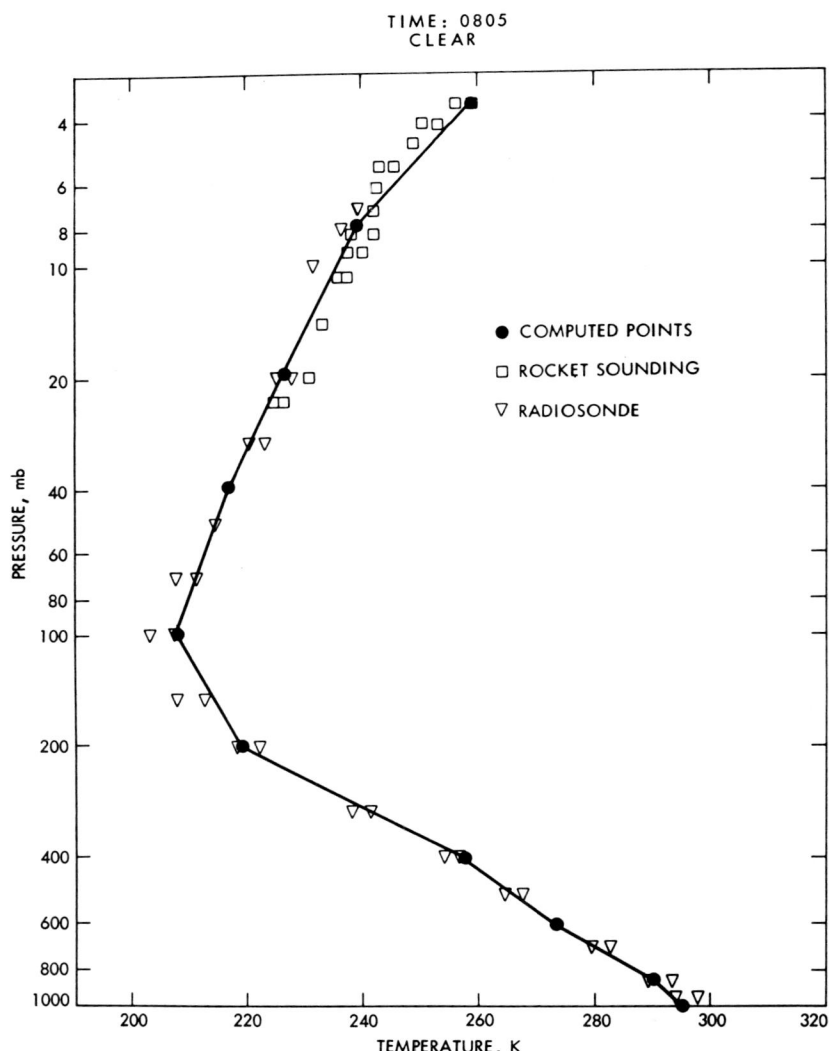

Fig. 6. Temperature profile recovered from measured radiance data in the 4.3 μm region. (From Ref. 10.)

according to the values of the corresponding random errors in observations. For zero random errors, the asymptotic value is equal to the quadrature errors.

Thus, in the presence of random errors in measurements the residuals do not give a false indication of convergence; they will not tend toward zero. The residuals first decrease and then approach an asymptotic value of the same order of magnitude as the errors in measurements. This property is the result of the partial (nonlinear) dependence of several sounding frequencies on temperature variations at one pressure level, and suggests that the iterative process should be terminated when $R^{(n)}$ becomes equal to the value of rms errors in measurements. The effect of continued iterations beyond this point does not increase the amount of information extracted from the radiance observations; it simply increases the rate of accumulation of errors in the reconstructed temperature values.

The solid circles in Fig. 4 correspond to the terminal orders of iteration at which the average absolute error in the reconstructed temperature values $<\Delta T>_{av}$ is within ± 0.1 K of the minimum. The solid circles occur always in the region of maximum curvature of the variation of $R^{(n)}$ with respect to n.

c. *Systematic errors*. Certain transmittance errors resulting from an approximate knowledge of the composition and the spectral properties of the atmosphere are systematic errors. The effect of these errors on the behavior of the residuals is qualitatively similar to the effect of random errors; that is to say, the larger the error in transmittance, the larger the corresponding asymptotic value of the residuals. And, inversely, the residuals tend to their minimum value when the errors in transmittance are minimum. We will show in the following section that by investigating the consequences of adopting a criterion which we shall call minimization of the residuals, we can develop a satisfactory method for the determination of other meteorological parameters, such as the constant mixing ratio of absorbing gases.

The mapping transformation applied in this section can be adapted to different data requirements as shown by Conrath (Ref. 8), Smith (Ref. 9), Shaw, et al. (Ref. 10), and Taylor (Ref. 11). Figure 7 is an example taken from Ref. 10 and shows a comparison between the temperature profile recovered from real data with colocated radiosonde and rocketsonde data. More recently, the technique has been applied by Jastrow and Halem (Ref. 12) to interpret the infrared radiance data from the 15 μm sounding on the National Oceanic and Atmospheric Administration (NOAA) satellites for numerical weather prediction purposes.

IV. DETERMINATION OF COMPOSITION PROFILES

The dependence of the radiative transfer equation on the concentration of absorbing gases $\rho_s(z)$ appears in the kernel of the equation as shown in Eqs. (17), (19), and (23). If the mixing ratio profile is a constant q_s, the dependence of $K(\nu, z)$ on q_s is simple and the retrieval of q_s can be obtained by a minimization process. However, if the mixing ratio is a function of z, the unknown $\rho_s(z)$ will appear as a *functional* in the integral equation, and the retrieval of $\rho_s(z)$ in this case becomes more difficult. We will examine these two cases in this section.

A. Case of the Constant Mixing Ratio

The property of the residuals given in Eq. (9) and Fig. 6 has been used in Ref. 4 to derive the constant mixing ratio q of

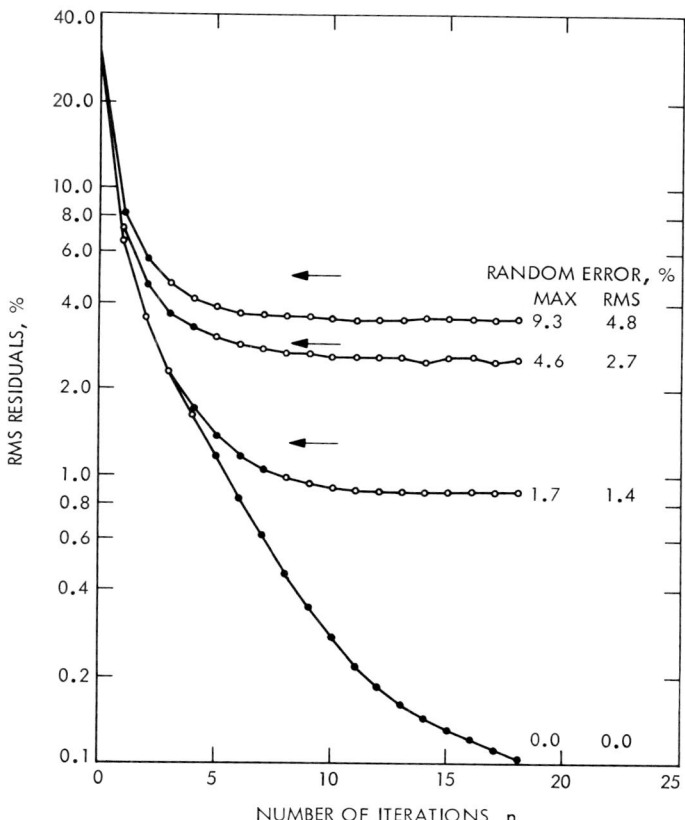

Fig. 7. Variations of the rms residuals with respect to the number of iterations for different noise levels in the radiance data. (From Ref. 4.)

absorbing gases, such as the constant mixing ratio of carbon dioxide in the terrestrial atmosphere. Here, an error in the value of the mixing ratio q_{CO_2} used in computing the kernel introduces an error in $K(\nu_j, p)$ which will prevent the residuals from converging to near zero. The residuals will reach an absolute minimum value, only when the correct kernel is used, i.e., when the correct mixing ratio is known, assuming all other sources of error to be relatively small.

As an illustration, we applied this method to the synthetic data of Fig. 3, assuming, however, the value of q_{CO_2} (which is equal to 462×10^{-6}) to be unknown. The results in Fig. 8 of the twelfth iteration clearly show that the residuals have one minimum at the correct value of the mixing ratio. The results of the third iteration, for which $T(z)$ is far from having converged, show that a good approximation to the value of the mixing ratio can be obtained with just a rough knowledge of the temperature profile.

B. Case of the Variable Mixing Ratio

The determination of the composition profile $\rho_s(z)$ can be obtained by applying a mapping transformation similar to the one used for $T(p)$. However, the relaxation equation required to transform the I axis into the ρ_s axis may, in some cases, be hard to express analytically. According to Eq. (7), the relaxation approach can be generalized to solve for any function or functional under the sign of integration. If $g(z_j)$ is the temperature profile then α_j can be obtained directly from Eq. (26). But in the case of the composition profile, the determination of α_j is more difficult because $\rho_s(z)$ appears as a *functional* in the kernel,

$$K(\nu_j, z, <\rho_s(z)>, \ldots) = \frac{\partial \tau(\nu_j, z, <\rho_s(z)>, \ldots)}{\partial z}$$

since τ and K depend on the distribution of $\rho_s(z)$ between $\bar{z}$ and z. The notation $<\rho_s(z)>$ indicates that the transmittance and the kernel are functionals of $\rho_s(z)$.

1. The General Approach

To determine $\rho_s(z)$, let us first integrate Eq. (23) by parts and write the result as in Ref. 5

$$I(\nu_j) = B[\nu_j, T(\bar{z})] - \int_{z_0}^{\bar{z}} \tau[\nu_j, z, <\rho_s(z)>, \ldots] \frac{\partial B}{\partial z} dz \quad (31)$$

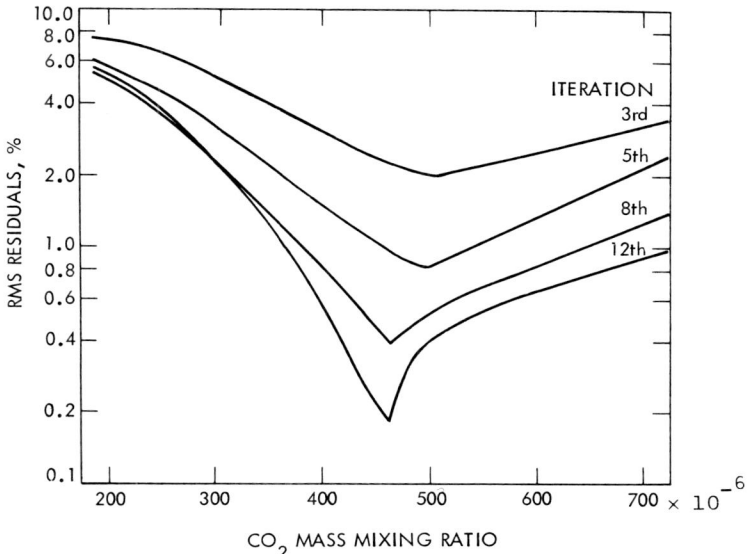

Fig. 8. Determination of the constant mixing ratio of an absorbing gas by the criterion of minimization of the residuals. (From Ref. 4.)

To determine $\rho_s(z_j)$ from a given set of radiance measurements, assuming that the temperature profile is known and is not isothermal, we map the ν_j axis into the z_j axis according to Eq. (5) then make an initial guess $\rho_s^{(n)}(z_j)$ and solve the equation

$$I(\nu_j) - B[\nu_j, T(\bar{z})] = \int_{z_0}^{z} \tau[\nu_j, z, <\alpha_j^{(n)} \rho_s^{(n)}(z_j)>, \ldots] \frac{\partial B}{\partial z} dz \quad (32)$$

to obtain a set of scaling factors $\alpha_j^{(n)}$, for $j = 1, 2 \ldots J$. We generate the next iteration through the relaxation equation

$$\rho_s^{(n+1)}(z_j) = \alpha_j^{(n)} \rho_s^{(n)}(z_i) \quad (33)$$

This iteration process is repeated until each value of the scaling constants approaches unity, which is equivalent to satisfying the residuals in Eq. (9). This relaxation method of solution leads to accurate determination of composition profiles without any *a priori* information for the expected solution, as shown in Fig. 9.

The relaxation method can be applied in conjunction with any interpolation formula. The extent of interpolation is dictated by the quadrature requirements, and by the need to optimize the quality of solutions obtained from a finite set of sounding frequencies. Additional details on this subject can be found in section 2 of Ref. 5 and in section 4 of Ref. 6.

2. *Approximate Relaxation Equation*

The determination of $\alpha_j^{(n)}$ can be rather time consuming because it requires reevaluation of τ many times. Two approximate relaxation equations have been derived in Eqs. (13) and (17) of Ref. 5. We give here one approximation to the relaxation equation

$$\alpha_j^{(n)} = 1 - \frac{\tilde{I}(\nu_j) - I^{(n)}(\nu_j)}{\int_{z_o}^{Z} \tau^{(n)} \ln \tau^{(n)} \frac{\partial B}{\partial z} dz} > 0 \qquad (34)$$

where τ and B are functions of both ν and z. Equation (34) proved to be very useful particularly when α is close to unity.

V. REMOTE SOUNDING IN THE PRESENCE OF CLOUDS

Equation (23) is derived for the case of plane, parallel, *homogeneous* and *nonscattering* atmospheres. This is an ideal case which does not usually apply to observations in the presence of clouds or other horizontal inhomogeneities. In this section, we treat the problem of remote sounding of cloudy atmospheres for the determination of the "clear column" vertical profiles, i.e., the vertical temperature profiles in the clear portions of the fields of view. We will separate the problem into two parts. The first part deals with the simple case of a single cloud layer or a single degree of horizontal inhomogeneity and the second part deals with the general case of multiple cloud layers. The treatment of this

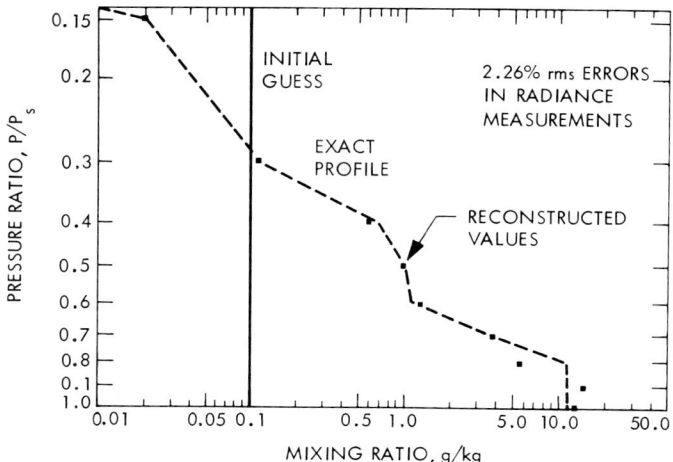

Fig. 9. Determination of the variable water vapor profile from synthetic data with random noise. (From Ref. 5.)

problem in this section will be brief and will describe only the results which have been published in Refs. 13 to 16. Additional work on this problem is still continuing.

A. Observations in the Presence of a Single Cloud Layer

In this part, we treat two cases of clouds. The first case is general and requires no *a priori* knowledge of the radiative properties of the clouds. The second case is for clouds with known properties, such as black or gray clouds.

1. *Clouds with Spectrally Unknown Characteristics*

We consider two adjacent fields of view having different fractional cloud covers at the same height z_c as shown in Fig. 10. We express the outgoing radiance $\tilde{I}_1(\nu)$ and $\tilde{I}_2(\nu)$ from the first and second fields of view as

$$\left. \begin{array}{l} \tilde{I}_1(\nu) = \left[I_1(\nu) \right]_{clear} - N_1 G(\nu, z_c, t, r, e, \ldots) \\ \tilde{I}_2(\nu) = \left[I_2(\nu) \right]_{clear} - N_2 G(\nu, z_c, t, r, e, \ldots) \end{array} \right\} \quad (35)$$

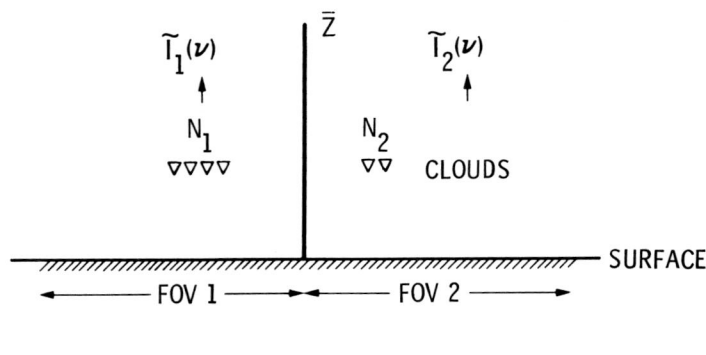

$$I_{CLEAR}(\nu) - \tilde{I}_1(\nu) = N_1 \, G\,(\nu, Z, t, e, P, \ldots)$$

$$I_{CLEAR}(\nu) - \tilde{I}_2(\nu) = N_2 \, G\,(\nu, Z, t, e, P, \ldots)$$

Fig. 10. The single cloud layer.

Equation (35) makes no assumption about the radiative transfer properties of clouds; it simply states that the observed radiance $\tilde{I}_k(\nu)$ is equal to the clear column radiance which would have been measured in the absence of clouds minus the radiance G "obscured" by the presence of a fractional cloud cover N_k. G is unknown and depends on ν, z_c, and the spectral properties of the clouds.

If the two fields of view are small and contiguous, we can assume that

$$\bar{I}(\nu) = \bigl[I_1(\nu)\bigr]_{clear} = \bigl[I_2(\nu)\bigr]_{clear}$$

and substitute into Eq. (23) to eliminate G and get

$$\bar{I}(\nu) = \tilde{I}_1(\nu) + \eta\,[\tilde{I}_1(\nu) - \tilde{I}_2(\nu)] \tag{36}$$

where η = unknown = $\dfrac{N_1}{N_2 - N_1}$

Thus, if η is known, we can reconstruct the clear column radiance according to Eq. (36) for all frequencies and proceed to recover $T(z)$ as described earlier.

According to Eq. (36), η can be determined from a knowledge of $I(\nu)$ at any frequency, say ν', as

$$\eta = \frac{I(\nu') - \tilde{I}_1(\nu')}{\tilde{I}_1(\nu') - \tilde{I}_2(\nu')} \qquad (37)$$

However, the exact value of the clear column radiance $I(\nu')$ is actually unknown because $I(\nu')$ itself depends on $T(z)$. Thus, the determination of η and $T(z)$ should be carried out simultaneously and the selection of ν' should satisfy certain convergence criterion in order to ensure uniqueness of the solution.

The argument for the selection of ν' is as follows: if $T^{(n)}(z)$ is any guess on the solution and $I^{(n)}(\nu')$ is the corresponding radiance according to Eq. (23), we get from Eq. (37).

$$\eta^{(n)} = \frac{I^{(n)}(\nu') - I_1(\nu')}{\tilde{I}_1(\nu') - \tilde{I}_2(\nu')} \qquad (38)$$

and by adding and subtracting $I(\nu)$ from the numerator we get

$$\eta^{(n)} = \eta + \frac{I^{(n)}(\nu') - I(\nu')}{\tilde{I}_1(\nu) - \tilde{I}_2(\nu)} = \eta + \frac{\Delta I^{(n)}(\nu)}{\tilde{I}_1(\nu') - \tilde{I}_2(\nu')} = \eta + \delta^{(n)}(\eta)$$

Now, in order to minimize $\delta^{(n)}(\eta)$, Eqs. (26) and (16) require that $\nu' < \nu_j$ where ν_j is the set of temperature sounding frequencies. By simple differentiation of the Planck function with respect to T and by substitution into Eq. (26), we can show that $\delta^{(n)}(\eta)$ is directly proportional to ν for a given temperature error $\Delta T = T(z) - T^{(n)}(z)$, as

$$\frac{\Delta I^{(n)}(\nu')}{I(\nu')} \approx \frac{\Delta B(\nu', T)}{B(\nu', T)} = \frac{b\nu'}{1 - e^{-\frac{b\nu'}{T}}} \frac{\Delta T}{T^2}$$

At the same time, the frequency range ν' should be cloud dependent so that $\tilde{I}_1(\nu') \neq \tilde{I}_2(\nu')$.

We can translate this into a condition for convergence and select ν' to ensure that

$$E = \left| \frac{I(\nu_j) - \bar{I}^{(n)}(\nu_j)}{I(\nu_j) - I^{(n)}(\nu_j)} \right| < 1 \quad (j = 1, \ldots J) \tag{39}$$

The basic steps of the method of solution for η and $T(z_j)$ are shown in Fig. 11.

1. Make an initial guess ($n = 0$) for $T^{(n)}(z)$.
2. Substitute $T^{(n)}(z)$ into Eq. (23) and compute $I^{(n)}(\nu')$ and $I^{(n)}(\nu_j)$; $j = 1, 2, \ldots J$.
3. Substitute $I^{(n)}(\nu')$ into Eq. (38) and compute $\eta^{(n)}$.
4. Substitute $\eta^{(n)}$ into Eq. (36) and reconstruct $\bar{I}^{(n)}(\nu_j)$; $j = 1, \ldots J$.
5. If the convergence criterion E is satisfied, then $\bar{I}^{(n)}(\nu_j)$ is closer to the exact clear column radiance $I(\nu_j)$ than the computed $I^{(n)}(\nu)$ and the relaxation equation

$$\frac{B[\nu_j, T^{(n+1)}(z_j)]}{B[\nu_j, T^{(n)}(z_j)]} = \frac{\bar{I}^{(n)}(\nu_j)}{I^{(n)}(\nu_j)} \quad (j = 1, \ldots J) \tag{40}$$

is used to obtain the new guess $T^{(n+1)}(z_j)$.

6. Go back to step 2 and repeat until

$$R^{(n)}(\nu_j) = \frac{\bar{I}^{(n)}(\nu_j) - I^{(n)}(\nu_j)}{\bar{I}^{(n)}(\nu_j)} \to 0 \tag{41}$$

Application of this method to recover $T(z)$ in the presence of clouds has shown that the solution exhibits the same characteristics as in the case of clear fields of view. See Ref. 6 for additional details.

2. *The Single Layer of Black Clouds*

For the case of black clouds, the term $G(\nu, z_c, \ldots)$ in Eq. (35) can be expressed analytically as

$$G(\nu, z_c) = - B[\nu, T(z_c)]\tau(\nu, z_c) + B[\nu, T(z_o)]\tau(\nu, z_o)$$

$$+ \int_{z_o}^{\bar{z}} B[\nu, T(z)]K(\nu, z)\, dz \qquad (42)$$

where z_c is the cloud-top height. By substituting Eq. (42) into Eq. (35), and using Eq. (23), we can express the measured radiance in any field of view $\tilde{I}_k(\nu)$ as an integral function of the clear column temperature profile $T(z)$, the cloud-top height z_c and the fractional cloud cover N, according to Eq. (1) in Ref. 13, as

$$\tilde{I}(\nu) = N\, \{\, B[\nu, T(z_c)]\tau(\nu, z_c)$$

$$+ \int_{z_c}^{\bar{z}} B[\nu, T(z)]K(\nu, z)\, dz\, \}$$

$$+ (1 - N)\, \{\, B[\nu, T(z_o)]\tau(\nu, z_o)$$

$$+ \int_{z_o}^{\bar{z}} B[\nu, T(z)]K(\nu, z)\, dz\, \} \qquad (43)$$

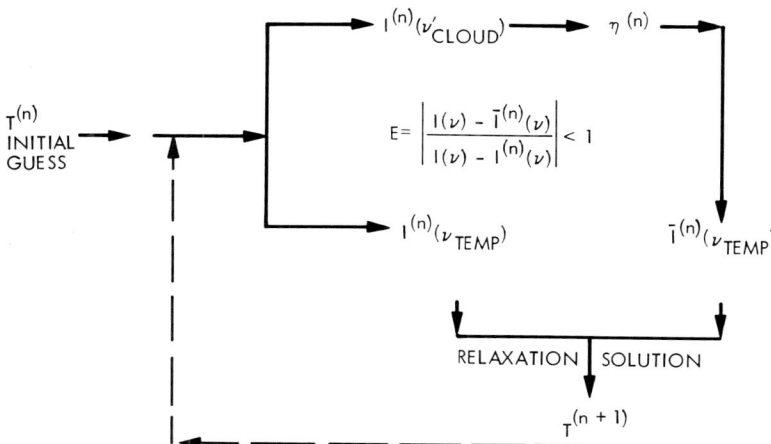

Fig. 11. Flow diagram of the iteration method of solution for the determination of the clear-column temperature profiles. (From Ref. 14.)

If N and z_c are known, it is obvious that the problem reduces to the case described earlier in this paper. In practice, however, neither z_c nor N are known, and a method of solution should be derived to account for (or eliminate) the effects of clouds. When we solve Eq. (23) for measurements made in the presence of clouds (i.e., this is equivalent to assuming N = 0 in Eq. (43)), we obtain an apparent temperature profile $\tilde{T}_k(z)$ which is different from the true clear-column profile $T(z)$. The author (Ref. 13) derived a simple relationship between $T(z)$ and $\tilde{T}(z)$ in the form of

$$B[\nu_j, T(z_j)] = B[\nu_j, \tilde{T}(z_j)] \qquad z_j \geq z_c$$

$$B[\nu_j, T(z_j)] = B[\nu_j, \tilde{T}(z_j)]$$
$$+ a\{B[\nu_j, \tilde{T}(z_j)] - B[\nu_j, T(z_c)]\} \qquad (z_j \leq z_c) \quad (44)$$

where $a = N/(1 - N)$ and B is the Planck function given in Eq. (16).

Let us consider next two adjacent fields of view having different amounts of clouds, N_1 and N_2, at the same height z_c. We can eliminate $B[\nu_j, T(z_c)]$ from Eq. (44) and write (see Eq. (12) in Ref. 13)

$$B[\nu_j, T(z_j)] = B[\nu_j, \tilde{T}_1(z_j)]$$
$$+ \eta\{B[\nu_j, \tilde{T}_1(z_j)] - B[\nu_j, \tilde{T}_2(z_j)]\} \qquad (45)$$

The practical benefits of Eq. (45) are obvious. The clear column temperature profile $T(z)$ can now be recovered directly by a simple transformation of the apparent temperature profiles $\tilde{T}_1(z)$ and $\tilde{T}_2(z)$ of two adjacent fields of view. η is an unknown constant which can be determined from an additional frequency as described earlier in this section.

3. *Determination of the Amount and Height of Clouds*

The value of the fractional cloud cover N and the height z_c of clouds can be determined when the radiative transfer properties of the clouds in $G(\nu, z_c, t, r, e, ...)$ are given.

For the case of black clouds, the author (Ref. 4) showed that by substituting Eq. (43) into Eq. (35) and by taking the ratio for two different sounding frequences ν_1 and ν_2, we get

$$\frac{I^{(n)}(\nu_1) - \tilde{I}_k(\nu_1)}{I^{(n)}(\nu_2) - \tilde{I}_k(\nu_2)} = \frac{G^{(n)}(\nu_1, z_c)}{G^{(n)}(\nu_2, z_c)} \qquad (46)$$

where Eq. (46) is now a function of one unknown z_c, assuming that $T^{(n)}(z)$ has been determined as described earlier. The solution for z_c can be obtained by minimization techniques as described in section 6 of Ref. 6. The corresponding value of N_k is obtained directly from Eq. (35) as

$$N_k = \frac{I^{(n)}(\nu_1) - \tilde{I}_k(\nu_1)}{G(\nu, z_c)} \qquad (47)$$

Applications of this approach to the infrared data from the Vertical Temperature Profile Radiometer (VTPR) sounder on NOAA 4 are illustrated in Fig. 12.

B. Observations in the Presence of Multiple Cloud Formations

The two methods described earlier for the elimination of the effects of a single layer of clouds can be extended to multiple cloud formations for the cases of clouds with spectrally unknown characteristics and for black clouds. Derivation of the required equations can be found in Ref. 14. A brief summary of the final results will be given next.

1. *Clouds with Spectrally Unknown Characteristics*

Let us reexamine Eq. (35) now and look at NG, on the right-hand side of the equation, as a one term expansion of the difference between the clear column radiance $I(\nu)$ and the radiance measured in the presence of clouds $\tilde{I}(\nu)$. In the case of Eq. (35), N is just a coefficient of expansion and $G(\nu, z, t, r, e, ...)$ is the expansion function. The mathematical form of G need not be

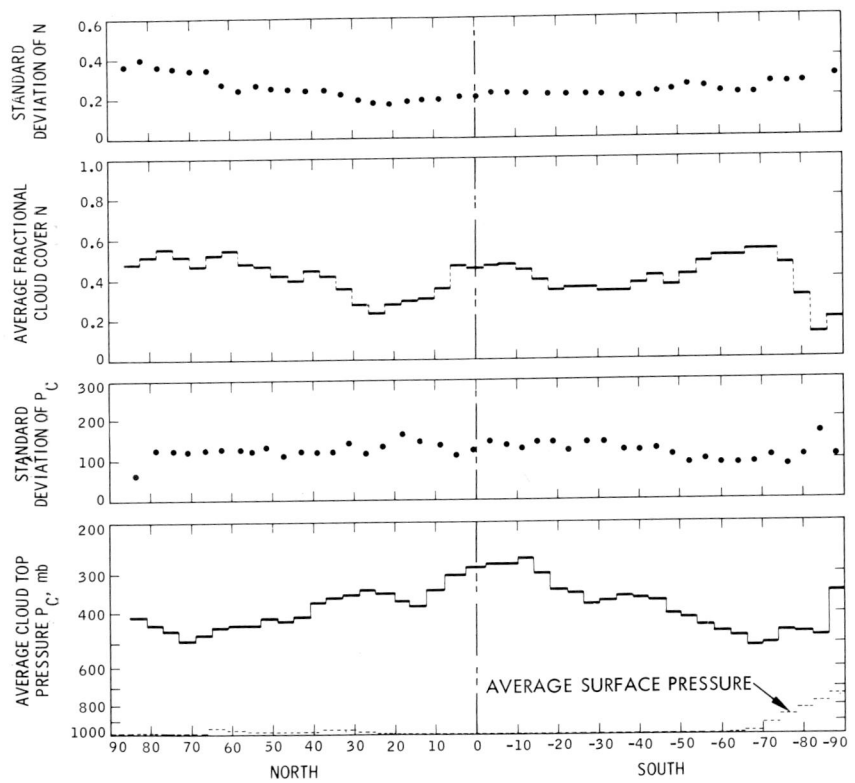

Fig. 12. Meridional profiles of zonally averaged cloudiness, from the VTPR sounder on the NOAA 4 satellite, for the period of January 1-7, 1975.

defined because it will be eliminated by using measurements from adjacent fields of view as shown in Eq. (35). A knowledge of the form of the function G is necessary, however, if we need to determine the amount of the corresponding fractional cloud covers.

The one-term expansion of Eq. (35) may not be sufficient in the presence of multiple cloud formations. Therefore, we propose to use a three-term expansion of the form

$$I^{clear}(\nu) - I_k(\nu) = N'_k G' + N''_k G'' + N'''_k G''' \qquad (48)$$

with $G' = G'(\nu, z, t', r', e', \ldots)$ and similarly for G'' and G'''. The expansion functions G depend on ν and z as well as on the cloud transmissivity t, reflectivity r and emissivity e. N', N'' and N''' are the expansion coefficients. The next step now is to eliminate G from Eq. (48) and express $I(\nu)$ as a function $\tilde{I}_k(\nu)$ and N'_k, N''_k and N'''_k.

In order to eliminate the expansion functions, we consider observations over four adjacent fields of view, K = 1, 2, 3, and 4, and use the first three equations to express G', G'' and G''' as functions of the N_k, $I(\nu)$ and $\tilde{I}_k(\nu)$. We substitute the results into Eq. (48) for k = 4 and write

$$I(\nu) = \tilde{I}_1(\nu) + \sum_{\ell = 1}^{3} \eta_\ell [\tilde{I}_1(\nu) - \tilde{I}_{\ell + 1}(\nu)] \qquad (49)$$

Details of the substitution are given in Appendix A of Ref. 15. As expected, we note that the first term, $\ell = 1$, of Eq. (49) corresponds to the case of a single cloud layer as given in Eq. (36).

The determination of $T(z)$ and η_ℓ is carried out simultaneously by iterations according to the steps given in Fig. 11. A typical illustration of the accuracy of this method is shown in Fig. 13, using synthetic radiance data. Additional details on the accuracy and stability of the solution are given in Ref. 14. Experimental

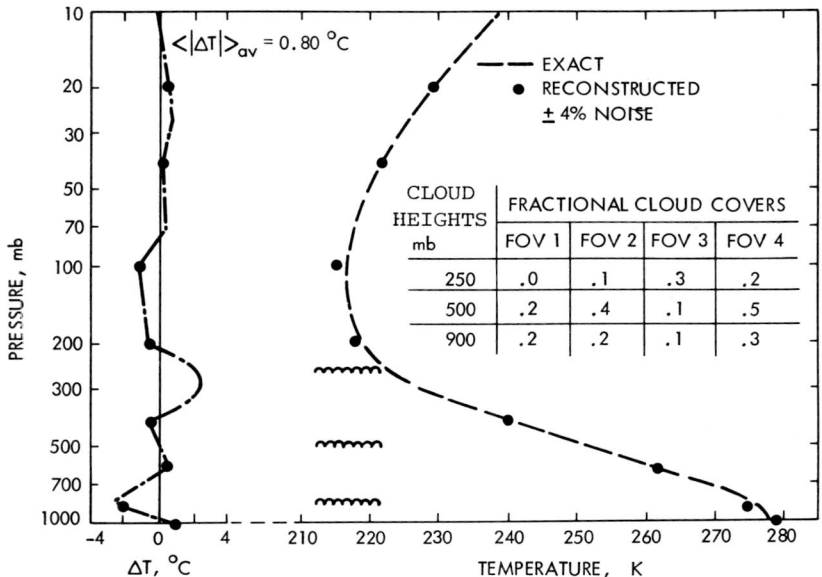

Fig. 13. Determination of the clear-column temperature profile in the presence of three cloud layers. (From Ref. 14.)

verification of the expansion approach has been obtained and discussed in Refs. 15 and 16.

We conclude the discussion here by indicating that the use of more than four expansion terms is feasible from a mathematical point of view, but from an experimental point of view the expansion may not converge always because of the effects of noise in the data and uncertainties in the accuracy of the computed kernels.

2. *Multiple Layers of Black Clouds*

In the case of black clouds, Eq. (45) can be extended to read

$$B[\nu_j, T(z_j)] = B[\nu_j, \tilde{T}_1(z_j)]$$

$$+ \sum_{\ell=1}^{3} \eta_\ell \{B[\nu_j, \tilde{T}_1(z_j)] - B[\nu_j, \tilde{T}_{\ell+1}(z_j)]\} \quad (5$$

where $\tilde{T}_k(z_j)$ is the apparent temperature profile, over the kth field of view (obtained without accounting for the effects of clouds). The coefficients η_ℓ should be determined from additional radiance data measured at different frequencies over the same k fields of view or from *a priori* knowledge of certain properties of the solution, such as the value of $T(z)$ or the lapse rate dT/dz at given heights z.

VI. APPLICATIONS AND EXTENSIONS

The mapping transformation method described in this paper has been presented specifically as a method of solution of the nonlinear radiative transfer equation. However, it must be obvious that the mapping transformation is general and can be applied to a wide class of nonlinear as well as linear integral equations, as in the case of the analytical example. The only requirement is for the kernel $K(\nu, z)$ to be a rapidly decaying function with maxima occurring at different values of z for different values of ν.

Extensions of this method and applications to other linear and nonlinear problems can be found in the works of Barcilon (Ref. 7), Menzies, et al. (Ref. 17), Twomey, et al. (Refs. 18 and 19), Twitty (Ref. 20), Grassl (Ref. 21), Gautier, et al. (Ref. 22), Encrenaz, et al. (Ref. 23), and Gille, et al. (Ref. 24).

SYMBOLS

a	constant defined in Eq. (16)
b	constant defined in Eq. (16)
B	Planck function
c	subscript denoting clouds
E	convergence criterion
g	function to be determined
G	radiance from cloudy portion of fields of view
I	radiance function

I	outgoing radiance measurement
k	absorption coefficient
K	kernel of integral equation
n	number of iterations
N	fractional cloud covers
$\mathcal{N}$	integral operator
p	Pressure
R	residual function defined in text
s	subscript denoting absorbing gas
T	temperature
$\tilde{T}$	apparent temperature profile
z	geopotential height
z, z_o	lower and upper limits of the range of integration
W	relaxation weight
α	scaling factor Eq. (7)
$\bar{\alpha}$	weighted scaling factor Eq. (10)
α'	line half width
ε	random error
η	cloud coefficients
ν	frequency, cm^{-1}
ρ	air density
τ	atmospheric transmittance
ϕ	instrument function

ACKNOWLEDGMENT

This paper presents the results of one phase of research carried out at the Jet Propulsion Laboratory, California Institute of Technology, under Contract Number NAS 7-100, sponsored by the National Aeronautics and Space Administration.

REFERENCES

1. L. D. Kaplan, Inference of atmospheric structure from remote radiation measurements, *J. Opt. Soc. Am.* 49, 1004 (1959).

2. M. T. Chahine, "Methods in Computational Physics," Vol. 4, p. 83. Academic Press, New York, 1965.

3. M. T. Chahine, Determination of the temperature profile in an atmosphere from its outgoing radiance, *J. Opt. Soc. Am. 58,* 1634 (1968).

4. M. T. Chahine, Inverse problems in radiative transfer: Determination of atmospheric parameters, *J. Atmos. Sci. 27,* 960 (1970).

5. M. T. Chahine, A general relaxation method for inverse solution of the full radiative transfer equation, *J. Atmos. Sci. 29,* 741 (1972).

6. M. T. Chahine, Remote sounding of cloudy atmospheres, I: The single cloud layer, *J. Atmos. Sci. 31,* 233 (1974).

7. V. Barcilon, On Chahine's relaxation method for the radiative transfer equation, *J. Atmos. Sci. 32,* 1626 (1975).

8. B. J. Conrath, R. A. Hanel, V. G. Kunde, and C. Parbhakara, The infrared interferometer experiment on Nimbus 3, *J. Geophys. Res. 30,* 5831 (1970).

9. W. L. Smith, Iterative solution of the radiative transfer equation for the temperature and absorbing gas profile of an atmosphere, *Appl. Opt. 9,* 1993 (1970).

10. J. H. Shaw, M. T. Chahine, C. B. Farmer, L. D. Kaplan, R. A. McClatchey, and P. W. Schaper, Atmospheric and surface properties from spectral radiance observations in the 4.3 micron region, *J. Atmos. Sci. 27,* 773 (1970).

11. F. W. Taylor, Temperature sounding experiments for the Jovian Planet, *J. Atmos. Sci. 29,* 950 (1972).

12. R. Jastrow and M. Halem, Accuracy and coverage of temperature data derived from the IR radiometer on the NOAA 2 satellite, *J. Atmos. Sci. 30,* 958 (1973).

13. M. T. Chahine, An Analytical transformation for remote sensing of clear-column atmospheric temperature profiles, *J. Atmos. Sci. 32,* 1946 (1975).

14. M. T. Chahine, Remote sounding of cloudy atmospheres, II: Multiple cloud formations, *J. Atmos. Sci. 34,* 744 (1977).

15. H. H. Aumann and M. T. Chahine, An infrared multidetector spectrometer for remote sensing of temperature profiles in the presence of clouds, *Appl. Opt. 15,* 2091 (1976).

16. M. T. Chahine, Remote sounding of cloudy atmospheres, III: Experimental verifications, *J. Atmos. Sci. 34,* 758 (1977).

17. R. T. Menzies and M. T. Chahine, Remote atmospheric sensing with an airborne laser absorption spectrometer, *Appl. Opt. 13,* 2830 (1974).

18. S. Twomey, Comparison of constrained linear inversion and an iterative nonlinear algorithm applied to the indirect estimation of particle size distributions, *J. Comput. Phys. 18,* 188 (1975).

19. S. Twomey, B. Herman and R. Rabinoff, An extension to the Chahine method of inverting the radiative transfer equation, *J. Atmos. Sci.,* 1977.

20. J. T. Twitty, The inversion of aureole measurements to derive aerosol size distributions, *J. Atmos. Sci. 32,* 584 (1975).

21. H. Grassl, Determination of aerosol size distributions from spectral attenuation measurements, *Appl. Opt. 10,* 2534 (1971)

22. D. Gautier and I. Ravah, Sounding of planetary atmospheres: A fourier analysis of the radiative transfer equation, *J. Atmos. Sci. 32,* 881 (1975)

23. Th. Encrenaz and D. Gautier, An iterative method to infer the Jovian atmospheric structure from infrared measurements, *Astronaut. Astrophys. 26,* 143 (1973).

24. J. D. Gille and F. B. House, On the inversion of limb radiance measurements, I: Temperature and thickness, *J. Atmos. Sci. 28,* 1427 (1972).

DISCUSSION

Rodgers: Can you produce an estimate of the error in your final solution using the relaxation method? I mean the total error.

Chahine: This is a nonlinear equation and it is very difficult to determine explicitly the effects of noise in the data on the accuracy of the final solution. By trying several cases of noise levels, it becomes possible to describe the growth of errors in the solution as a function of noise in the data. Professor Barcilone[1] has shown that the growth of errors in the final solution is cyclical. In other words, he has shown that the growth of errors in the final solution does not vary linearly with the growth of noise in the given function.

Rodgers: This isn't quite what I meant. I meant the departure of your solution from the real profile. I didn't mean the sensitivity of your solution on noise in the measurement.

Chahine: That depends on the degree of discretization of the given function or in other words on the information content of the given data. It depends also on any *a priori* information available about the expected solution.

Gille: When you iterate and drive the radiance residuals down, they drop monotonically. Does the error of your temperature solution also drop monotonically?

Chahine: During the first few iterations, the errors in the solution decrease monotonically with the radiance residuals. However, when the decrease in the radiance residuals $R^{(n)}$ becomes slow[2] the accuracy of the solution becomes dependent not only on the $R^{(n)}$ but also on its first and second derivatives. Thus, at higher iterations, beyond the point of maximum curvature in the Figure, the errors in the solution do not necessarily decrease monotonicall with $R^{(n)}$; and very often they don't.

Fleming: I would like to take exception to your statement that your solution does not depend on the initial profile. If the Chairman will permit me, I have a slide to illustrate my point.

---

[1] See Ref. 7 in this paper.

[2] Pointing to Fig. 7.

GENERALIZATION OF THE RELAXATION METHOD

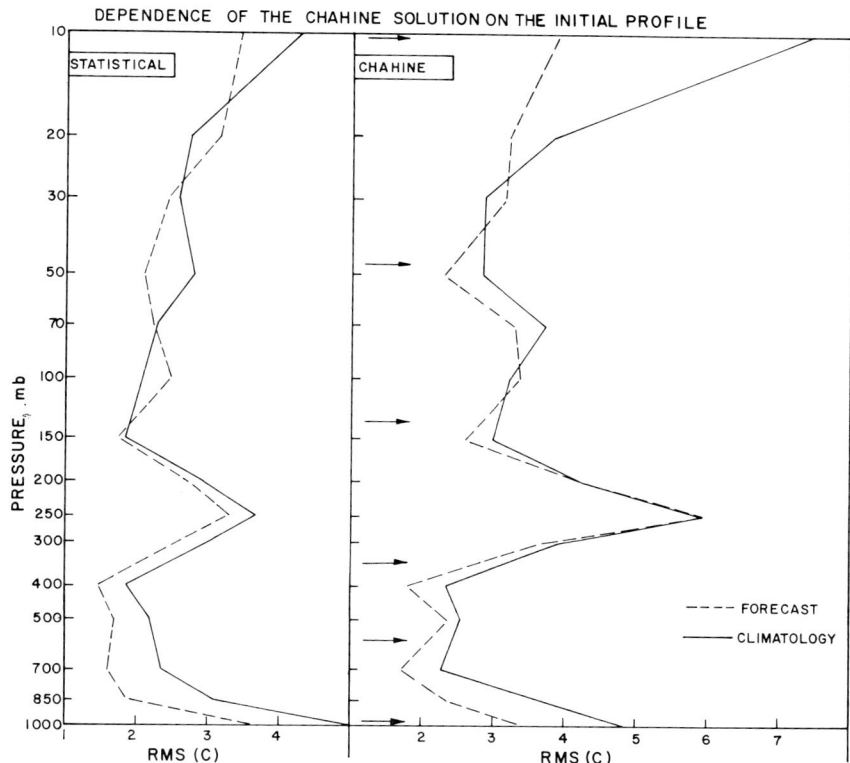

This is a VTPR simulation study in which we have 107 retrievals by the Chahine method on the right-hand side. The arrows show the locations of the peaks of weighting functions and the dashed line is the RMS error of the solutions when forecasts were used for the initial approximation. The solid line is the RMS error of the solutions when climatology is used for the initial approximation. Notice that there is a definite dependence of the solution on what is used as the initial approximation. On the left is the same situation, except the retrieval method is the Rodgers-Strand-Westwater statistical solution. Clearly, the Chahine method shows a dependence on the first guess virtually of the same magnitude as the Rodgers-Strand-Westwater method.

Chahine: I agree with you here because the slide I presented on the VTPR showed very strong dependence on the initial guess. I agree with your results.

Fleming: But you just said and your abstract said it is not dependent on the initial guess.

Chahine: That depends on the properties of the kernel and the given functions. In my abstract I was not dealing specifically with the VTPR data. I said from a complete set of data you should be able to get a complete solution independent of the initial guess. But for the case of the VTPR, the given data are incomplete and, therefore, the solution is incomplete and depends on the initial guess. For the specific result you have shown on the VTPR I agree with your conclusion.

Susskind: I would like to make a comment on the past two comments. When we talk about the dependence of the initial guess, there are two different things which would cause dependence on the initial guess that Dr. Chahine was just referring to. First, if we think about the effects of clouds and the ability to retrieve surface temperatures, if you only have VTPR soundings and you want to filter out the effects of clouds, then you have to return the guess at the surface, because there is sufficient information in the observations to construct the clear column radiances. That's one dependence of the guess. You get a surface temperature that looks like the guess, then that's one kind of problem. Even if you have the clear column radiance and you don't have to worry about the clouds, the shape of the profile to some extent is what is important. You have information at, say, the peaks of the weighting functions of the channel, that much information, and you have to make some assumptions as to what's happening in between those points. If the shape of the initial guess is wrong, the shape of the solution will be wrong. The fine vertical structure of the solution has to follow the guess. If the points come closer and closer together so that you have more information, then you could say to better approximation that the solution does become independent of the guess because you're talking about errors in shape between points that are very close together and it's not--well, it's essentially linear in there. It doesn't matter what's happening. So there are two different things that cause dependence on the guess.

Staelin: Does your method apply to certain other situations? You began by saying you might associate each frequency with that altitude where the weighting function peaks. There are certain problems where all the weighting functions peak at the same altitude, however. For example, if you are attempting to sound temperature by looking up from the ground every weighting function peaks at the surface. Other problems involve weighting functions that have two peaks. Does your method apply to these cases?

Chahine: Yes it does, but the results are not as satisfactory as for the case where the weighting function has distinct peaks at different values of the physical axis. If you cannot discriminate between the information received from different heights, you cannot have a satisfactory inverse solution. Professor Twomey has introduced a modification to the relaxation equation which applies to the case you have just described.[1]

Twomey: You showed one of the last graphs and there was the residual decreasing to the error level and then flattening out. I have seen this same behavior in algorithms, and it's kind of spookey, really. The thing I want to ask though is the following: This seems to hold with random errors and I have the feeling that there must be some kind of error that would be disastrous which the thing would try to invert. Have you found out anything about the character of that kind of error?

Chahine: Yes. But first, regarding the property of the residuals shown here[2], I thought at first when I noticed that the residuals did not decay down to zero in the presence of noise in the data, this behavior is a property of the nonlinear equations I was studying. However, in examining the residuals obtained by Dr. Twitty[3] for his linear equation, I found out that the residuals in his case also decreased rapidly at first and then approached an asymptotic value nearly equal to the errors in the data. Because of this, I now believe that this behavior of the residuals is due to the types of *overlapping* kernels under consideration. Now when does this property of the residuals break down? Usually it does not fail in the presence of systematic noise alone or random noise alone. But it sometimes breaks down when random and system noise are both large and are both present in the data.

Westwater: On your statement that the residuals converged to the noise level, does this convergence depend on the method of interpolation that you use to fit the function "f"? You would think that linearly interpolating between estimated points might conceivably give a different residual than a quadratic interpolation or a higher order interpolation.

Chahine: Yes, because in the absence of noise in the data, the residuals decrease rapidly and approach an asymptotic level equal to the quadrature error. Now, different interpolation methods will introduce different quadrature errors and, therefore, the residuals will approach different values reflecting the level of quadrature errors in the solution.

Westwater: That would imply to me that you're really not converging to the noise level, because if you're converging to a

---

[1] See Ref. 19 in this paper.
[2] See Fig. 7 in this paper.
[3] See Ref. 20 in this paper.

quantity that depends upon method of interpolation that you're using, that really does not bear a direct relationship to the noise level.

Chahine: The noise level in the system of equations is the sum of the systematic noise, the random noise as well as the quadrature noise.

STATISTICAL PRINCIPLES OF

INVERSION THEORY

C. D. Rodgers
Clarendon Laboratory

The accuracy of solutions to the inverse problem of radiative transfer is a topic that has received very little attention from the theoretical point of view in the meteorological literature. Many of the sources of error are statistical in nature, and statistical methods must be used to deal with them. Such methods give considerable insight into both the nature of the problem and the nature of the solution.

All the available information about an unknown profile can be expressed in the form of values of functions of that profile and error estimates of these values. Estimation theory shows how these values are combined to give an estimate of the unknown profile and its error covariance. Many inversion methods can be expressed in this form, although the error estimate is not usually carried out. Practical applications are described, both for inversion of individual profiles, and the global analysis of satellite data.

I. INTRODUCTION

This paper is based, to a large extent, on a review paper (Ref. 1), which covers many of the topics in much more detail. The subject is not new by any means. However, the implications of a correct statistical analysis of meteorological inverse problems do not seem to be properly appreciated, so it seemed reasonable to try to state them clearly at this Workshop. All physical problems involving measurements of continuous variables must be analyzed by statistical methods simply because measurement error is statistical in nature. We can never measure an exact value, we can only say that the value lies in such and such a range, or belongs to a population with a known probability density function. Any quantity derived from the measurement is also a random variable, and in developing methods for finding derived quantities, we must take account of this fact. There are standard statistical tools available (e.g., Ref. 2); therefore, it is not necessary to invent anything new--it is just a matter of recognizing when a particular tool is applicable to the problem in hand.

The kind of question that we would like to answer is something like this: Given a set of measurements of radiation emitted (reflected, scattered) by the atmosphere, our current understanding of the physics of the atmosphere, and any other measurements that may be relevant, what can we say about the state of the atmosphere? The question is usually expressed in terms of profiles of unknown quantities, such as temperature and composition. Given a measurement of a known function of a profile, estimate that profile when there are experimental errors in the measurement, and errors in our knowledge of the function. An essential part of the measurement is a proper characterization of both kinds of errors. If we do not know the magnitude of the error in a measurement or in the theory, the measurement is not worth making.

A little thought shows that this question can only be answered in a statistical sense. Given a probability density function representing our knowledge of some function of a profile, find a probability density function representing our knowledge of the profile itself. This applies to almost any physical measurement of almost anything, but it is a little more complicated in our case because the result is a profile, rather than a single number to which we can assign error bars.

If we are going to discuss general profiles, we should use the algebra of Hilbert space. It seems to me that the use of Hilbert space notation is an unnecessary complication in this subject. This is not because the algebra is more difficult than the algebra of matrices, but simply that many physicists are unfamiliar with the mathematical jargon, and in the end computer programs that implement inversion methods will use the algebra of matrices. In practice, it is not possible to specify an absolutely general profile because this would require an uncountable infinity of numbers. We must, therefore, deal with profiles that can be specified by a finite number of numbers, using some representation or discretization of the profile. An important consideration in the choice of representation is that it must be indistinguishable from the true profile for all practical purposes. In our case, the radiation emitted by an atmosphere with the true profile must be the same as that emitted by an atmosphere with a profile given by the corresponding representation well within experimental error.

At this point, we run up against the main problem of all inversion methods. The problem is ill posed (underconstrained). We need a relatively large number of numbers to construct a reasonable representation, and our measurement of radiation is usually made in a relatively small number of spectral intervals (or scan angles in the case of limb sounding). Even if there were no experimental error, there are usually not enough measurements to determine a profile uniquely. The presence of experimental error simply makes the problem worse.

II. A STATISTICAL APPROACH

The only way of solving the problem is by making use of some kind of extra *a priori* information. (The alternative is to find a different problem to solve.) The source of *a priori* information may be the physics of the problem, statistics of other measurements, arbitrary restrictions, prejudice, etc. The information, itself, can take many forms, such as representations with smaller number of parameters, smoothness, least squares deviation from some *a priori* profile, climatology, etc., but whatever the form, they are all similar in nature. They contain (or purport to contain) information about the unknown profile, just as the measurements do. An enlightening way of describing *a priori* information is to call it "virtual measurements." They provide values of known functions of the unknown profile with specified errors just as the real measurements do. To solve an inverse problem, the real measurements and the virtual measurements must contain enough pieces of information to determine all the parameters of the profile with an accuracy adequate for the application, i.e., there must be enough virtual measurements to make the problem well posed.

One way of considering the problem in order to develop some physical insight is to express it in terms of N-dimensional geometry, with N = 3 for the purpose of imagining what might happen. If the representation of the profile requires N parameters, we can treat it as a point in an N dimensional "profile space" with the parameters as coordinates. *A priori* information can be regarded as giving a value and uncertainty (including covariance) to these parameters, or equivalently, can be regarded as specifying a region of profile space within which we believe the solution to lie. This *a priori* constraint region may or may not be infinite in extent. In the example in Fig. 1, the *a priori* region is represented by the large ellipsoid.

The measurement consists of M numbers, each of which is a known function of the profile. They represent a point in an M dimensional "measurement space," which is a mapping of profile space. Typically, M << N, and a point in measurement space maps on to a region of profile space. This region is the class of profiles which are consistent with the measurements. In Fig. 1, such a region is represented by the axis of the infinite cylinder. The finite width of the cylinder expresses experimental error. The class of solutions which is consistent with both the measurements and the constraints within the error bounds is represented by the

STATISTICAL PRINCIPLES OF INVERSION THEORY

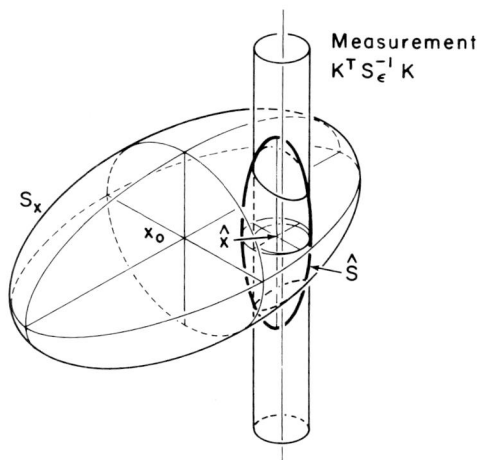

Fig. 1. Illustrating the relationship in profile space between the a priori information (x_o, S_x), the measurement $(K^T S_\epsilon^{-1} K)$ and the solution $(\hat{x}, \hat{S})$.

small ellipsoid in Fig. 1. We can see that the main effect of the constraints is to determine those components of the profile which are not determined by the measurement--in this case, position along the cylinder. The aim of statistical inverse theory is to determine the position and extent of the solution region of profile space, i.e., to find the solution and its uncertainty, or to characterize the class of profiles consistent with both the real and the virtual measurements.

III. LINEARITY

We can classify inverse problems according to their degree of nonlinearity in terms of the methods used to solve them and to carry out the error analysis. One such classification is

1. Linear: Linear problems can be solved explicitly, provided that there are enough measurements, both real and virtual,

so that the problem is well posed. Few real problems are linear.

2. Nearly linear: This class of problems can be linearized and solved with a few iterations.

3. Moderately nonlinear: These are sufficiently nonlinear for an ad hoc method to be required to find a solution efficiently, but are linear enough within the error bounds to ensure that the error analysis can be carried out with linear theory.

4. Grossly nonlinear: These are nonlinear, even within the error bounds of the solution.

In many cases, considerable improvement can be gained by not solving for the profile directly, but by solving for some nonlinear function of it. For example, if we solve for a Planck function profile rather than a temperature profile, in the case of 4.3 μm band or 15 μm band temperature sounding, the problem becomes nearly linear rather than moderately nonlinear.

IV. LINEAR ERROR ANALYSIS

The purpose of error analysis is to characterize the class of profiles which are consistent with the measurement. Linear error analysis applies to all except grossly nonlinear problems. We will assume that we have somehow found a profile which is a solution within the error bounds. We will use it as a linearization point.

The measurement y is a known function $F(x)$ of the unknown profile represented by the vector of parameters x, with error covariance S_y

$$y = F(\underset{\sim}{x}) + \varepsilon \qquad \text{Covariance } (\varepsilon) = S_y$$

We assume that within the error bounds of the solution, this equation can be linearized about our initial solution to

$$y = F(x_0) + \frac{\partial F}{\partial x} (x - x_0) + \varepsilon$$

STATISTICAL PRINCIPLES OF INVERSION THEORY

Denote the Frechet derivative $\partial F/\partial x$ by the matrix $\underset{\sim}{K}$, and we can write:

$$y - F(x_o) = \underset{\sim}{K}(x - x_o) + \varepsilon \tag{1}$$

The *a priori* information can be linearized similarly. For simplicity, we will only consider the case where $\bar{x}$ is an *a priori* value for the profile x, with covariance S_x.

$$x = \bar{x} + \underset{\sim}{\xi} \qquad \text{Covariance } (\underset{\sim}{\xi}) = S_x. \tag{2}$$

where $\underset{\sim}{\xi}$ is a random vector.

There are standard statistical methods for combining independent measurements of the same quantity, such as Eqs. (1) and (2). They are based on expected value, minimum variance or maximum likelihood estimators. The one everybody should be familiar with is the combination of two scalar measurements. If x_1 and x_2 are two independent measurements of an unknown x, with standard deviations σ_1 and σ_2, respectively, then the best estimate of x is $\hat{x}$ with standard deviation $\hat{\sigma}$ where

$$\hat{\sigma}^2 = (\sigma_1^{-2} + \sigma_2^{-2})^{-1}$$

$$\hat{x} = \hat{\sigma}^2(x_1/\sigma_1^2 + x_2/\sigma_2^2)$$

i.e., the measurements are weighted inversely with their variances. In our multivariate case, this generalizes to

$$\underset{\sim}{\hat{S}} = (S_x^{-1} + \underset{\sim}{K}^T S_y^{-1} \underset{\sim}{K})^{-1} \tag{3}$$

$$\hat{x} - x_o = \underset{\sim}{\hat{S}}\{S_x^{-1}(\bar{x} - x_o) + \underset{\sim}{K}^T S_y^{-1}(y - F(x_o))\} \tag{4}$$

which is in effect a weighted mean of two independent measurements of $x - x_o$, weighted with inverse covariance matrices. One estimate is $\bar{x} - x_o$, and the other is $\underset{\sim}{K}^*(y - F(x_1))$ where $\underset{\sim}{K}^*$ is any matrix such that $\underset{\sim}{K}\underset{\sim}{K}^*$ is a unit matrix. The second estimate is not unique because $\underset{\sim}{K}^*$ is not unique. For the purposes of this section, $\underset{\sim}{\hat{S}}$ is the more important quantity, being the covariance matrix of the solution $\hat{x}$. It contains information about the uncertainty in $\hat{x}$, it bounds the class of acceptable solutions.

Equations (3) and (4) comprise the complete solution in the case of linear problems and define the iteration to be carried out in the case of nearly linear problems. At each stage of the iteration x_o is replaced by $\hat{x}$ from the previous stage. Apparently different solutions (e.g., statistical, Twomey-Tichenov, etc.) only differ in the form of $\bar{x}$ and S_x, which express the nature of the *a priori* constraints. In the case of moderately nonlinear problems, it may be necessary for efficiency to find x_o by using some *ad hoc* procedure, such as Chahine's method, but a final stage of linearization and error analysis should be carried out according to Eqs. (3) and (4).

To understand the meaning of the covariance matrix $\hat{S}$, we can examine the diagonal elements, which comprise the "residual variance" or accuracy of the individual elements of the solution profile. This examination is illustrated in Fig. 2 for an idealized ca Residual variance gives a useful rough estimate of the accuracy of the solution profile, but it is not a complete description. Off diagonal elements of $\hat{S}$ are generally nonzero, so that there are correlations between the errors at different levels.

The form of the errors may be more easily understood if we diagonalize the error covariance matrix by finding its eigenvalues and vectors. The vectors form a set of "error patterns" whose coefficients are statistically independent from each other, and the values are the variances of these coefficients. An illustration of such a set of error patterns is given in Fig. 3 and Table 1. The solution is uncertain to the extent of adding each pattern to it with a coefficient which is random and has a variance given by λ_n. Note that all the patterns have only fine scale structure which is on the same kind of space scale in all cases.

STATISTICAL PRINCIPLES OF INVERSION THEORY

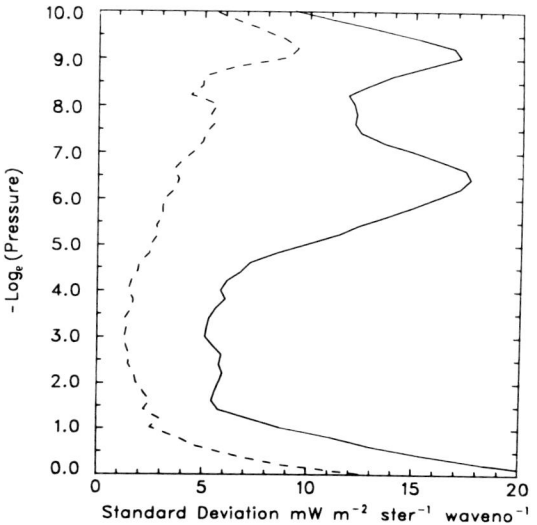

Fig. 2. A comparison of total variance (solid line) and residual variance (dashed) for a synthetic case using eight weighting functions which peak on the range from 1.7 to 7.5 scale heights, and a realistic statistical covariance matrix.

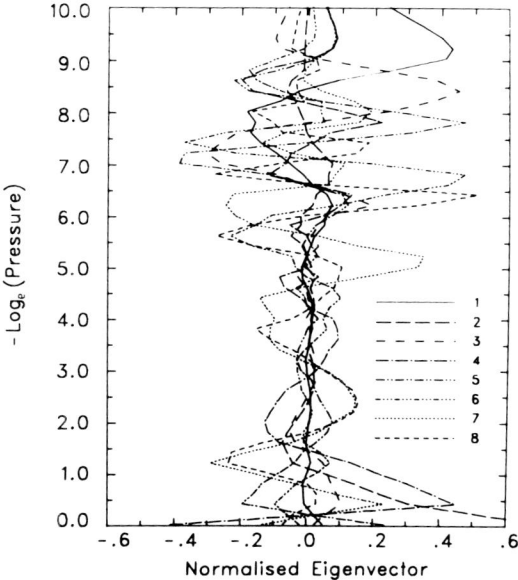

Fig. 3. Error patterns for the same case as Fig. 2.

TABLE 1

Eigenvalues of the Solution Covariance for an Idealized Case

n	1	2	3	4	5	6	7	8
λ_n	469	274	84	47	40	36	21	19

V. NONLINEAR ERROR ANALYSIS

There is no general method of error analysis that can apply to all nonlinear problems, just as there is no general solution. However, we can outline a strategy based on the linear error analysis which will give a good indication of the solution error and the degree of nonlinearity.

We must first define what is meant by a solution in the nonlinear case. This will be something like the maximum of a likelihood, or the minimum of a quadratic risk function, for example $\hat{x}$ might be the minimum of

$$(\bar{x} - x)^T S_x^{-1} (\bar{x} - x) + (y - F(x))^T S_y^{-1} (y - F(x)) \qquad (5)$$

where we have used linear *a priori* information $\{\bar{x}, S_x\}$ and a nonlinear observing system--$F(x)$. More general expressions can be used. The minimization is possible in principle, but I do not wish to discuss how it should be done. A linearization is possible about $\hat{x}$, and we can define a matrix $\hat{S}$ as in Eq. (3). The linearization may not be valid in the region of profile space defined by $\hat{S}$, and the only way to find out is to explore that region empirically, or to examine higher order terms in the expansion of y in terms of $x - \hat{x}$. It is usually much easier to explore empirically than to expand to quadratic and cubic terms. A sensible strategy is to explore in the direction of the eigenvectors of $\hat{S}$ to a distance from $\hat{x}$ given by the square root of the

STATISTICAL PRINCIPLES OF INVERSION THEORY

corresponding eigenvalue. In terms of the illustration in Fig. 1, we are exploring in the directions of the axes of the error ellipsoid, to a distance of one σ. At these new points in profile space, we evaluate either y or the risk function, both according to the linear approximation and according to the proper expression given in Eq. (5). A comparison of the two values will give a measure of the nonlinearity within the error bounds and an estimate of the true error bounds.

Using this technique, we can distinguish between moderately nonlinear and grossly nonlinear problems. This should be done in the development stages of an inversion method, even if it is not done when the method is in operational use.

Similarly, the only way of distinguishing between nearly linear and moderately nonlinear problems is to try a linearization algorithm, and to make some kind of quality judgment on the number of iterations required for convergence.

VI. SEQUENTIAL ESTIMATION

The basic concept of statistical inversion methods is one of updating information. We start with some *a priori* knowledge of a profile, together with a measure of uncertainty, as may be expressed in a covariance matrix. We then measure something related to the profile, enabling us to make a new, updated estimate of the profile, and an updated covariance matrix.

Once this simple principle has been grasped, a host of possibilities become apparent. I will briefly describe three of them.

1. It is not necessary to update the estimate with a whole vector of observations at once. The measurements can be included one at a time as scalars. The updating can then be carried out without inverting a matrix, because the estimation Eqs. (3) and (4) can be manipulated to give

$$\hat{\underset{\sim}{S}} = \underset{\sim}{S}_x - \underset{\sim}{S}_x K^T (K \underset{\sim}{S}_x K^T + \underset{\sim}{S}_y)^{-1} K \underset{\sim}{S}_x \qquad (6)$$

$$\hat{x} - \bar{x} = \underset{\sim}{S}_x K^T (K \underset{\sim}{S}_x K^T + \underset{\sim}{S}_y)^{-1} (y - F(x_o) + \underset{\sim}{K}(\bar{x} - x_o)) \qquad (7)$$

If y is a scalar, then S_y is a scalar and K is a vector: thus, the inverse becomes a scalar reciprocal. After updating with one component of y, $\hat{S}$ and $\hat{x}$ take the place of S_x and x_o for the next stage. The matrix inverse is eliminated, but, in fact, the total number of operations is similar so that there is no direct computational advantage. However, if the order in which the data is used is chosen correctly, problems of nonlinearity can be reduced so that the number of iterations is reduced. This is simply a matter of using the more linear measurements first: thus, when the time comes to incorporate the nonlinear measurements, the current estimate is nearer to the solution.

2. We can make use of the horizontal homogeneity of the atmosphere in a very simple way. If we have inverted a profile at position n, obtaining $\hat{x}_n$ and $\hat{S}_n$, then we can use this in conjunction with any other available estimate to construct the *a priori* estimate at position n + 1. We could, for example, use a very simple model of the horizontal behavior of the atmosphere:

$$x^o_{n+1} = \hat{x}_n + (\bar{x}_{n+1} - \bar{x}_n)$$

$$S_{n+1} = \hat{S}_n + \Delta S$$

where $\{x^o_{n+1}, S^o_{n+1}\}$ is the *a priori* estimate at n + 1, $\bar{x}$ is a climatology, and ΔS is a measure of the horizontal correlations. This simple model is something like a random walk process. The effect, in the case of temperature sounding in cloudy cases, for example, is to propagate information into a cloudy region from a surounding clear region, taking proper account of the growing uncertainty as we go further from the clear region.

3. We can, in principle, do a global analysis of fields of

STATISTICAL PRINCIPLES OF INVERSION THEORY

temperature and composition using sequential estimation and updating the whole analysis at every measurement time. In practice, the covariance matrices required would be prohibitively large. However, we can use the principle to carry out two-dimensional analyses and then combine the analyses to construct a three-dimensional field. The method can be used to reconstruct missing measurements or to interpolate objectively on to synoptic times. I will describe in some detail how this method has been used to analyze Nimbus 5 SCR radiances.

Nimbus 5 is in a near polar orbit, and the SCR measures radiances from every latitude between 80°S and 80°N twice each orbit. We assume that the radiance at any latitude can be represented by a Fourier series in longitude whose coefficients vary with time:

$$R(\lambda,t) = \bar{R}(t) + \sum_1^6 a_n(t)\cos(n\lambda) + b_n(t)\sin(n\lambda) \qquad (8)$$

where $R(\lambda,t)$ is the radiance (or other quantity to be analyzed) at longitude λ and time t. $\bar{R}(t)$ is the zonal mean; $a_n(t)$ and $b_n(t)$ are the Fourier coefficients. Six wave numbers have been used and 13 coefficients are given because the satellite has 13.4 orbits per day: thus, on a time scale of a day resolution of a finer structure cannot be expected.

We may write Eq. (8) in a vector product form

$$R(\lambda,t) = \underline{K}^T(\lambda) \times \underline{x}(t)$$

where $\underline{x}$ is a column vector of Fourier coefficients, and $\underline{K}^T$ is a row vector of sines and cosines. We operate on each latitude independently. At time t_n, we make a measurement of $R(\lambda(t_n),t_n)$ at some longitude $\lambda(t_n)$. Let us call that measurement y, with error variance σ^2. The available *a priori* information at t_n is the estimate of the Fourier coefficients made at time t_{n-1}, the last time a measurement was made at the same latitude. Let us call this estimate $\hat{x}_{n-1}$. We must make some assumption about the statistics of the time evolution of the Fourier coefficients. The simplest assumption is a random walk, in which case the *a priori* estimate for time t_n is

$$x_n^o = \hat{x}_{n-1} \qquad (9)$$

$$s_n^o = \hat{s}_{n-1} + (t_n - t_{n-1})\Delta S \qquad (10)$$

where ΔS is a measure of the increase of our uncertainty per unit time. We now combine the measurement y_n of $K(\lambda_n) \times K(t_n)$ according to the linear version of Eqs. (6) and (7).

$$\hat{s}_n = s_n^o - s_n^o K_n (K_n^T s_n^o K_n + \sigma^2)^{-1} K_n^T s_n^o$$

$$\hat{x}_n = x_n^o + s_n^o K_n (K_n^T s_n^o K_n + \sigma^2)^{-1} (y_n - K_n^T x_n^o)$$

where the notation has been simplified in an obvious way. Note that the inverse is a scalar reciprocal, so that the arithmetic to be performed at each stage is not excessive. If a measurement is missing at time t_n, then x_n^o and s_n^o are the best estimate.

This sequential estimator permits an estimate of the time development of the Fourier coefficients x, and ensures knowledge of the accuracy of the estimation. It can be used to interpolate missing data and to smooth existing data in an objective manner. (Incidentally, the *a priori* information and its covariance allows us to detect bad data by means of a 3σ test.) We can also use it to reconstruct global fields at synoptic times, simply by evaluating the Fourier series at the appropriate times.

As it stands, the method is one sided. We can only use the measurements which precede t_n to estimate at t_n. If the method is being used operationally, this is the best we can do, because the future is not available to us. In this case, more attention should be paid to the "forecast" from $\hat{x}_{n-1}$ to x_n^o (Eqs. 9 and 10). However, if we are analyzing a long run of data for research purposes, we would like to use both sides of the time axis. This can be done quite easily by analyzing the data backwards in time, and combining the forwards and backwards estimates in the proper statistical manner--by using the reciprocal of the covariance as a weight.

Figures 4 and 5 illustrate the effect of this kind of estimator in reconstructing missing data and smoothing existing data, when operating on a time series of measurements about 10 days long at a latitude of $40°N$.

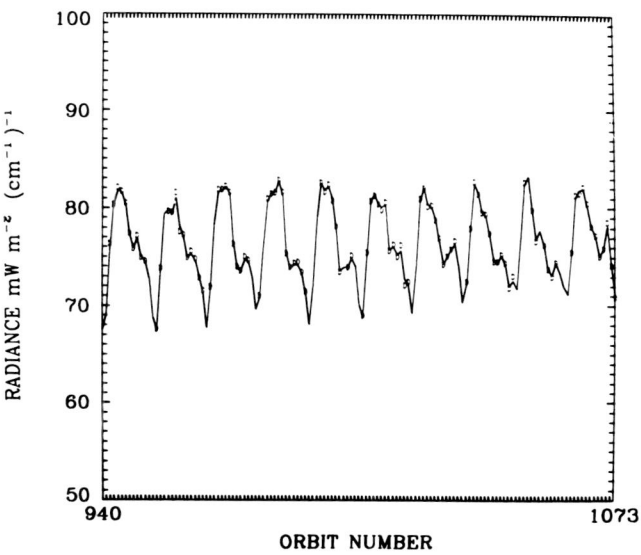

Fig. 4. Reconstruction and smoothing of Nimbus 5 SCR channel B12 at $40°N$--a typical case. The data points are marked D; the reconstructed data is a continuous line.

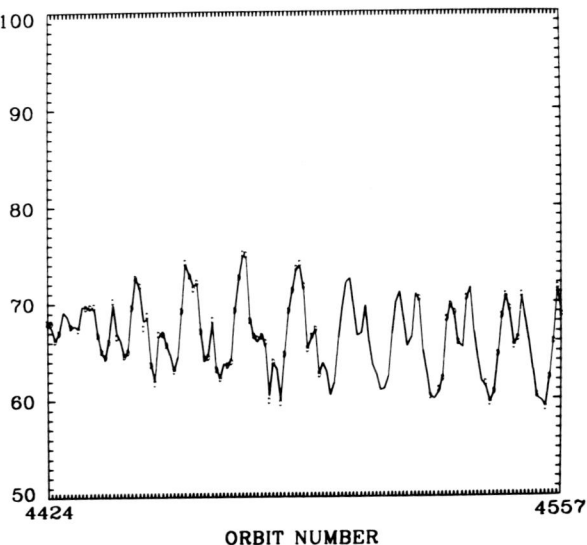

Fig. 5. Illustrating reconstruction of about 1-½ days of missing data.

VII. DISCUSSION

All the previous discussion is a special case of the Kalman filter (Ref. 3), which is a particularly powerful and general approach to the analysis of multivariate time series. Some generalizations of the problem, such as the case where it is not the profile, but some function of the profile (e.g., thickness) which is of interest have not been discussed. Problems of this type are also amenable to Kalman filtering.

STATISTICAL PRINCIPLES OF INVERSION THEORY

SYMBOLS

a_n, b_n	Fourier coefficients
$F(x)$	known function of the unknown profile
$\underset{\sim}{K}$	matrix denoting the Frechet derivative $\partial F/\partial X$
$\underset{\sim}{K}^*$	any matrix satisfying the relation $\underset{\sim}{KK}^*$ = unit matrix
K_n	value of K at time t_n
R	radiance
$\bar{R}$	zonal mean
S_y	error covariance
$\underset{\sim}{\hat{S}}$	covariance matrix of x
$\hat{S}_n$	values of $\hat{S}$ at position n
S^o_{n+1}	a priori estimate of $\hat{S}_n$ at position n + 1
t	time
t_n	time at nth position
T	transpose of a matrix
x	parameters of $F(x)$
x_1, x_2	two independent measurements of x
$\hat{x}$	best estimate of x
$\hat{x}_n$	values of x for an idealized case
x_o	initial value of x
x^o_{n+1}	a priori estimate of $\hat{x}_n$ at position n + 1
$\bar{x}$	a priori value for the profile x
y	measured quantity
Y_n	measurement of $\underset{\sim}{K}(\lambda_n) \times \underset{\sim}{K}(t_n)$
ΔS	measurement of the horizontal correlations
ε	measurement error
$\underset{\sim}{\xi}$	random vector
λ	longitude
λ_n	nth eigenvalue of the solution covariance
σ	standard deviation
$\hat{\sigma}$	standard deviation of $\hat{x}$
σ^2	error variance
σ_1, σ_2	standard deviations for x_1 and x_2, respectively

REFERENCES

1. C. D. Rodgers, Retrieval of atmospheric temperature and composition from remote measurements of thermal radiation, *Rev. Geophys. Space Phys. 14,* 609 (1976).

2. R. Deutsch, "Estimation Theory." Prentice Hall, Inc., Englewood Cliffs, New Jersey, 1964.

3. R. E. Kalman, A new approach to linear filtering and prediction problems, *J. Basic Eng. 82D,* 35 (1960).

STATISTICAL PRINCIPLES OF INVERSION THEORY

DISCUSSION

Susskind: I just wanted to ask something quickly about what you have just done. These radiances--are these the observed radiances, or the radiances taking out the effects of clouds, or what? I mean, if you have clouds coming into the picture, it is certainly going to foul up.

Rodgers: These are actually stratospheric radiances.

Susskind: Okay, so you're not worrying about that.

Rodgers: This isn't the Nimbus 5 selective chopper. It's about 45 kilometers.

Green: On your first slide, you said "given some knowledge of a function." Should you have replaced the word "function" by "functional"?

Rodgers: Sorry, yes, it is functional.

Green: Thus, the problem is how to get a function from a functional.

Rodgers: I was just not being quite rigorous.

Chahine: Clive, you made a statement that we should pay just as much attention to our actual measurement as we do to virtual measurements. I know how to improve my actual measurements. I have the physics. How can I improve my virtual measurements and be sure of that?

Rodgers: By the same sort of techniques as your actual measurements. If you have no virtual measurements, you just can't solve the problem. You just have to go into some other problem. The only way of producing a profile is by having enough virtual measurements from somewhere. It may be physics. I can produce a virtual measurement off the top of my head immediately. I can say the temperature in the atmosphere anywhere is going to lie between zero and 500°. I know it's not going to help you very much; it reduces the variance a bit. It just makes it noninfinite at least. But it still means the errors on all the points are going to be 250°.

Chahine: For the nonlinear method, I assume the temperature to be positive and real. But in your case, you are using *a priori* statistics.

Rodgers: This isn't only statistics. This applies to any kind of virtual measurement.

Chahine: Only real physical data should be used to judge virtual data. That's because of the high variability of the atmosphere and clouds. How can you be sure that the virtual measurements that you have now are good tomorrow or are good in the presence of a front?

Rodgers: I don't know. But in the method you described you have virtual measurements which you haven't stated explicitly and this is your interpolation rule. That's virtual measurement, which is a very unrealistic one in fact.

Chahine: Not for temperature. Professor Kaplan might like to comment on the assumption of constant lapse rate between two levels in the atmosphere.

Kaplan: That is assuming you know where the lapse rate changes. And anybody who has worked on recording radiosonde measurement of temperature knows that the temperature can increase in the atmosphere in general linear with height up to the point where the lapse rate changes. Of course, the lapse rate changes at arbitrary levels and you have to be able to pick the point at which this comes in. I mean, there are assumptions. You assume if you pick the height at which you attribute a frequency, measurements at a frequency, this is a point at which the lapse rate changes or you fit to a polynomial. But there are these constraints.

Rodgers: But you have got to recognize that it is a virtual measurement of some kind, with some variance.

Kaplan: Yes, and to the extent to which your vertical resolution gets worse as you get larger, this is more and more an error. As you narrow your resolution this becomes much less important.

Chahine: Have you done an analysis to determine how often you hit and how often you miss?

Rodgers: What do you mean?

Chahine: With your probability approach, how often do you end up with the correct answer in applying your technique to real data?

Rodgers: Well, this doesn't really apply. This is not really within the scope of what I was trying to describe. I wasn't going into techniques of how you find $\hat{x}$, I was going into techniques of how you find $\hat{S}$.

Chahine: I wanted to see, in applying the statistical approach to real data, how often you hit?

Rodgers: You mean how good can your statistics be?

STATISTICAL PRINCIPLES OF INVERSION THEORY

Chahine: Yes.

Rodgers: That's another question which I haven't really tried to touch on here. I know it is very difficult to get *a priori* information. Statistics is only one way. But whatever your *a priori* information is, you have to work on it as hard as you can.

Wark: Have you ever given any consideration to other aspects of the radiance field? Namely, that we usually try to solve for profiles as though they existed as the only profile in nature, and then we go on to another set of radiances and try to solve them. In addition to this, we have gradients which exist in the radiance field and these gradients are very strongly tied to the dynamics of the atmosphere. That is, the same set of radiances in the same location at the same time of year can be associated with quite different profiles, mainly because of the gradients which occur, which are the physical dynamical processes in the atmosphere.

Rodgers: This is why I have been recently getting interested in doing global analysis of radiances to try and make an estimate of the global distribution rather than individual profiles.

Wark: But have you tried to associate this with the gradients in the temperature field?

Rodgers: Not yet. My feeling about analyzing meteorological data is that it should be the meteorological analyst's job, not ours. We shouldn't go through this interface of profiles or anything like it. We should get him good calibrated radiances. He's already doing an inverse problem. He's solving for the field of whatever it is, given certain measurements of something different. Radiances are just another thing.

Wark: That's right. We should be giving radiances to the meteorologists and let them inject them into their analyses.

Chahine: What you are asking for is a simultaneous solution of the radiative transfer equation and the equations of motions. This is a great aim. It isn't easy.

Wark: But I wanted to emphasize the point that the solution for profiles is not necessarily our aim here in a meteorological sense.

Rodgers: Sure. There are lots of things you can get out of this stuff other than profiles.

Malchow: A bit of a detailed question about your $\hat{S}$. In non-linear iterative processes, how is it supposed to be handled? It seems to be unstable if it is iterated within the iterative process.

Rodgers: It certainly isn't unstable. No.

Malchow: I was wondering if you had any experience with that problem?

Rodgers: Well, $\hat{S}$ is trying to find the small ellipsoid in Figure 1 here. Intuitively, it is not going to be unstable providing you have got the right kind of *a priori* information. If $\hat{S}_x$ is the wrong shape compared with this cylinder, then perhaps $\hat{S}$ is going to stretch way up the cylinder. It is a matter of getting the right information in and if you've got enough information, $\hat{S}$ is easy. If S is not easy, you haven't got enough information.

Deirmendjian: This is not a question but just a comment. Since this is an interactive Workshop, may I interject some non-scientific thoughts about Dr. Rodgers' introduction of the word "aesthetics." I like it because "aesthetics" derives from a Greek verb meaning "to perceive with the senses." It is a scientist's prerogative to introduce--and to be governed a little by--aesthetics in his work. This implies things like restraint, non-exaggeration, nonreliance on innumerable assumptions, criteria or data banks, and so on. I would like to make an analogy, if I may, between sailing, about which I know some things, and the use of mathematical inversion techniques, about which I know very little. Some people want us to use more and more instruments and electronic gadgetry which are supposed to help us sail better, on the assumption that we have no senses--seeing, hearing, sensory feeling--or judgment or "sea-sense." A good sailor does use all these things to advantage for a successful voyage. So, in analogy to this, I feel that sometimes we tend to resort to inversion techniques too blindly, without using our judgment or "feel" about handling a given problem, which may lead to "anti-aesthetic" excesses.

INVERSE SOLUTION OF THE PSEUDOSCALAR TRANSFER

EQUATION THROUGH NONLINEAR MATRIX INVERSION

Jean I. F. King
Air Force Geophysics Laboratory

The upwelling radiance from a plane-parallel planetary atmosphere viewed either in a limb or frequency scan depends on the internal scattering and thermal state of the atmosphere. Each upwelling profile is the solution of a uniquely specified, but generally unknown, pseudoscalar transfer equation.

Nonlinear matrix inversion operators have been developed which, applied to observed radiances, infer maximal information regarding atmospheric scattering parameters and vertical distribution of radiant sources and sinks. The algorithm has the attractive feature of noise discrimination, attributing instrumental errors to extra-atmospheric sources.

I. THE MILNE PROBLEM ⟨MDO MODE⟩ IN CODON LANGUAGE

Assume the upwelling intensity $I(0,\mu)$ to be exactly representable in $\mu = \cos\theta$ space by a linear expression and $n-1$ hyperbolic components.

$$I(0,\mu) = \text{const} \left[\mu + Q + \sum_{\alpha=1}^{n-1} \frac{L_\alpha}{1 + K_\alpha \mu} \right] \tag{1}$$

Substituting the more convenient transform variable $\kappa = \sec\theta = 1/\mu$, we have

$$\frac{I(0,1/\kappa)}{\kappa} = \text{const} \left\{ \frac{1}{\kappa^2} + \frac{Q}{\kappa} + \sum_{\alpha=1}^{n-1} \frac{L_\alpha}{\kappa + K_\alpha} \right\} \tag{2}$$

The positive definite character of $I(0, 1/\kappa)$ requires the hyperbolic poles, $\kappa = -K_\alpha (\alpha = 1,2,\ldots,n-1)$, to be negative.

By "clearing of fractions" the upwelling intensity can be alternatively expressed as the quotient polynomial with n roots $\kappa = -1/\mu_i$ $(i = 1,2,\ldots,n)$. Thus,

$$\frac{I(0,1/\kappa)}{\kappa} = \text{const} \frac{\Pi_{i=1}^{n}(\mu_i \kappa + 1)}{\kappa^2 \Pi_{\alpha=1}^{n-1}\left(\frac{\kappa}{K_\alpha} + 1\right)}$$

$$= \text{const } \Pi_{i=1}^{n} \mu_i \, \Pi_{\alpha=1}^{n-1} K_\alpha \frac{H(1/\kappa)}{\kappa} \quad (3)$$

where we have defined the H-function as

$$H(1/\kappa) = \frac{\Pi_{i=1}^{n}\left(\kappa + 1/\mu_i\right)}{\kappa \, \Pi_{\alpha=1}^{n-1}(\kappa + K_\alpha)} \quad (4)$$

Again, the positive definiteness of the upwelling intensity requires the polynomial roots, $\kappa = -1/\mu_i (i = 1,2,\ldots,n)$, to be negative.

The quotient polynomial H-function can be characterized by its sequence of roots and poles along the real axis in the transform plane. We shall call such a linear array a *codon*. The importance of this representation lies in the fact, as we shall see, that all the radiation physics is implicit in the codon root-pole morphology.

Using partial fraction analysis, we are able to express the coefficients of the upwelling intensity expansion as *residues (or pole-strengths)* of the H-function codon at the poles $\kappa = -K_\alpha$. Thus, we find from Eqs. (3) and (4)

$$\frac{I(0,1/\kappa)}{\kappa} = \text{const} \frac{H(1/\kappa)}{H_o \kappa}$$

$$= \text{const} \left[\frac{1}{\kappa^2} + \frac{Q}{\kappa} + \sum_{\alpha=1}^{n-1} \frac{L_\alpha}{\kappa + K_\alpha}\right] \quad (5)$$

PSEUDOSCALAR TRANSFER EQUATION

where the residues

$$L_\alpha \equiv \frac{1}{H_o} \lim_{\kappa \to -K_\alpha}(\kappa + K_\alpha)\frac{H(1/\kappa)}{\kappa} = \frac{\Pi_{i=1}^{n}(1/\mu_i - K_\alpha)}{H_o K_\alpha^2 \Pi_{\substack{\beta=1 \\ \neq \alpha}}^{n-1}(K_\beta - K_\alpha)} \tag{6}$$

and

$$H_o \equiv \lim_{\kappa \to 0}\kappa H(1/\kappa) = \frac{1}{\Pi_{i=1}^{n}\mu_i \, \Pi_{\alpha=1}^{n-1}K_\alpha}$$

The Q-constant, the residue of the double pole at the origin, is evaluated after some algebra as

$$Q = \sum_{i=1}^{n}\mu_i - \sum_{\alpha=1}^{n-1}\frac{1}{K_\alpha} \tag{7}$$

II. CODON TRANSFER THEORY

The upwelling profile can be considered the externally sensed solution of an internal transfer problem. We proceed now to construct, i.e., to infer, the unique transfer equation, source function, constitutive relation, and characteristic function which are implied by the observed intensity.

We need first an equation of transfer. Now the transfer equation can be viewed as a conservation condition imposed on the intensity under steady-state. We shall show that the transfer equation has a deeper origin as a codon identity in transform space.

From Eq. (2) we see that the α component of the upwelling intensity obeys the *split codon identity*

$$\frac{\kappa}{\kappa + K_\alpha} + \frac{K_\alpha}{\kappa + K_\alpha} = 1 \tag{8}$$

Let us consider this identity as a codon operator, viz.

$$\frac{\kappa}{\kappa + K_\alpha} J_\alpha(\tau) + \frac{K_\alpha}{\kappa + K_\alpha} J_\alpha(\tau) = J_\alpha(\tau) \tag{9}$$

and seek the internal (scalar) τ-space eigenfunction which converts this codon equation into the equation of transfer.

By defining the internal (vector) intensity by

$$I_\alpha(\tau, 1/\kappa) = I_\alpha(0, 1/\kappa) J_\alpha(\tau) = \frac{\kappa}{\kappa + K_\alpha} J_\alpha(\tau) \tag{10}$$

we see by inspection that the required eigenfunction form is

$$J_\alpha(\tau) = L_\alpha e^{-K_\alpha \tau} \tag{11}$$

for then we have the *equation of transfer*

$$\frac{dI_\alpha(\tau, 1/\kappa)}{\kappa d\tau} = I_\alpha(\tau, 1/\kappa) - J_\alpha(\tau) \tag{12}$$

We note in passing from Eqs. (2) and (11) that the eigenfunction $J(\tau)$ is the *inverse Laplace transform* of the upwelling intensity

$$J(\tau) = L^{-1} \left| \frac{I(0, 1/\kappa)}{\kappa} \right|$$

$$= \text{const} \left[\tau + Q + \sum_{\alpha=1}^{n-1} L_\alpha e^{-K_\alpha \tau} \right] \tag{13}$$

Equation (12), as it stands, is a relation between the two dependent variables $I(\tau, 1/\kappa)$ and $J(\tau)$. To eliminate one of these, a second or constitutive relation is needed. This expresses the source function, the radiation emitted at level τ in the direction $\theta = \cos^{-1}\mu$, as the sum of radiation incident in all directions which is scattered into θ, viz.

$$J(\tau, \mu) = \frac{1}{2} \int_{-1}^{1} \psi(\mu, \mu') I(\tau, \mu') d\mu' \tag{14}$$

For local thermodynamic equilibrium, the source emission is uncorrelated with the directions of the incident beams $\theta' = \cos^{-1}\mu'$, in which event

PSEUDOSCALAR TRANSFER EQUATION

$$J(\tau,\mu) \to J(\tau) \tag{15}$$

so that the eigenfunction becomes the source function

$$J(\tau) = \frac{1}{2}\int_{-1}^{1} \psi(\mu')I(\tau,\mu')d\mu' \tag{16}$$

We can write, therefore, as the *integro-differential equation of transfer*

$$\frac{dI(\tau,1/\kappa)}{\kappa d\tau} - I(\tau,1/\kappa) = \frac{1}{2}\int_{-1}^{1} \psi(\mu)I(\tau,\mu)d\mu \tag{17}$$

The finite exponential sum character of Eq. (13) demands that the source function $J(\tau)$ be compounded of radiation restricted to a finite number of fixed directions. This is mathematically accommodated by expressing the characteristic function as a finite delta-function sum, i.e.,

$$\psi(\mu) = \sum_{i} \psi_i \delta(\mu - \mu_i) \tag{18}$$

Equation (18) implies that the characteristic function acts as a filter and permits only the beams in the directions $\mu = \mu_j = \pm\mu_i$ to participate in the source function. The ID transfer equation then becomes

$$\frac{dI(\tau,1/\kappa)}{\kappa d\tau} - I(\tau,1/\kappa) = J(\tau) = \frac{1}{2}\sum_{i} \psi_i I(\tau,\mu_i) \tag{19}$$

The determination of the strength and direction of the incident beams is facilitated by taking the derivative of the constitutive relation

$$J'(\tau) = \frac{1}{2}\sum_{i} \psi_i \frac{dI(\tau,\mu_i)}{d\tau} \tag{20}$$

Substituting from Eqs. (10) and (13), we can write

$$1 - \sum_{\alpha=1}^{n-1} K_\alpha L_\alpha e^{-K_\alpha \tau} = \frac{1}{2}\sum_{i}\psi_i \left(1 - \sum_{\alpha=1}^{n-1} K_\alpha \frac{L_\alpha e^{-K_\alpha \tau}}{1 + K_\alpha \mu_i}\right) \tag{21}$$

The constant terms are equal in the conservative case $\frac{1}{2}\sum \psi_i = 1$. Interchanging the order of summation leads to

$$\sum_{\alpha=1}^{n-1} K_\alpha L_\alpha e^{-K_\alpha \tau} = \sum_{\alpha=1}^{n-1} K_\alpha L_\alpha e^{-K_\alpha \tau} \left(\frac{1}{2} \sum_i \frac{\psi_i}{1 + K_\alpha \mu_i} \right) \quad (22)$$

which in turn requires

$$\frac{1}{2}\sum_{i=1}^{n} \frac{\psi_i}{1 + K_\alpha \mu_i} + \frac{1}{2}\sum_{i=1}^{n} \frac{\psi_{-i}}{1 - K_\alpha \mu_i} = \sum_{i=1}^{n} \frac{\psi_i}{1 - K_\alpha^2 \mu_i^2} = 1 \quad (23)$$

where we have assumed $\mu_{-i} = -\mu_i$ and $\psi_{-i} = \psi_i$.

We shall now demonstrate that ψ_i is the residue of reciprocal H-functions. Consider the bilaterally symmetric T-function defined by

$$T(1/\kappa) \equiv \frac{1}{H(1/\kappa)H(-1/\kappa)} = \frac{\kappa^2 \prod_{\alpha=1}^{n-1}\left(\kappa^2 - K_\alpha^2\right)}{\prod_{i=1}^{n}\left(\kappa^2 - \frac{1}{\mu_i^2}\right)}$$

$$= 1 + \sum_{i=1}^{n} \frac{\psi_i}{\mu_i^2 \kappa^2 - 1} \quad (24)$$

Clearly, we have for $\kappa = \pm K_\alpha$

$$T(\pm 1/K_\alpha) = 0 = 1 - \sum_{i=1}^{n} \frac{\psi_i}{1 - K_\alpha^2 \mu_i^2} \quad (25)$$

which shows that the T-function satisfies the requirement of Eq. (23).

The strength of the beam at $\mu = \mu_i$ is given now as the residue of the reciprocal H-functions

PSEUDOSCALAR TRANSFER EQUATION

$$\psi_i = \kappa \xrightarrow{\lim} \pm \frac{1}{\mu_i} \mu_i^2 \left(\kappa^2 - \frac{1}{\mu_i^2}\right) T(1/\kappa)$$

$$= \kappa \xrightarrow{\lim} \pm \frac{1}{\mu_i} \frac{(\kappa^2 - 1/\mu_i^2)}{\kappa^2 H(1/\kappa) H(-1/\kappa)} = \frac{\prod_{\alpha=1}^{n-1}(1/\mu_i^2 - K_\alpha^2)}{\prod_{\substack{j=1 \\ \neq i}}^{n}(1/\mu_i^2 - 1/\mu_j^2)} \quad (26)$$

The deep reciprocity inherent in the codon formulation of the radiative transfer problem is seen by comparing L_α and ψ_i. The coefficients L_α of the exponential source function expansion, as defined in Eq. (13), namely,

$$J(\tau) = \text{const} \left[\tau + Q + \sum_{\alpha=1}^{n-1} L_\alpha e^{-K_\alpha \tau}\right]$$

are identified as residues at the poles of the H-function codon, from Eq. (6), namely,

$$H_o L_\alpha = \kappa \xrightarrow{\lim} - K_\alpha (\kappa + K_\alpha) \frac{H(1/\kappa)}{\kappa} = \frac{\prod_{i=1}^{n}(1/\mu_i - K_\alpha)}{K_\alpha^2 \prod_{\substack{\beta=1 \\ \neq \alpha}}^{n-1}(K_\beta - K_\alpha)} \quad (27)$$

In contrast, the weights of the filter function ψ_i which enter the constitutive relation, from Eq. (19), namely,

$$J(\tau) = \frac{1}{2} \sum_i \psi_i I(\tau, \mu_i),$$

are residues at the poles of the *reciprocal* paired H-function codons

$$\psi_i = \kappa \xrightarrow{\lim} \pm \frac{1}{\mu_i} \frac{\kappa^2 - 1/\mu_i^2}{\kappa^2 H(1/\kappa) H(-1/\kappa)} = \frac{\prod_{\alpha=1}^{n-1}(1/\mu_i^2 - K_\alpha^2)}{\prod_{\substack{j=1 \\ \neq i}}^{n}(1/\mu_i^2 - 1/\mu_j^2)}$$

It is further seen that a knowledge of the codon structure, i.e., its roots and poles, serves to specify all the radiation functions and parameters of the problem. The codon concept shifts the emphasis and changes the character of transfer theory. Rather than seeking the solution for a particular radiation parameter, one develops algorithms for determining the underlying codon structure from the given data. Thus, in the *forward* problem in which the internal scattering characteristic function $\psi(\mu)$ is known, Gaussian quadrature is used to determine the weights ψ_i and directions μ_i. These determine, in turn, the constants K_α and L_α and thus the upwelling intensity. On the other hand, in the *inverse* problem we must construct the codon structure from the observed upwelling intensity profile. This requires a nonlinear matrix inversion algorithm, developed by the author (Ref. 1), which solves uniquely and exactly the following problem: Given 2n upwelling intensities $I_j = I(0, \mu_j)$ sensed at the arbitrary nadir angles $\theta_j = \cos^{-1}\mu_j$ (j = 0,1,..., 2n - 1), find the unique 2n constants C, Q, and the n - 1 pairs (L_α, K_α), which are specified by the measurements. This is equivalent to the inversion of the following nonlinear equation set

$$I_j = C \left[\mu_j + Q + \sum_{\alpha = 1}^{n - 1} \frac{L_\alpha}{1 + K_\alpha \mu_j} \right], \quad j = 0,1,\ldots,2n - 1 \quad (28)$$

The Planck intensity source function follows readily from Eq. (13); namely,

$$B(\tau) = C \left[\tau + Q + \sum_{\alpha = 1}^{n - 1} L_\alpha e^{-K_\alpha \tau} \right] \quad (29)$$

The inference of the internal scattering parameters becomes a mere evaluation of μ_i and ψ_i from the inferred constants L_α, K_α.

III. SUMMARY, GENERALIZATION, AND PROSPECTUS

In this paper, we have discovered three features: first, that underlying the interaction of radiation and matter, i.e., radiative transfer theory, is a simple code; second, that the code consists of linear arrays of roots and poles along an axis in a complex transform space (which we have called "codons"); and third, that the code translates easily into observable and inferable radiation parameters. Stated succinctly, we have found a code, we have determined its structure, and we have broken the code.

In assessing the importance and implications of this work, we must set forth how transfer theory differs from conventional differential analysis. The specification of density distributions is of great interest in physics, for example, the fluid density in the Euler-Lagrange equation or the ψ-probability density in Schrödinger wave mechanics. In these and other cases, the densities, specified as solutions of partial differential equations, are considered primal and, hence, not further analyzable into more basic component parts. In contrast, the transfer equation through its linkage between two spaces relates the source function, and energy density, to the solution of an integro-differential equation. In this relation, the density $J(\tau)$ is analyzed and dissected as the discrete sum of a more primitive concept, the beam field $I(\tau; \mu, K)$ quantized in the directions μ_i and e-folding lengths K_α^{-1}.

We have seen that the physics of the generalized transfer problem is completely determined by the root-pole morphology of the codon. Further, the alternate root-pole structure along the negative real axis of the H-codon in classical theory is a relatively restricted grouping. It is natural to inquire into the physical systems implied by more general codon patterns.

We find that there are codon identities other than the linear expression, Eq. (8), which can generate transfer equations. In particular, the quadratic identity can be used to generate a wave transfer equation. The Planck intensity, derived from first principles, is such an example. We may ask what is the condition for the source function to obey a differential equation? We find this occurs if, and only if, the associated codon exhibits the group property of invariance under displacement. This, in turn, occurs if, and only if, the codon roots and poles are equally spaced. We are able to express all the finite polynomials (Legendre, Laguerre, Hermite, Jacobi) entering into the ψ-functions of simple quantum systems as inverse of such equally spaced (canonical) codons (Ref. 2). Finally, it proves possible to infer the relativistic radial (Dirac) wave equation as the inverse of a split quadratic codon equation.

SYMBOLS

B	Planck intensity source function
H_o	H-function residue at origin
$H(\mu)$	discrete ordinate H-function of Chandrasekhar
$I(\tau, \mu)$	radiant intensity at optical depth τ at an angle $\theta = \cos^{-1} \mu$ with the zenith
J	source function
K	reciprocal e-folding depth of radiant beam
L_α	coupling constant between the forcing function and the K_α medium modes
Q	$Q = q(\infty)$, where $q(\tau)$ is the Hopf q-constant
T	discrete ordinate T-function of Chandrasekhar
α, i	dummy variables specifying particular values of K and μ, respectively
κ	independent variable specifying the complex transform plane
ψ	characteristic function

REFERENCES

1. J. I. F. King, Remote Temperature Inference and Diagnosis of Instrumental Error Using Padé Inversion, *Proc. of the Inter. Radiation Symp., IAMAP (IUGG), Sendai, Japan,* 1972.

2. Jean I. F. King, Schrödinger equation and quantum state condons and discrete transform space, *J. Math. Phys. 15,* 1849 (1974).

DISCUSSIONS

Green: You used quantum mechanics as an illustration for your problem. In most applications of quantum mechanics, the Schrödinger equation or the Dirac equation does not serve as the goal of the search. For example, in the nuclear force problem, you can assume that the Schrödinger equation or the Dirac equation is a basic law of nature or the basic equation of motion and you might be attempting from scattering data, experimental accelerator data, to infer a basic law force so that the Schrödinger equation or the Dirac equation has the same role as the radiative transfer equation and the scattering data has the role of some experimental observation from which you hope to find a basic nuclear force or atomic force. Thus, it seems that if in your analogy of your codons you arrive at the Schrödinger equation or the Dirac equation you do not have a parallelism to the real world use of the Schrödinger equation or the Dirac equation as something which connects scattering observations to some physical aspect of the description of your system. Which is the law of force governing your system which you would then insert into the Schrödinger equation?

King: Let me try to answer that. Perhaps our conventional view of the Schrödinger and Dirac equations needs reassessing. Let us think generally. A codon identity quite literally is a separation of unity into two or more parts. The radiative transfer equation is the transcription into our space of this codon separation. The radiative heat exchange represents the flow resulting from this nonequilibrium partition arising from the nonisothermal temperature distribution. The Schrödinger and Dirac equations may be similarly viewed as representing unequal partitioning of the Psi-function, concentrated in regions of high electron expectancy and dilute elsewhere. At a deeper level, this separation is represented in the Dirac equation by a triply split codon identity. This identification of the Dirac equation as an inverse codon identity has the important conceptual consequence that nature, at least in the radial Dirac equation describing the Kepler atom, is contrained by a *law of form*, rather than a law of force.

Unidentified Speaker: Could you predict something that isn't known?

King: Yes. At present no theory or model exists for the fine structure constant. It currently enters into field theory as an *ad hoc,* empirically determined, externally imposed coupling constant between the electromagnetic field and the electron. I hope to model the fine structure constant as a codon counting algorithm involving the harmonic sum of codon poles.

Irvine: I am a little bit perplexed at your unhappiness with the transfer equation, Jean. You say there are ambiguities in it, but

PSEUDOSCALAR TRANSFER EQUATION 151

that is true of all physics. In every physical equation we are making a mathematical approximation to reality. The best description of radiation we think we have now would be quantum field theory. Of course, there are aspects of that that are not present in Maxwell's equations. Likewise, in the transfer equation, we are negelcting information which is present in Maxwell's equations.

King: Yes, I agree with you. What I should have said was that a linear transfer equation is satisfied only by exponential-type source functions. In other words, only restricted classes of source functions satisfy the linear transfer equation. We agree to that.

Irvine: You also expressed unhappiness about the fact that one uses a local thermodynamic equilibrium approximation for the source function. Of course, that is not necessary. That is simply because we are ignoring any possible information on the microphysics at a given point.

Twitty: I guess I am either missing something or I don't understand this relationship between the codons and the equations you have derived. There are an infinite set of equations you could write down that are identities, and one could do these kind of transforms on all of those and produce some infinite set of differential equations which hopefully would not describe anything we know about physics. What is so special about the ones you have derived?

King: The special feature is that the transfer equation is the only inverse statement of a *linear* codon identity. And I would say that quadratic codon identities lead to sinusoidal source functions and quadratic transfer equations.

Twitty: But then the Dirac equation looks far more complicated.

King: It is still a wave equation.

Twitty: No, I am not referring to the Dirac equation. Your identity that gives it is far more complicated.

King: Yes, the reason for that is the additional constraint which codons must obey in quantum mechanics. The differential equation format of wave mechanics demands that the roots and poles of the corresponding codons in the transform plane be equally spaced. These codon patterns associated with various quantum systems are developed in the paper cited.

Twitty: So it seems to me that somewhere in this identity, mathematical identity is what you're really talking about, there is the basic physics and you have used some additional knowledge about the physics to say that this identity corresponds to what we know in terms of the wave equation describing the problem.

King: Yes. That requirement is that the Psi-function codon be invariant under translation in the transform plane. This invariance, in turn, demands equally spaced roots and poles, namely, canonical codons, for superposition.

Fymat: I have a question and a comment. The question is: Have you carried on this work for the case of scattering and with contributions from single and multiple scattering in the source function?

King: No, I have not. The formalism appears indifferent to whether or not the impinging photon is singly- or multiply-scattered. Perhaps this is not a disadvantage inasmuch as it is difficult to conceive any measurement which would discriminate between the two.

Fymat[1]*:* The comment is that in his book, Professor van de Hulst has emphasized that we should use what he called the scattering amplitude matrix, a 2 x 2 matrix which contains information both on the amplitude and the phase of the wave. He carried the theory only for single scattering. On the other hand, Fano has shown the analogy between this treatment and quantum mechanics. Dr. Vasudevan and myself have pursued this analogy further in a paper we have recently published.[2] This article provides the complete theory of multiple scattering for both amplitude and phase by exploiting the close analogy between the quantum mechanical states of half-spin systems and the polarization states of electromagnetic radiation. The interest of this new formulation is that it enables you to carry the multiple scattering with the phase information. Although we are not observing the phase in the visible, this is possible in other regions of the spectrum. The phase also contains

---

[1] Dr. Fymat's post-Workshop comment: "I did not say that the scatterer molecule or particle conserved any memory of the scattering order. Rather, I was concerned with the formalism that your codon approach would take had you considered a particular scattering process and the consequent polarization it induces in the radiation field. Here, the radiative transfer equation would become four-dimensional (using, for example, Stokes' representation of the polarization state), and the source-function would be far more complicated than Planck's function as it would receive contributions from both single and multiple scattering."

[2] Published in Astrophysics and Space Science, Vol. 38, pp. 95-124, 1975.

information about the atmosphere. I thought you might wish to use this work were you interested in extending your codon approach to scattering and polarization.

King: Yes, I see. All of these methods must ultimately rest on some conservation principle.

BACKUS-GILBERT THEORY AND ITS APPLICATION TO RETRIEVAL OF OZONE AND TEMPERATURE PROFILES

Barney J. Conrath
Goddard Space Flight Center

The analytical methods of Backus and Gilbert were originally formulated for application to inverse problems associated with the physics of the solid earth. However, the theory is sufficiently general to be applicable to many types of inverse problems, and, in particular, constitutes a useful tool for analyzing the information content of atmospheric profile retrievals. Basically, the method provides a quantitative evaluation of the trade-off between vertical resolution of a retrieved profile and formal root-mean-square (rms) error due to measurement noise propagation. As one example of an application of the theory, the problem of retrieving the topside ozone profile from backscattered ultraviolet (BUV) measurements is considered. For measurements of the type currently being obtained with the Nimbus 4 and AE-E BUV experiments, it is found that a vertical resolution of approximately 0.75 scale height can be achieved for a formal volume mixing ratio profile error of 10%. Other examples include treatments of the retrieval of temperature profiles from measurements in the 15 μm CO_2 absorption band for both the terrestrial and Martian atmospheres. Finally, the method is applied to the problem of retrieving temperature profiles of the Jovian planets from measurements in the far infrared pressure induced H_2 lines to be obtained from the Mariner Jupiter/Saturn fly-by missions. In the latter example, the results of the Backus-Gilbert analysis are compared with an analysis by Gautier and Revah, based on more conventional information theory techniques.

I. INTRODUCTION

The method of Backus and Gilbert was originally developed for application to inverse problems encountered in the physics of the solid earth. The theory was formulated primarily from a physical point of view in a series of three papers by Backus and Gilbert (Refs. 1, 2, and 3) and summarized within a more formal framework by Backus (Ref. 4). In addition to providing an inversion algorithm, the method also provides diagnostic information which can be used in assessing the value of a given set of measurements. Although the original applications were in the field of seismology, the theory is quite general and can be readily applied to inversion problems encountered in atmospheric physics. Examples of such applications include those of Conrath (Ref. 5), Westwater and Cohen (Ref. 6), Fleming (Ref. 7), Wang (Ref. 8), and Rodgers (Ref. 9).

In the present paper, a review of the method as applied to profile retrieval in planetary atmospheres is given. The basic theory is discussed, and certain aspects are illustrated through the use of examples. This review will be followed by the presentation of results of a recent application of the method to the problem of retrieving high level ozone profiles from satellite measurements of back-scattered ultraviolet radiation. Finally, applications to problems of temperature profile retrieval from remote infrared measurements of the earth's atmosphere as well as the atmospheres of Mars, Jupiter, and Uranus are considered. In the case of Jupiter, results are compared with those obtained by Gautier and Revah (Ref. 10) who employed a different theoretical approach.

II. THEORY

Basically, the method of Backus and Gilbert treats a general set of integral equations of the form

$$g_i = \int K_i(z) f(z) dz \qquad i = 1, 2, \ldots m \qquad (1)$$

BACKUS-GILBERT THEORY

It is assumed that there exist measurements of the quantities g_i and that the $K_i(z)$ are known functions of the independent variable z. The problem is to infer information on the unknown function $f(z)$. In the applications considered in this paper, the g_i are generally related to radiance measurements the kernels $K_i(z)$ are determined by the radiative transfer process (scattering, absorption, etc.), $f(z)$ is an atmospheric profile, and z is some measure of height within the atmosphere.

The finite number of kernels $K_i(z)$ do not constitute a complete set, in general, so it is not possible to obtain an exact specification of $f(z)$ from measurements of g_i. Nevertheless, it still may be possible to specify certain useful properties of $f(z)$ from the available measurements. Let $\hat{f}(z)$ be an integral property of the profile associated with level z, and assume it is to be obtained by a linear estimate of the form

$$\hat{f}(z) = \sum_{i=1}^{m} a_i(z) g_i \qquad (2)$$

where the z-dependent coefficients $a_i(z)$ are determined by the inversion method chosen. A relation between $\hat{f}(z)$ and $f(z)$ is obtained by substituting Eq. (1) into Eq. (2), i.e.,

$$\hat{f}(z) = \int A(z, z') f(z') dz' \qquad (3)$$

where

$$A(z, z') = \sum_{i=1}^{m} a_i(z) K_i(z') \qquad (4)$$

Thus, the nature of the estimate $\hat{f}(z)$ is controlled by the behavior of $A(z, z')$, usually called the averaging kernel. The essence of the Backus-Gilbert method is to attempt to control the shape of $A(z, z')$ through the choice of the coefficients $a_i(z)$. Originally, effort was directed primarily toward making $A(z, z')$ resemble a delta function as nearly as a given set $K_i(z)$ would permit. In other words, the goal was to achieve in some sense the best resolution possible. However, there are some applications for which it is desirable to force $A(z, z')$ toward some other predetermined shape. For example, it may be required that $\hat{f}(z)$ approximate a uniformly weighted average of $f(z)$ over some range of the independent variable centered on z, in which case $A(z, z')$ should be made to approximate a rectangular function. In any event, one approach to controlling the shape of $A(z, z')$ is to choose the coefficients $a_i(z)$ such that they minimize some quadratic form for each value of z, for example,

$$Q(z) = \int J(z, z') [A(z, z') - D(z, z')]^2 \, dz' \tag{5}$$

where the weight $J(z, z')$ and the function $D(z, z')$ are chosen to produce the desired behavior in $A(z, z')$.

Another aspect of the problem which must be considered is measurement noise. Any measurements of g_i will have associated with them errors of unknown magnitude. However, if the statistical properties of the measurement errors are known, the resulting statistical properties of the errors in $\hat{f}(z)$ can be found. By assuming that the measurements possess zero mean error and have an error covariance matrix $\underset{\sim}{E}$, it is easily shown that $\sigma_{\hat{f}}^2$, the error variance in $\hat{f}$, is given by

$$\sigma_{\hat{f}}^2(z) = \underset{\sim}{a}^T(z) \, \underset{\sim}{E} \, \underset{\sim}{a}(z) \tag{6}$$

where $\underset{\sim}{a}(z)$ is the column vector of coefficients $a_i(z)$, and the superscript T denotes matrix transposition. From the point of view of controlling the propagation of measurement noise, it is desirable to choose the coefficients $a_i(z)$ such that $\sigma_{\hat{f}}^2(z)$ is minimized at each level z. However, it is not possible, in general, to minimize both $Q(z)$ and $\sigma_{\hat{f}}^2(z)$ simultaneously; therefore, a compromise is reached by minimizing a linear combination. This relationship can be written as

$$R(z) = w \, Q(z) + (1 - w) \, r \, \sigma_{\hat{f}}^2(z) \tag{7}$$

BACKUS-GILBERT THEORY

where the factor r ensures that both terms have the same physical dimensions. By varying the weight w, emphasis can be shifted from minimizing the error to maximizing control over the shape of $A(z, z')$. Thus, there is a tradeoff between the two considerations, and the best choice for w must be determined by the nature of the particular application.

The value of $\underline{a}(z)$ which minimizes $R(z)$ can be readily calculated; as a result, an expression for $\hat{f}(z)$ of this form is obtained

$$\hat{f}(z) = \underline{a}^T \underline{g} = \underline{v}^T(z) \left[\underline{s}(z) + \left(\frac{1-w}{w}\right) r \underline{E} \right]^{-1} \underline{g} \tag{8}$$

where

$$V_i(z) = \int K_i(z')D(z, z')J(z, z')dz' \tag{9}$$

and

$$s_{ij}(z) = \int K_i(z')K_j(z')J(z, z')dz' \tag{10}$$

An interesting special case of this form of solution results when $J(z, z') = 1$ for all z and z', and $D(z, z') = \delta(z - z')$. Then

$$s_{ij} = \int K_i(z)K_j(z)dz \tag{11}$$

and

$$V_i(z) = K_i(z) \tag{12}$$

Thus, Eq. (8) has the form of the "minimum information" solution (Ref. 11) for the continuous case

$$\hat{f}(z) = \underline{K}^T(z) \left[\int \underline{K}(z')\underline{K}^T(z')dz' + \gamma \underline{E} \right]^{-1} \underline{g} \tag{13}$$

where

$$\gamma \equiv \left(\frac{1-w}{w}\right) r$$

Thus, viewed from within the framework of Backus-Gilbert theory, the minimum information solution is that solution for which the averaging kernel lies closest to a delta function subject to a constraint on error propagation controlled by the value of γ. Note that for the general form of Eq. (8), the matrix to be inverted is a function of z so a matrix inversion is required for each value of z considered. However, in the special case when $J(z, z')$ is independent of z, a single matrix inversion is required for all values of z. This may be a nontrivial point of consideration when an inversion method is being chosen for processing a large quantity of data.

A second form of $Q(z)$ which has been used extensively is that obtained with

$$D(z, z') = \delta(z - z') \quad \text{and} \quad J(z, z') = 12(z - z')^2$$

This form generally results in an averaging kernel with a broader central peak but smaller sidelobes compared with the averaging kernel obtained by using $J(z, z') = 1$. This choice from Eq. (5) yields

$$Q(z) \equiv s(z) = 12 \int (z - z')^2 A^2(z, z') dz' \tag{14}$$

where $s(z)$ is called the "spread." It has units of z and is a measure of the spread of $A(z, z')$ about $z' = z$. The unusual normalizing factor 12 is chosen so a rectangular A of unit area has a value of s equal to its width. Although this choice is quite arbitrary, it can be demonstrated that s is a good approximation to the usual measures of the widths of other well-known functions. Some examples are:

(a) Gaussian

$$A(z) = \frac{1}{\sqrt{2\pi}\, b} e^{-z^2/2b^2}$$

$$s = \frac{3}{2\sqrt{\pi}} \times 2b = 0.85 \times 2b$$

(b) Triangular

$$A(z) = \begin{cases} (1 - z/2b)/2b & |z| \leq 2b \\ 0 & |z| > 2b \end{cases}$$

$$s = 1.2 \times 2b$$

(c) Lorentzian

$$A(z) = \frac{1}{\pi} \frac{b}{z^2 + b^2}$$

$$s = \frac{3}{\pi} \times 2b = 0.95 \times 2b$$

where b is the half-width at half-maximum. Thus, it appears that s can be taken as a reasonable though somewhat unconventional measure of the resolution in z.

As before, $\underline{a}$ is determined by minimizing a linear combination of s and $\sigma_{\hat{f}}^2$; however, in this case, it is necessary to impose an additional constraint to obtain a nontrivial solution. The constraint chosen by Backus and Gilbert is that $A(z, z')$ be unimodular, i.e.,

$$\int A(z, z')dz' = 1 \qquad (15)$$

This is a reasonable choice if $\hat{f}(z)$ is interpreted as some average value of f. However, it should be noted that, in general, $A(z, z')$ can be negative for some values of z'; thus, the analog with a weighted average is not complete unless one is willing to admit negative weights. Carrying out the minimization process yields

$$\underline{a}(z) = \frac{\underline{\underline{W}}^{-1}(z)\underline{u}}{\underline{u}^T \underline{\underline{W}}^{-1}(z)\underline{u}} \qquad (16)$$

where

$$\underline{\underline{W}}(z) = w \underline{\underline{s}}(z) + (1 - w) r \underline{\underline{E}} \qquad (17)$$

$$u_i = \int K_i(z)dz \qquad (18)$$

and

$$s_{ij} = 12 \int (z - z')^2 K_i(z') K_j(z')dz' \qquad (19)$$

Thus, by varying the weight w a tradeoff between error and resolution as measured by s can be obtained. If s is plotted against $\sigma_{\hat{f}}$ as w varies from 0 to 1, a "tradeoff curve" is obtained which can be used to pick the appropriate value of w for the application at hand. One such tradeoff curve is obtained for each value of z considered. Examples are given in the following sections.

Other parameters can be defined which are useful in characterizing the behavior of $A(z, z')$. One such parameter is the "center" defined as

$$c(z) = \int z' A^2(z,z') dz' \bigg/ \int A^2(z, z') dz' \qquad (20)$$

A "resolving length" can then be defined as the spread about the center

$$\ell(z) = 12 \int [c(z) - z']^2 A^2(z, z') dz \qquad (21)$$

Obviously, if $\hat{f}(z)$ is to represent a weighted average of f over a region centered on z, then we would like to have $c(z) \approx z$. Note that the spread also can be written as

$$s(z) = \ell(z) + 12[z - c(z)]^2 \int A^2(z, z') dz' \qquad (22)$$

Thus, $s(z)$ has contributions due both to the width of $A(z, z')$ and the displacement of its center from z. Therefore, minimizing $s(z)$ has the desirable property of reducing both the width of A and the departure of its center from the value of z being considered.

It is of interest to note that the minimum information form (Eq. (13)) does not, in general, result in a unimodular averaging kernel. However, the derivation leading to Eq. (13) can be modified to incorporate a unimodular constraint, resulting in the solution

$$\hat{f}(z) = \left\{ \underset{\sim}{K}^T(z) + \left[\frac{1 - \underset{\sim}{K}^T(z) \underset{\sim}{\omega}^{-1} \underset{\sim}{u}}{\underset{\sim}{u}^T \underset{\sim}{\omega}^{-1} \underset{\sim}{u}} \right] \underset{\sim}{u}^T \right\} \underset{\sim}{\omega}^{-1} \underset{\sim}{g} \qquad (23)$$

where

$$\underset{\sim}{\omega} = \int \underset{\sim}{K}(z) \underset{\sim}{K}^T(z) dz + (1/w - 1) \, r \, \underset{\sim}{E}$$

BACKUS-GILBERT THEORY

III. APPLICATION TO OZONE PROFILE RETRIEVAL

The problem of retrieving information on the vertical distribution of ozone within the earth's atmosphere from measurements of backscattered ultraviolet radiance using a satellite borne sensor can be analyzed from the point of view of Backus-Gilbert theory. Although a considerable body of literature on the treatment of this type of inversion problem exists (e.g., Refs. 12, 13, and 14), no attempt will be made here to review the various methods. Rather, we shall be concerned with more general questions concerning the information content of the measurements.

The extraction of ozone information from backscattered radiance measurements is in principle a straightforward process. The incident solar radiation is scattered back to the satellite sensor from various levels within the atmosphere and from the lower boundary surface. In addition, if the measurement is made within an ozone absorption band, the radiation is attenuated by absorption along the total path. To a first approximation, the majority of the radiation is backscattered from an effective scattering layer. If the distribution of scatterers (atmospheric molecules and aerosols) is assumed known, then a measurement of the ratio of the backscattered radiance to the incident solar flux permits the attenuation due to ozone absorption to be inferred. From a knowledge of the ozone absorption coefficient, the total ozone above the effective scattering layer can be inferred. For an estimate of total ozone, the effective scattering layer should be located in the troposphere. To obtain profile information, measurements at several different wavelengths are required, corresponding to scattering layers covering a range of heights in the stratosphere.

The problem of extracting ozone information divides naturally into two distinct problems: retrieval of high level profiles and retrieval of total column abundance. Since the retrieval of total column abundance does not require a profile inversion in the usual sense, only the upper level profile retrieval will be considered here. Most investigations of profile inversion have been limited to the "topside" or region above the ozone peak so that only single scattering need be considered. For example, the Nimbus 4 BUV data have been used to retrieve profiles only above approximately the 10 mb level.

For a single scattering Rayleigh atmosphere, the backscattered spectral radiance I_i measured by a nadir viewing sensor at wavelength λ_i can be written as

$$I_i = \frac{3}{16\pi}(1+\cos^2\theta)\beta_i F_i \int_0^{p_o} e^{-(1+\cos\theta)[\beta_i p + k_i X(p)]} dp$$

where F_i is the solar spectral flux, θ is the solar zenith angle, β_i is the scattering extinction coefficient, k_i is the absorption coefficient, $X(p)$ is total ozone (cm-kPa) above pressure level p, and p_o is lower boundary pressure level. It is assumed that the integrand in Eq. (1) approaches zero as $p \to p_o$. Given measurements of I_i/F_i, the retrieval problem then is to solve the integral equation for $X(p)$.

In forming a linear inversion problem, most methods consider perturbations about some starting profile or first guess $X^o(p)$. For this discussion, the atmosphere is divided into discrete layers of uniform thickness in $\ln p$. Further, because of the large range of variation of I with λ, a logrithmic scaling is found convenient. Then the deviation of $\ln I_i$ from the value it would have for an ozone profile $X^o(p)$ can be written to first order

$$\delta \ln I_i = \sum_j \frac{\partial \ln I_i}{\partial \ln x_j} \delta \ln x_j$$

where x_j is the amount of ozone in the jth layer. The partial derivatives are evaluated at $X = X^o(p)$ with the integral written in numerical quadrature form and provide a measure of the sensitivity of the radiance at the ith wavelength to changes in the ozone content in the jth layer. In the notation of Section II, the partial derivatives correspond to a discrete form of $K_i(z)$, whereas $\delta \ln I_i = g_i$ and $\delta \ln x_j = f_j$. A set of kernel functions for seven wavelengths in the Hartley-Huggins band is shown in Fig. 1. (These kernels were provided by C. L. Mateer.)

Using the methods of Section II, with the spread as the parameters characterizing the averaging kernel, tradeoff curves were calculated. An example of a tradeoff curve is shown in Fig. 2 for the 3.76 mb level. In this case, a random noise in the measurement comparable to that achieved with the Nimbus 4 BUV experiment ($\approx$ 1%)

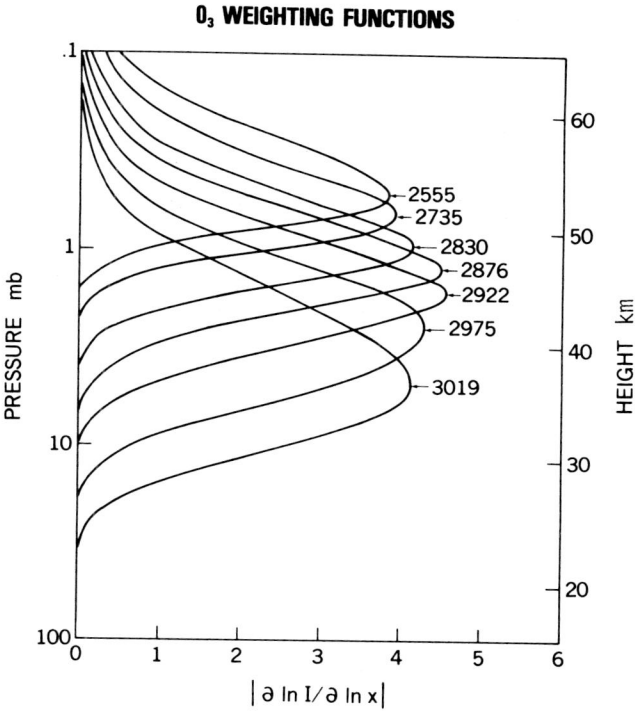

Fig. 1. Kernel functions for topside ozone profile retrieval using backscattered ultraviolet measurements. Each curve is labeled by the wavelength (Angstroms) to which it pertains. (1 bar = 100 kPa.)

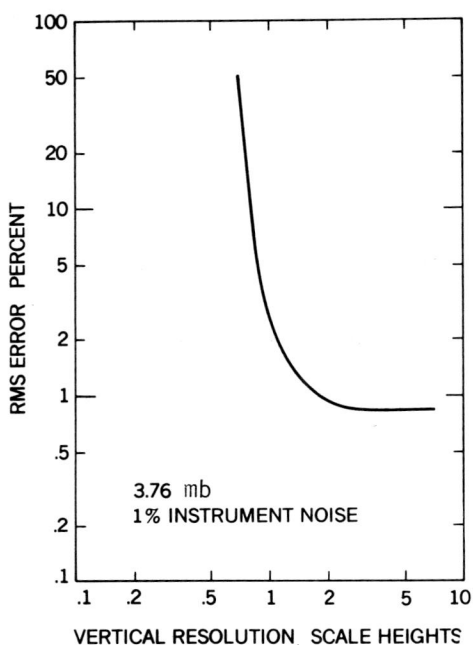

Fig. 2. Tradeoff curve at the 3.76 mb level for ozone profile retrieval from backscattered ultraviolet measurements. The spread is used here as a measure of vertical resolution, and an rms measurement error of 1% was assumed. (1 bar = 100 kPa.)

was assumed. The L-shape of the curve is characteristic of remote sensing profile retrieval techniques in general. If an attempt is made to improve the resolution as measured by the spread much beyond 1 scale height, the rms profile error increases rapidly. By combining information from curves such as these from many different levels, the spread as a function of height was calculated for several different rms errors (Fig. 3). If a 10% error is accepted, then a vertical resolution of one scale height or better can be achieved between 0.5 and 10 mb. If, however, an error no larger than 1% is demanded, virtually no vertical structure information can be obtained.

BACKUS-GILBERT THEORY 167

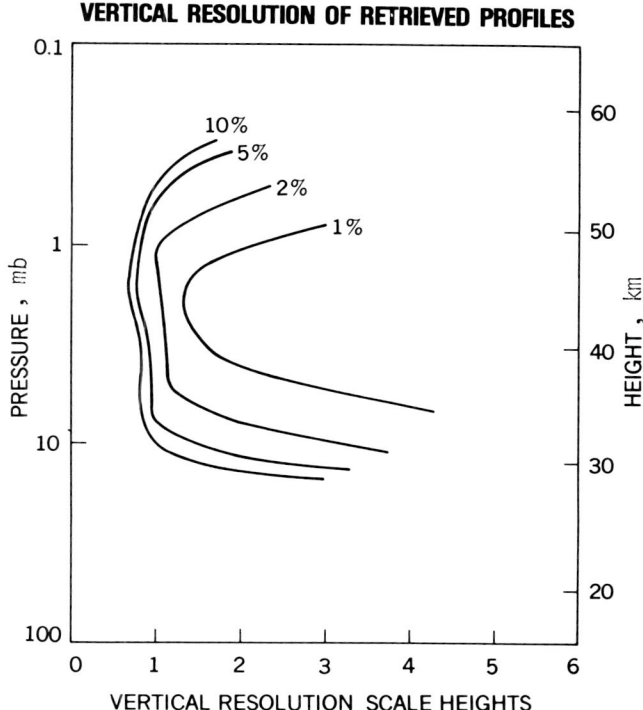

Fig. 3. Spread as a function of height for various ozone profile errors. An rms measurement error of 1% was assumed. (1 bar = 100 kPa.)

IV. APPLICATION TO TEMPERATURE PROFILE RETRIEVAL

Backus-Gilbert theory has found considerable use in the analysis of problems associated with the retrieval of atmospheric temperature profiles from remotely measured, thermally emitted, infrared radiation. The method has been applied with considerable success to the sounding of the terrestrial atmosphere; however, it has proven to be an especially useful tool in analyzing the potential of various types of measurements for sounding the atmospheres of other planets. In the case of the Earth's atmosphere, there is usually information on the temperature profile available in addition to the radiance measurements. For planetary atmospheres, the radiance measurements are frequently all that is available for estimating a profile. Examples of both types will be considered.

For purposes of this discussion, consider a nonscattering atmosphere in local thermodynamic equilibrium. Then the spectral radiance in the ith spectral interval as measured with a sensor located above the atmosphere can be written as

$$I_i = B_i(T_o)\tau_i(0) + \int_0^\infty B_i[T(z)] \frac{\partial \tau_i(z)}{\partial z} dz \qquad (24)$$

In this expression, $z = -\ln(p/p_o)$, i.e., z would be the geometric height expressed in units of scale height for an isothermal atmosphere. $B_i(T)$ is the Planck function for temperature T within spectral inverval i (assumed to be narrow), and $\tau_i(z)$ is the atmospheric transmittance between level z and the sensor. The term $B_i(T_o)\tau_i(0)$ is the contribution from the lower boundary located at pressure level p_o here assumed to be that of a blackbody at temperature T_o. This term is usually specified from measurements in transparent spectral intervals.

The problem then is to use measurements of I_i to retrieve information on the temperature profile T(z). The problem is first linearized by expansion about an appropriately chosen reference profile $T^o(z)$; this results in the set of linear integral equations

$$\Delta I_i = \int_0^\infty K_i(z)\Delta T(z) dz \qquad (i = 1, 2, \ldots m) \qquad (25)$$

where m is the number of spectral intervals for which measurements exist. The radiance difference is $\Delta I_i = I_i - I_i^o$ where I_i^o is calcula from Eq. (24) using the reference profile, and $\Delta T(z) = T(z) - T^o(z)$. The kernel functions $K_i(z)$ are given by

$$K_i(z) = \frac{d B_i[T^o(z)]}{dT} \frac{\partial \tau_i(z)}{\partial z} \qquad (26)$$

These equations are now of the form of Eq. (1) so the methodology of Section II can be applied.

A. Earth

Remote temperature sounding in the Earth's atmosphere has been carried out using measurements within the 15 μm and 4.3 μm CO_2 bands as well as measurements in O_2 microwave lines. Only soundings within the 15 μm band will be considered.

Conrath (Ref. 5) analyzed measurements within the 15 μm band in terms of the tradeoff between averaging kernel spread and profile error, assuming an error covariance matrix of the form

$$\underline{E} = \sigma_\varepsilon^2 \, \underline{1} \qquad (27)$$

where σ_ε^2 is the variance of the measurement noise and is assumed to be the same in all spectral intervals and $\underline{1}$ is the unit matrix. With these assumptions, it was possible to calculate tradeoff curves in terms of spread as a function of $\sigma_T^2/\sigma_\varepsilon^2$. Two cases were considered, one for a set of seven spectral intervals and the other for a set of 16 intervals. Kernel functions for the seven-interval set are shown in Fig. 4. Typical tradeoff curves for the 49-mb level are shown in Fig. 5. Again, the characteristics L-shape curves result. Averaging kernels from selected points on the tradeoff curve for the seven-interval set (broken curve in Fig. 5) are shown in Fig. 6. The first averaging kernel corresponds to a point near the minimum error-maximum spread end of the tradeoff curve while the last is from the maximum error-minimum spread end of the curve. Note that the latter has a substantially more narrow central peak than the former but has developed sidelobes with negative excursions.

The center as a function of height calculated using Eq. (20) is shown in Fig. 7. The same value of $\sigma_T^2/\sigma_\varepsilon^2$ was used at all levels. Ideally, c(z) should lie on the broken diagonal line shown in the figure. However, above approximately 10 mb, c(z) stops increasing. This level corresponds to the peak of the uppermost kernel function,

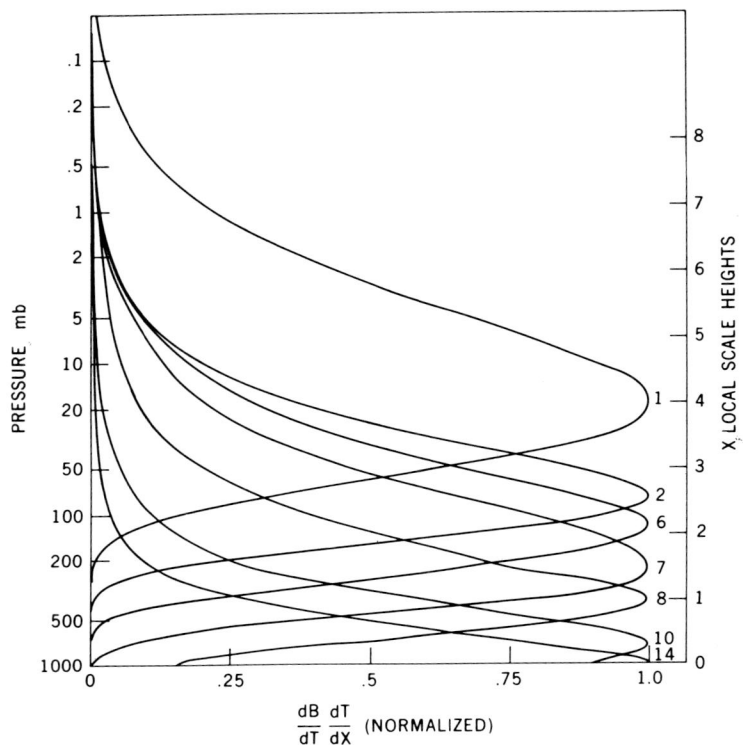

Fig. 4. Kernel functions for the 15-μm CO_2 band in the terrestrial atmosphere. The labels 1, 2, 6, 7, 8, 10, and 14 refer to frequencies 667.5 cm^{-1}, 677.5 cm^{-1}, 697.5 cm^{-1}, 702.5 cm^{-1}, 707.5 cm^{-1}, 727.5 cm^{-1}, and 747.5 cm^{-1}, respectively. (1 bar = 100 kPa.)

and essentially no useful information is obtained above that level. The corresponding resolving length as a function of height is given in Fig. 8. The values shown correspond to formal rms temperature errors σ_T of 1 to 2 K for instrument noise levels typical of those achievable with existing spectrometers operating in the 15-μm band. Thus, the resolution as measured by the resolving length is approximately 0.5 scale height in the lower troposphere, but degrades to in excess of two scale heights at 10 mb.

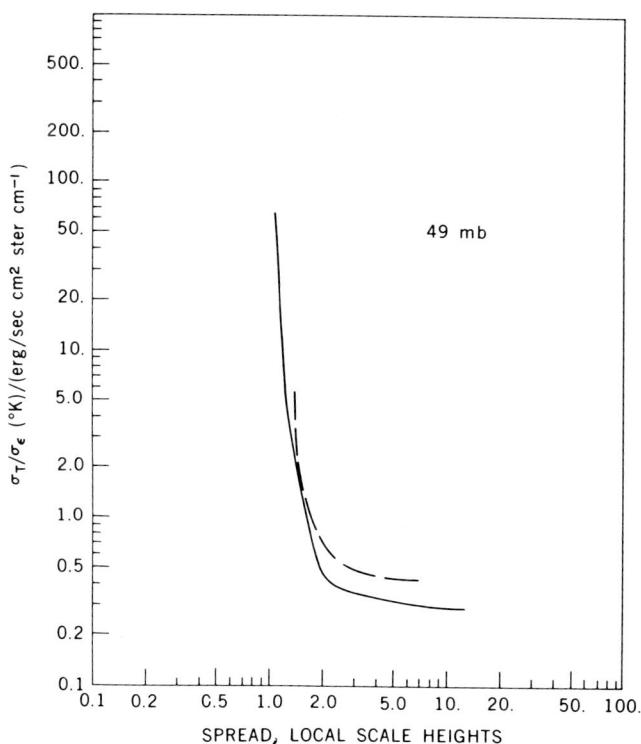

Fig. 5. Tradeoff curves at the 49-mb level for temperature profile retrieval from measurements within the 15-μm CO_2 band in the terrestrial atmosphere. The broken curve is for a set of seven spectral intervals while the solid curve is for a sixteen interval set. (1 bar = 100 kPa ; 1 erg = 10^{-7} J.)

Fleming (Ref. 7) has considered the problem of attempting to construct averaging kernels which are approximately rectangular in shape. This effort was motivated by the fact that in meteorological applications, the thickness of atmospheric layers between constant pressure surfaces is frequently the desired quantity. Since the thickness is proportional to the mean temperature of a layer, a rectangular averaging kernel permits direct estimation of

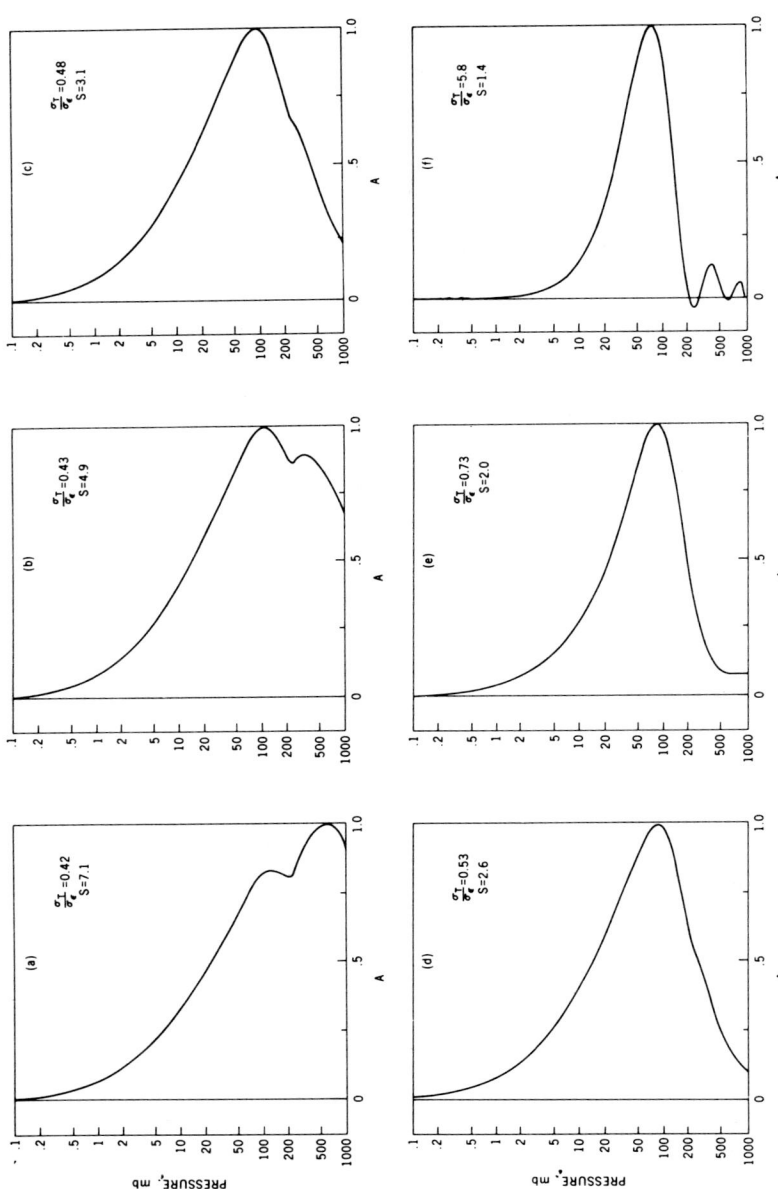

Fig. 6. Averaging kernels corresponding to various points along the tradeoff curve for the seven spectral interval set shown in Fig. 5. (1 bar = 100 kPa.)

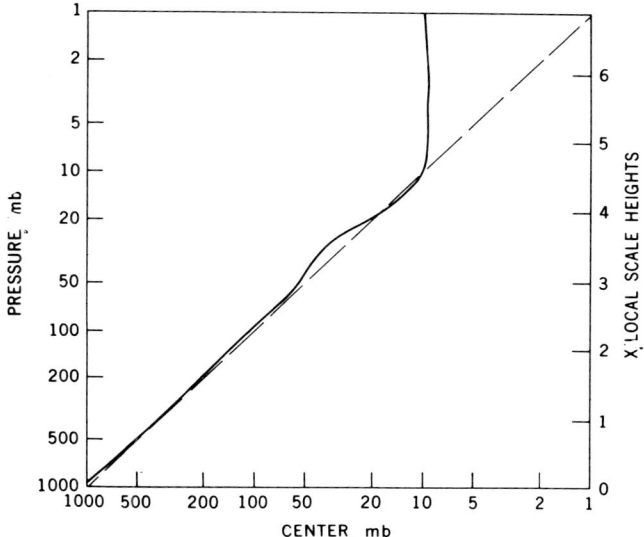

Fig. 7. Center as function of scale height for the seven spectral interval set whose kernel functions are shown in Fig. 4. (1 bar = 100 kPa.)

thicknesses. For this purpose, $D(z, z')$ in Eq. (5) was taken as a rectangular function, and the weight used was

$$J(z, z') = [1 - D(z, z')]^2 \tag{28}$$

This choice of weight acts as a penalty function which tends to suppress the amplitude of $A(z, z')$ outside the region delineated by $D(z, z')$. Fleming's results showed that a slightly better thickness estimate could be obtained in this way than by integrating a retrieved temperature profile, although the improvement was considered marginal.

Recently, Rodgers (Ref. 9) has employed the Backus-Gilbert formulation to analyze the vertical resolution achievable with statistical estimation techniques. The basic approach used was to treat the available *a priori* statistics as a measurement of the

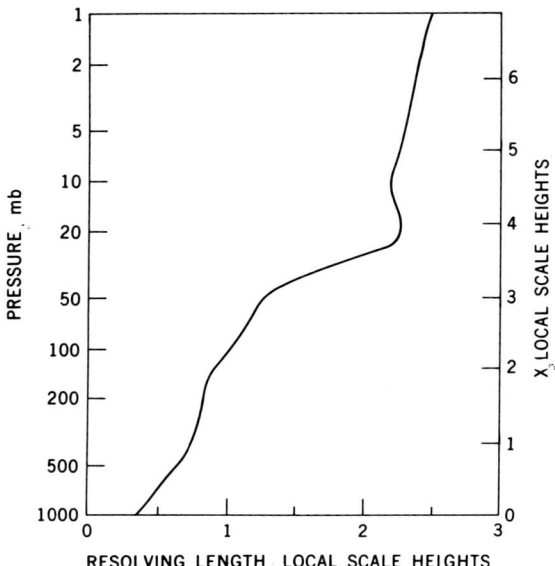

Fig. 8. Resolving length as a function of scale height for the seven spectral interval set whose kernel functions are shown in Fig. 4. The assumptions used in the calculations are described in the text. (1 bar = 100 kPa.)

temperature profile with high vertical resolution but large error. The results of the study indicate that the use of statistics along with the radiance measurements considerably improves the vertical resolution over that achieved with the radiance measurements alone. For example, by using statistics representative of forecast errors, it was found that an improvement in resolution by about a factor of two can be achieved at a profile error level of 0.5 K.

B. Mars

The Mariner 9 spacecraft, launched into orbit about Mars in November 1971, carried a Michelson interferometer operating in the infrared. Measurements within the 15-μm CO_2 band were used to retrieve temperature profiles on a near-global basis. Although the average basal pressure of the atmosphere is only about 5 mb, the fact that it is almost pure carbon dioxide permitted a large portion of the lower atmosphere to be sounded.

Because of the virtual nonexistence of *a priori* information on the Martian thermal structure, only the radiance measurements were available for retrieval purposes. Kernel functions for a 5-interval set of measurements are shown in Fig. 9. An analysis was carried out with this set of functions in terms of a tradeoff between spread and error. The resulting vertical resolution as measured by the resolving length is shown in Fig. 10 as a function of height for a formal rms temperature profile error of 2 K. The center as a function of height is shown in Fig. 11. For purposes of comparison with Figs. 7 and 8, the Martian pressure scale height is approximately 10 km. Not surprisingly, the results for Earth

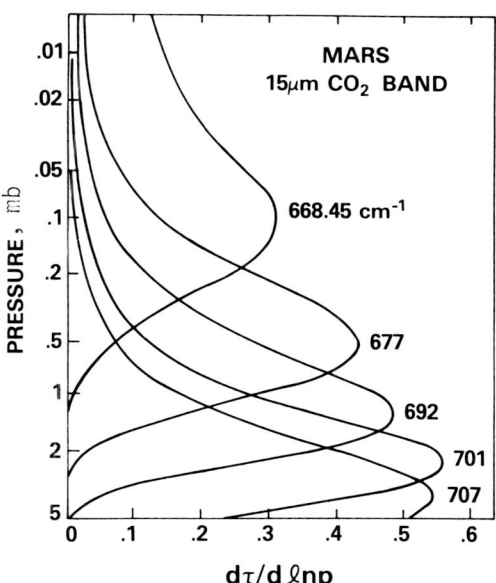

Fig. 9. Set of 15-µm CO_2 band kernel functions used for temperature sounding in the Martian atmosphere. (1 bar = 100 kPa.)

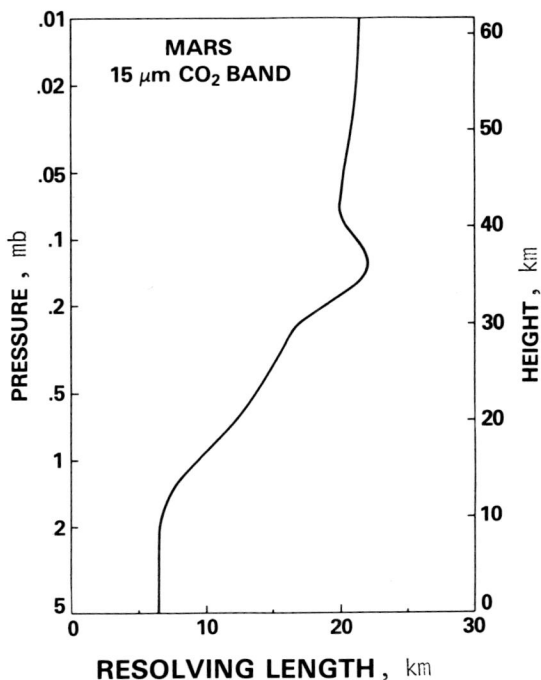

Fig. 10. Resolving length as function of height in the Martian atmosphere obtained with the kernel functions shown in Fig. 9. (1 bar = 100 kPa.)

and Mars are quite similar. The behavior of c(z) indicates that little useful information is obtained above 0.1 mb, and the resolution varies from slightly greater than 0.5 scale height in the lower atmosphere to in excess of 2 scale heights at the upper limit of the sounding region. Unlike the Earth, however, the vertical resolution for Mars cannot be significantly improved at the present time because of a lack of additional information. Nevertheless, almost 20 thousand Martian temperature profiles have been retrieved (Refs. 15 and 16) and used to study the dynamic regime of the lower atmosphere (Refs. 17, 18, and 19).

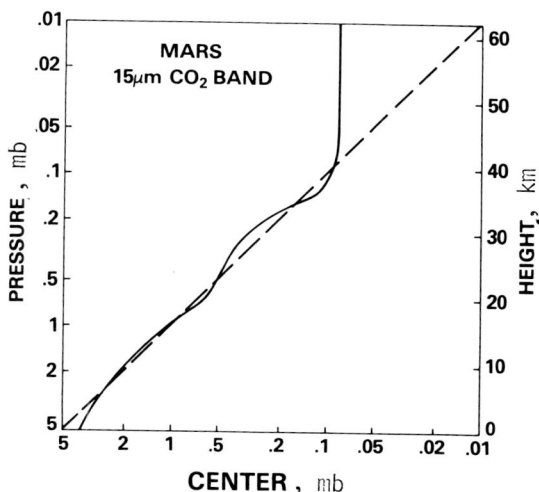

Fig. 11. Center as function of height in the Martian atmosphere obtained with the kernel functions shown in Fig. 9. (1 bar = 100 kPa.)

C. The Jovian Planets

The atmospheres of the Jovian planets (Jupiter, Saturn, Uranus, and Neptune) are composed primarily of hydrogen and helium with small admixtures of methane, ammonia, and other gases. The hydrogen spectrum includes pressure induced lines, two of which are centered at ≈ 360 cm^{-1} and ≈ 600 cm^{-1}. These lines are quite broad (≈ 100 cm^{-1}) and can be used for temperature sounding even with measurements of moderate spectral resolution. In addition, the ν_4 CH$_4$ band can be used for temperature retrieval, provided that measurements with a spectral resolution of the order of a few cm are available. Kernel functions for Jupiter for a set of spectral intervals including both hydrogen and methane absorption features are shown in Fig. 12. The uppermost kernel requires a spectral resolution of 1 cm^{-1}, while the remaining kernels are for 5 cm^{-1} resolution.

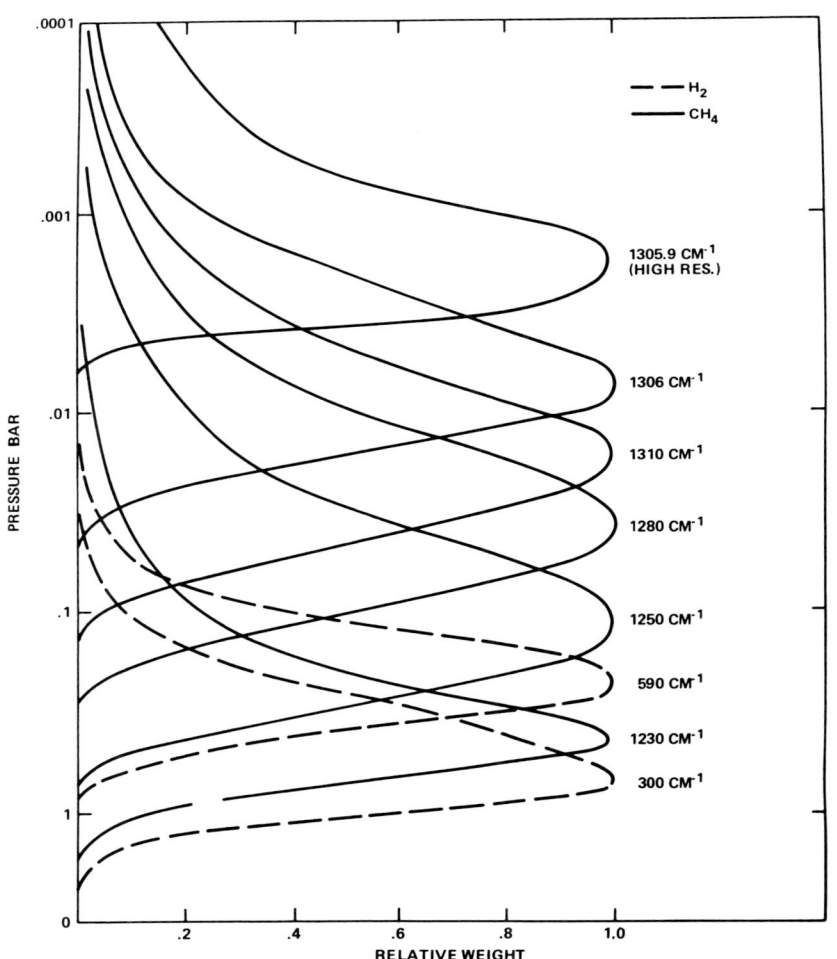

Fig. 12. Kernel functions for temperature profile retrieval in the atmosphere of Jupiter. Spectral intervals within the ν_4 methane band and pressure induced hydrogen lines are included. (1 bar = 100 kPa.)

Earth-based measurements of the Jovian thermal emission spectrum have been used to obtain disk-averaged temperature profiles (see, e.g., Ref. 20). In addition, the Mariner Jupiter/Saturn (MJS) 1977 Mission will carry a Michelson interferometer capable of giving high quality spectral measurements at good spatial resolution during flybys of Jupiter, Saturn, and possibly Uranus. Therefore, there is considerable interest in analyzing the capabilities of measurements of this type. The set of kernels shown in Fig. 12, excluding the uppermost kernel, were employed in a Backus-Gilbert analysis. The results are summarized in Fig. 13

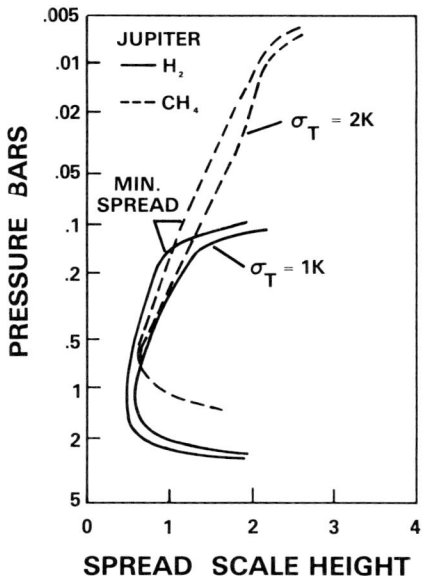

Fig. 13. Spread as a function of height in the Jovian atmosphere obtained with the kernel functions shown in Fig. 12. Both the minimum spread obtainable and the spread obtainable for the temperature profile errors indicated are shown. Measurement errors consistent with current state-of-the-art instrumentation were assumed. (1 bar = 100 kPa.)

where spread as a function of height is plotted. The hydrogen and methane regions were treated separately, and two curves are shown for each; one represents the minimum spread achievable and the second is the spread obtainable with the profile errors indicated. An instrument noise level equal to that anticipated for the MJS 1977 interferometer was assumed. The rapid increase of the spread as either end of a sounding region is approached is due to the fact that $c(z)$ does not increase outside the region from which useful information is obtained (see Eq. (22)). Thus, the total vertical range covered by the combined spectral regions is slightly over two pressure decades, although the lower limit may be set in practice by the presence of cloud decks. As a final example, an analysis for Uranus using only the hydrogen absorption features is presented in Fig. 14. The results are similar to those for Jupiter except that the pressure range over which information can be obtained is somewhat displaced.

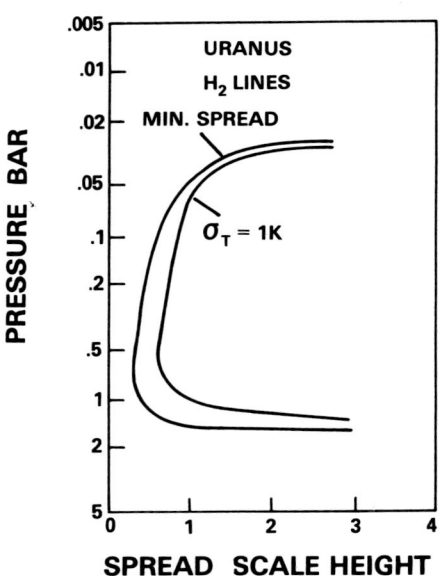

Fig. 14. Spread as a function of height in the atmosphere of Uranus obtained with the kernel functions shown in Fig. 12. (1 bar = 100 kPa.)

Recently, Gautier and Revah (Ref. 10) have provided an analysis of the Jovian temperature sounding problem from a more conventional information theory point of view, and it is of interest to compare the results of that study with those obtained with Backus-Gilbert theory. In the study of Gautier and Revah, it was assumed that the radiative transfer Eq. (24) could be written in a linearized form

$$G(\nu) = \int_0^\infty B[\nu_o, T(p)] \, e^{-t(\nu,p)} \frac{\partial t}{\partial p} \, dp \tag{29}$$

where $G(\nu)$ is a function of the measured radiance at frequency ν, and ν_o is a reference frequency. The Jovian atmosphere was assumed to be infinitely deep so there is no boundary term. For measurements within the hydrogen lines, the optical depth has the functional form

$$t(\nu, p) = k(\nu) p^2 \tag{30}$$

By introducing the new variables,

$$\eta = \ell n \, t(\nu_o, p) \tag{31}$$

$$\xi = -\ell n \, k(\nu)/k(\nu_o) \tag{32}$$

Eq. (29) can be rewritten in the convolution form

$$g(\xi) = \int_{-\infty}^\infty K(\xi - \eta) f(\eta) d\eta \tag{33}$$

where $g(\xi)$ is obtained from the measurements, $f(\eta)$ is related to the temperature profile, and the kernel has the functional form

$$K(\xi - \eta) = \exp[-(\xi - \eta) - \exp(\xi - \eta)] \tag{34}$$

By an appropriate continuation of $g(\xi)$ outside the range of ξ for which measurements are possible, the convolution theorem can be applied to Eq. (33) to obtain

$$g^*(k) = K^*(k) f^*(k) \tag{35}$$

where the asterisks denote a Fourier transformation and k is interpreted as a spatial frequency in the vertical. It is assumed that the measurements are contaminated by noise whose Fourier spectrum $N^*(k)$ is essentially constant over the range of interest. In general, $|g^*(k)|$ will tend to be a monotonically decreasing function of k. At some value of k, such as k_m, $|g^*(k)|$ will become equal to the noise $|N^*(k)|$. Therefore, for higher spatial frequencies, no useful information can be obtained. The minimum sampling interval $\delta\eta$ required to reproduce all the profile information contained in the measurements is obtained from Shannon's sampling theorem

$$\delta\eta = \frac{\pi}{k_m} \tag{36}$$

Gautier and Revah define the vertical resolution to be $\delta\eta$, which is one-half of the shortest wavelength retrievable. By assuming hydrostatic equilibrium and using Eqs. (30) and (31), Eq. (36) can be rewritten

$$\delta z = \frac{\pi}{2k_m} \tag{37}$$

where, as before, z is height in units of scale height. Thus, if k_m can be established, the vertical resolution can be estimated. To calculate k_m, $g^*(k)$ must be known. Gautier and Revah estimated this quantity by using a model Jovian atmosphere to which white noise of approximately 2 K standard deviation was added to simulate fine scale structure. This permitted δz to be calculated as a function of measurement signal to noise ratio.

It is obvious that the Gautier-Revah and Backus-Gilbert approaches are philosophically somewhat different, as the former makes use of the anticipated properties of the profile to be retrieved while the latter depends solely on the behavior of the kernel functions. Nevertheless, it is of interest to compare the results obtained with the two methods. For signal to noise ratios comparable to those anticipated in the hydrogen lines with the

MJS interferometer, the calculations of Gautier and Revah indicate a vertical resolution of approximately 0.5 scale height which is consistent with the Backus-Gilbert results shown in Fig. 14.

V. CONCLUDING REMARKS

The Backus-Gilbert theory has proven to be a useful tool for analyzing the potential of various types of remote radiation measurements for the retrieval of atmospheric profiles. While it has generally come to be regarded as a means of studying the tradeoff between vertical resolution and profile error, its applicability is, in fact, considerably broader. It can be used to construct averaging kernels whose shapes are dictated by the requirements of a particular application. The extent to which the method is successful can be judged on the basis of whether the resulting averaging kernels are better approximations to what is required than are the kernel functions of the original set of integral equations.

Examples from several widely differing applications were presented in this review. However, the results were quite similar in all cases. The broad, smooth kernels associated with radiative transfer processes whether it is backscattered solar radiation or thermally emitted radiation cannot be combined into sharply peaked averaging kernels without incurring strong propagation of measurement errors. However, examination of each case reveals that for reasonable error levels, averaging kernels can be obtained which are narrower than the original radiative transfer kernels. In this sense, inversion of the data sets is judged to be worthwhile. In the case of temperature profiles in the Earth's atmosphere, a large body of information is available in addition to radiance measurements and this information can be used to improve the quality of the retrievals. For other planetary atmospheres, frequently little other information is available. In such cases, analyses of the type presented are particularly useful in establishing the value and limitations of retrieved profiles.

The Backus-Gilbert theory is, of course, not unique in its ability to analyze vertical resolution. Another approach, which is of interest because of its use of more conventional information theory techniques, was briefly reviewed and found to give essentially the same vertical resolution, at least for the one example considered.

SYMBOLS

$\underset{\sim}{a}$	column vector of coefficients (Eq. (6))
$a_i(z)$	coefficients used in linear estimation
$A(z, z')$	averaging kernel
b	half width at half maximum
$B_i(T)$	Planck radiance at temperature T and ith frequency
$c(z)$	center of averaging kernel defined in Eq. (20)
$D(z, z')$	desired function in definition of $Q(z)$
$\underset{\sim}{E}$	measurement error covariance matrix (Eq. (6))
E_{ij}	matrix elements
$f(z)$	unknown profile
$f_j(z)$	value of f corresponding to jth layer
$\hat{f}(z)$	inferred property of unknown profile
F_i	solar flux at ith frequency or wavelength
$\underset{\sim}{g}$	vector of measurement quantites (Eq. (8))
g_i	measured quantities
$G( )$	function of measured radiance at frequency
I_i	radiance at ith frequency or wavelength
$J(z, z')$	weight factor in definition of $Q(z)$
k_i	absorption coefficient
$\underset{\sim}{K}(z)$	kernel functions
$K_i(z)$	kernel elements
$\ell(z)$	resolving length of kernel defined in Eq. (21)
p	atmospheric pressure level
p_o	lower boundary pressure level

$Q(z)$	quadratic form used to control shape of averaging kernel
r	factor used in $R(z)$
$R(z)$	linear function of $Q(z)$ and $\sigma_{\hat{f}}^2(z)$
$\underset{\sim}{S}$	matrix defined in Eq. (8)
$s(z)$	spread of averaging kernel defined in Eq. (14)
$s_{ij}(z)$	matrix elements defined in Eq. (10)
$t(\nu, p)$	optical depth from level p to top of the atmosphere at frequency
$T(p)$	atmospheric temperature as a function of pressure
$T^o(z)$	reference temperature profile
$\underset{\sim}{u}$	vector defined in Eq. (16)
$u_i(z)$	vector elements
$\underset{\sim}{V}$	vector defined in Eq. (8)
$V_i(z)$	vector elements
w	weight used in $R(z)$
$\underset{\sim}{W}$	matrix defined in Eq. (17)
$W_{ij}(z)$	matrix elements
x_j	amount of ozone in jth layer
$X(p)$	amount of ozone in an atmospheric column above pressure level p
$X^o(p)$	first guess for $X(p)$
z	height-related independent variable
β_i	scattering extinction coefficient
η	height-related variable defined in Eq. (31)
ν	frequency
ξ	frequency related variable defined in Eq. (32)
$\sigma_{\hat{f}}^2$	error variance in $\hat{f}(z)$
σ_ε^2	measurement error variance
σ_T	error variance in T
$\tau_i(z)$	atmospheric transmittance from level z to the top of the atmosphere at ith frequency
$\underset{\sim}{\omega}$	matrix defined following Eq. (23)

ω_{ij} matrix elements
δ delta function
γ $= (1 - w)r/w$
θ solar zenith angle
ν_o reference frequency

REFERENCES

1. G. E. Backus and J. F. Gilbert, Numerical applications of a formalism for geophysical inverse problems, *Geophys. J. R. Astron. Soc.* 13, 247 (1967).

2. G. E. Backus and J. F. Gilbert, The resolving power of gross earth data, *Geophys. R. J. Astron. Soc.* 16, (1968).

3. G. E. Backus and J. F. Gilbert, Uniqueness in the inversion of inaccurate gross earth data, *Philos. Trans. R. Soc. London,* A266, 123 (1970).

4. G. E. Backus, Inference from inadequate and inaccurate data, I-III, *Proc. Nat. Acad. Sci.* 65, 1, 281; 67, 282 (1970).

5. B. J. Conrath, Vertical resolution of temperature profiles obtained from remote radiation measurements, *J. Atmos. Sci.* 29, 1262 (1972).

6. E. R. Westwater and A. Cohen, Application of Backus-Gilbert inversion technique to determination of aerosol size distributions from optical scattering measurements, *Appl. Opt.* 12, 1340 (1973).

7. H. E. Fleming, A method for calculating atmospheric thicknesses directly from satellite radiation measurements, *Conf. on Atmos. Radiation, Fort Collins, Colorado, August 7-9* [Preprint Volume published by AMS, Boston, Massachusetts, 1972.]

8. J. Y. Wang, On the estimation of low-altitude water vapor profiles from ground-based infrared measurements, *J. Atmos. Sci.* 31, 513 (1974).

9. C. D. Rodgers, The vertical resolution of remotely sounded temperature profiles with *a priori* statistics, *J. Atmos. Sci.* *33*, 707 (1976).

10. D. Gautier and I. Revah, Sounding of planetary atmospheres: A Fourier analysis of the radiative transfer equation, *J. Atmos. Sci.* *32*, 881 (1975).

11. M. Foster, An application of the Wiener-Kolmogorov smoothing to matrix inversion, *J. Soc. Ind. Appl. Math.* *9*, 387 (1961).

12. S. Twomey, On the deduction of the vertical distribution of ozone by ultraviolet spectral measurements from a satellite, *J. Geophys. Res.* *66*, 2153 (1961).

13. D. N. Yarger, An evaluation of some methods of estimating the vertical atmospheric ozone distribution from the inversion of spectral ultraviolet radiation, *J. Appl. Meteorol.* *9*, 921 (1970).

14. C. L. Mateer, A review of some aspects of inferring the ozone profile by inversion of ultraviolet radiance measurements, in "Mathematics of Profile Inversion" (L. Colin, Ed.), NASA TM X-62150, 1972.

15. R. A. Hanel, et al., Investigation of the Martian environment by infrared spectroscopy on Mariner 9, *Icarus*, *17*, 423 (1972).

16. B. J. Conrath, et al., Atmospheric and surface properties of Mars obtained by infrared spectroscopy on Mariner 9, *J. Geophys. Res.* *78*, 4267 (1973).

17. J. A. Pirraglia, Polar symmetric flow of a viscous compressible atmosphere: An application to Mars, *J. Atmos. Sci.* *32*, 60 (1975).

18. J. A. Pirraglia and B. J. Conrath, Martian tidal pressure and wind fields obtained from the Mariner 9 infrared spectroscopy experiment, *J. Atmos. Sci.* *31*, 318 (1974).

19. B. J. Conrath, Influence of planetary-scale topography on the diurnal thermal tide during the 1971 Martian dust storm, *J. Atmos. Sci.* *33*, 2430 (1976).

20. L. Wallace and G. R. Smith, On Jovian temperature profiles obtained by inverting thermal spectra, *Astrophys J.* *203*, 760 (1976).

DISCUSSIONS

Chahine: Do you really need to use the Gilbert-Backus method to find that the optimum spacing is reduced to half width?

Conrath: It depends on what you are trying to do. It is not always true. The optimum resolution turns out to be of the order of the half-width of the original kernel. However, you can improve on it a bit by moving around on the trade-off curve, particularly in the vicinity of the elbow. It depends on how you wish to trade off between resolution noise.

Chahine: The degree of resolution is not only a function of the kernel and the data, but it is also a function of the structure of the solution profile. To make my point clear, I need to draw a very brief diagram on the blackboard.

Conrath: I think the point is that you have to be very careful by what you mean by "resolution." Go ahead.

Chahine: If you are trying to "resolve" the tropopause from remote sensing data you will find that this is an extremely difficult task no matter how narrow the weighting functions are. This is a result of the fact that the outgoing radiance is a weak function of the value of temperature at the tropopause. On the other hand, one can resolve the stratopause with the same set of weighting function. Thus, my question is: How can you get the resolving power without taking the structure of the solution into account?

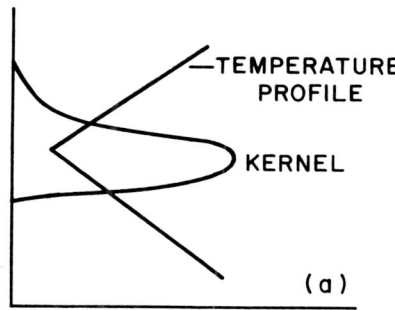 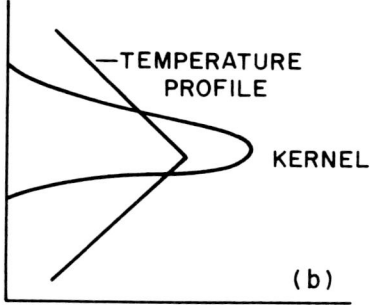

Fig. D-1.

Conrath: Again, I think it depends on how you define the resolving power of the resolution essentially. In the case of Backus-Gilbert, if you use the spread of the resolving length as the parameter, it is in the same sense that you would use the width of a spectrometer slit function as a measure of your resolution. Obviously, what sort of features you can see depend completely on the details of the spectrum. You can see or detect a spectral line much narrower than your slit function provided it's strong enough. You'll see it spread out, of course.

Chahine: Then would I change my resolving power if my solution happened to be like in Fig. D-1(a) instead of being like in Fig. D-1(b)?

Conrath: Well, not in the definition I am using. So you are free to define it however you prefer.

Chahine: Okay.

Conrath: So you have to take it in the context in which it is used.

Chahine: So the resolving power that I wanted to ask is, is the function really of the solution you are after?

Conrath: If you want to define it your way, yes.

Kaplan: But he is not defining it that way.

Conrath: I think it is arbitrary.

Rodgers: The situation that Dr. Chahine has drawn on the board, the resolving power is identical in both cases. There is no distinction between those two profiles. It is just a matter of sign, providing it is a reasonably linear problem and the 15 micron band is reasonably linear in that case. There is quite a distinction between information and signal and it must be realized. The eye may see a low-wattage bulb because the eye is a logarithmic device. The detectors we use are linear devices.

Chahine: If the solution happened to be the stratopause, I can resolve it; if it is the tropopause, I cannot. What is the resolving power then?

Rodgers: The resolving power is something that doesn't have any kind of meaning in your relaxation method.

Chahine: No, I am not thinking in terms of the relaxation method.

Rodgers: In Backus-Gilbert method, those two situations are identical. The sign is irrelevant. Absolute value is irrelevant.

Drayson: I think the point of nonlinearity here is very important. If you work in the 4.3 micron band the temperatures, the Planck function is a very nonlinear function of temperature. But if one is holding for the Planck function and gets a temperature from that

BACKUS-GILBERT THEORY

then you do have a linear problem. So, I think the point Clive is making about the instrument senses in a linear sort of way is a very good point here. But if you are trying to get temperature directly, what Chahine says about the--if you are thinking on a 4.3 micron band, then you can't see that. You very much enhance the positive temperature there because the signal is much much stronger, because of a nonlinearity of the Planck function.

Kaplan: Of course, nobody is interested in the Planck function unless it's a game to play. It's really the temperature that we are after, and so the nonlinearity does come in and it is crucial. I'll talk more about this later.

Westwater: I really was going to amplify Dr. Rodgers' comments. But what the Backus-Gilbert technique really does is estimate one functional from another and the functional that you estimate is the convolution of the averaging kernel with the particular profile that you are trying to retrieve. And, of course, that functional will vary depending on the radiance that your observations have sensed. But the measure of resolution itself, mainly the spread of the averaging kernel, does not depend on the profile you are trying to retrieve. And I think this is what Clive Rodgers also has said.

Conrath: Yes, again in the sense of the analogy with the instrument slit function of a spectrometer in sensing a spectrum; what you actually see depends on the spectrum.

Twomey: I simply wanted to comment that the possibilities of getting a trade-off curve and the scanning function or slit function or whatever, this can be applied with any linear method at all. But in the Gilbert-Backus procedure it is looked at explicitly. The nonlinear method, you produce a combination of your measurements, your g's, whatever they may be called. And so provided it is a linear method, your solution is a convolution type operation or is gotten by a convolution type operation on your inaccessible f(x). And you can always calculate this function and you can, of course, look at the spread of it. It is not a specific property of the Gilbert-Backus procedure.

Conrath: That's right. In fact, you can derive various other algorithms from the Backus-Gilbert point of views simply by minimizing a different quadratic form.

Barkstrom: I would like to make a comment and see if the procedure that I am suggesting is correct on the Backus-Gilbert, that in the procedure if you go back to the original papers, for example, Backus' paper and Backus-Gilbert and Parkers exposition, there's a suggestion in there that you would go ahead and perform an inversion and then the trade-off curve is computed quite separately.

It would suggest an inversion procedure where, in fact, you performed the original inversion procedure and then smoothed after you have done the retrieval. Is that a . . .?

Conrath: That is not my impression of what Backus and Gilbert originally had in mind. I suppose you could approach it from that point of view. If you use it as an inversion method, the retrieval you get depends on where you are on the trade-off curve. You have to choose your value of what I call "w" here.

Barkstrom: How constant is the value of "w"?

Conrath: What do you mean "how constant"?

Barkstrom: If you choose a given tolerance on your retrieved profile, how constant is the value of "w" across the atmosphere?

Conrath: You mean in going from one sounding to another?

Barkstrom: Going from one region of the atmosphere to another.

Conrath: Horizontally?

Barkstrom: Presumably vertically.

Conrath: Vertically, you can choose "w" different at different levels. In the type of Backus-Gilbert method I have outlined here, you essentially do the analysis at each individual level independent of all other levels.

Barkstrom: Yes, and then if you set the tolerance on the retrieve profile at some particular value . . .

Conrath: No, "w" would change with height in general.

Fleming: Just to go back to Roland Drayson's comments, I think it is a matter of how you treat your kernel function. If you linearize the Planck function, you get a factor derivative of the Planck function with respect to temperature. In that case, if you simply redefine your kernel functions to include the original waiting function times that factor and call that new quantity the kernel function, then, indeed, the kernel is temperature dependent and then the resolving power will be a function of temperature. But, if you use the method of Dr. Chahine, he specifically objects to linearizing the Planck function, in which case, he must live with the original weighting function which is temperature independent, except for there is a dependence but in another sense. In which case, I would have to agree with Dr. Conrath.

BACKUS-GILBERT THEORY

Conrath: I think what you're talking about essentially is a kernel of the form ($d\tau/dx \times dB/dT$).

Fleming: Yes. In that case, it becomes temperature dependent. If you actually define your kernel functions that way, then you have included a temperature dependence and I would, of course, have to agree with Dr. Chahine.

Susskind: I'm a little confused about what you mean by "resolution." I just want to ask a very simple question. Say we have a range of atmosphere 10 scale heights and you have 10 measurements. Is it possible to have a resolution of better than one scale height? Can you by this Backus-Gilbert theory show the optimum is better than one scale or you'd be showing my 10 measurements aren't really all that independent and maybe I can only get two scale heights? I mean, is it possible to have more resolution than you have measurements in the range in which you are trying to measure things?

Conrath: I think the answer is no.

Susskind: I should think so too, but I wasn't quite clear. So in other words you could only be degrading what you think you have?

Conrath: That's right.

Susskind: Not doing any better than what it appears to be. All right, so you're talking about the optimum number of independent things you can think you measure, and not do any better. Good.

INVERSION OF INFRARED LIMB EMISSION

MEASUREMENTS FOR TEMPERATURE AND

TRACE GAS CONCENTRATIONS

John C. Gille and Paul L. Bailey
National Center for Atmospheric Research

Limb emission measurements are characterized by sharp weighting functions at high altitudes, and for temperature determinations, strongly nonlinear dependence of the weighting function on the temperature. Several methods for inverting this type of measurement have been described and used, including iterative, statistical, nonlinear and approximate direct approaches. These approaches will be described; advantages and disadvantages of each will be outlined.

I. INTRODUCTION

In techniques utilizing limb emission, often called the limb scanning approach, the data for inversion are measurements of the radiance emitted by the atmosphere, made while a passive radiometer scans from the planet to space across the planetary limb. The change in geometry changes the inversion problem from that for downward or nadir viewing instruments, most importantly by making the problem of temperature determination strongly nonlinear, but more agreeably by yielding narrower weighting functions.

The geometry of limb radiance measurements is shown in Fig. 1. The advantages of the technique, which follow from the geometry, have been described by Gille and House (Ref. 1) [herein referred to as GH]. They need be summarized only briefly here. First,

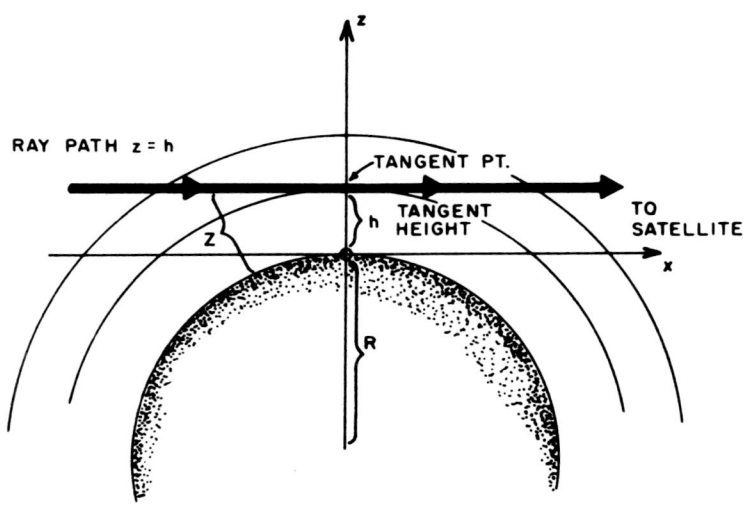

Fig. 1. The geometry of limb scanning.

because of the geometry, there is no contribution from atmospheric layers below the lowest point along the ray path. Because of the rapid drop of atmospheric density above the lowest point, called the tangent point, very narrow weighting functions result. Secondly, the long slant path leads to more emitter along a path through a given altitude, and, thus, sensitivity to higher altitudes. A third advantage is that the cold background of space means that all the signals originate in the atmosphere, and no variability can be attributed to the background.

Because the vertical resolution is coming from the geometric effects in a real instrument, the vertical field of view must be as narrow as practicable to obtain the advantages of the inherent high vertical resolution. However, the spectral width can be made very broad in order to get more signal. This illustrates a general tendency for geometric and spatial effects to play the role in limb scanning that spectral effects do in nadir sounding. In operation, the radiometer samples the atmospheric signal many times during the scan across the limb.

The equation from which the outgoing limb radiation is calculated is given as

$$N(h) = \int_{-\infty}^{\infty} B(x) \frac{d\tau(h, x)}{dx} dx \qquad (1)$$

where N(h) is the radiance received when looking at tangent point h, B is the Planck function, x is the coordinate along the ray path, and $\tau(h, x)$ is the transmittance between the spacecraft and the point x along a path through h. By converting from a horizontal to a vertical integral, and rearranging terms, one obtains

$$N(h) = \int_{h}^{\infty} B(z) c \rho(z) \left[\left(\frac{d\tau}{da}\right)_a - \left(\frac{d\tau}{da}\right)_p \right] \left(\frac{dx}{dz}\right) dz \qquad (2)$$

where z is the vertical coordinate, c is the concentration of the absorber gas, ρ the atmospheric density, a the amount of absorber along the ray path between the atmospheric level and the satellite, and subscripts a and p refer to points on the ray path anterior and posterior to the tangent point. Equation (2) can be written

$$N(h) = \int_{h}^{\infty} B(z) W(z, h) dz \qquad (3)$$

where W is the weighting function which tells how much the level z contributes to radiation observed along a path whose lowest point is h.

Weighting functions for an ideal instrument with an infinitesimal field of view have been presented for a broad carbon dioxide channel by GH. They typically have widths at the half-power points in the order of 3 km. These are for an ideal instrument with an infinitesimal field of view. For a real instrument, with a nominal 2 km field of view, the weighting functions are as shown in Fig. 2. Note that as a result of convolving the infinitesimal weighting functions with a finite field of view, the weighting function is now about 5 km wide.

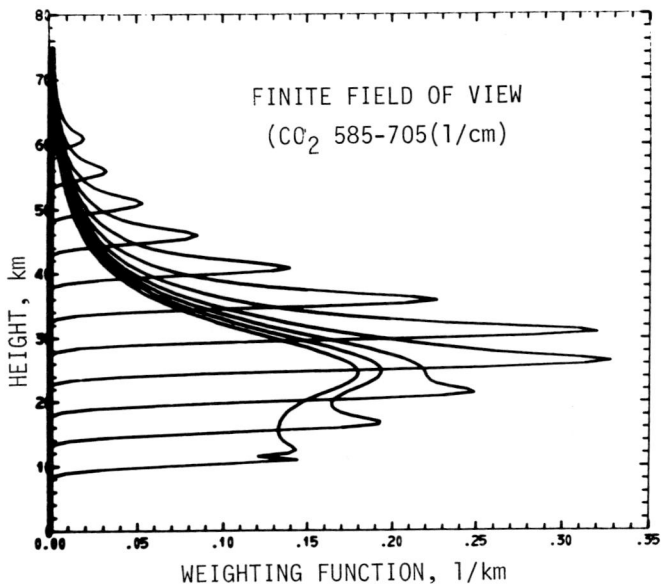

Fig. 2. Weighting functions for limb scanning, with a finite field of view. The field of view has a Gaussian shape, and nominal 2 km width.

The important thing to note is that the weighting function depends strongly on the density at the tangent point. This is the reason for the major nonlinearity; through the hydrostatic equation, temperature at levels below a geometric level will affect the pressure and density at the level, and, therefore, change the weighting function. This is what makes the temperature determination from limb radiance measurements the strongly nonlinear problem that it is.

Weighting functions for trace constituents, such as ozone, have been calculated by Gille and others (Ref. 2) and by Russell and Drayson (Ref. 3). They also display the narrow width expected in the limb geometry. Once the temperature profile is determined, the trace gas concentration measurements are a straightforward, nearly linear problem as in the nadir viewing case.

INFRARED LIMB EMISSION MEASUREMENTS

First to be presented in this paper will be results and thoughts on the information content of limb radiance measurements. This will be followed by a description of four inversion techniques currently being actively pursued.

At this time, an operational method for inverting the data obtained from the Nimbus 6 Limb Radiance Inversion Radiometer (LRIR) is being developed. In a few months, it is expected that much more definitive comparisons between the different techniques and quantitative results confirming the technique, and establishing the accuracy of the results, will be available. In the concluding section, some preliminary results of the recovery of temperature and ozone profiles from the data are shown.

II. INFORMATION CONTENT OF LIMB RADIANCE MEASUREMENTS

It is desirable to be able to assess the amount of information in a set of measurements, in order to guide algorithm development, and to ensure that there is a reasonable match between the effort that goes into the inversion and the information contained in the radiance data. A technique to determine the information content has been described by Gille and Bailey (Ref. 4). It will be summarized very briefly here.

In that reference, it was pointed out that three effects have the potential to degrade the information that is contained in a set of radiance measurements. First, radiative transfer itself tends to obliterate some detail. Second, the characteristics of the measuring instrument will add further smoothing and reduction of information. Finally, the retrieval may also reduce the information that still remains. In the reference cited, the first two effects were studied.

The method was to calculate the change in the outgoing radiance due to sinusoidal temperature perturbations with 2 K amplitude and vertical wavelengths from 2 to 14 km. It became clear that the features with shorter wavelengths are greatly attenuated compared with larger disturbances just by the radiative transfer process. This is what one would expect when considering the width of the weighting functions, even for an infinitesimal field of view.

The instrument field of view response may be approximated by a Gaussian, for which one can calculate analytically a smoothing of the signal. One finds that small vertical scales are further reduced. By comparing these with the effects of instrumental noise and the effects of imprecise sample spacing, one can determine what vertical wavelengths are just at the noise level. (This can also be used to determine how frequently the radiance profile needs to be sampled for given instrument characteristics.) Finally, because there is a trade-off between instrument field of view and noise, it is possible to determine the optimal field of view for a given application.

Another way of looking at this is to perform a spectral decomposition of limb scanning measurements. An example of such a spectrum is shown in Fig. 3, in which the amplitude of the components are plotted against spatial frequency. For the Nimbus 6 geometry, one cycle per milliradian is a wave of 4 km vertical extent; 0.5 cycle per milliradian is, therefore, a wave with 8 km wavelength.

From the previous discussion, it will be recognized that the amplitude of the signal at high frequencies contains only noise, and that there is no information about the atmosphere contained at these frequencies. One may then apply an optimal filter (Ref. 5) to determine what the real amplitude might be. In Fig. 3, the circles represent the unfiltered signal, whereas the symbol F indicates the result of applying an optimal filter. If one knows something about the modulation transfer function of the instrument, then one can do a certain amount of boosting of the middle frequencies (to compensate for the smoothing introduced by the instrument). It is even possible that the atmospheric smoothing can be partially compensated. Clearly, however, one must have a high

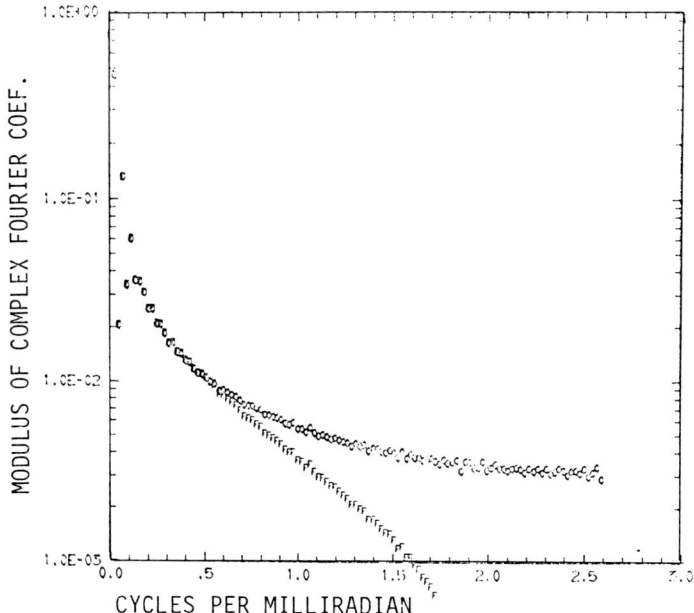

Fig. 3. Amplitude spectrum of the narrow CO_2 channel limb radiance profile. Circles indicate the original spectrum; F's show the spectrum after optimal filter is applied.

signal to noise ratio in order to be reasonably confident that one is actually increasing the realism of the reconstructed radiance profile.

III. ITERATIVE INVERSION TECHNIQUE

The equation for the limb radiance profile can be transformed to (GH)

$$N(h - z_o) = p_o c R^{-1} \int_{h-z_o}^{\infty} B[T(\tau)] \frac{dx}{dz} [(\frac{\partial \tau}{\partial a})_a - (\frac{\partial \tau}{\partial a})_p]$$
$$\times \exp[-R^{-1} \int_{z_o}^{z} g\, T(z')dz']\, T(z)^{-1}\, dz \qquad (4)$$

where the pressure p_o is at the reference level z_o. The exponential terms show the effect of the hydrostatic equation on the pressure, and the nonlinear dependence of radiance on the temperature profile. Because of this, and the sharply peaked weighting function, GH utilized the iterative inversion technique of the Chahine type. This is very fully described in GH and will not be gone into in greater detail here. Some features of the iterative scheme are presented in Ref 6. Perhaps most interesting is that for the implementation used by GH, although radiance residuals continue to drop with continued iteration, temperature errors went through a minimum and then began to grow. Another feature was that in early stages of the iteration radiance residuals oscillated as a function of iteration number, presumably as the atmosphere "sloshed" while it adjusted itself to the hydrostatic equation.

This method gives good results, which are nearly independent of the initial guess. The major problem as far as reducing a large amount of data, however, is the large number of iterations required to drive the residuals below the root mean square (rms) radiance error. An iterative approach must be made very fast in order to process large amounts of data.

Gille and others (Ref. 2) also applied the iterative inversion to constituent profiles. For this problem, the results were fairly fast in terms of the number of iterations; only two to four iterations were generally required to get the residuals below the radiometric error. However, the calculation of the radiance is still a very time-consuming process. Two variations of this approach might be noted; Gille and others (Ref. 2) corrected the profile at each level, based only on the measurement at that level. Tallamraju (Ref. 7) used a corrector for a level based on all the radiances.

IV. DIRECT INVERSION

In the early studies of the limb radiance problem, it was recognized that a simple first-order inversion method was possible due to the highly localized nature of the weighting function (Ref. 8). This behavior provided the motivation for the corrector equation in the iterative scheme used by GH. More recently, a similar principle has been suggested for use with pressure modulated limb scanning radiometer proposed for Nimbus G.

As described by House and Ohring (Ref. 8), the radiance at a tangent height h may be approximated by

$$N(h) \simeq \bar{B}(T(h)) \varepsilon(h) \qquad (5)$$

where $\varepsilon(h)$ is the effective atmosphere emittance. If the emittance is known, the temperature may be recovered from the Planck function. If the constituent concentration is sought, the temperature must be known as well as a relationship between emittance and mixing ratio.

For an isothermal, constant mixing ratio atmosphere with constant absorption, the emittance is given by Burn and Upplinger (Ref. 9)

$$\varepsilon(p(h), W, T) = 1 - \exp\left(\frac{-K\, p(h)\, W}{\sqrt{T}}\right) \qquad (6)$$

where p(h) is the pressure at tangent height h, W is the mixing ratio, T is the temperature, and K is the absorption coefficient.

The localization of the radiance contribution to the vicinity of the tangent point is such that virtually all the radiance comes from within two scale heights of the tangent point for moderate or weak absorption. On the order of 80 to 90% comes from within one scale height of the tangent point. In a more realistic situation, where the absorption coefficient is directly proportional to pressure, the degree of localization will be enhanced since the largest pressures occur at the tangent point. The radiance at tangent height h should then be approximated by the form

$$N(p(h)) \simeq \bar{B}(\bar{T}(h)) \left| 1 - \exp \frac{(-K(p(h))p(h)\bar{W}(h)}{\sqrt{\bar{T}(h)}} \right| \quad (7)$$

where the variables $\bar{T}(h)$ and $\bar{W}(h)$ represent temperature and mixing ratio averaged in some manner over the region of the atmosphere from which the radiance arises. By assuming the radiance all comes from within one scale height of the tangent point,

$$\begin{aligned} \bar{T}(h) &= \int_h^{h+H} T(z)A(z)dz \bigg/ \int_h^{h+H} A(z)dz \\ \bar{W}(h) &= \int_h^{h+H} W(z)A(z)dz \bigg/ \int_h^{h+H} A(z)dz \end{aligned} \quad (8)$$

where $A(z)$ is an arbitrary weight depending upon the optical characteristics of the atmosphere in the vicinity of h.

The effective absorption coefficient may be difficult to accurately specify and could most easily be obtained empirically from detailed forward radiance calculations. The most useful form, however, is to have the effective emittance specified in terms of the atmospheric variable of interest

$$\varepsilon(p, \bar{T}, \bar{W}) = \frac{N(p)}{\bar{B}(\bar{T})} \quad (9)$$

In the case of CO_2, the mixing ratio profile is assumed to be globally uniform and does not need to be considered explicitly. Temperature variability also appears to be of second order importance and can be neglected as an explicit parameter. This results in a simple relationship

$$N(p) = \bar{B}(\bar{T}(p))\varepsilon(p) \quad (10)$$

where $\bar{B}(\bar{T})$ is the instrument response weighted Planck function, which may be parameterized and inverted to give $\bar{T}(\bar{B})$. The $\varepsilon(p)$ are determined from Eq. (9) by calculating $N(p)$ and $\bar{T}$ (using Eq. 8) for a set of atmospheres. The effective emissivity for LRIR

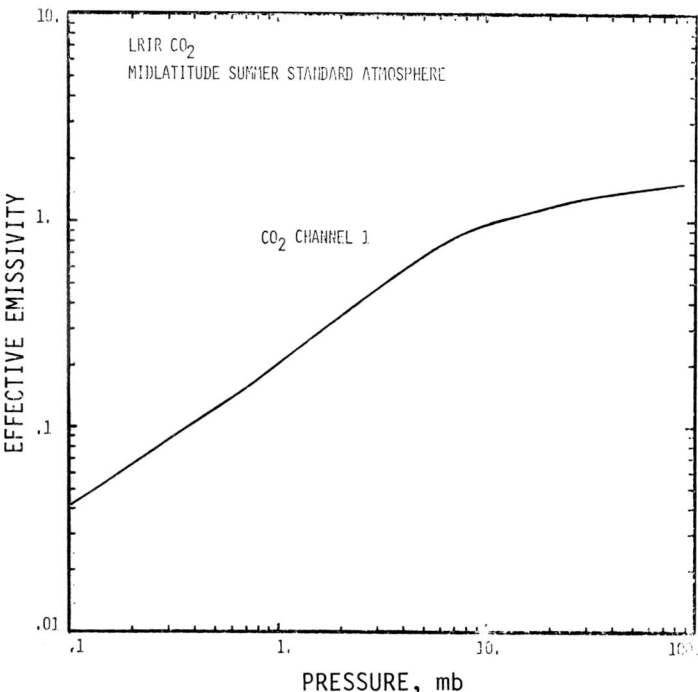

Fig. 4. Effective emissivity of the narrow carbon dioxide channel for midlatitude summer standard atmosphere, as a function of pressure. (1 bar = 100 kPa.)

Channel 1 (narrow CO_2) is shown in Fig. 4 for the summer midlatitude standard atmosphere. The effective emissivities for other atmospheres fall very close to the same curve. The effective emissivity for Channel 2 (broad CO_2) has a similar shape, but is displaced toward lower emissivity and higher pressures.

It can be seen that values for the emissivity become greater than one as the channel becomes opaque. In reality, the effective temperature for these pressures should be derived from a much deeper layer of the atmosphere which is appreciably warmer than the region near the tangent point.

For nonuniformly mixed constituents, Eq. (9) is solved for $\varepsilon(p, \bar{T}, \bar{W})$, which is then interpreted to give $\bar{W}$ as a function of pressure. Inversion for temperature using a uniformly mixed constituent like CO_2 can proceed in several ways. If the radiance $N(p)$ is known at pressure p, the emissivity and hence, $B(\bar{T}(p))$ and $\bar{T}(p)$ can be obtained directly.

If the radiance is known to correspond to some weighted temperature, the pressure corresponding to the resulting emissivity can be obtained from the emissivity curve. For constituents such as ozone, the mixing ratio can be obtained if the radiance is known to correspond to a specified temperature and pressure. In this situation, it is convenient to fit emittance in the form shown in Eq. (7)

$$\varepsilon(p, \bar{T}, \bar{W}) = \varepsilon(\frac{p\bar{W}}{\sqrt{\bar{T}}}) \tag{11}$$

In application, CO_2 radiance will not be known as a function of either temperature or pressure. Both parameters must be determined from observations made in two different spectral channels in the 15-μm CO_2 band having different optical properties.

If both channels are looking at a level in the atmosphere where one of the channels is becoming opaque, the effective radiating temperature of the opaque channel is very close to the atmospheric temperature at that level. By using this temperature, the emissivity for the transparent channel and hence the pressure at that level may be determined. If the angular separations between samples on the radiance profiles are known, the entire profile of effective temperature against pressure may be reconstructed by applying the hydrostatic equation, the spectrally weighted Planck function, and the curves of emissivity against pressure. This is shown schematically in Fig. 5.

Depending upon the averaging function $A(z)$ used in Eq. (8), the $\bar{T}(p)$ profile may be shifted an appropriate amount in P to

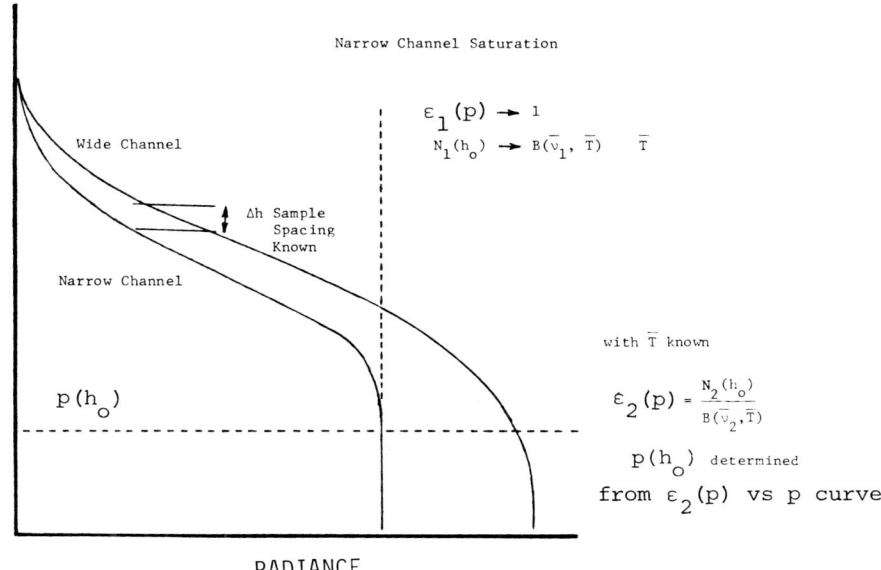

Fig. 5. Schematic of the direct inversion scheme. Point where narrow channel is saturated allows determination of the temperature, from which the broad channel emissivity $\varepsilon_2(p)$, and, therefore, $p(h_o) = p_o$ may be determined.

provide an estimate of the T(p) profile. Thus, the technique has great potential for providing initial profiles for more exact inversion schemes.

Once the temperature and pressure profiles have been determined, a straightforward application of Eqs. (9) and (12) can be used to obtain $\bar{W}(p)$. Again, a shift in pressure compatible with the weights used in Eq. (8) will result in an approximate W(p) profile.

V. STATISTICAL RETRIEVALS

Another approach which we have explored is the use of statistical relationships. If one writes the measured radiances as a column vector, the radiative transfer equation may be written

$$N(h_i) = C_{ij} B_j \tag{12}$$

where the B_j's are the Planck functions, subscripts indicate levels and $\underset{\sim}{C}$ contains the information about the distribution of material (including the hydrostatic equation) and transmittance of the gases. In this form, $\underset{\sim}{C}$ is required to include most of the nonlinearity. This coefficient matrix can be calculated from synthetic data, calculated in turn from a wide range of atmospheric profiles. Once it is known, it can be applied to real data to determine the Planck function and, therefore, the temperature. It is a rather empirical approach, but it appears to work moderately well and it was very fast.

A problem with this approach is that by putting all the nonlinearity into the coefficient matrix, one may be left with larger errors than desired. A more interesting idea to expand the true limb radiance profile as a linear combination of approximate (nonlinear) radiance profiles

$$N_T(p_{hi}) = D_{ij} N_A(p_{hj}) = D_{ij} B(T_j) \varepsilon(p_j) \tag{13}$$

Here the tactics are to incorporate the nonlinearity in a known, interpretable way into N_A and to keep D_{ij} linear or nearly so. A special case may be noted--when $D_{ij} = \delta_{ij}$, we again have the direct inversion, described earlier. When the elements of D are found by regression, then we have a full linearized statistical inversion.

Now, by applying D^{-1}, $B\varepsilon$ is determined. By knowing p, ε and B are immediately known. The nonlinearity is now incorporated in the $\varepsilon(p)$. The trick is to find the best way to incorporate a large part of the nonlinearity in an analytically manageable form so that, after a linear inversion, the nonlinear form may be analytically interpreted with small errors.

Some preliminary results indicate that Eq. (13) does indeed give somewhat better results than Eq. (12) for temperature, and considerably better for ozone. One of the problems that tends to

INFRARED LIMB EMISSION MEASUREMENTS

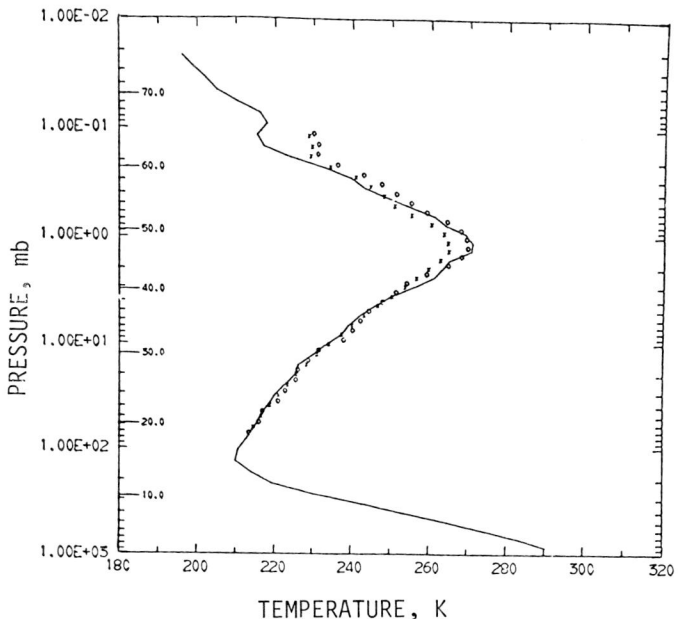

Fig. 6. *Preliminary temperature retrieval for Wallops Island for 29 July 1975 compared with in-situ measurements. The solid line is the rocketsonde plus radiosonde measurement, made at 1803 Greenwich mean time (GMT). The points are from two retrievals of limb radiance for the LRIR over pass at 1737 GMT. x's and o's are for retrievals before and after a correction for atmospheric non-sphericity. (1 bar = 100 kPa.)*

arise, however, is that although one can deal with 75% of the atmospheric cases fairly well, the other 25% (which includes things like stratospheric warmings and other interesting cases) are not handled well. It is very desirable to develop a technique that is general enough to handle all these cases without requiring separate treatment.

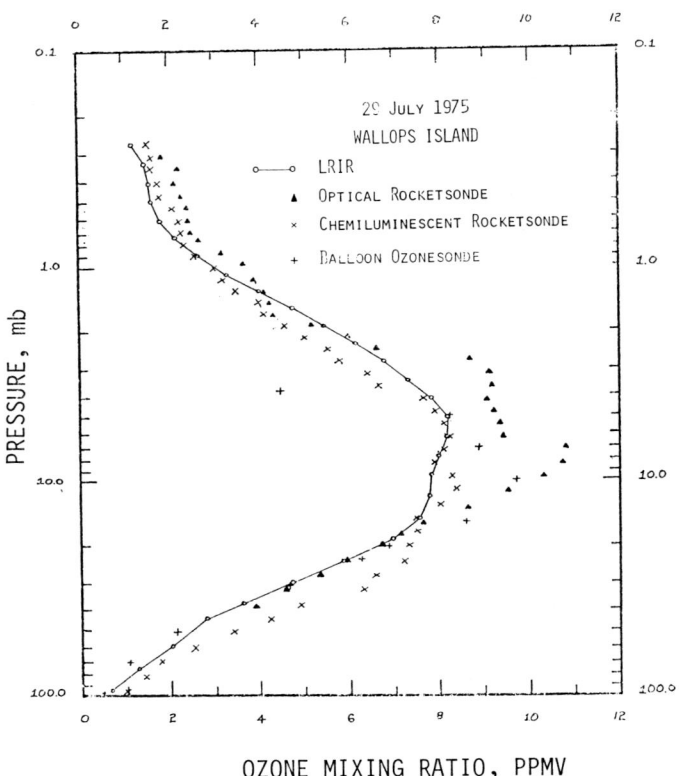

Fig. 7. Preliminary ozone retrieval for Wallops Island for 29 July 1975, compared with in-situ measurements. The points are based on a Krueger optical rocketsonde launched at 1717 GMT, a Hilsenrath chemiluminescent rocketsonde launched at 1920 GMT and a balloon ozonesonde launched at 2051 GMT. The solid line is from the retrieval of LRIR radiances measured at 1737 GMT. (1 bar = 100 kPa.)

VI. CONCLUDING REMARKS

As noted above, many months of real data are now in the process of being analyzed. Much of the past year has been spent in processing the raw data in order to have good, calibrated radiances with which to work. At present, a search is being conducted for an inversion method that will have the ability to extract the large amount of information contained in the limb scans, but with acceptable computer time.

As an indication of the progress to date, and the information obtainable by limb scanning, Figs. 6 and 7 show temperature and ozone[1] retrievals, respectively, at Wallops Island in July, 1975. They are compared with rocket soundings taken within three hours of the satellite overpass. The temperature profile and the rocket are in reasonably good agreement from 25 to 55 km. The root-mean-square (rms) error is about 3 K which is close to the experiment objective. Similarly, the ozone profile in Fig. 7 looks very reasonable. If one takes the percentage difference between the ozone retrieval and the optical rocket measurement at each level, and then takes the rms value of that percentage difference, one gets 14%. This is close to the claimed accuracy of the rocket instrument.

These results to date give us great confidence that the infrared limb scanning approach is very powerful and technically feasible. They also indicate that the data quality from LRIR is such that large amounts of new data will be available to refine limb scanning techniques and to study the upper atmosphere in hitherto unobtainable detail.

---

[1] See Footnote 1 in Discussions.

SYMBOLS

g	gravitational acceleration
N_T	true limb radiance
N_A	approximate limb radiance
R	gas constant for air
T	temperature
x	coordinate along the ray path

ACKNOWLEDGMENT

The National Center for Atmospheric Research is sponsored by the National Science Foundation.

REFERENCES

1. J. C. Gille and F. B. House, On the inversion of limb radiance measurements I: Temperature and thickness, *J. Atmos. Sci.* *28*, 1427 (1971).

2. J. C. Gille, F. B. House, R. A. Craig, and J. R. Thomas, Nimbus-F Limb Radiance Inversion Experiment, vol. I Technical, Honeywell, Inc. Aerospace Division, Rep. 65-D-47, 1970.

3. J. M. Russell, III and S. R. Drayson, The inference of atmospheric ozone using satellite horizon measurements in the 1043 cm^{-1} band, *J. Atmos. Sci.* *29*, 376 (1972).

4. J. C. Gille and P. L. Bailey, On the Information Content of Limb Radiance Measurements, *Conference on Atmospheric Radiation*, Am. Meteorol. Soc., Boston, Mass., Aug. 1972, pp. 13-15.

5. J. W. Brault and O. R. White, The analysis and restoration of astronomical data via the fast Fourier transform, *Astron. Astrophys.* *13*, 169 (1971).

6. J. C. Gille, Limb Radiance Inversion: Iterative Convergence for a Nonlinear Kernel, in "Mathematics of Profile Inversion (L. Colin, ed), NASA TM X-62, 1972, p. 150.

7. R. K. Tallamraju, Inference of Stratospheric Minor Constituents from Satellite Limb Radiant Intensity Measurements, University of Michigan Technical Report, ORA Project 011023, University of Michigan, Ann Arbor, 152 pp.

8. F. B. House and G. Ohring, Inference of Stratospheric Temperature and Moisture Profiles from Observations of the Infrared Horizon, NASA CR-1419, 1969.

9. J. W. Burn and W. G. Uplinger, The Determination of Atmospheric Temperature Profiles from Planetary Limb Radiance Profiles, NASA CR-1513, 1970.

DISCUSSIONS

King: As you know the geometry of limb viewing truncates the kernel which has the effect of imposing the tangent height as the bound of your intensity integral. This finite bound converts the integral equation from a *Fredholm* to a *Volterra* type. Now for fairly general classes of kernels of this truncated type, the Volterra equation has direct solutions which do not involve matrix inversions. One could progressively work down to retrieve the temperature, if the true atmospheric kernel is sufficiently approximated by a soluble class of Volterra kernels. Have you tried this?

Gille: Not explicitly. I think one of the major problems is, first of all, knowing where you are starting. If you have that point, then working up and down from there is what I described in the case of what we called the direct method. If one already has radiance as a function of pressure (which is the hard part), then working down would be the "onion peeling" approach, which is equivalent to inverting a diagonal matrix. We have done a little work on this, and expect to do more. There is a problem with sensitivity to noise at the top levels. We have not tried anything like an analytic approach because the transmittances are not accurately approximated by closed form expressions. It sounds like something worth looking into, however.

Chahine: I have two questions and one comment. Your weighting functions are the most beautiful weighting functions that I have seen today. They might not be aesthetic but they are mathematically beautiful. I would like to have seen a comparison between the results obtained by the three or four methods you have described. If you tried the relaxation and the emissivity approach, would you get large variations in your solution?

Gille: I think the variations would be rather small. I think one of the major differences would be computer time, which might be rather different for different kinds of approaches. We don't have that kind of comparison. That's one of the things I was alluding to when I said in two or three months I think we will have a good deal more. I think many of these are now getting to the point where we can try them not only on synthetic data but on some of the real data and see how well they do.

Chahine: The second question is on the emissivity coefficients you have described. And you have said they are fairly independent of temperature. I am surprised because in the stratosphere you can have large variations in temperature, from day to day or season to season. Were you able to determine that the emissivity coefficient really was a weak function of temperature even for variations of 20 degrees?

INFRARED LIMB EMISSION MEASUREMENTS

Gille: The first order effect is that if you are looking at a geometric altitude, the atmospheric pressure or density varies as the temperature changes, through hydrostatic adjustment. That is the thing you take out by plotting versus pressure. Then it is things like the variation of line intensities over the band, and changes in the intensities of overtone bands that matter. While these effects make a difference, they don't add a lot of spread to the emissivities.

Susskind: Is it broad banded?

Gille: Yes, the channels are quite broad, of the order 100-200 cm^{-1}.

Mateer: I was a little disappointed that you didn't show any results of ozone profiles, John.[1]

Gille: The focus of this meeting has been on methods and I felt I should talk about methods rather than results, Carl. Also, I was involved in a NASA meeting at NCAR until just before I left, and some slides I meant to bring were inadvertantly left behind. Our ozone profiles, although I can't show you one, look quite reasonable. We have results from about 20 km up to 55 km, with a maximum in mixing ratio between 30-35 km in mid-latitudes. We have a few cases of simultaneous rocket data for comparison, but have only looked hard at a comparison from Wallops Island where there was a simultaneous flight by two rocket ozone sensors. The satellite and rocket retrievals had an rms percentage difference between 20 and 50 km of 14%. That is about what is claimed for the accuracy of the rocket measurements.

Kaplan: I am surprised, Carl, that you haven't asked him about the variation from day and night and at twilight and dawn! Do you have any results on ozone changes?

Gille: The ozone top at the moment is at about 55 kilometers and we don't see any extremely large effects there. There could be effects of perhaps 20 percent. We're still checking it out.

Kaplan: Do you see, or don't you want to say yet, dawn and twilight effects at all?

Gille: We appear not to see them. In fact, I can tell you one problem that does occur. We look at the day side, looking back not along the orbit plane but 30° off it, north to south at a

---

[1] In response to Dr. Mateer's comment, the authors included, after the Workshop, their ozone retrieval results (Fig. 7).

given altitude. Going south on the night side, we are looking south to north. Some differences between day and night are due to the fact that the viewing geometry is slightly different and one can see small effects of gradients, but at the altitudes we are talking about those effects are not there. By and large, up to 55 there does not seem to be any very significant effect. What we really want to do is push up a little bit higher and I think the signal to noise ratio will allow us to do it. I think there we ought to be able to do it.

INVERSION OF SCATTERED RADIANCE HORIZON

PROFILES FOR GASEOUS CONCENTRATIONS

AND AEROSOL PARAMETERS

Harvey L. Malchow and Cynthia K. Whitney
Charles Stark Draper Laboratory, Inc.

This paper presents techniques that have been developed and used to invert limb scan measurements for vertical profiles of atmospheric state parameters. The parameters which can be found are concentrations of Rayleigh scatterers, ozone, NO_2, and aerosols, and aerosol physical properties including a Junge-size distribution parameter and real and imaginary parts of the index of refraction. The novel techniques developed for this problem should be of interest for nonlinear numerical search problems in general.

I. INTRODUCTION

There is growing scientific opinion that physical processes in the earth's stratosphere are of vital importance to man and the biosphere, and that far too little is understood about these processes. Examples include the interactions and the resulting balance between stratospheric ozone, NO_2 and aerosols, which affect the ultraviolet radiation and the temperature environment at the Earth's surface. It is widely appreciated that a proper understanding of these phenomena and their consequences will require a significant body of new experimental data, and that the scope of the requirement is such that remote sensing by satellite offers the most practical approach.

A variety of remote sensing techniques based on radiometric data is available for stratospheric monitoring. It is possible to consider any combination of extinction, emission, and scattering as potential signal generating phenomena. Furthermore, in the cases of emission and scattering, it is possible to consider any combination of vertical, horizontal, or inclined scan directions for obtaining the information.

This paper reports on new data processing techniques that were developed for one of the many possible stratospheric monitoring techniques. It is anticipated, however, that some of the techniques presented here will be useful for the other types of experiments as well.

In the problem addressed here, the measurements consist of multispectral limb scans of visible scattered sunlight. Figure 1 illustrates the experiment geometry, and Fig. 2 illustrates a typical *simulated* "measurement data" set. These measurements are inverted for the *atmospheric state,* which is comprised of vertical profiles of atmospheric parameters, within the altitude regime of

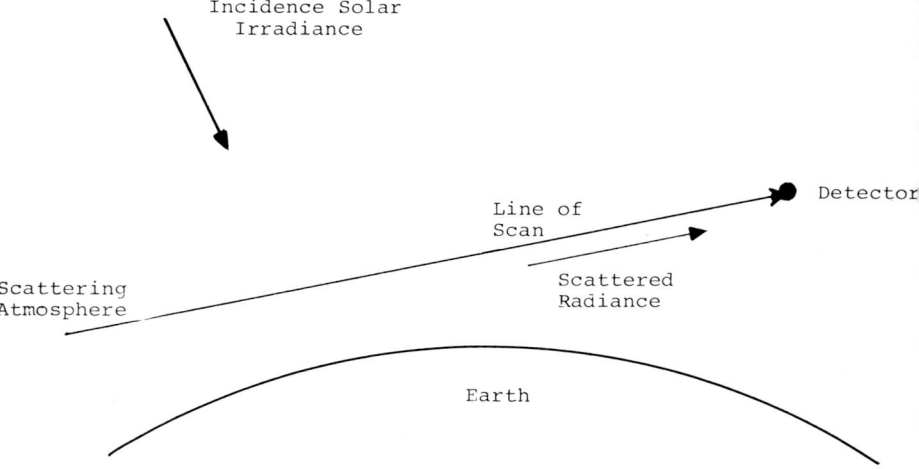

Fig. 1. Scattered radiance limb scan geometry.

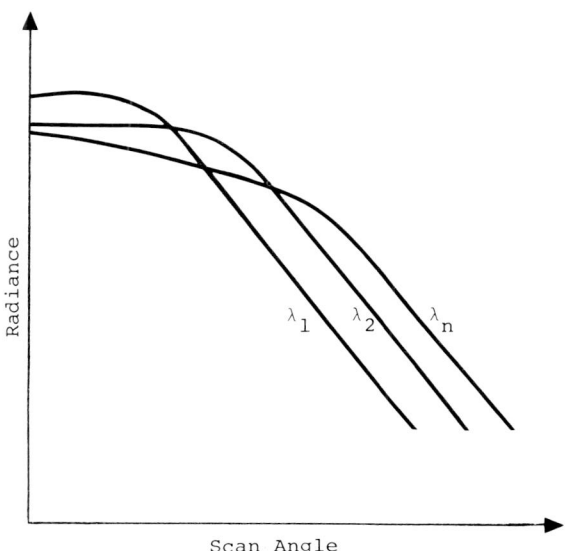

Fig. 2. *Multispectral Limb Scan Measurements.*

observations. The parameters are the concentrations of Rayleigh scatterers, ozone, NO_2 and aerosols, and the aerosol physical properties, such as Junge size distribution parameter (within the optically active size range) and the real and imaginary parts of index of refraction.

Inclusion of so many state parameters may seem ambitious, but this is necessary because they are all optically active over the same band of visible wavelengths. Therefore, it is not possible to invert for a subset of these parameters except by assuming values for the remaining ones. The ability to treat them all is, in fact, a significant advantage of the limb scan experiment. Examples of the inversion results for simulated experiments are presented here to demonstrate the viability of the inversion technique.

II. GENERAL FRAMEWORK

The limb scan inversion problem has some features which are similar to, and some which are different from, other inversion problems in atmospheric sciences. The purpose of this section is to set a framework for discussing the limb scan problem by focussing on those features which are shared with other inversion problems, and to note the standard techniques which are applicable to them.

A major similarity between the limb scan and other inversion problems is the inevitable presence of noise in the measurements. Let x_a represent the actual atmospheric state, and $z(x_a)$ represent the measurement as predicted from the state. The actual measurement observed is

$$z = z(x_a) + n$$

where n represents noise. The above is called the measurement equation. If linearized about some x_0 by Taylor expansion with the partial derivative H, it takes the same form as that which occurs in many other inversion problems where a Fredholm integral is replaced by a quadrature:

$$z - z(x_0) = (\underset{\sim}{H}) [x_a - x_0] + n$$

The problem of extracting the state x_a from this equation has been approached in various ways by Deutsch (Ref. 1), Twomey (Ref. 2), Mateer (Ref. 3), Westwater and Strand (Ref. 4), Rodgers (Ref. 5) and others.

The presence of noise suggests that a stochastic approach to the problem is appropriate. In fact, one can argue that not only are the measurements random variables due to the noise, but also that the state is a random variable drawn from an ensemble of possible states. The well-known theory of optimal estimation offers a general framework for addressing most such inversion problems, and it is, in fact, the approach adapted here to the limb scan inversion problem. This common point of departure is reviewed and related to other work in the remainder of this section.

A way of defining an optical state estimate is suggested by the Gauss-Markov theorem, which states that a minimum variance, unbiased estimate of x_a can be obtained as a linear function of measurements by minimizing a Euclidian distance. Various authors use various names for the construct playing this role. One common name is "cost function," expressing the desirability of minimizing it. Defining a Euclidian distance generally requires two basic choices: the quantities to measure the distance between, and the positive definite metric to use. This means choosing the terms to put in the cost function, and the weights to give them. These choices are resolved by considering what is done with the cost function.

A typical way to minimize the cost function is to differentiate with respect to the state estimate, set the derivative to zero, and solve for the estimate. To provide an estimate that can actually be evaluated, it is necessary that the cost function involve only *known* variables (and not, for instance, the unknown actual state). A well-defined estimate is guaranteed by requiring that each term in the cost function involve the deviation between a known value of a quantity and the known value that quantity would be expected to have if the state were known to be equal to the estimate. Typically, each such term is multiplied by the inverse of the variance of that quantity. That is, distance is measured between known values and expected values that depend on the state estimate and variances define the metric coefficients.

In the limb scan problem, there are many atmospheric parameters in the state and many wavelength channels and altitudes for measurement. It is appropriate to use vectors for state and measurement, and covariance matrices for expressing uncertainties of each. With

q_a = actual vector quantity

$\langle q \rangle$ = vector quantity expected if state = estimate

C_q = quantity covariance

the typical cost function term is

$$(q_a - \langle q \rangle)^T C_q^{-1} (q_a - \langle q \rangle)$$

where T indicates transpose.

Generally, the measurement vector contributes the leading term in the cost function. Some authors include various additional terms as well. Often there is a term representing a prior estimate of the state vector. It should be noted that exclusion of such a term is only a special case: zero prior estimate and infinite prior covariance. The fact that the prior estimate is always implicitly present makes the optimal estimation procedure potentially recursive. The optimal state estimate and its covariance obtained after one batch of data could be used as the prior estimate and covariance before another batch of data.

This idea of recursion leads naturally to consideration of cases where the state is evolving with some running variable v. (Examples of such a variable include scattering particle radius, scattering angle, altitude, longitude, latitude, and time.) When there is evolution, it may be appropriate that the prior estimate at v be determined or at least modified by the posterior estimate at $v - \Delta v$. This is accomplished by some authors by including terms in the cost function representing one or more derivatives of the state with respect to the running variable. These terms have the same effect as modifying the prior estimate at v to align better with the posterior estimate at $v - \Delta v$; thus, the posterior estimate at v is also closer to that at $v - \Delta v$, and the whole function is more smoothly behaved. In the case of the limb scan inversion, the point of the experiment is to find excursions *from* rather than

SCATTERED RADIANCE HORIZON PROFILES

parameters of a smooth model, so no smoothing is involved, and no derivative terms appear in the cost function.

The following notation will be a convenient basis for subsequent discussion. Let the vector x be the state estimate and the matrix $\underset{\sim}{P}$ be its covariance. The prior state estimate is x_0 with a covariance $\underset{\sim}{P}_0$. Because of noise, the measurement vector z has a covariance matrix $\underset{\sim}{R}$. If the state were, indeed, known to be equal to x, then the expected values of x_0 and z would be x and $z(x)$. The cost function is thus

$$\text{Cost function} = (z - z(x))^T \underset{\sim}{R}^{-1} (z - z(x))$$
$$+ (x_0 - x)^T \underset{\sim}{P}_0^{-1} (x_0 - x)$$

With $z(x)$ linearized about x_0 and with partial derivative matrix $\underset{\sim}{H}$, the optimal state estimate is found to be

$$x = x_0 + \underset{\sim}{K}(z - z(x_0))$$

where

$$\underset{\sim}{K} = \underset{\sim}{P}_0 \underset{\sim}{H}^T (\underset{\sim}{H} \underset{\sim}{P}_0 \underset{\sim}{H}^T + \underset{\sim}{R})^{-1}$$

Intuitively reasonable behavior for $\underset{\sim}{K}$ is demonstrated by noting that for large noise covariance $\underset{\sim}{R}$, $\underset{\sim}{K}$ is small and x is largely determined by x_0. The measurement has influence only in proportion to its trustworthiness.

The gain matrix $\underset{\sim}{K}$ is sometimes called the Kalman gain because the procedure being discussed here is a special case of the well-known Kalman filter (Ref.6). In the typical Kalman filter problem, the state evolves with a running variable v, and the prior estimate $x_0(v)$ is obtained from the posterior estimate $x(v - \Delta v)$ by using a dynamic model for the evolution over Δv. Use of such a dynamic model can have the effect of smoothing, so that aspect of the Kalman filter formalism has not been applied to the limb scan problem.

Once the optimal state estimate x is determined, its covariance $\underset{\sim}{P}$ follows by simply substituting the expression for x into the definition

$$\underset{\sim}{P} = \langle (x - x_a)^2 \rangle$$

and evaluating the expectations. The result is

$$\underset{\sim}{P} = (\underset{\sim}{I} - \underset{\sim}{K}\,\underset{\sim}{H})\underset{\sim}{P}_0$$

The covariance update provides small $\underset{\sim}{P}$ whenever $\underset{\sim}{R}$ is small. For $\underset{\sim}{R} = 0$,

$$\underset{\sim}{K}\,\underset{\sim}{H} = \underset{\sim}{P}\,\underset{\sim}{H}^T (\underset{\sim}{H}\,\underset{\sim}{P}\,\underset{\sim}{H}^T)^{-1}$$

Premultiplying by

$$\underset{\sim}{I} = (\underset{\sim}{H}^T\underset{\sim}{H})^{-1} (\underset{\sim}{H}^T\,\underset{\sim}{H})$$

and rearranging parentheses establishes $\underset{\sim}{K}\,\underset{\sim}{H} = \underset{\sim}{I}$ and $\underset{\sim}{P} = 0$ in the limit. That is, there is no uncertainty at all concerning the state after a noiseless measurement. The $\underset{\sim}{P}$, in general, represents residual state uncertainty which remains because the measurement was not noise free.

In the case of highly nonlinear problems, the Taylor series expansion $z(x) = z(x_0) + (\underset{\sim}{H}) (x - x_0)$ is not very accurate and as a result, the posterior state estimate may not even approximately reproduce the observed measurement. The standard technique for overcoming such a difficulty is to make many iterations on the same data, with the partial derivatives and the state estimate (but not the covariance) updated at each iteration. The process is commonly called multiple local iteration. Iteration is to be distinguished from recursion, iteration being applied to overcome nonlinearity and recursion being applied to smooth over noise.

The above remarks review the standard aspects of nonlinear optimal estimation that are applicable to the limb scan inversion problem. There are, however, a number of features of the limb scan

SCATTERED RADIANCE HORIZON PROFILES

inversion problem which tend to distinguish it from other inversion problems and require special techniques. These are the subjects of the following sections.

III. OVERCOMING COMPUTATIONAL PROBLEMS ASSOCIATED WITH LIMB SCANS IN PARTICULAR

Computational problems arise in the limb scan inversion problem because of requirements to (1) perform matrix inversions in computing the optimal state estimate, and (2) perform radiative transfer simulations to compute expected measurements and partial derivatives.

The required matrix inversions are difficult first because of the inherent large dimensionality of the problem. Typically there may be from ten to one hundred scan positions, with several state variables and several measurements per scan position. Rigorously, each measurement depends on every one of the state variables. This is because at each scan position, all the lower atmosphere provides a source for multiply-scattered photons, whereas all the upper atmosphere contributes to and damps the received signal. Thus, in a completely rigorous inversion, the dimensionality of covariance and partial derivative matrices can be of the order of hundreds by hundreds.

The solution to the problem of large dimensionality lies in approximations that replace the one large inversion, with its many measurements and many state variables, by many small inversions, each involving just a few measurements and a few state variables. The convenient partitioning is by scan position; the state variables at the tangent altitude of a given scan are found from the multispectral radiance measurements at that scan. The effect of the upper atmosphere can be fully accounted for by making the state a function of a running variable v = altitude, starting the inversion at the top of the atmosphere, and working downward from there, so that accurate estimates of the atmospheric parameters above each position are available. The lower atmosphere can be accounted for at least approximately by using the prior estimate for it. The overall approach is colloquially called "onion peeling."

The onion peeling technique described above is remarkably efficient because most of the signal received at each scan position comes from near the tangent altitude. This is the case because the atmosphere is maximally dense near the tangent altitude, and because the geometric length of the line of sight through altitudes near tangent is large. Figure 3 plots density of signal contribution against slant range along the scan. For a nearly exponentially decaying atmosphere, the curve is nearly Gaussian, and the width of that Gaussian corresponds to a small altitude increment because of the limb scan geometry.

Even with the dimensionality of the problem reduced by the onion peeling described above, there is still a potential difficulty due to the dynamic range of the variables. It is possible, in particular, for the diagonal entries of covariance matrices to be of significantly different magnitudes, the result being that the required matrix inversions are numerically difficult. This

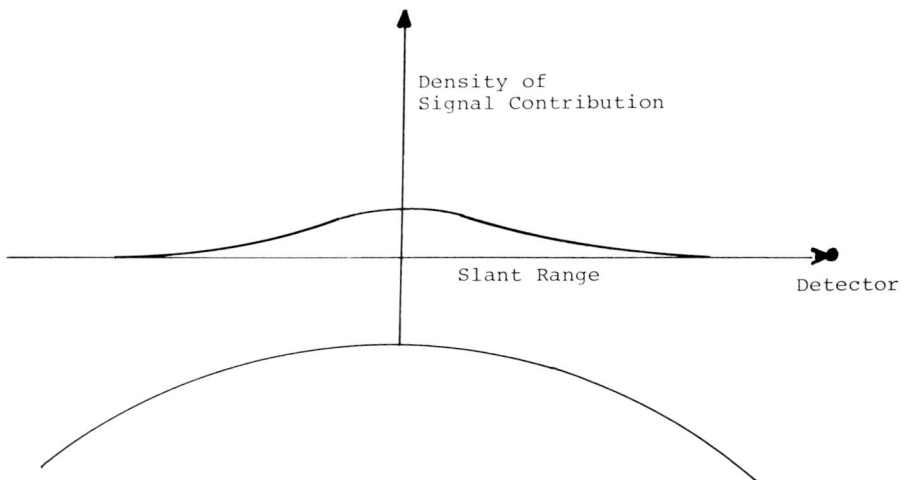

Fig. 3. Scan through Earth's Atmosphere

difficulty is, however, easily circumvented by redefining the variables. We let the measurement be log (radiance) instead of radiance, and correspondingly we let the state elements be log (constituent density) or log (aerosol physical property), as the case may be.

Radiative transfer simulations are required in the limb scan inversion problem to produce the predicted measurement $z(x)$ and the partial derivative matrix H. A first problem to be faced in performing a simulation is that the aerosol physical properties (Junge size parameter and complex refractive index) enter the radiative transfer problem only indirectly, through the resulting optical properties (phase functions and cross sections). In an inversion procedure, it is not practical to take a conventional approach, calculating the aerosol optical properties from the physical properties with a Mie code. Instead, a precalculated aerosol model can be used. The model used here consists of polynomial coefficients obtained from multivariate regression of a large data base of aerosol optical properties calculated from physical properties by a Mie code. In milliseconds of computation, the resultant model yields values of phase function, scattering cross section and absorption cross section, given Junge-size parameter and complex index of refraction.

For the limb scan situation, the calculation of measurement given optical state is still a nontrivial problem. The process is governed by a complicated integro-differential equation of radiative transfer requiring elaborate computer simulation. The curved geometry inherent in the problem restricts the number of applicable computer models, and at present only two accurate techniques are available: Monte Carlo simulation as described by Collins and Wells (Ref. 7) and DART simulation as described by Whitney (Ref. 8). Other than these, there is only a simple but unrealistic single scattering approximation.

The choice of which of the available radiative transfer models to use is dictated by speed requirements. In the course of a limb scan inversion, the radiative transfer model is called many thousands of times. The number of calls grows multiplicatively with the number of wavelengths, number of state variables, number of iterations, and number of scan positions, and typically approaches 10^5. Therefore, only the single scattering and DART models are really practical to use.

Although the computational difficulties associated with limb scan inversion are largely circumvented by the steps described above, there remain difficulties associated with nonlinearity. These are of a fundamental nature not particularly limited to the limb scan problem. Their solution is described in the next section.

IV. ESTIMATING PARTIAL DERIVATIVES FOR NONLINEAR INVERSIONS IN GENERAL

It was found that even the combined application of all the techniques described in the last section was insufficient to produce a scattered limb scan inversion. The reason is the extreme nonlinearity of the radiative transfer over the dynamic range of the state variables normally encountered. Because nonlinearity is a characteristic of many inversion problems, and not just the limb scan, the technique for overcoming it presented here may be fruitfully applied to other nonlinear problems as well.

The heart of the problem is in the calculation of the partial derivative matrix $\underset{\sim}{H}$. The radiative transfer is sufficiently nonlinear over the state variable excursions required that the usual idea of partial derivative is not suitable. The local slope of a curve of radiance against state variable simply does not follow the curve far enough to be used as an accurate basis for updating the state.

Given the nonlinearities of the problem, we are likely to

encounter the situations illustrated in Figs. 4a and 4b. The radiance as a function of number density of aerosols is plotted as a sigmoid curve. The left asymptote is found at aerosol densities so low that the radiance is entirely determined by the other atmospheric constituents. The right asymptote is found at aerosol densities so high that the atmosphere is essentially opaque. A measured value of radiance is indicated by the dashed lines in the figures. In Fig. 4a, the prior estimate of the state is in a region where the local slope is so small that the updated state estimate resulting from that slope is vastly larger than the actual state. In Fig. 4b, the prior estimate is in a region where the

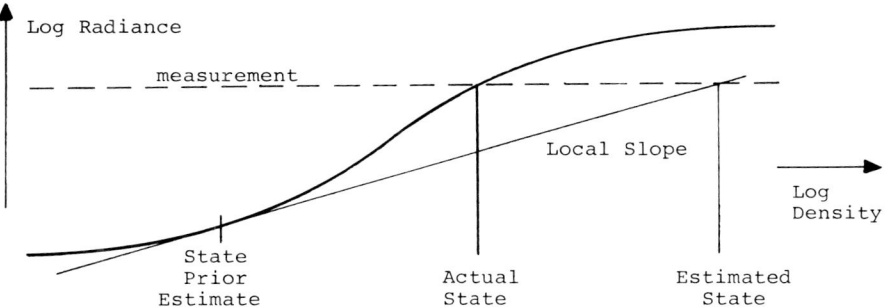

(a). Overshoot due to partial derivative.

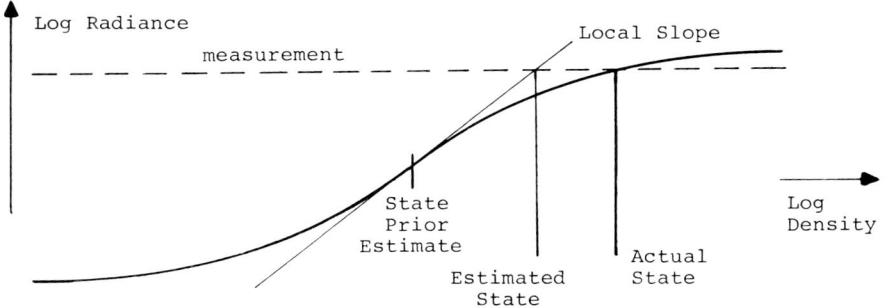

(b). Undershoot due to partial derivative.

Fig. 4. Radiance as a function of number density of aerosols.

local slope is so large that the updated state estimate resulting from that slope is much too close to the prior estimate. It is possible for the overshoot, in the case of Fig. 4a, to throw the problem outside the numerical operating range of the computer, and it is possible for the undershoot, in the case of Fig. 4b to interminably delay convergence, in either case making the problem unsolvable in a practical sense.

The remedy for the problems described is to use a derivative-like construct that is more global in its meaning than the local partial derivative. Such a construct is determined by first formulating a mathematical model for the curves of radiance as a function of the state parameter. To acknowledge the changing slope, the radiance is modeled as a quadratic function of the state parameter. Figure 5 illustrates how such a model can be fairly accurate over a region substantial enough to include the required state excursions. The quadratic model is actually constructed by the following steps:

1. Establish the prior estimate of radiance, based on the state prior estimate.

2. Perturb the state ± one standard derivation, and calculate

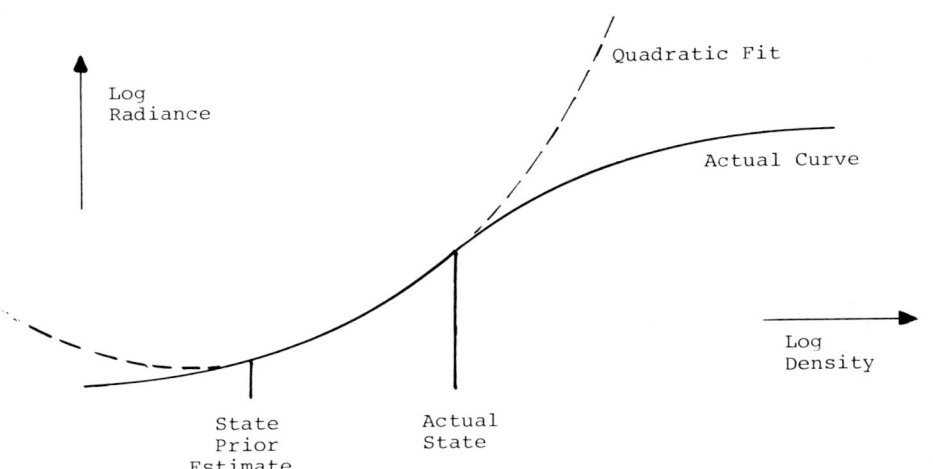

Fig. 5. Quadratic fit.

the perturbed radiances.

3. Fit a quadratic curve through the three points.

Thus, the model is:

$$z_i = a + b(x_i - x_2) + c(x_i - x_2)^2$$

where $i = 1, 2, 3$, and the coefficients are found as:

$$a = z_2$$

$$b = \frac{1}{D}\left[(z_3 - z_2)\frac{(x_2 - x_1)^2}{2} + (z_2 - z_1)\frac{(x_3 - x_2)^2}{2}\right]$$

$$c = \frac{1}{D}\left[(z_3 - z_2)(x_2 - x_1) - (z_2 - z_1)(x_3 - x_2)\right]$$

$$D = \frac{1}{2}\left[(x_3 - x_2)(x_2 - x_1)^2 + (x_2 - x_1)(x_3 - x_2)^2\right]$$

The quadratic model is the basis for evaluating a derivative like construct that is more global in its meaning then the local partial derivative is. The construct is evaluated by the following algorithm.

1. If the measurement line intersects the quadratic model curve, replace the partial derivative by the slope of the line from prior estimate point to the nearest intersection.

2. If the measurement line does not intersect the quadratic model curve, replace the partial derivative by the slope of the line from prior estimate point to the extremum of the quadratic.

V. SIMULATION RESULTS

This section presents inversion results for a simulated limb scan experiment. In the simulation, a set of limb radiances is calculated from a known atmospheric state. These radiances are treated as "measurements" and inverted to retrieve the known state.

Seven state parameters are inverted, namely: (1) concentration of Rayleigh scatterers; (2) concentration of ozone; (3) concentration of nitrogen dioxide; (4) aerosol extinction; (5) aerosol Junge-size parameter; (6) aerosol refractive index, real part; and (7) aerosol refractive index, imaginary part. Eight wavelengths are used in the inversion: 0.3500, 0.4000, 0.4500, 0.5000, 0.5500, 0.6750, 0.7770, and 0.8630 µm. These were chosen to fall within the sensitive spectral range of a silicon diode detector, and at the same time to involve only the chosen parameters as optically active constituents. The specific problem geometry has the scan direction, Earth centroid, and satellite lying in the same plane with the Sun nearly behind the detector so that the scattering angle is approximately $164°$. For aerosol inversion, this is a moderately good geometry, in that the aerosol angular scattering function is roughly midway between its highest and lowest values.

For each state parameter, three graphs have been constructed. The first displays the state parameter as a function of altitude, the second shows the evolution of the inverted state parameter as the inversion process is recycled at a particular scan altitude, and the third displays error bars for the completed inversion.

The particular graphs presented were constructed with input noise statistics, but without actual noise values. Ideally, they should be representative of mean performance with zero mean noise, and the error bars should describe the spread to be expected. Later runs with actual random noise included required an artificial scaling down of the gain by a factor of 5 to produce the anticipated behavior.

The atmospheric state is defined at altitude intervals of 1 kilometer, and three varieties of state are represented in the first graph. These are:

1. Prior Estimate State--The first guess, constructed from various standard sources which are compiled in Malchow (Ref. 9). This state is represented by relatively smooth functions of altitude.

2. True State--This is the state that is used to produce the simulated measurements. It is constructed from various sources of real data, except for the aerosol parameters, which are chosen by random selection from what is considered a likely range of variation.

3. Inverted State--This is the state found by the inversion process. Since no noise has been added to the simulated measurements, this state is identical to the true state when the inversion is perfect.

The scan range extends from 22 to 13 kilometers. For 22 kilometers and above, the true state and prior estimate state are set equal in order to focus attention on the inversion process itself and not the consequences of initial errors. It has been found in other simulations that such initial errors can be overcome in approximately three scans.

The second graph of each set shows the path followed by each parameter through state space en route to convergence. Since the limb problem is highly nonlinear, many iterations of the inversion procedure are carried out at each scan position. The convergence curve represents only the 20 kilometer scan altitude. For this particular altitude it can be seen from the curves that a stable solution is achieved in 10 iterations. However, more iterations are needed at the higher altitudes.

The third graph of each set shows the 1σ error bars. The error bar plotted at 23 kilometers represents the covariance input to the problem. The error bars at lower altitudes are generally smaller, having been reduced by the information gained in the inversion process. In cases where even further reduction is desired, avaraging over data sets can be employed.

The first set of data (Figs. 6, 7, and 8) is for Rayleigh scatterers (also called neutral density). Since this constituent has little variability relative to its range with altitude, the prior

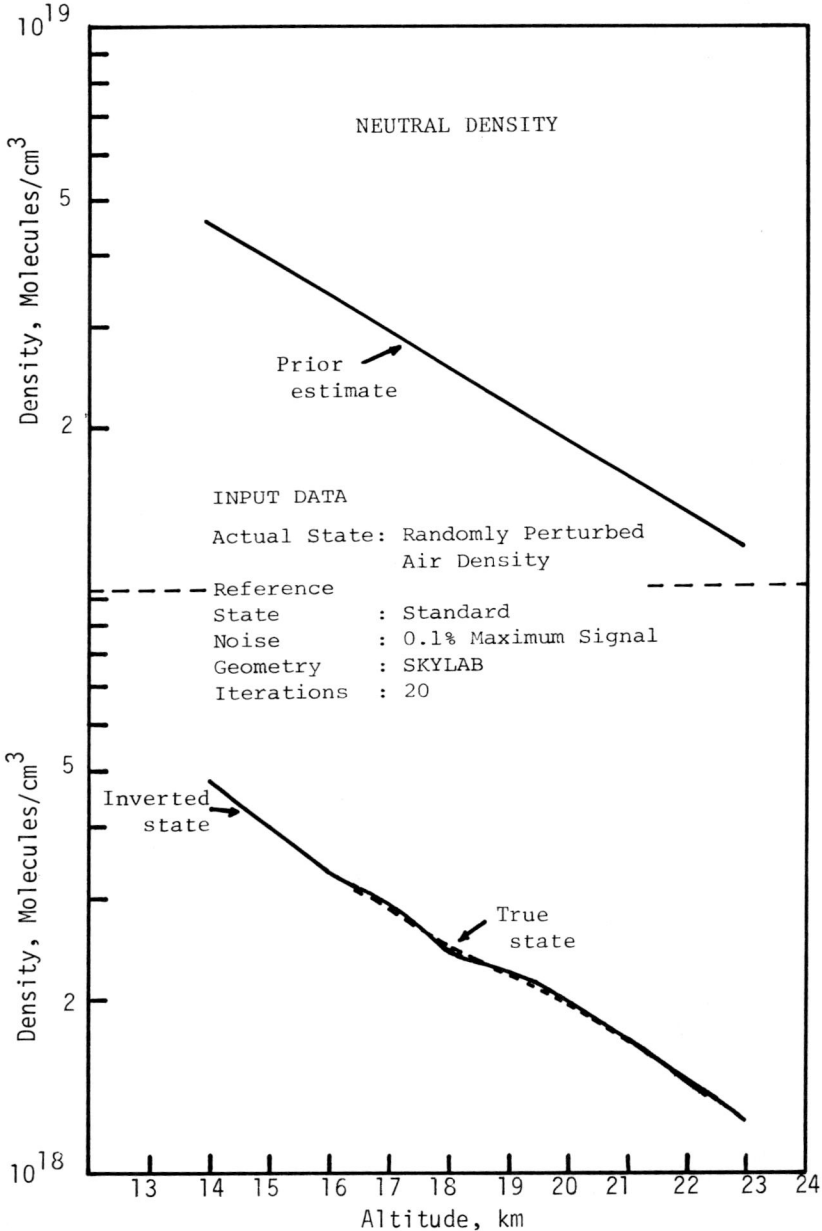

Fig. 6. Neutral density solution.

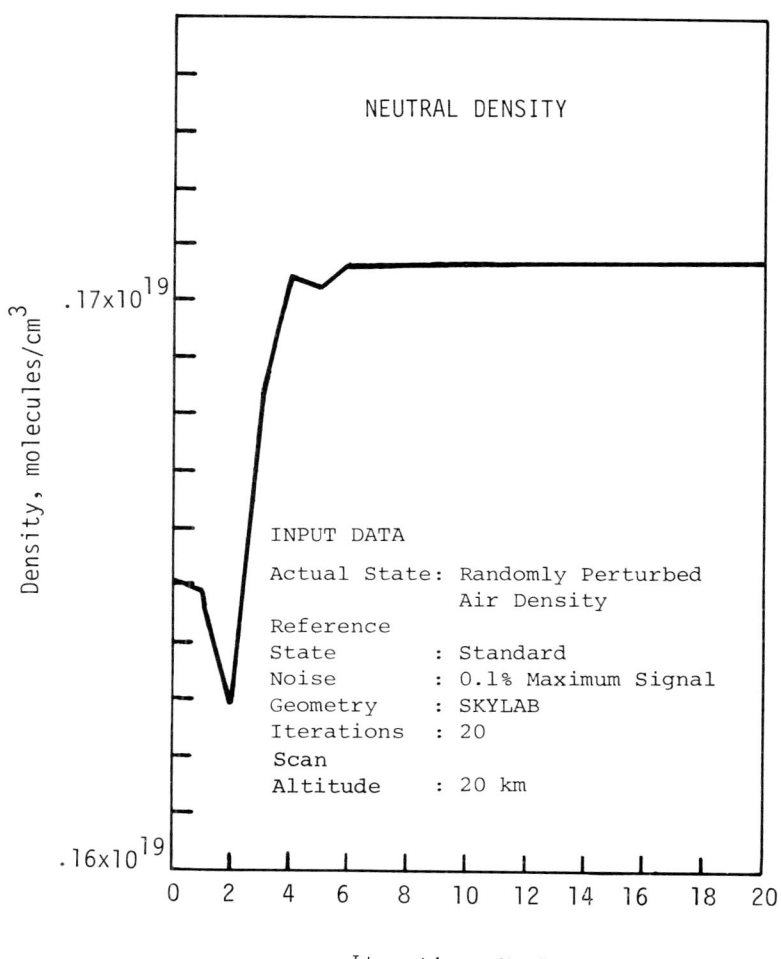

Fig. 7. Neutral density convergence.

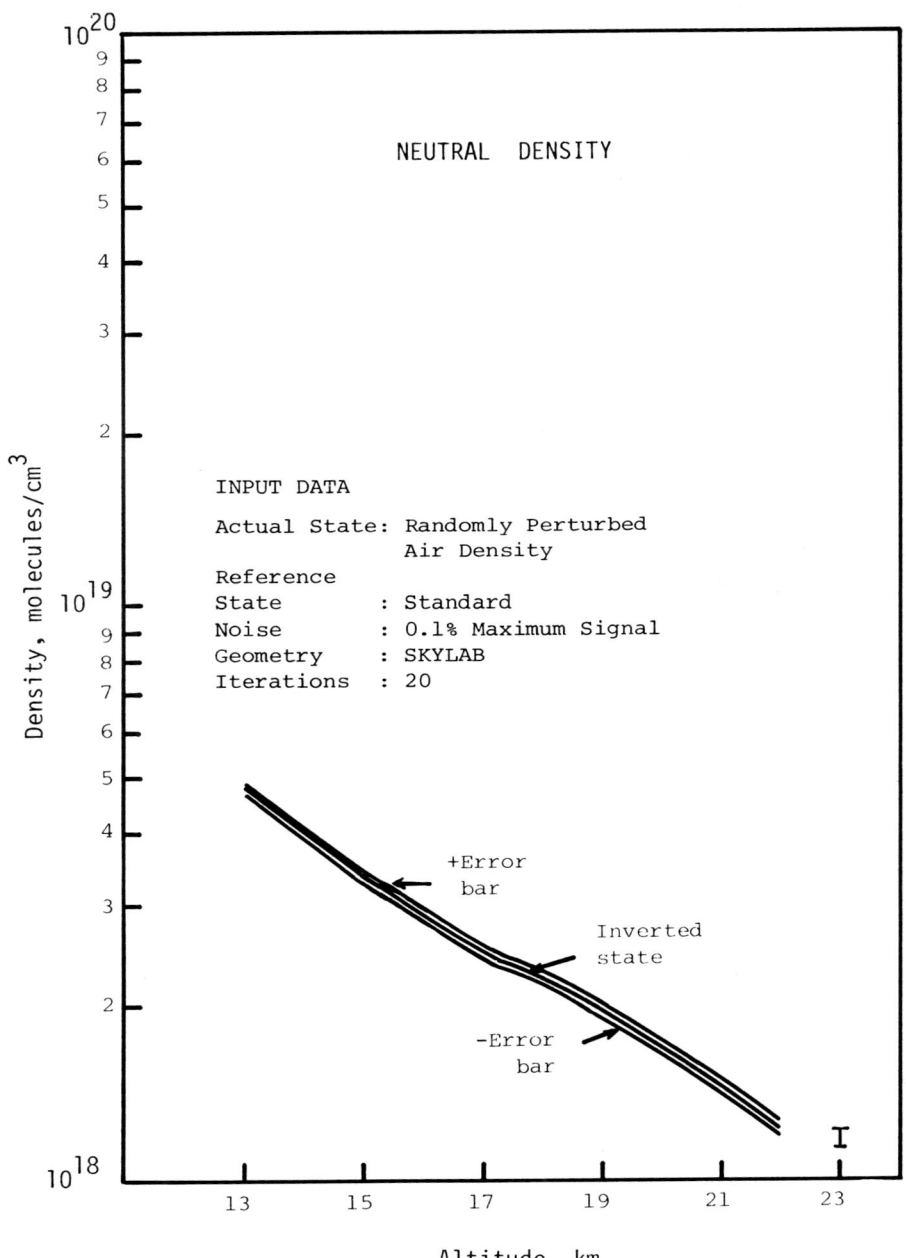

Fig. 8. Neutral density error bars.

SCATTERED RADIANCE HORIZON PROFILES

estimate is plotted separately from the inverted and true states. The first graph (Fig. 6) shows a good tracking of the true state by the estimated state. At 20 kilometers the true and prior estimate states differ by 3% initially. This is representative of a 1σ excursion. After 8 iterations, the second graph (Fig. 7) shows close convergence to the true state. Error bar results for this constituent (Fig. 8) show a small reduction of initial input uncertainty.

The second set of data (Figs. 9, 10, and 11) is concerned with ozone concentrations. The true ozone state (based on data from Ref. 10), is marked by strong perturbations both up and down from the prior estimate. However, ozone is strongly represented in the simulated measurements at the chosen scan altitudes, with the result that this constituent can be inverted quite accurately as the inverted and true states show. At 20 kilometers, the initial perturbation was about -50%. The initial uncertainty assumed for ozone is $1\sigma = 60$%, and the inversion reduces this by a factor of 3 or 4.

The next set of graphs (Figs. 12, 13, and 14) is concerned with NO_2. This constituent has relatively small prior estimate (Ref. 11) optical depths for the chosen scan altitude regime and the true state (based on our SKYLAB measurements) is even smaller. Therefore, one expects and gets less accuracy for NO_2 in the inversion than for ozone. The error bars show a modest reduction from the assumed initial uncertainty of $1\sigma = 300$%.

Aerosol extinction is the subject of the next set of graphs (Figs. 15, 16, and 17). In this case, the prior estimate curve is based on Elterman's data (Ref. 12). The true state is constructed from lidar data (Ref. 13), and shows layered structure. Since Since aerosol extinction exhibits a great deal of spatial dynamics, it is important that the inversion process can deal with a large range of extinction values. This particular inversion problem shows that order of magnitude changes in aerosol extinction over a 1

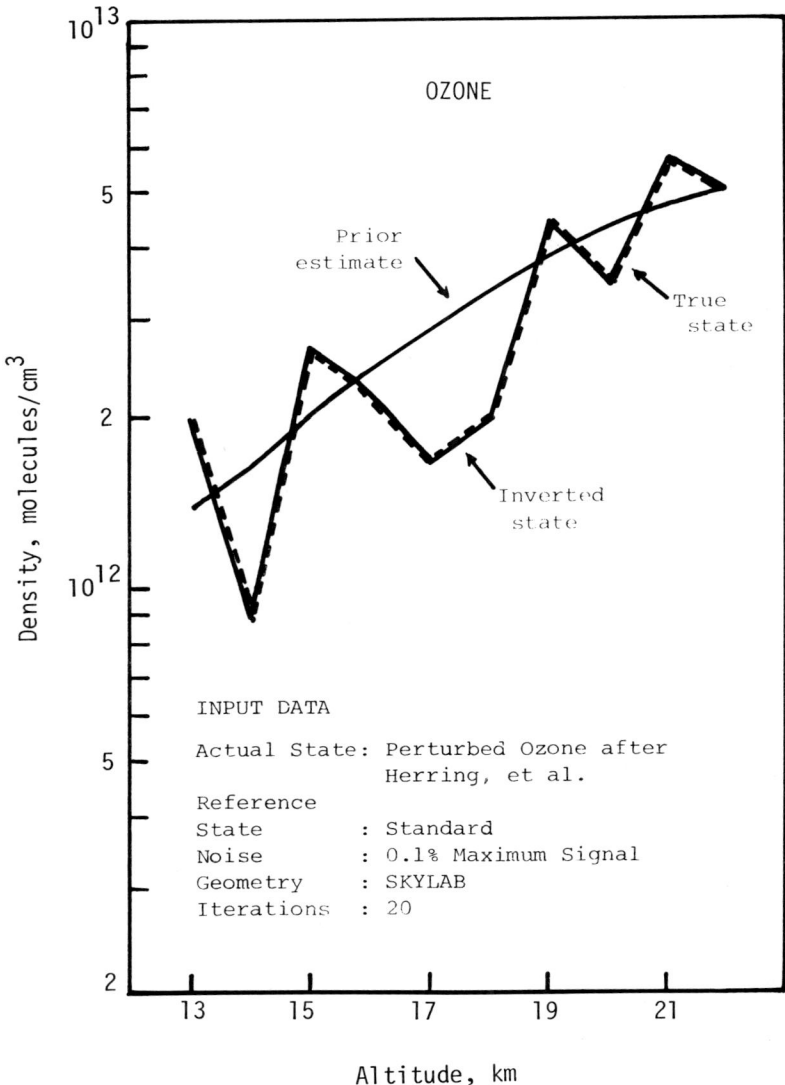

Fig. 9. Ozone solution.

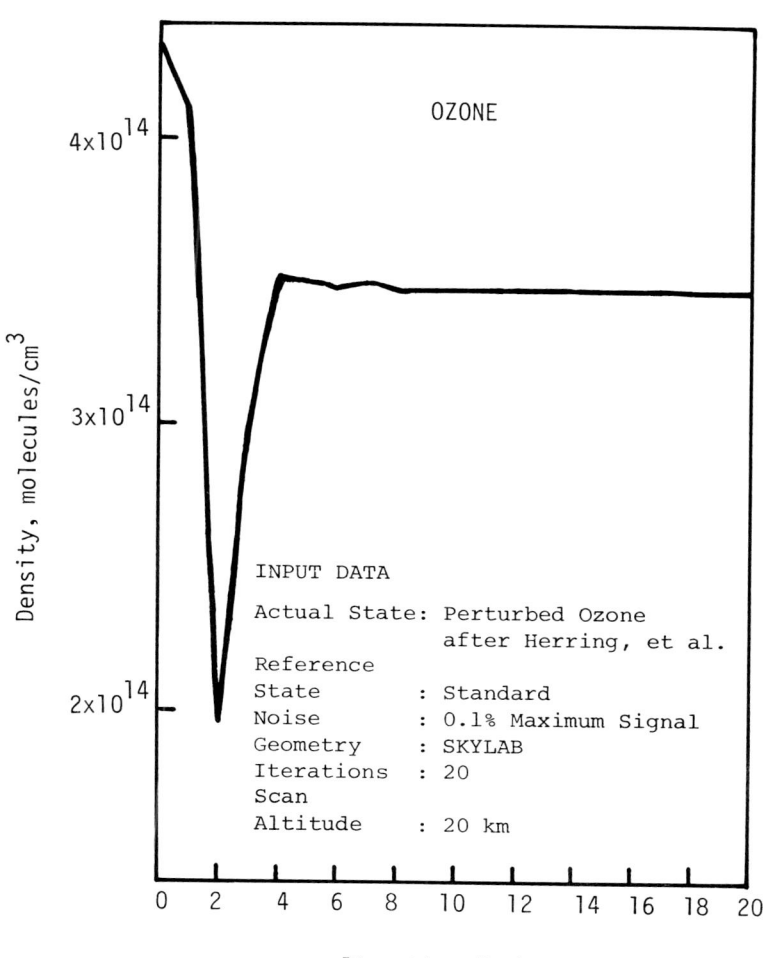

Fig. 10. Ozone convergence.

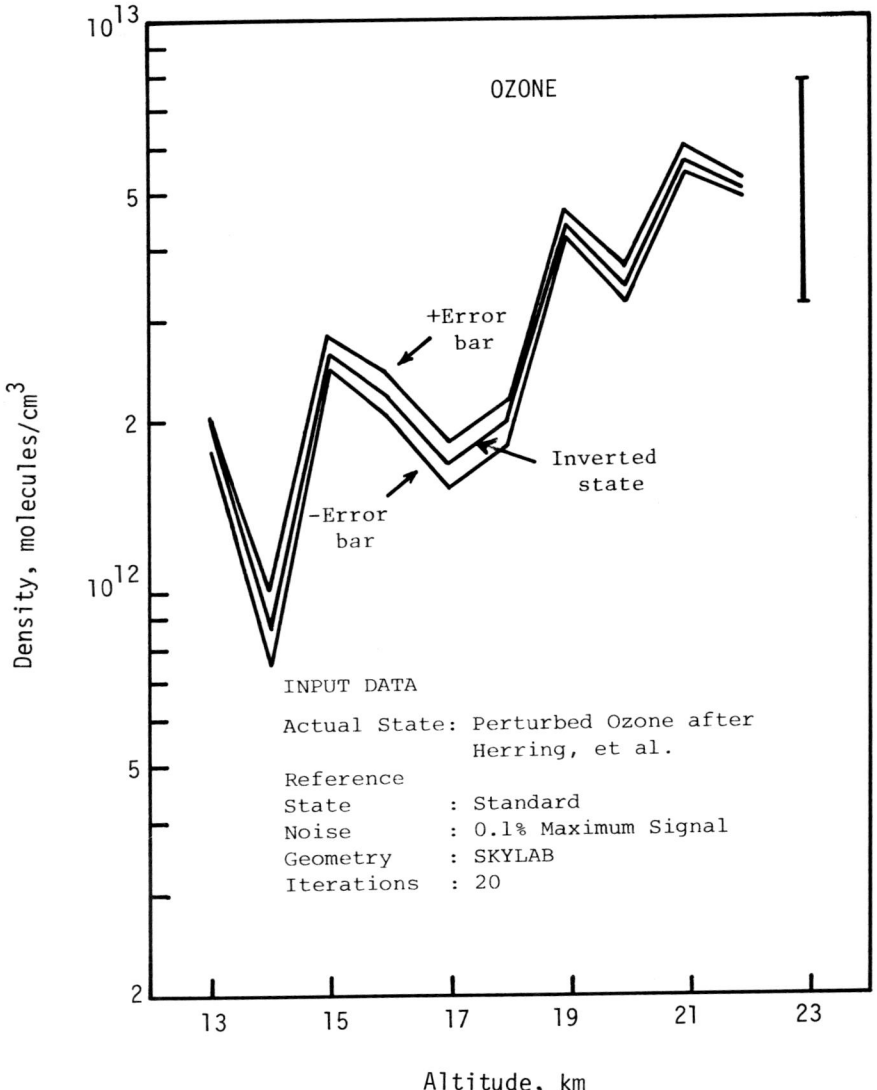

Fig. 11. Ozone error bars.

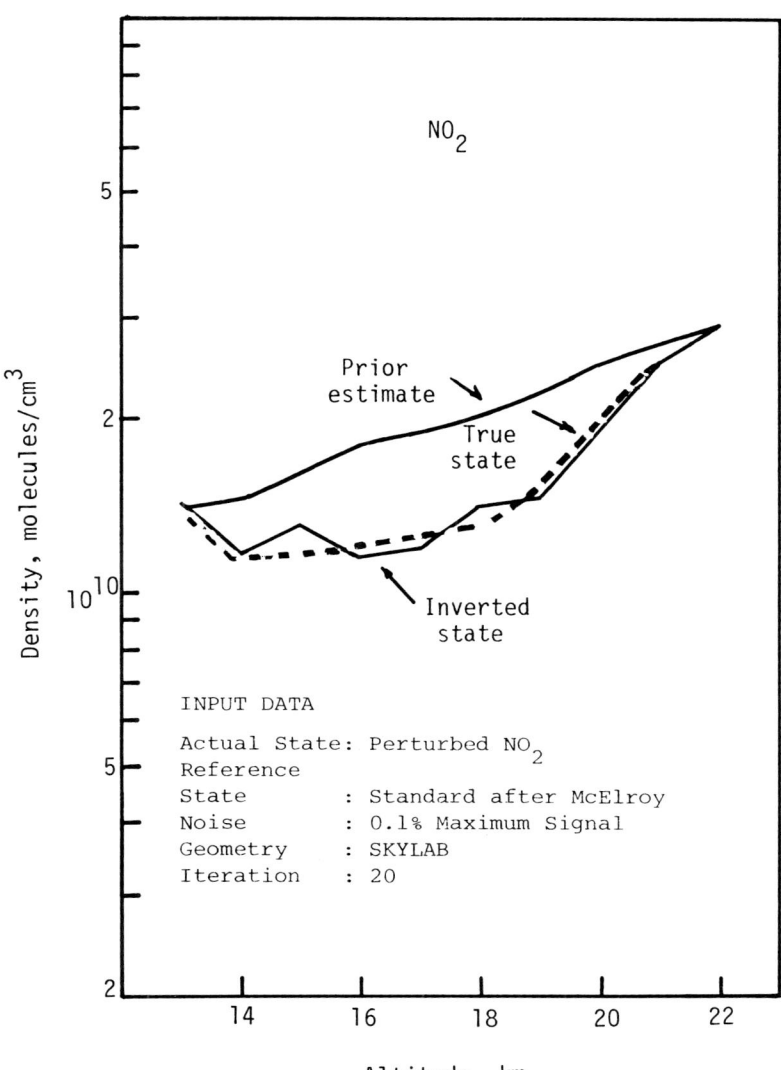

Fig. 12. Nitrogen Dioxide Solution.

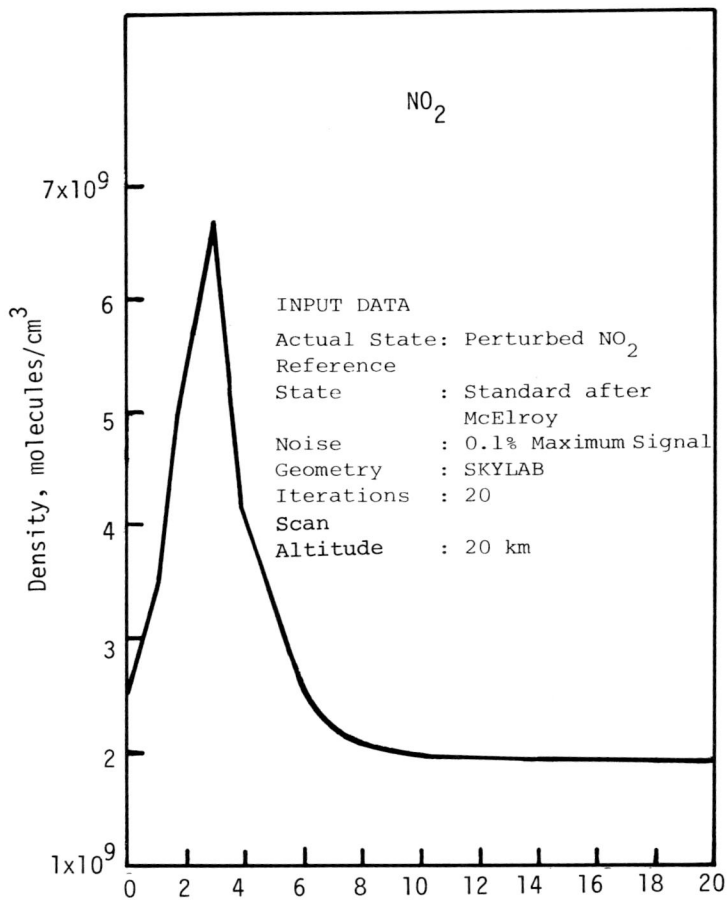

Fig. 13. Nitrogen dioxide convergence.

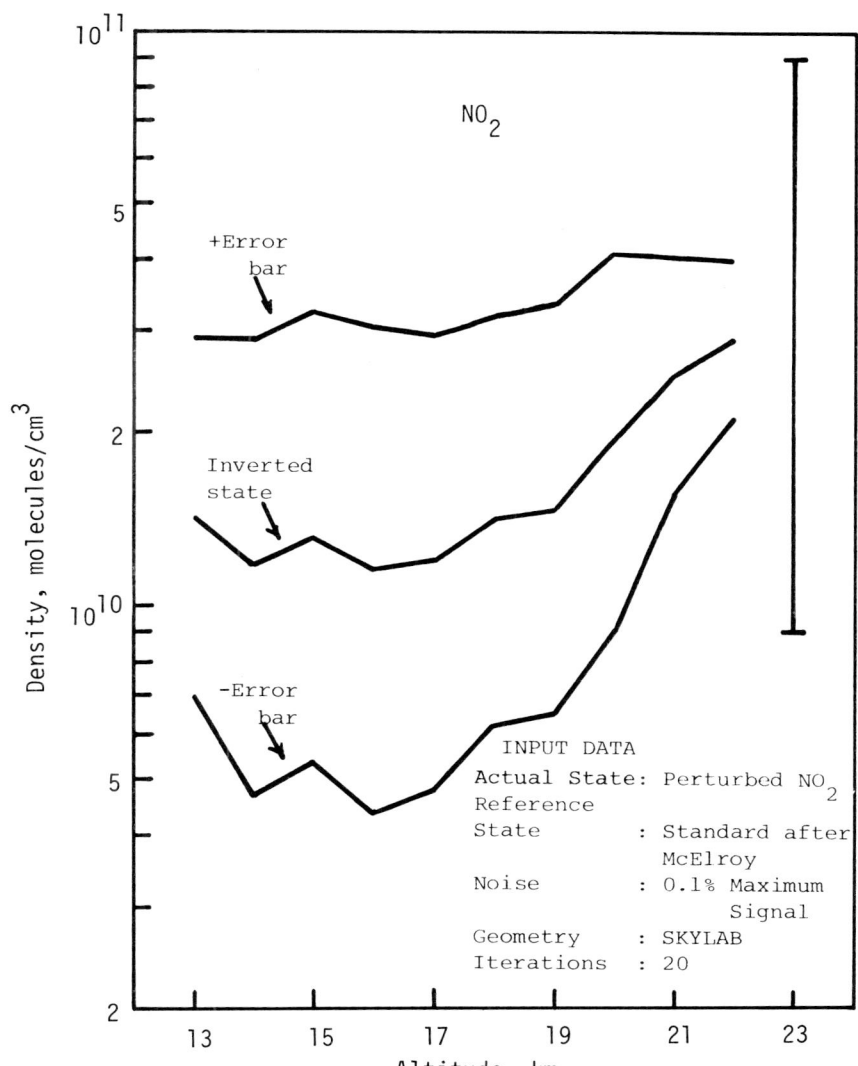

Fig. 14. Nitrogen dioxide error bars.

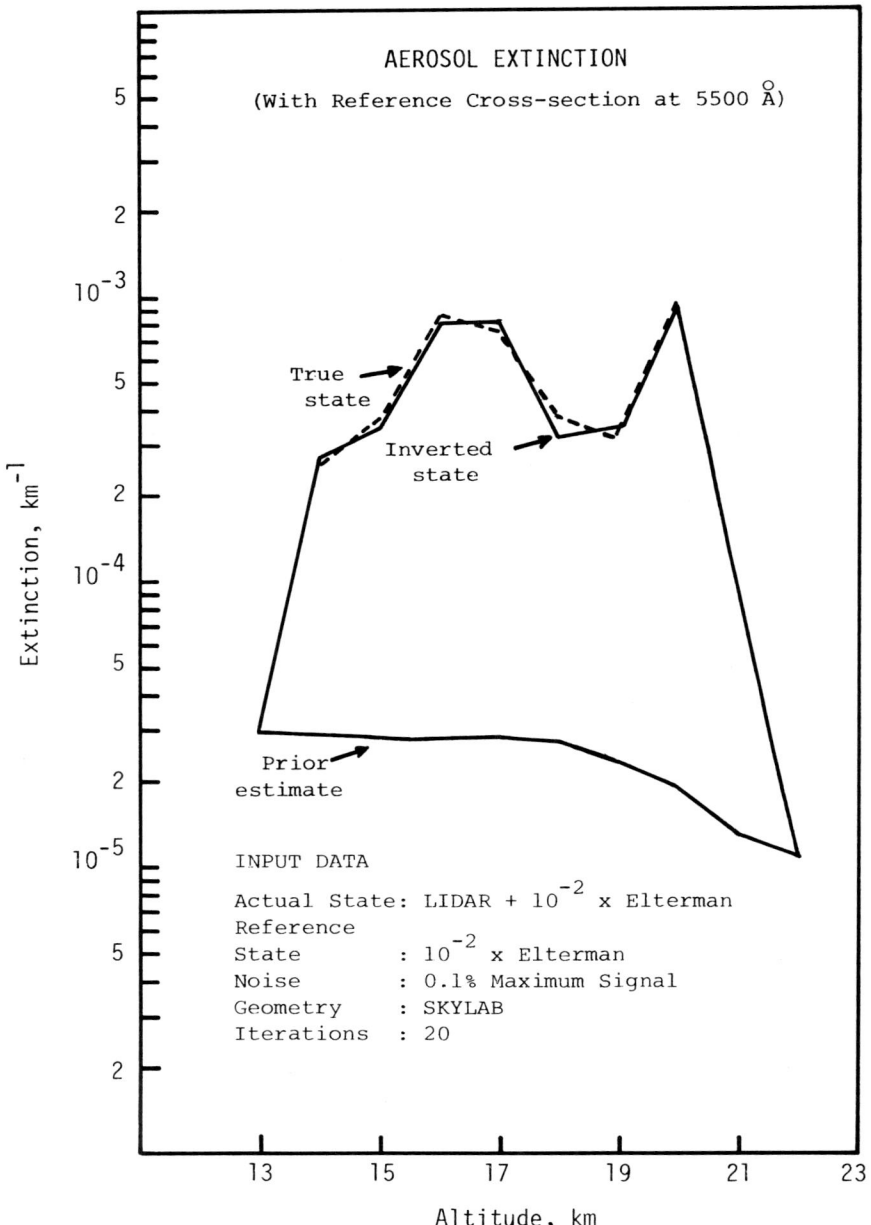

Fig. 15. Aerosol extinction solution.

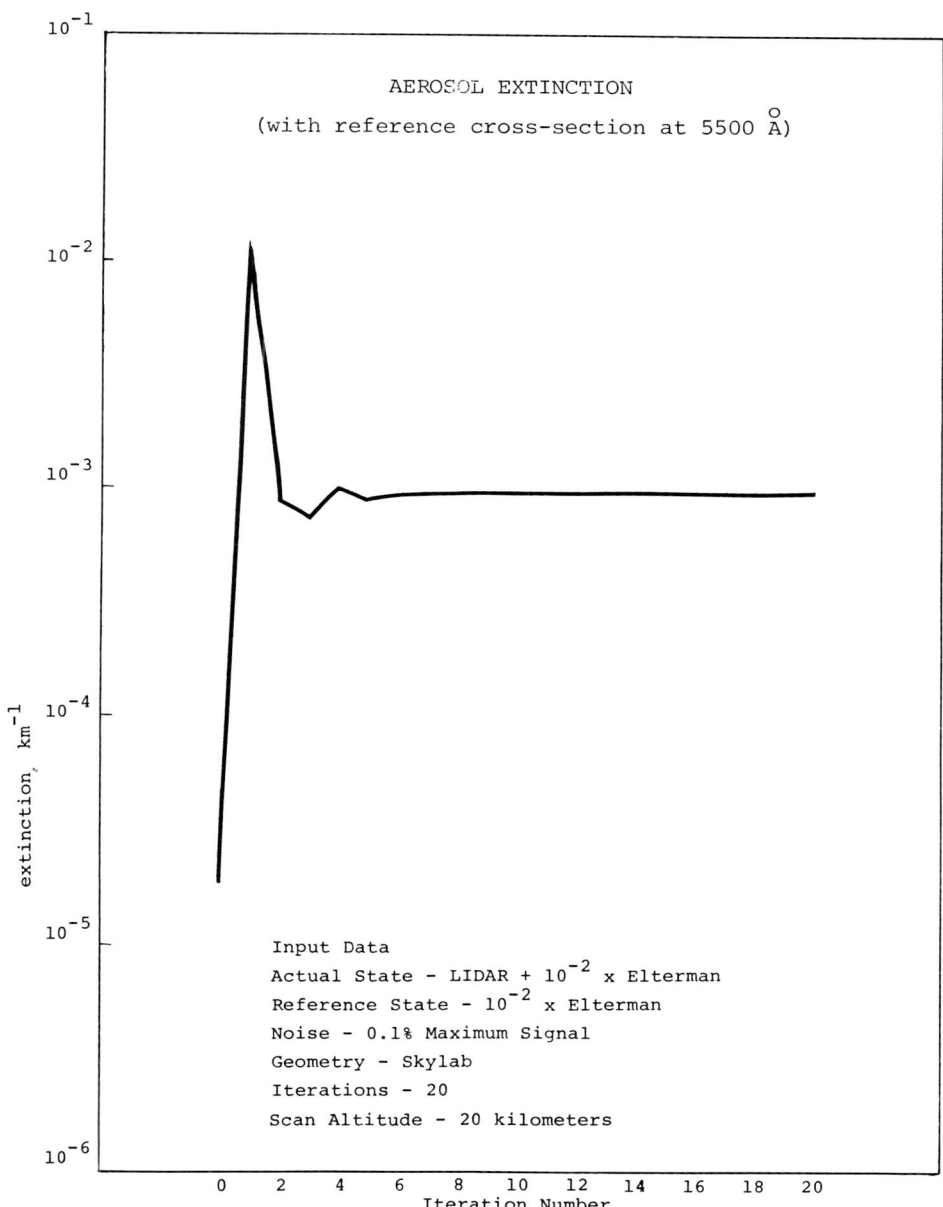

Fig. 16. *Aerosol extinction convergence.*

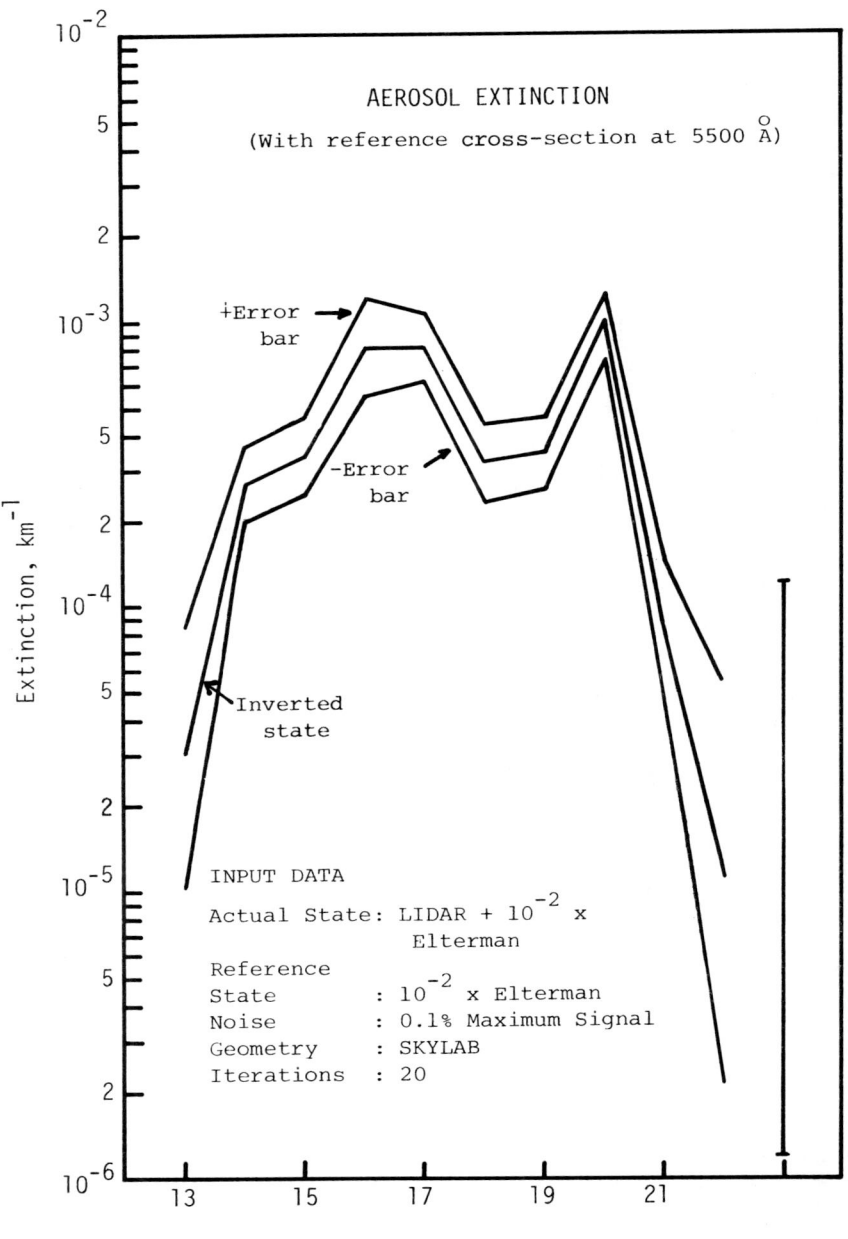

Fig. 17. Aerosol extinction error bars.

SCATTERED RADIANCE HORIZON PROFILES

kilometer interval can be dealt with by the nonlinear algorithm. The error bars for aerosol extinction show a reduction of the initial order of magnitude uncertainty by as much as a factor of 50.

The final three sets of graphs (Figs. 18 to 26) are concerned with inversion of aerosol physical parameters. A noteworthy feature of these graphs is that the inversion accuracy for these parameters is proportional to the aerosol extinction. In each case, the closest agreement between the true and inverted states occurs at the peaks in aerosol extinction. Graphs of aerosol extinction have been superposed on the error bar graphs to illustrate this point. The results show that when moderate aerosol concentrations are present, useful information about the aerosol physical characteristics can be obtained from limb scans.

In general, the simulated measurement calculations displayed here demonstrate that sufficient information is present in the considered wavelength set to separate and invert the studied constituents. An important next step in the simulation studies will be to relate the expected inversion accuracy to the quality of the measurements. Other potential error sources that will require analysis include albedo uncertainties, modeling errors, such as those from quadrature in the radiative transfer calculations, and errors in the constituent cross-section averages over finite wavelength intervals.

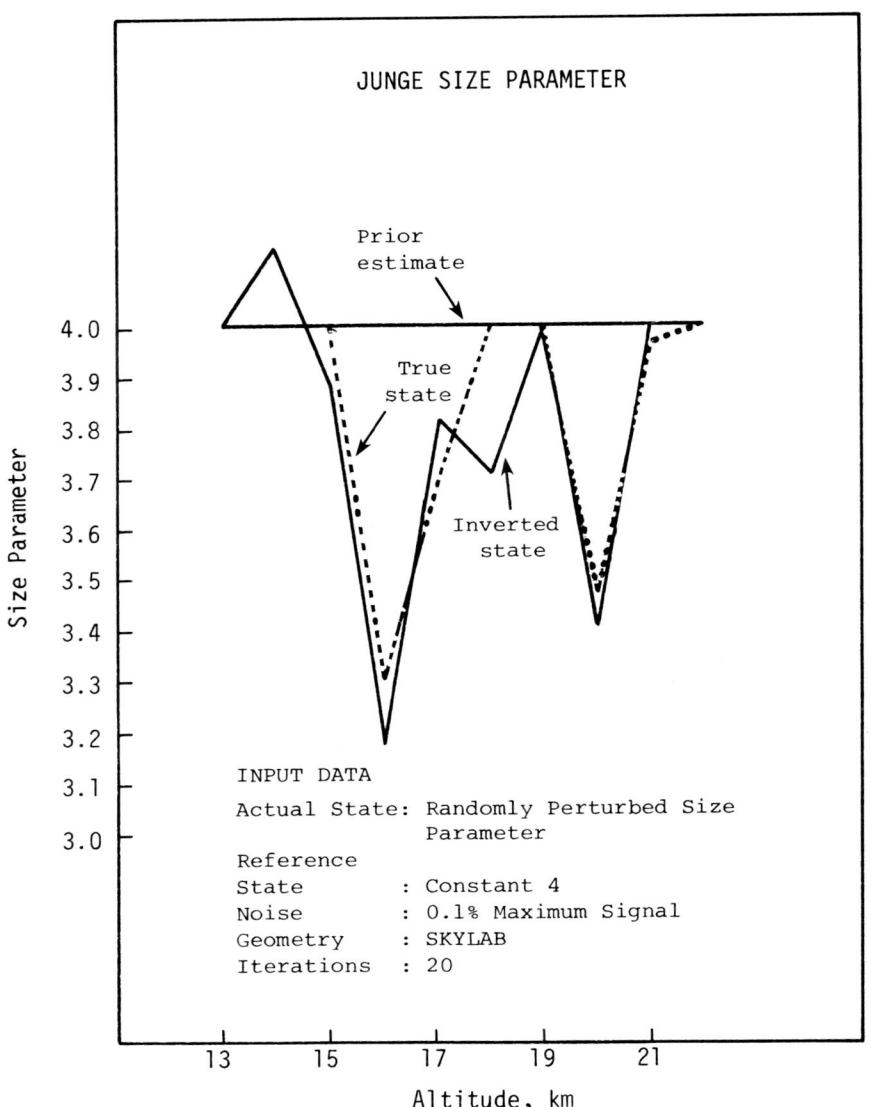

Fig. 18. Junge size parameter solution.

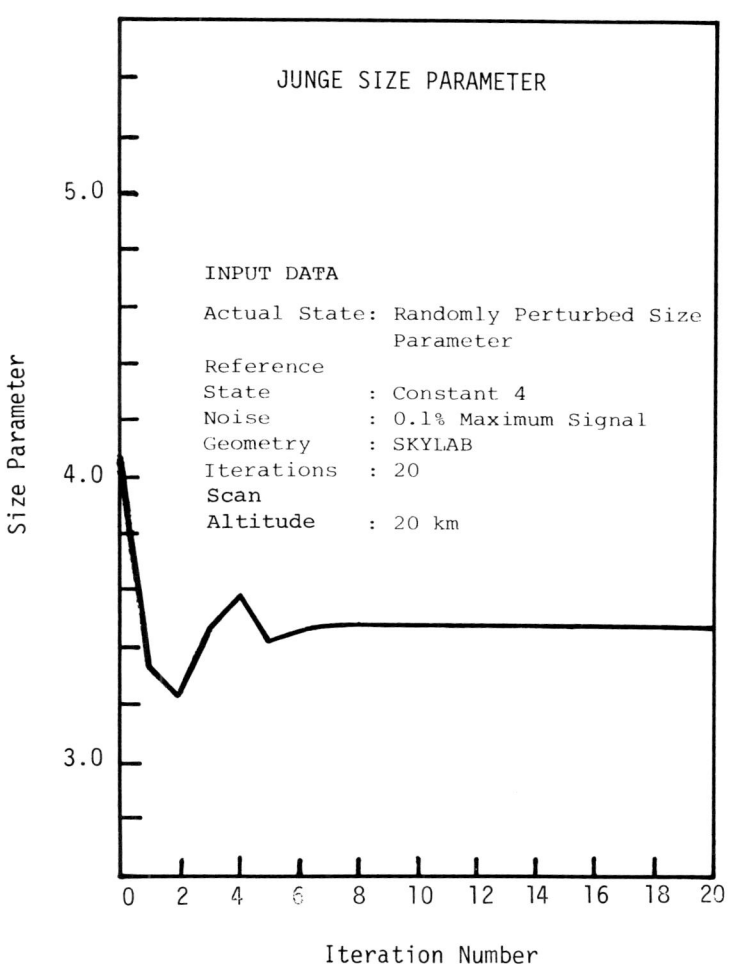

Fig. 19. Junge size parameter convergence.

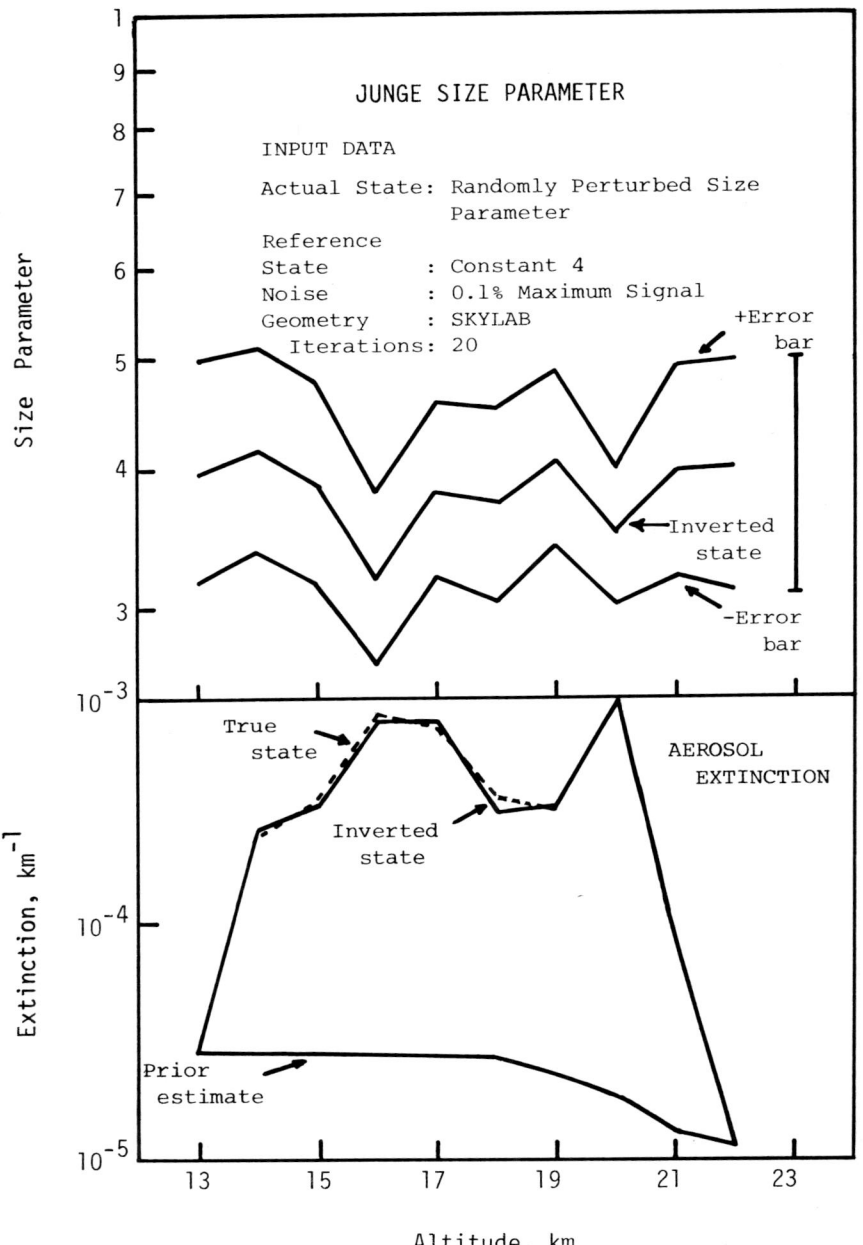

Fig. 20. Junge size parameter error bars.

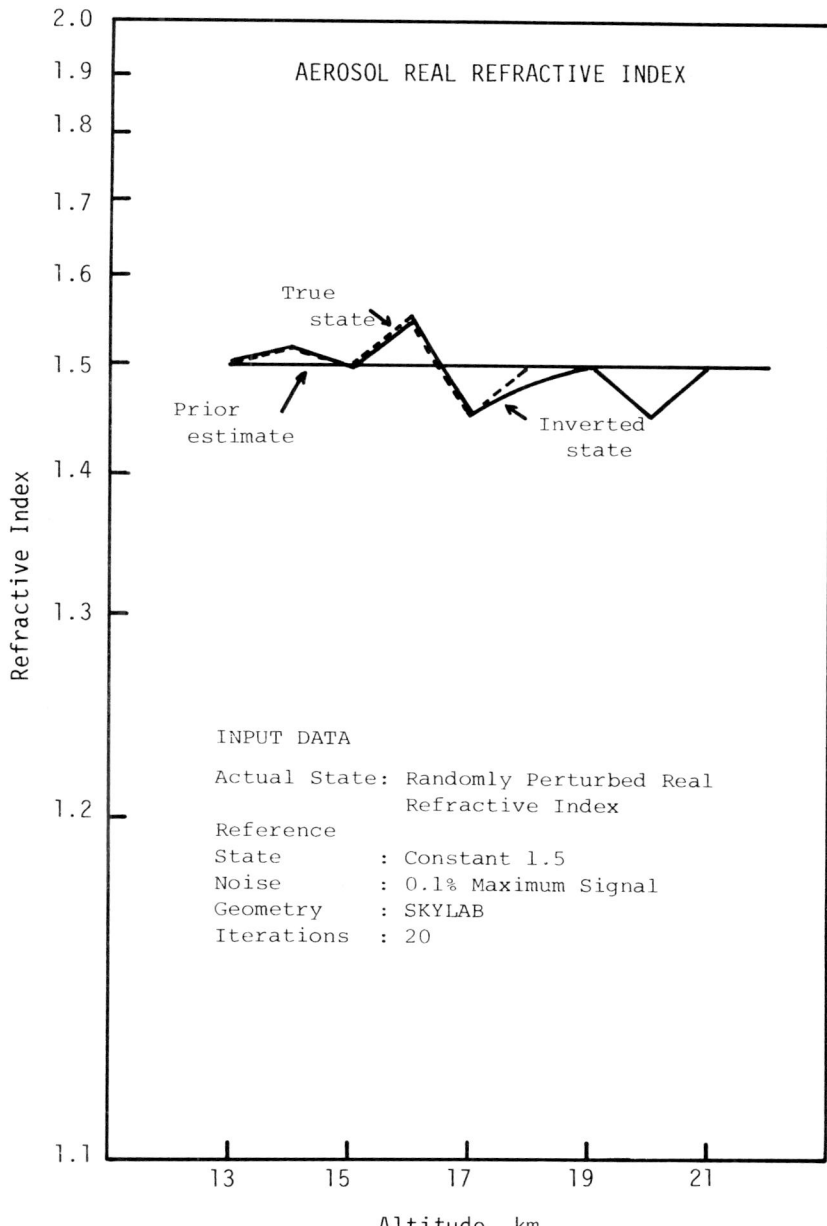

Fig. 21. Aerosol real refractive index solution.

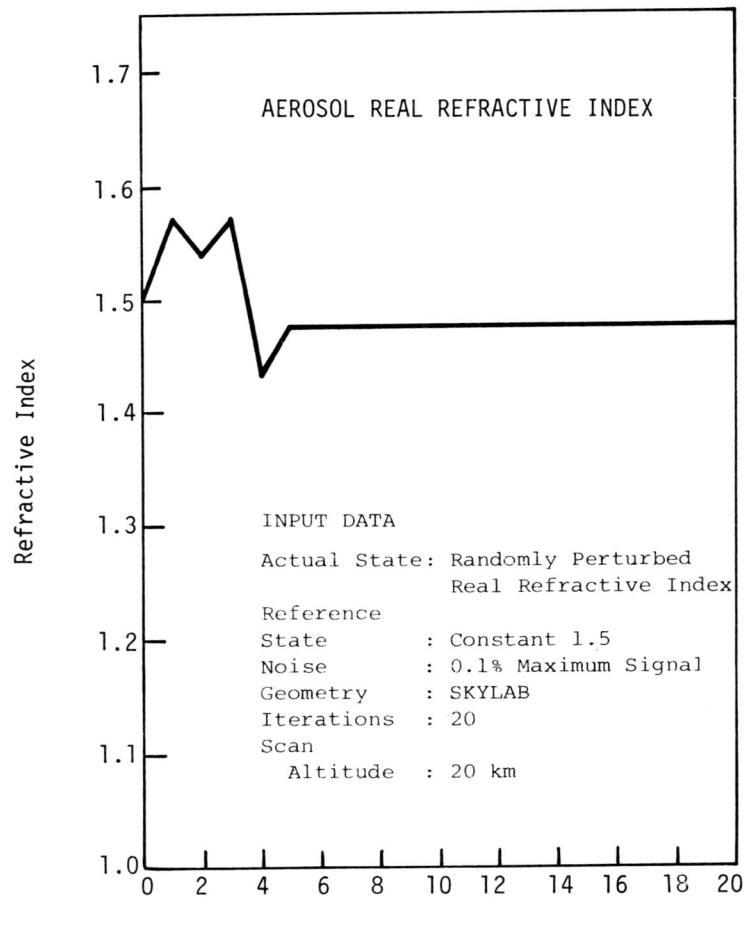

Fig. 22. Aerosol real refractive index convergence.

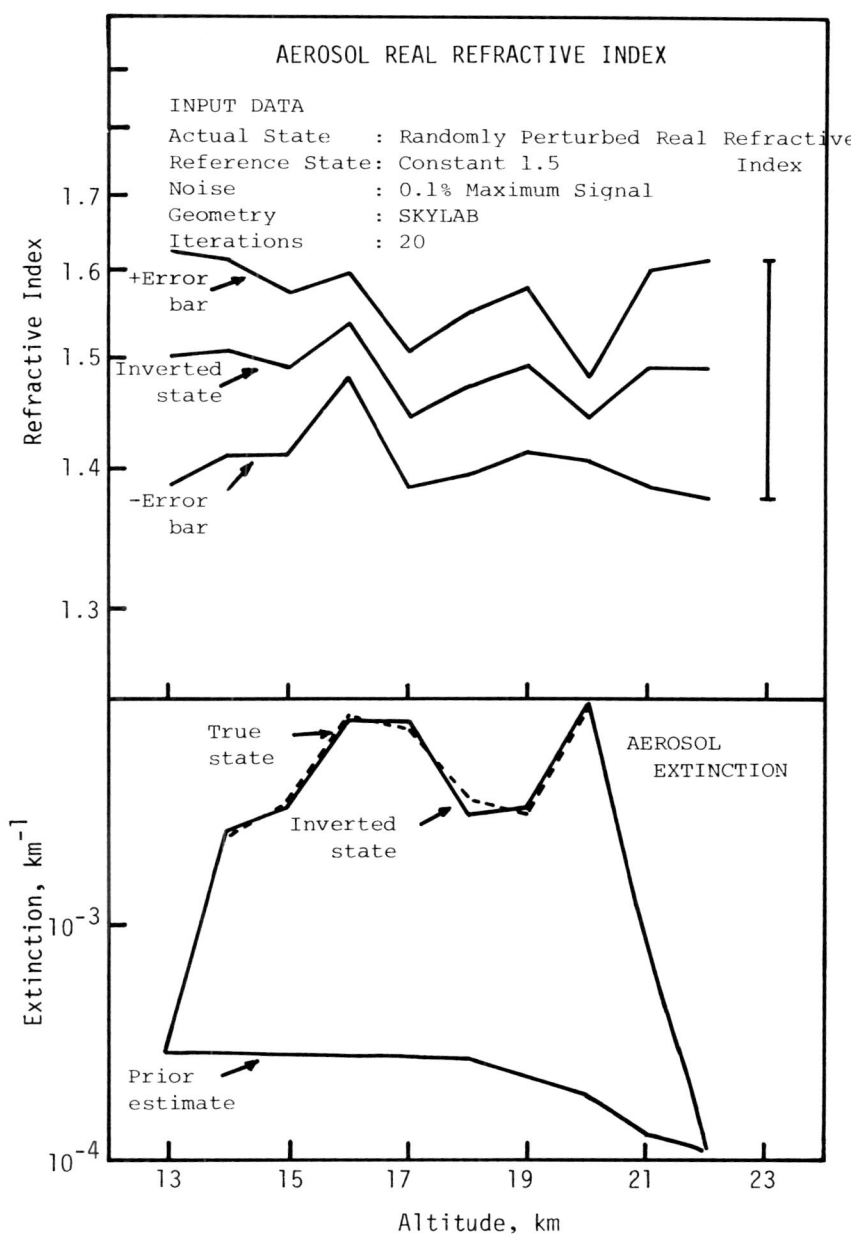

Fig. 23. Aerosol real refractive index error bars.

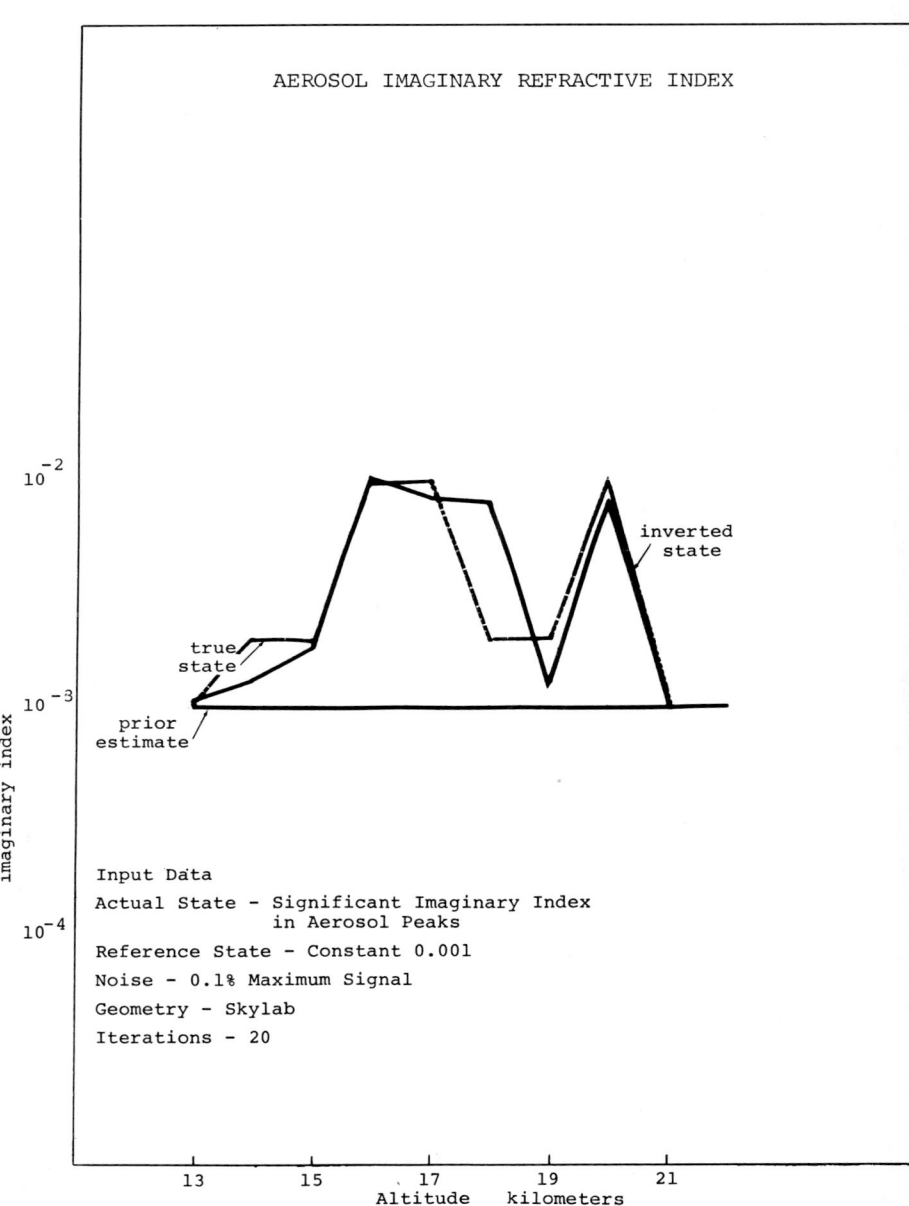

Fig. 24. Aerosol imaginery refractive index solution.

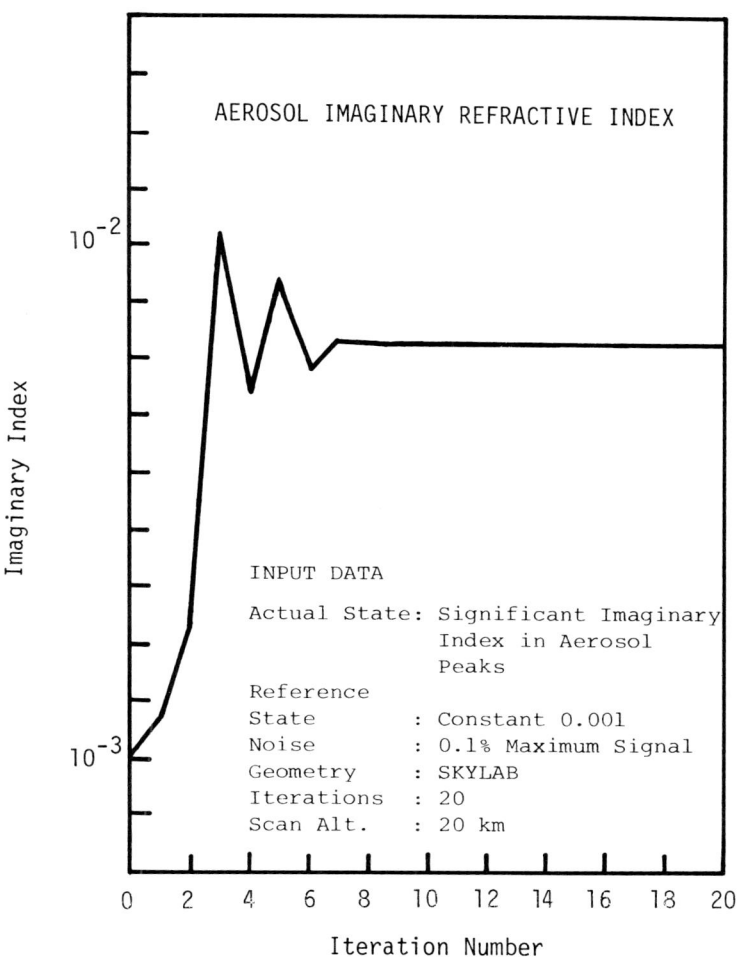

Fig. 25. Aerosol imaginery refractive index convergence.

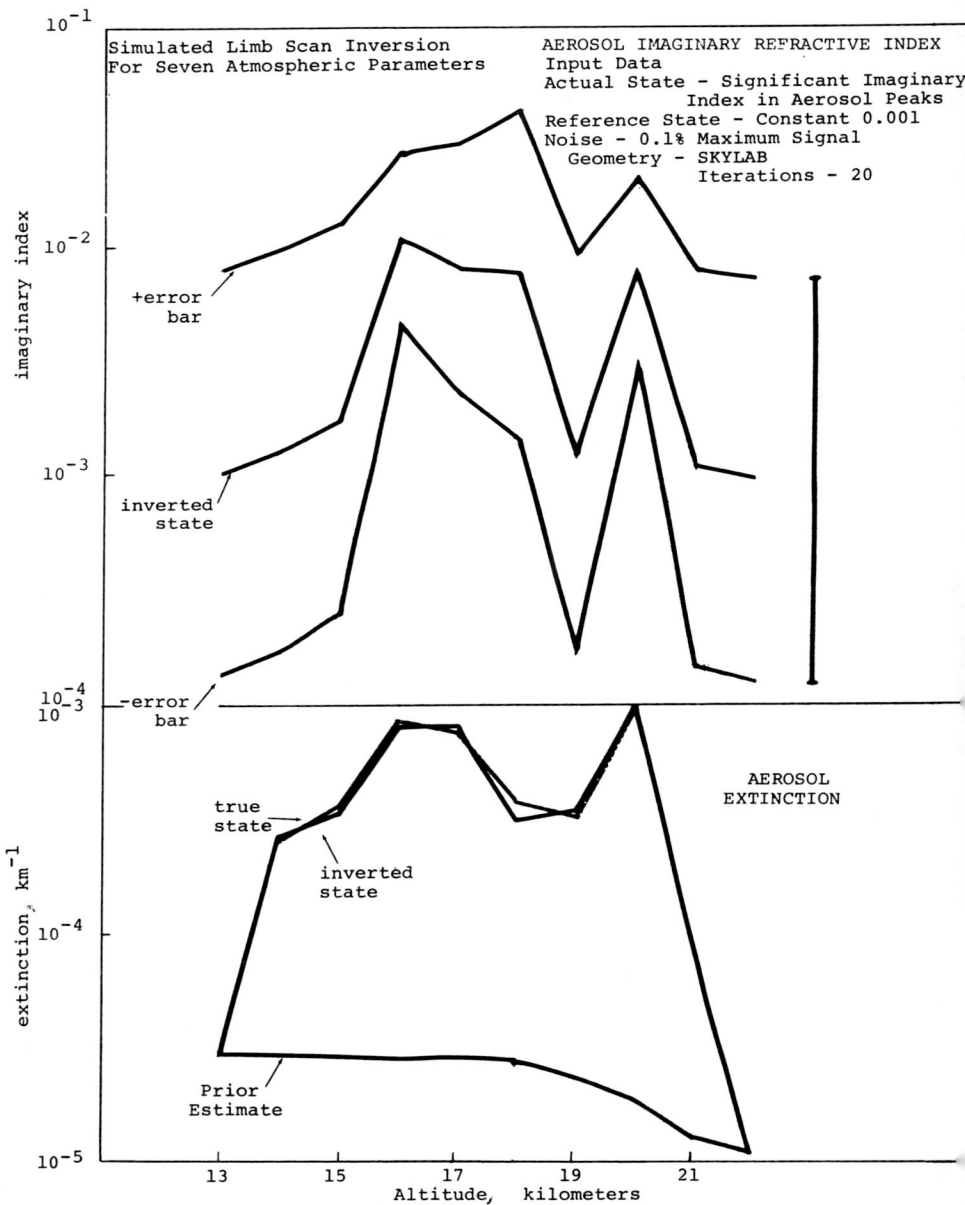

Fig. 26. Aerosol imaginary refractive index error bars.

SCATTERED RADIANCE HORIZON PROFILES

VI. CONCLUDING REMARKS

This paper has demonstrated the feasibility of inverting multispectral horizon profile radiance data at altitude intervals of 1 kilometer for number densities of Rayleigh scatterers, ozone, NO_2, and aerosols, and for aerosol physical properties including a size distribution parameter and the complex index of refraction. This is the first demonstration of a remote sensing technique that would provide all this information simultaneously on a global basis and with all the correlations.

The inversion technique itself extends standard nonlinear estimation theory in order to accommodate the limb scan problem. A crucial aspect of the extension is the use of a derivative-like construct which is more global in its meaning than the local partial derivative. Nonlinear estimation theory extended in this way apparently constitutes the first numerical search technique based on a construct other than the local partial derivative.

The multi-spectral horizon profile remote sensing technique and the extended nonlinear inversion technique together imply data requirements and, hence, specifications for a high quality radiometer. In conjunction with the analysis and simulation effort reported here, Charles Stark Draper Laboratory has designed an instrument specifically to meet the implied requirements.

SYMBOLS

a	scalar coefficient
b	scalar coefficient
c	scalar coefficient
C_q	covariance of q
D	determinant
$\underset{\sim}{H}$	partial derivative matrix
$\underset{\sim}{I}$	identity matrix

$\underset{\sim}{K}$	gain matrix
n	noise
$\underset{\sim}{P}$	state covariance
$\underset{\sim}{P}_0$	prior estimate of state covariance
q	quantity
q_a	actual value of quantity
<q>	expected value of quantity
$\underset{\sim}{R}$	noise covariance
T	transpose
v	running variable
Δv	increment of running variable
x	atmospheric state estimate
x_a	actual state
x_0	prior estimate of state
z	radiance measurement
$\lambda_1, \lambda_2, \lambda_n$	wavelength
σ	standard deviation

ACKNOWLEDGMENT

The work presented in this paper was sponsored by NASA-Langley Research Center under an Advanced Applications Flight Experiment program, NAS 1-14150 with the Charles Stark Draper Laboratory, Inc.

REFERENCES

1. R. Deutsch, "Estimation Theory." Prentice Hall, Inc., New York, 1965.

2. S. Twomey, Comparison of constrained linear inversion and an iterative nonlinear algorithm applied to the indirect estimation of particle size distribution, *J. Comput. Phys.* 18, 188 (1975).

3. C. L. Mateer, On the information content of Umkehr observations, *J. Atmos. Sci.* 22, 370 (1965).

4. E. R. Westwater and O. N. Strand, Inversion techniques in "Remote Sensing of the Environment," (V. E. Derr, Ed.), NOAA Report, 1972.

5. C. D. Rodgers, Some theoretical aspects of remote sounding in the Earth's atmosphere, *J. Quant. Spectrosc. Radiat. Transfer* *11*, 767 (1971).

6. R. E. Kalman and R. Bucy, New results in linear filtering and prediction, *J. Basic Eng. 83D*, 95 (1965).

7. D. Collins and M. Wells, Flash, a Monte Carlo procedure for use in calculating light scattering in a spherical shell atmosphere, AFCRL Report No. 70-0206, 1970.

8. C. K. Whitney, The DART Method in "Standard Procedures to Compute Atmospheric Radiative Transfer in a Scattering Atmosphere"(J. Lenoble, Ed.), I.U.G.G., 1975, pp. 80-83.

9. H. L. Malchow, Standard models of atmospheric constituents and radiative phenomena for inversion simulation, MIT Aeronomy Program Internal Report No. AER 7-1, 1971.

10. W. S. Hering and T. R. Borden, Ozonesonde observations over North America (4), AFCEL-64-30, 1967.

11. J. C. McConnell and M. B. McElroy, Odd Nitrogen in the Stratosphere, *J. Atmos. Sci. 30,* 1465 (1973).

12. L. Elterman, An atlas of aerosol attenuation and extinction profiles for the troposphere and stratosphere, AFCRL Report 66-828, 1966.

13. M. P. McCormick and W. H. Fuller, Lidar measurements of two intense stratospheric dust layers, *Appl. Opt. 14,* 4 (1975).

DISCUSSION

Twitty: First, your results are really incredible. I missed somewhere just what went into this. What is your input from which you are extracting all of these parameters? What spectral inputs and what scattering information?

Malchow: We are assuming that there is a multi-channel photometer that has eight wavelength channels of information ranging from 4000 to 7000 Å. The band width of each channel is rather narrow, something like 20 or 50 Å. It depends somewhat on the instrument design.

Chahine: What are you measuring?

Malchow: We are doing scans of the sunlit limb.

Twitty: So this is through the limb with the Sun in one specific place?

Whitney: It doesn't matter where the Sun is. Sun can be anywhere.

Twitty: For the results you show here, the Sun was way overhead?

Malchow: It was somewhat behind and off to the side.

Whitney: It happened to be what was in that Skylab experiment. We constructed the geometry accordingly.

Gille: Do you have any data from Skylab or other similar kind of experiments that you can test this on?

Malchow: We have Skylab data but it is really bad. First of all, the channels are not calibrated relative to one another because there was a rush to get the instrument onto the spacecraft. They were calibrating it at Woods Hole or somewhere when the calibration light bulb burned out. They could not get another one before the instrument had to go on the spacecraft. So they did try to calibrate it but with about 10% uncertainty in one channel or another. We cannot really work with that much noise. If the noise levels were lower, we could probably fit it to a clean atmosphere at higher altitudes, but the present noise levels are just too large.

Herman: I think what is puzzling some people here is the question of whether these were simulated measurements or not. They were simulated, am I correct?

Malchow: Yes.

SCATTERED RADIANCE HORIZON PROFILES

Herman: Now to do this, apparently you must have performed calculations using, let's say, a Junge size distribution?

Malchow: Yes.

Herman: Now suppose the actual size distribution is non-Junge. That would throw everything off, am I correct?

Malchow: Yes, indeed.

Herman: Now the other thing is you were able to extract more this way than we can from ground-based measurements and I am trying to figure out why. There must be something in your computer simulation that is enabling you to reproduce all these various variables more accurately than can be achieved in practice.

Malchow: I think one of the advantages over the ground-based data problem is that in the stratosphere we are not looking through as much atmosphere as on the way in or out. So we're getting more signal from the particular layer that we are looking at relative to the total signal that we are getting. As to the question of whether the Junge-size parameter is relevant or not, in any of these inversion processes you find what you are looking for. If you put in six parameters to describe the state, you find six parameters. All we have done is an exercise with the simplest analytic model that gave us some kind of information about the aerosol size distribution.

Unidentified Speaker: Where are the errors involved?

Whitney: There were error statistics in the measurement, which did not actually perturb the measurements. That would be the desirable thing to do, namely, to run through that exercise.[1]

Malchow: Yes, all we have shown is a closed problem in effect.

Reagan: To amplify a little bit on what Dr. Herman was talking about, I am curious about the globality or the global ability for the optimization to home in and along with that I would like to ask a question about the sequence that you went about for say determining refractive index and size. Was it sequential? Was it simultaneous?

Whitney: Yes, simultaneous.

---

[1] Post-Workshop comments added by the authors--"We were able to do that soon after the Workshop. We found that preserving the stability required cutting the gain by a factor of five."

Reagan: Also, if you changed things a bit and you still came up with the same results, that would, it seems, say something about the globality of it.

Irvine: You have assumed that you have values of the intensity measured in a two-dimensional altitude and frequency plane. How much do the parameters that you are looking for influence the intensity in different portions of that plane so that you get some separation there?

Whitney: Well, for instance, the NO_2 tends to be in a layer that is at a different altitude than the ozone, and aerosols have their own characteristics with altitudes.

Irvine: And there is different wavelength dependence too?

Whitney: Correct. The various constituents are active at different wavelengths. We have not yet conducted a channel optimization and that is an exercise that must be done. I believe that the only basis for working with those particular wavelengths shown here is that they correspond to what was in the Skylab data.

Irvine: Can I ask you something else? What percentage abundance of NO_2 can you hopefully detect? What mixing ratio?

Malchow: As you saw in our results, we are getting into difficulties with the levels that we used in this experiment. The numbers that appear on those graphs are getting down to about the lower limit of what you can deal with.

Irvine: And was that something like one part in 10^{10}?

Malchow: I don't remember what the mixing ratio was. I can show you what the actual density values were. But I would say that for nominal NO_2 it is probably shaky in the altitude region where we did the inversion.

Whitney: That's pretty far away from where it is peaked.

Irvine: Does one have any hope of detecting some of the compounds that are related to the ozone abundance--the fluorocarbons and the ClO and things like that?

Whitney: I tend to be doubtful with the wavelengths we used because there is a limit to how many independent wavelengths there really are. And you cannot look for more bits of information than are really there.

Shettle: I have a question regarding the relation of your error bars to the accuracy of your convergence. Your fitted values seem to converge within a small fraction of the error bars you show.

Whitney: That is because we did not actually perturb the measurement values. We simply processed the statistics. We did not do a Monte Carlo type of problem where you would actually put the perturbations on in accordance with those statistics.

Thomas: I feel that the quality of your results in regard to the aerosols might arise from the fact that you only had one parameter in the size distribution. We have taken a look at the possible inversion of all of the matrix elements, using a three-parameter model, and we found three orders of magnitude separation between the first and second eigenvalues. So limiting the model to one parameter might be of great value to you.

Malchow: Yes, I think it was. Two parameters would be really tough, and more parameters than that I cannot really see being handled easily.

Editorial Footnote: The following statement was submitted by the authors soon after the Workshop--"Some of the questioners of our results expressed concern that we had not produced simulations that included random noise inputs. Within the week following the meeting, we did successfully run simulations with additive random noise, and we obtained results that were quite satisfactory. To produce these results, we found it necessary to introduce a small amount of gain damping."

INVERSION OF SOLAR AUREOLE MEASUREMENTS

FOR DETERMINING AEROSOL

CHARACTERISTICS

Adarsh Deepak
Old Dominion University
and
Institute for Atmospheric Optics and
Remote Sensing (IFAORS)

Solar aureole is the region of enhanced sky brightness within about $20°$ around the Sun's disk, mainly because of the predominantly forward scattering of aerosol particles. It is shown that the solar aureole radiance is very sensitively dependent on the aerosol size distributions. The photographic solar aureole isophote (PSAI) measurement technique for determining the aerosol size distribution n(r) and other characteristics takes advantage of this sensitivity. Single scattering theory of the solar aureole is given. The assumptions and conditions imposed on the single scattering theory to make it tractable to inversion are discussed. The important role of the almucantar measurements is also discussed. Efforts that need to be performed in the near future are also stated.

I. INTRODUCTION

As man makes more and more of an impact on his environment because of the rapidly expanding technology, it becomes increasingly imperative to study the background level of these aerosols in order to monitor how man is affecting the balance in the atmosphere and what effects these aerosols will have on such things as climate, air quality, solar radiation dosage, man's health, food production, etc.

Remote sensing techniques based on extinction and scattering of electromagnetic radiation by aerosols are perhaps the most practical and economical way of diagnosing and monitoring the atmospheric aerosols on a long-term basis. The aerosol characteristics of special interest are: the size distribution n(r), the complex refractive index $\tilde{m} = m' - im''$, and the altitude distribution of these quantities. Of these, the size distribution perhaps plays the most important role in electromagnetic scattering phenomena and atmospheric processes. It is assumed for the sake of simplicity that the aerosols are spherical, r being the radius.

Nearly all the aerosol remote sensing methods, active and passive, are based on the measurement of extinction, scattered intensity, or polarization of the direct, backscattered, or multi-angle scattered radiation made through narrow bandpass spectral filters. It is by inverting the measurements obtained by one or a combination of these methods that the aerosol characteristics are usually determined. In this paper, simple and practical methods based on the measurement and inversion of the solar aureole radiance are described.

II. SOLAR AUREOLE MEASUREMENTS

The solar aureole is the area of enhanced brightness closely surrounding the Sun's disk (within about $20°$) because of mostly aerosol scattering of sunlight. Since the aerosols scatter predominantly in the forward direction, the contribution of atmospheric haze to the sky radiance for angles close to the Sun is roughly 10^2 to 10^3 times the contribution by molecular scattering. This is illustrated in Fig. 1. It is to take advantage of this large signal range that a simple, portable, photographic solar aureole measurement (PSAM) technique was developed at the University of Florida in 1970 (Ref. 1) and has since been used to diagnose the aerosol size-altitude distributions (Refs. 2 to 5) by using the aureole radiance measurements along the almucantar.

SOLAR AUREOLE MEASUREMENTS

The almucantar is a conical scan of constant solar zenith angle with the local zenith as the axis. A solar aureole measurement program was subsequently initiated at the NASA Langley Research Center Center in 1974 and the photographic solar aureole isophote (PSAI) method (Refs. 6 and 7) was developed. Isophotes are lines or curves of equal radiance. The PSAI method is an extension of the almucantar solar aureole radiance (ASAR) method. In the latter, the radiance measurements taken along the almucantar are used to infer the aerosol properties. In the PSAI method, in addition to making the almucantar measurements, one takes advantage of the fact, which emerged from our computer studies, that the shape of the solar aureole isophotes is sensitively dependent on the characteristics of the aerosol size distribution. Suggestions for the use of solar aureole measurements to determine aerosol properties had also been made earlier by Deirmendjian (Refs. 8 and 9) and other researchers (Refs. 10 to 13). Solar aureole measurements, taken with scanning photometers for determining aerosol properties

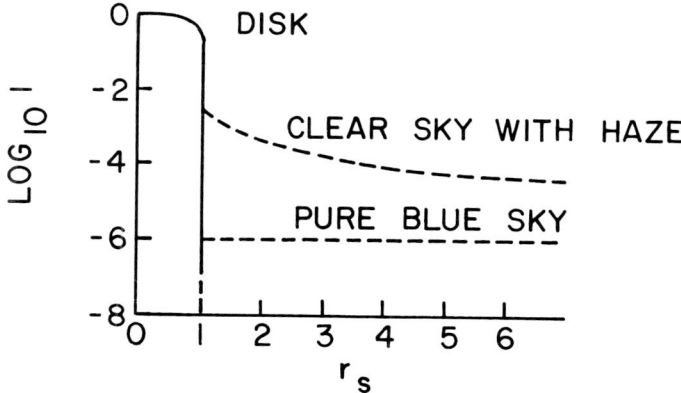

Fig. 1. Relative intensities of the sky brightness as compared to that of the Sun's direct light. r_s represents the distance from the Sun in solar radii. (From Ref. 14.)

have also been made by Shaw (Ref. 15) and Twitty, et al. (Ref. 16).

The following sections briefly describe the theory, photographic measurements, and results of the PSAI method. For details of some of the theoretical aspects discussed here, see Ref. 1.

III. SIMPLIFYING ASSUMPTIONS

The forward scattered radiance, being highly sensitive to $n(r)$, is relatively insensitive to effects due to aerosol refractive index, polarization, and multiple scattering (MS), a fact that helps in simplifying the inversion problem. In this paper, only the single scattering (SS) theory treatment is considered which should help in understanding the difficulties involved in the inversion of aerosol scattering measurements. Therefore, one makes the following reasonable simplifying assumptions:

(1) Particles are spherical so that results of the Mie theory can be used in computations.

(2) The atmosphere is horizontally homogeneous and vertically inhomogeneous.

(3) Absorption effects are ignored by selecting to work in spectral regions for which atmospheric absorption is nil.

(4) The polarization effects are small for forward scattered light and can be ignored.

(5) For relatively clear days (visibility > 15 km), the MS effects at the forward scattering angles are small compared with SS (Ref. 4) and can be ignored.

(6) An average value for the refractive index of all atmospheric aerosols is assumed for forward scattering.

(7) The atmosphere is treated as plane-parallel; the spherical Earth effects, which become significant for zenith angles ϕ larger than $75°$, are incorporated into the theory by the use of the generalized Chapman type functions $S(\phi)$ (Refs. 4 and 17) in place of the secant functions.

IV. SINGLE SCATTERING THEORY OF THE SOLAR AUREOLE

Figure 2 illustrates the geometry of the calculation. Shown is an acceptance cone $d\Omega$ originating at the detector and a solid angle $d\Omega'$ centered at the elemental scattering volume dV at altitude y(km). ϕ_s and ϕ are the zenith angles of the Sun and the narrow view cone and ω is the dihedral angle between the normals to the Sun zenith and view cone zenith planes intersecting at dV. The scattering angle ψ is then given by the relation

$$\cos \psi = \cos \phi \cos \phi_s - \sin \phi \sin \phi_s \cos \omega \tag{1}$$

The element dV is given by

$$dV = R^2 \, d\Omega \, S(\phi) \, dy \tag{2}$$

where the generalized Chapman type functions (Refs. 4 and 17)

$$S(\phi) = \sec \phi \qquad (\text{for } \phi \leq 75°) \tag{3}$$

The optical depth defined by

$$\tau_j(\lambda, y) = \int_y^\infty \beta_j'(\lambda, y) \qquad (j = M, A) \tag{4}$$

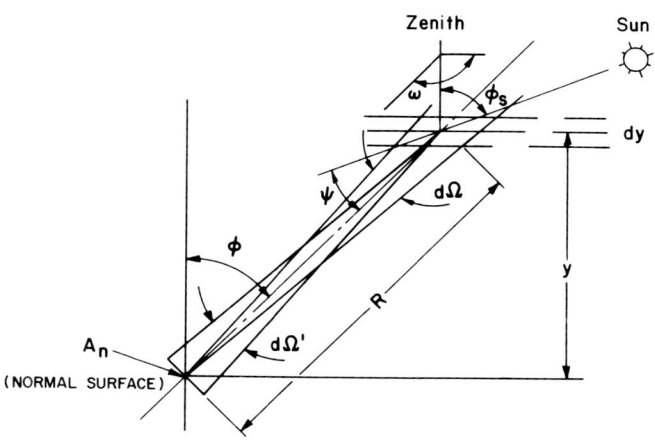

Fig. 2. The geometry of the sky single scattering problem.

for a ray traversing the distance from the Sun to the air mass element dV is given by

$$\tau_1 = \Sigma_j \tau_j(\lambda, y) S_j(\phi_s) \qquad (j = M, A) \qquad (5)$$

and from the air mass to the detector by

$$\tau_2 = \Sigma_j \{\tau_j(\lambda, 0) - \tau_j(\lambda, y)\} S_j(\phi) \qquad (6)$$

where M denotes air molecules; A, the aerosol species; λ, the wavelength; and β' the volume scattering coefficient (VSC) (km^{-1}) at altitude y for the jth constituent. In this paper, all quantities represented by τ_A, β'_A, β_A, F'_A, F_A, P'_A, and P_A are functions of $\tilde{m}$, even though their $\tilde{m}$ dependence is not indicated in their representation form. The primes denote the y-dependence of the quantities. β'_A and β'_M are defined by the following relations:

$$\beta'_A(\lambda, y) = \int_{r_2}^{r_1} \eta(r, y) \pi r^2 Q(x, \tilde{m}) dr \qquad (7)$$

where $Q(x, \tilde{m})$ is the efficiency factor (Ref. 18), $x = 2\pi r/\lambda$ is the particle size parameter, r_1 and r_2 are minimum and maximum values of r and $\tilde{m} = m' - im''$, the complex refractive index of aerosols.

$$\beta'_M(\lambda, y) = \beta_M(\lambda) \rho_M(y) \qquad (8a)$$

where the VSC for molecules is

$$\beta_M(\lambda) = \frac{8\pi^3 (n^2 - 1)^2}{N \lambda^4} \frac{(4 + 3d)}{(4 - 3d)} \qquad (8b)$$

In Eq. 8(b), N is the number of molecules cm^{-3}; n, the refractive index of the medium, $d = 4\Delta/(1 - \Delta)$; and Δ is the depolarization of scattered light at a scattering angle of 90° for a linearly polarized incident radiation with its electric vector perpendicular to the scattering plane. For unpolarized incident light, Δ is replaced by $\xi = 2\Delta/(1 + \Delta)$. Then the volume scattering function

(VSF) for air molecules ($j = M$) is

$$F'_M(\psi, \lambda, y) = \beta_M(\lambda) P_M(\psi) \rho_M(y), \quad km^{-1} sr^{-1} \tag{9}$$

where the molecular phase function is

$$P_M(\psi) = \frac{3}{16\pi} (1 + \cos^2\psi), \quad sr^{-1} \tag{10}$$

where $\rho_M(y)$ is the dimensionless function representing the altitude distribution of molecular density. The VSF for aerosols ($j = A$) is

$$F'_A(\psi, \lambda, y) = \beta'_A(\lambda, y) P'_A(\psi, \lambda, y) \tag{11}$$

where the aerosol volume phase function is

$$P'_A(\psi, \lambda, y) = \frac{1}{2k^2 \beta'_A(\lambda, y)} \int_{r_1}^{r_2} n(r, y) \{i_1(\psi, \tilde{m}, x) + i_2(\psi, \tilde{m}, x)\} dr \tag{12}$$

and i_1 and i_2 are the Mie intensity functions and $k = 2\pi/\lambda$.

The sky radiance due to the molecules and aerosols in the volume element dV is then given by

$$dB(\phi, \phi_s, \omega, \lambda) = H_0(\lambda) \{S_M(\phi) F'_M(\psi, \lambda, y) + S_A(\phi) F'_A(\psi, \lambda, y)\} e^{-(\tau_1 + \tau_2)} dy \tag{13}$$

Integrating over all such elemental volumes, the total single scattered sky radiance is

$$B(\phi, \phi_s, \omega, \lambda) = H_0(\lambda) e^{-\Sigma_j \tau_j S_j(\phi)} \left\{ S_M(\phi) \int_0^\infty e^{-\Sigma_j \tau_j(\lambda, y) D_j} F'_M(\psi, \lambda, y) dy \right. $$
$$\left. + S_A(\phi) \int_0^\infty e^{-\Sigma_j \tau_j(\lambda, y) D_j} F'_A(\psi, \lambda, y) dy \right\} \tag{14}$$

where

$$D_j = S_j(\phi_s) - S_j(\phi) \qquad (j = M, A) \qquad (15)$$

Before discussing the inversion problem, a brief description of the photographic solar aureole measurement (PSAM) technique will be in good order.

V. THE PHOTOGRAPHIC SOLAR AUREOLE MEASUREMENT TECHNIQUE

A photographic technique of making measurements of the solar aureole radiance is briefly described here. Figure 3 schematically illustrates the equipment. Photographs of the Sun's aureole are taken with a small format camera (35 mm or 70 mm) through a wavelength filter with the Sun occulted by a neutral density (ND) disc (ND = 4) held coaxially on a stem about 0.6 m to 1.3 m (2 to 4 ft.) in front of the lens. The ND filter attenuates the radiance of the direct sunlight by a factor of 10^4, so that the optical density of the Sun's image is of the same magnitude as the optical densities of the surrounding aureole, as shown in a typical photograph in Fig. 4a. The Sun's image not only enables

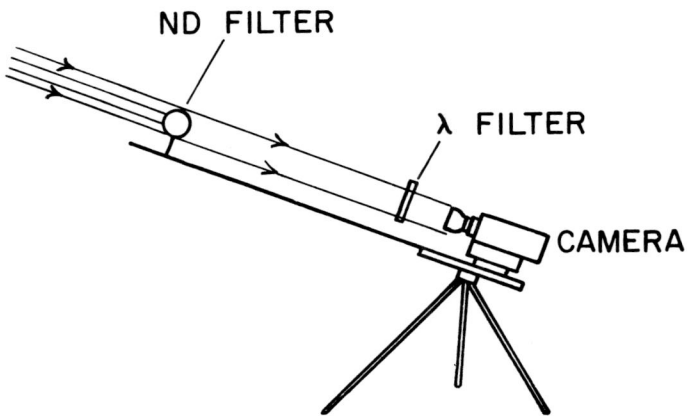

Fig. 3. Schematic illustration of the arrangement of the solar occulting disc and the camera for aureole photography.

SOLAR AUREOLE MEASUREMENTS

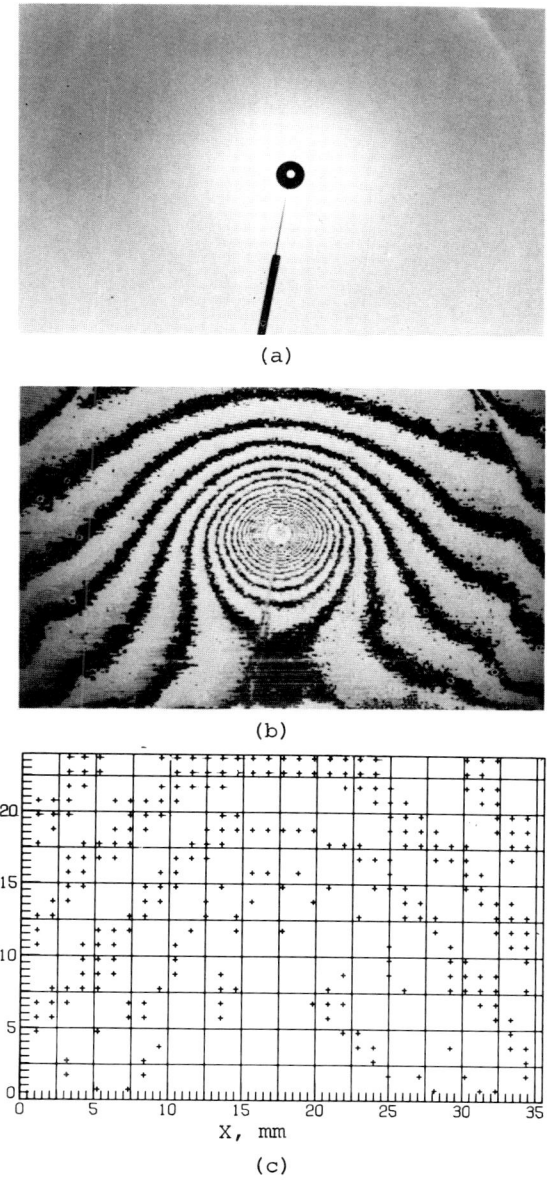

Fig. 4(a). A typical solar aureole 35 mm-photograph taken through a wavelength filter λ = 500 nm. (b). An isodensity tracing of the photograph. (c). The computed solar isophote mapping of the photograph.

one to calibrate the entire photograph relative to its radiance, but also enables accurate measurements to be made of the angular distances from the Sun. Photographs are taken through different wavelength filters. In addition, the direct solar irradiance measurements through the same filters are made with a photometer to determine $\tau(\lambda, 0)$.

A. Solar Aureole Isophotes

The optical density of the whole photograph is read with the help of a Joyce-Loebl Isodensitracer which gives digital data output on a magnetic tape and at the same time provides an isodensity tracing, such as shown in Fig. 4b. Isodensities are lines or curves of equal optical density in a photograph. Isophotes are then generated from the taped data output, as shown in Fig. 4c, where an economy-wise reduction has been made in the number of data points.

B. Almucantar Radiance

The photogrammetry of the solar aureole is presented in another paper submitted for publication, where it is shown that the almucantar projects on the film as a conic (Fig. 5). The shapes of the conics for three different values of the solar zenith angle ϕ_s are illustrated in Fig. 6. Accordingly, the measured almucantar radiance as a function of the scattering angle is shown in Fig. 7. The peak at $0°$ corresponds to Sun's direct light reduced in intensity by the ND disc.

VI. THE INVERSION OF SOLAR AUREOLE MEASUREMENTS

In order to make Eq. (14) simpler and amenable to inversion, measurement should be restricted to zenith angles ϕ_s or ϕ less than $75°$. Then the spherical Earth correction can be neglected and

SOLAR AUREOLE MEASUREMENTS

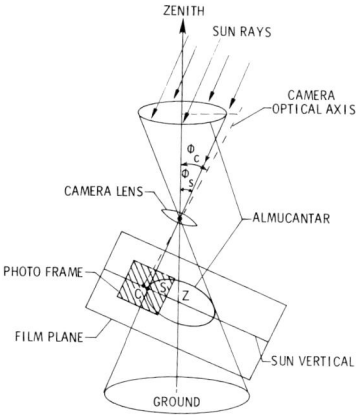

Fig. 5. Almucantar projection on the film as a conic.

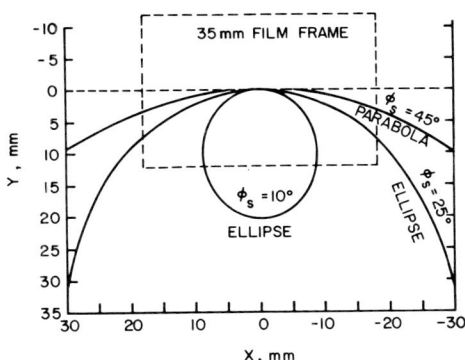

Fig. 6. The shapes of the conics for three solar zenith angles $\phi_s = 10°, 25°,$ and $45°$, for a lens of focal length 55 mm.

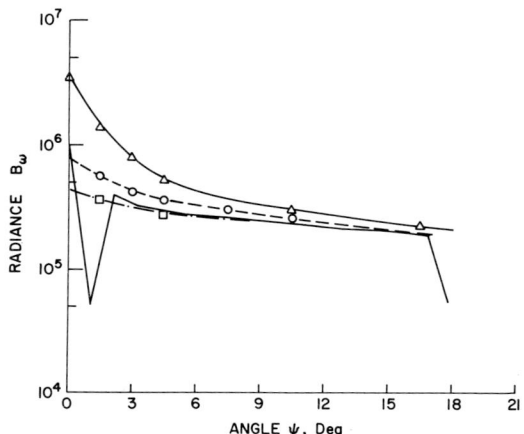

Fig. 7. Plot of the almucantar radiance as a function of the scattering angle (solid line). Symbols indicate theoretically computed radiance for different size distributions.

$$D_A = D_M = D = \sec \phi_s - \sec \phi \qquad (16)$$

Equations (14) and (15) reduce to

$$B(\phi, \phi_s, \omega, \lambda) = G \int_0^\infty \{F_M'(\psi, \lambda, y) + F_A'(\psi, \lambda, y)\} e^{-\tau(\lambda, y)D} dy \qquad (17)$$

where

$$G = H_o(\lambda) \sec \phi \, e^{-\tau(\lambda, 0)\sec \phi} \qquad (18)$$

With Eq. (17), it is possible to obtain the information about the size altitude distribution of aerosols from the multispectral measurements of (a) sky radiance $B(\lambda)$ as a function of ϕ_s, ϕ, and ω and (b) the total optical depth $\tau(\lambda, 0)$.

The altitude distribution of molecular density $\rho_M(y)$ can be obtained from tables or from radiosonde data. The functions $F_A'(\psi, \lambda, y)$ and $\tau_A(\lambda, y)$, both of which depend on the altitude size distribution $\eta(r, y)$, are the unknowns in Eq. (16). Therefore, in order to make Eq. (16) tractable to inversion, an assumption has to be invoked about the *separability* of the altitude-size distribution function $\eta(r, y)$ in the form

$$\eta(r, y) = n(r)\rho_A(y) \tag{19}$$

where the dimensionless quantity $\rho_A(y)$ is defined by

$$\rho_A(y) = N(0, y)/N(0, 0) \tag{20}$$

$N(0, y)$ being the number of particles cm^{-3} at altitude $y(km)$.

This assumption implies that the form of the particle size distribution $n(r)$ does not itself change with altitude y, but only the number density or concentration varies. This is a reasonable assumption in view of Junge's experimental observations that up to about 3 km, the aerosol size distribution $n(r)$ remains nearly constant. Since most of the aerosols are concentrated below 3 km, and according to Elterman's data, the aerosol density falls off nearly two to three orders of magnitude at an altitude of about 5 km, the error introduced due to extension of the assumption of constancy of $n(r)$ to regions above 3 km is small. Then, functions β_A', F_A', and P_A' reduce to

$$\beta_A'(\lambda, y) = \beta_A(\lambda)\rho_A(y) \tag{21}$$

$$F_A'(\psi, \lambda, y) = F_A(\psi, \lambda)\rho_A(y) \tag{22a}$$

and

$$P_A'(\psi, \lambda, y) = P(\psi, \lambda) \tag{22b}$$

where

$$\beta_A(\lambda) = \int_{r_1}^{r_2} \pi r^2 Q(x, \tilde{m}) n(r) dr \tag{23a}$$

and

$$F_A(\psi, \lambda) = \frac{1}{2k^2} \int_{r_1}^{r_2} (i_1 + i_2) n(r) dr \tag{23b}$$

Thus,

$$P_A(\psi, \lambda) = F_A(\psi, \lambda)/\beta_A(\lambda) \tag{23c}$$

In case two aerosol layers having different size distributions are present in the atmosphere, it is then convenient to choose a two-term model for $\eta(r, y)$, such as

$$\eta(r, y) = n_1(r)\rho_{1A}(y) + n_2(r)\rho_{2A}(y) \tag{24}$$

which is separable in r and y in each of its terms. This case is not discussed here, but will be treated in a subsequent paper.

In Eq. (17), both the molecular and aerosol scattering contributions depend on the unknown quantity $\tau_A(y)$. Let us define the "effective transmission" functions for air molecules and aerosols as

$$T_M(D) = \frac{1}{w_M(0)} \int_0^\infty dy\, \rho_M(y)\, e^{-\Sigma_j \tau_j(\lambda,\, y)D} \tag{25}$$

and

$$T_A(D) = \frac{1}{w_A(0)} \int_0^\infty dy\, \rho_A(y)\, e^{-\Sigma_j \tau_j(\lambda,\, y)D} \tag{26}$$

where the integrated thickness is given by

$$w_j(y) = \int_0^\infty \rho_j(y')dy' \qquad (j = M, A) \tag{27}$$

By using Eqs. (25) and (26), Eq. (17) can be written as

$$B(\phi, \phi_s, \omega, \lambda) = G\{P_M(\phi, \lambda)\tau_M(\lambda, 0)T_M(D) + P_A(\psi, \lambda)\tau_A(\lambda, 0)T_A(D)\}$$

The y-dependence on the right-hand side of Eq. (28) is now confined to the two factors, T_M and T_A, representing effective transmission

SOLAR AUREOLE MEASUREMENTS

factors. It can easily be shown that T_M is related to T_A as follows:

$$T_M(D) = \frac{1}{\tau_M(\lambda, 0)} \left\{ e^{-\tau_M(\lambda, y)D} \left[\frac{1 - e^{-\tau_A(\lambda, 0)D}}{D} \right] e^{-\tau_A(\lambda, y)D} \right.$$

$$\left. \left[\frac{1 - e^{-\tau_M(\lambda, 0)D}}{D} \right] \right\} - \frac{\tau_A(\lambda, 0)}{\tau_M(\lambda, 0)} T_A(D) \tag{29}$$

Granting we know $H_o(\lambda)$, $\tau(\lambda, 0)$ can be determined by direct radiance B_s of the half-degree Sun cone Ω_s given by

$$B_s = \frac{H_o(\lambda) e^{-\tau(\lambda, 0)\sec\phi_s}}{\Omega_s} = 1.67 \times 10^4 \, H_o(\lambda) e^{-\tau_o(\lambda, 0)\sec\phi_s} \tag{30}$$

$\tau_M(\lambda, 0)$ can be computed from radiosonde data for the observation site so that

$$\tau_A(\lambda, 0) = \tau(\lambda, 0) - \tau_M(\lambda, 0) \tag{31}$$

Many techniques for obtaining the aerosol characteristics suggest themselves in the light of this analysis. One of the simplest and the most elegant is the method based on the almucantar radiance measurements, as explained in the following section.

VII. ALMUCANTAR AS AN INDICATOR OF n(r)

The difficulty of obtaining information about aerosol characteristics from Eq. (28) lies in that the dependence of B on n(r), $\tilde{m}$ and $T_A(y)$ is not separable. One simple method of handling this problem is to make radiance measurements in the almucantar ($\phi = \phi_s$).

For $\phi = \phi_s$, $T_A = T_M = 1$ so that Eq. (28) reduces to

$$B(\phi = \phi_s, \omega, \lambda) = G\{P_M(\psi, \lambda)\tau_M(\lambda, 0) + P_A(\psi, \lambda)\tau_A(\lambda, 0)\} \quad (32)$$

which is independent of y. $P_A(\psi, \lambda)$ can then be obtained from almucantar radiance measurements by rewriting Eq. (32) as

$$P_A(\psi, \lambda) = \frac{1}{\tau_A(\lambda, 0)} \left\{ \frac{B(\phi = \phi_s, \omega, \lambda)}{G} - P_M(\psi, \lambda)\tau_M(\lambda, 0) \right\} \quad (33)$$

Figure 8 shows a plot of the experimental phase function curve normalized to unity at $3°$.

In the following is described an algorithm for obtaining $n(r)$ from $P_A(\psi, \lambda)$. If we assume for $\tilde{m}$ an average value of say 1.55 + i(0.0), it should then be possible to obtain $n(r)$ from Eq. (33) by either a numerical inversion scheme or a model-fitting approach. In either case, theoretical values of $P_A(\psi, \lambda)$ obtained from Eq. (23c) are compared with experimental values of $P_A(\psi, \lambda)$

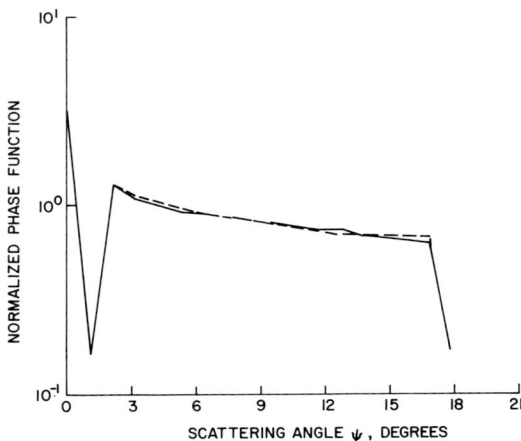

Fig. 8. Experimental phase function curve obtained from almucantar radiance curve shown in Fig. 7. Dashed curve corresponds to right-hand side of the Sun; solid curve to the left-hand side.

SOLAR AUREOLE MEASUREMENTS

obtained from Eq. (33). In the latter approach, a catalog of phase functions is generated by use of an analytic model for $n(r)$, such as

$$n(r) = \frac{p_3}{Q} \frac{e^{(p_1/r)^{\nu-3}}}{\left[p_2(e^{(p_1/r)^{\nu-3}} - 1) + 1\right]^2} \frac{1}{r^\nu} \qquad (34)$$

where p_1, p_2, p_3, and ν are adjustable constants and the efficiency factor, $Q = 2$, for large r. Typical size distribution curves for different values of the parameter ν are shown in Fig. 9. Any other realistic analytic model of $n(r)$ will do just as well.

A set of phase function curves corresponding to values of ν that range from 4.0 to 5.2 are shown in Fig. 10. By comparing the experimental phase function curve in Fig. 8 with one of the curves in the catalog, one obtains a reasonably good estimate of the $n(r)$. If, however, one wants to go a step further and obtain the $n(r)$ that gives the best fit to the experimental phase function curve, a least squares computer code is used. But the cost of such a computation is often prohibitive, especially when Mie theory is used, encouraging one to stop short and settle for the $n(r)$ obtained with a visual fit to one of the computed curves in the catalog.

VIII. SOLAR AUREOLE ISOPHOTES AS INDICATORS OF $n(r)$ and $\rho_A(y)$

An extensive parametric computation of the solar aureole isophotes as functions of $n(r)$ and $\rho_A(y)$ and to some extent, of $\tilde{m}$, has been carried out by using Eqs. (28) and (34) and the aerosol number density profile $N(0, y)$, such as the one obtained from lidar measurements and shown in Fig. 11. Everything else being the same, circumsolar isophotes corresponding to three values of the parameter ν of the size distribution $n(r)$ (Fig. 9) are shown in

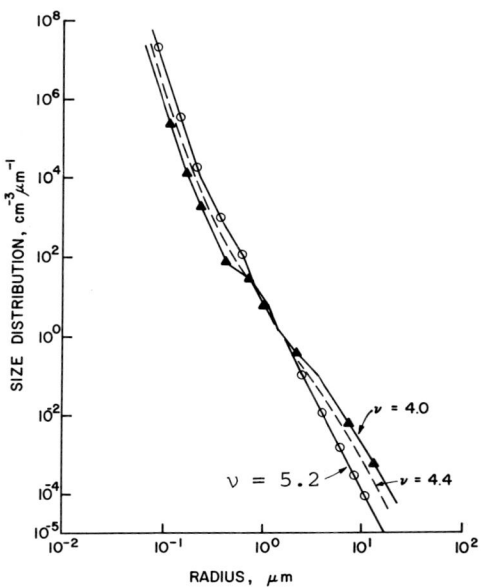

Fig. 9. Typical size distribution curves for parameter $\nu = 4.0, 4.4,$ and 5.2.

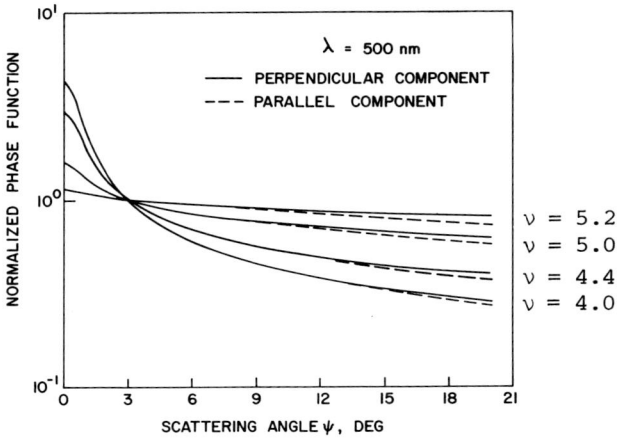

Fig. 10. A set of phase function curves corresponding to the size distribution curves shown in Fig. 9 for ν values in the range 4.0 to 5.2.

SOLAR AUREOLE MEASUREMENTS

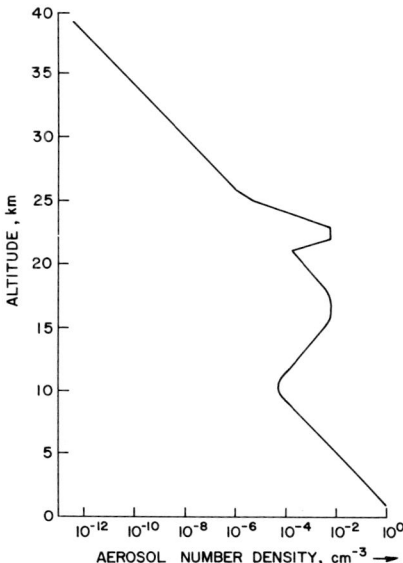

Fig. 11. Altitude distribution of particle number density (cm^{-3}) obtained from lidar observations.

Figs. 12a, 12b, and 12c. It is easy to see from Figs. 9 and 12 that by slightly increasing the number of smaller particles and decreasing the number of larger particles as ν becomes larger, the shape of the isophotes undergoes a dramatic change. As the value of ν increases upwards from a value of 4.0, the isophote pattern attains a shape increasingly similar to that of the experimentally obtained isophotes (Fig. 4c), until for a value of $\nu = 5.0$ the computer generated pattern (Fig. 12c) best resembles the latter. Thus, Fig. 12 illustrates the fine sensitivity of the patterns of the circumsolar isophotes to the size distribution $n(r)$. In contrast, in the case of the lidar backscattering ratio profile, defined by

$$R_A(\pi, \lambda, y) = 1 + \frac{F'_A(\pi, \lambda, y)}{F'_M(\pi, \lambda, y)} \tag{35}$$

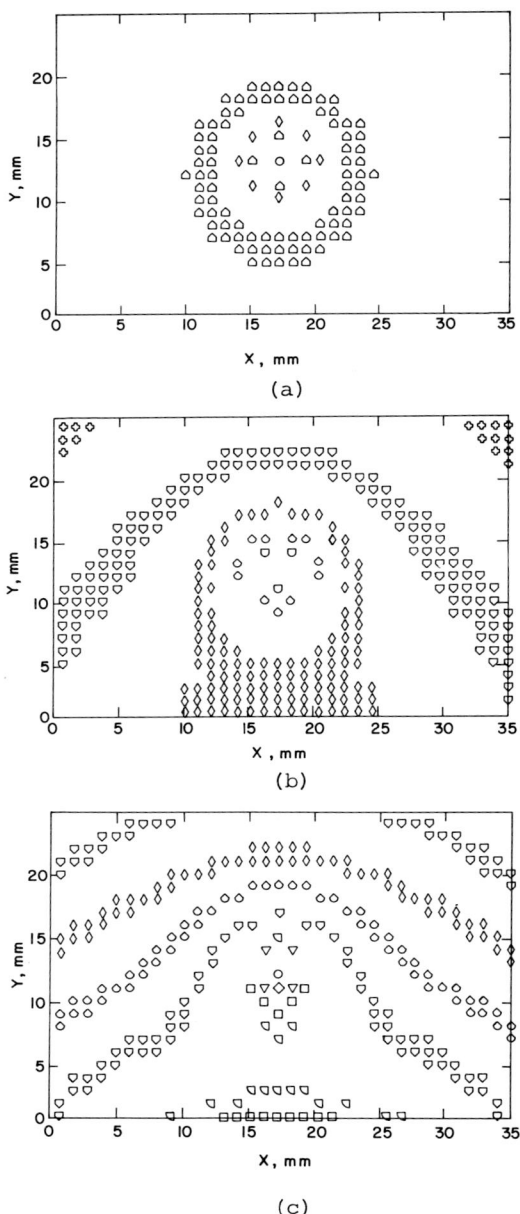

Fig. 12. Computer generated SS circumsolar isophotes for different values of the size distribution parameter ν: (a) $\nu = 4.0$, (b) $\nu = 4.4$, and (c) $\nu = 5.0$.

SOLAR AUREOLE MEASUREMENTS

The changes in R_A corresponding to the changes in the value of ν from 4.0 to 5.2 occur in the peak values of R_A at 16.5 km and 22.5 km (Fig. 13) which is a feature not too sensitive to $n(r)$ since these differences can be easily removed by adjusting the scaling parameter in $n(r)$.

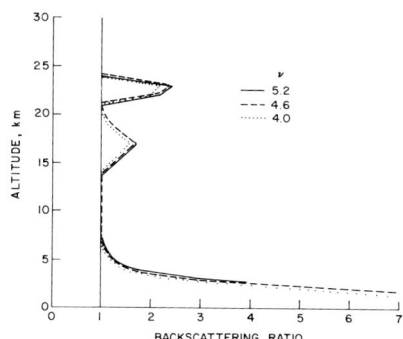

Fig. 13. Lidar profiles of the backscattering ratio for $\nu = 4.0$, $\nu = 4.6$, and $\nu = 5.2$ showing two layers of volcanic dust with peak concentrations at altitudes of 16.5 km and 22.5 km over Hampton, Virginia.

Attempts are underway to optimize computer programs that will obtain the best fit to the isophote pattern or numerically invert the isophote data by utilizing, to their advantage, the fact that radiance along each isophote remains constant. In this regard, it is important to keep in mind that even though isophotes depend on both $n(r)$ and $\rho_A(y)$, within the aureole region the isophote shape is more sensitively dependent on $n(r)$ than on $\rho_A(y)$. The sensitivity of the isophotes to $\rho_A(y)$ increases for larger angular distances from the Sun.

IX. NUMERICAL INVERSION METHODS

Whereas in the modeling approach one starts with a model judicially guessed at by experience, in the numerical inversion approach one starts with an initial estimate of parameters. A typical iterative scheme, such as the one described by Malchow and Whitney (Ref. 19), is illustrated in Fig. 14.

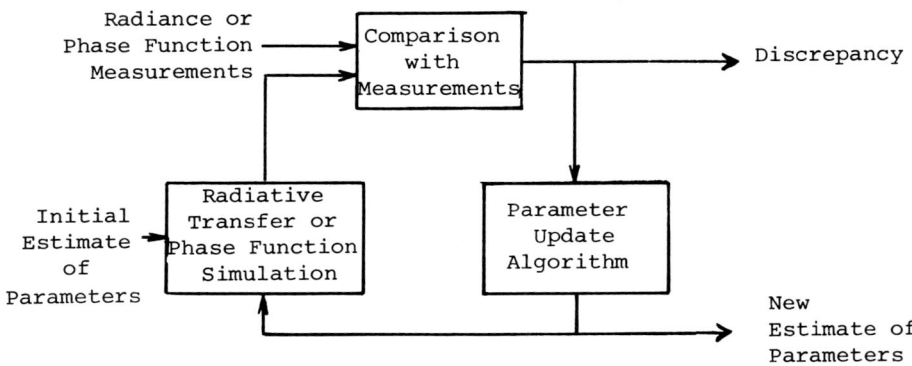

Fig. 14. Schematic representation of iterative inversion of measurements.

Simulations of the sky radiance isophote shape or the phase function are performed by using the radiative transfer equation (such as Eq. (28) for SS) or Eq. (23c), respectively, along with the initial set of parameters. Comparisons are made with the appropriate measurements. If the discrepancy is greater than a minimum prescribed value, then a parameter updating algorithm is used to obtain a new estimate of parameters. However, when the convergence criteria is satisfied, the final estimates of parameter are assumed to be accurate.

In such an iterative scheme, the simulation relation (e.g., the radiative transfer Eq. (28)) is called several times during

SOLAR AUREOLE MEASUREMENTS

each iteration. And here lies the problem, namely, the prohibitive costs of computation. One way to handle the problem is to use simple approximations for P_A and β_A in terms of the parameters in $n(r)$, $\tilde{m}$, and $N(0, y)$. Exact Mie theory computations of scattered radiance when used in an inversion scheme were found to be prohibitively expensive. Thus, it is *imperative* that in developing a radiative transfer code with the aim of using it in an inversion scheme, it must be as *computationally fast* as possible in order to be of practical use in numerical retrievals of aerosol characteristics. Work is in progress on the development and optimization of such inversion schemes and radiative transfer programs based on the exact Mie theory results.

X. CONCLUDING REMARKS

Under the present state of the art in inverting sky scattered radiance, it seems that for the present the modeling approach applied to the solar aureole measurements gives reasonably good estimates about the aerosol characteristics, particularly about the size distribution $n(r)$. In addition, such a parametric modeling approach provides not only a clear physical insight into the problem, but also understanding of the sensitivity of the individual aerosol parameters that are sought. On the other hand, catalog or modeling methods can become very unwieldy when more than one aerosol parameter is sought. It is precisely in such cases that when several parameters are simultaneously sought, the numerical inversion schemes have an added advantage over the modeling approach.

In spite of the tremendous upsurge in research activity in this field in recent years, much of our present knowledge of aerosol characteristics is either on a localized scale or is spatially and/or temporally averaged. To acquire a large scale or global view of aerosol characteristics, their monitoring must be carried from aboard space platforms. Greater endeavor needs to be made in the area of aerosol remote sensing techniques and, particularly, on the inversion aspects of the problem.

SYMBOLS

B	sky radiance
B_s	direct solar radiance
B_ω	almucantar radiance
d	$= 4\Delta/(1-\Delta)$, in Eq. (8b)
D	$= \sec \phi_s - \sec \phi$
D_j	$= S_j(\phi_s) - S_j(\phi)$, in Eq. (15)
F_j, F'_j	volume scattering functions. Prime denotes y-dependence
G	factor defined in Eq. (18)
$H_o(\lambda)$	unattenuated solar irradiance
i_1, i_2	Mie intensity functions
I	intensity of solar disk normalized to unity at its center
j	$= M$ for molecules and $= A$ for aerosols
k	$= 2\pi/\lambda$
$\tilde{m}$	complex refractive index of aerosols; $\tilde{m} = m' - im''$
m'	real part of $\tilde{m}$
m''	imaginary part of $\tilde{m}$
$n(r)$	differential size distribution, $cm^{-3} \mu m^{-1}$
n	molecular refractive index
$N(0, y)$	number density of aerosols at $y(km)$, cm^{-3}
N	number density of molecules, cm^{-3}
p_1, p_2, p_3	adjustable constants in Eq. (34)
P_j, P'_j	phase functions. Prime denotes y-dependence
Q	efficiency factor
r	radius, μm
r_s	solar radius
R	distance of scattering volume to the detector

SOLAR AUREOLE MEASUREMENTS

R_A	lidar backscattering ratio
T_j	"effective transmission" function, defined in Eqs. (25) and (26)
V	scattering volume of atmosphere
w_j	integrated thickness, defined in Eq. (27)
x	distance along x-direction on photograph
y	altitude, km
Y	distance along y-direction on photograph
β_j, β_j'	volume extinction coefficients for the jth constituent. Prime denotes y-dependence
η	two-dimensional function of r and y
λ	wavelength
ρ_j	dimensionless factor representing the altitude dependence of jth constituent
τ	total optical depth for all constituents
τ_j	optical depth for the jth constituent
τ_1, τ_2	optical depths from top of atmosphere to V and V to detector, respectively
ν	adjustable constant in Eq. (34)
ϕ, ϕ_s	zenith angles of the Sun and detector view cone, respectively
ψ	scattering angle
ω	dihedral angle between the normals to the Sun zenith and view cone zenith planes intersecting at the scattering volume
Ω, Ω'	solid angles of cones centered at the detector and the scattering volume
Ω_s	half-degree Sun cone

ACKNOWLEDGMENTS

This work was supported by NASA grant NSG-1252. The author wishes to gratefully acknowledge the excellent support in the experimental aspects of this research effort by R. R. Adams, NASA Langley Research Center.

REFERENCES

1. A. E. S. Green, A. Deepak, and B. J. Lipofsky, Interpretation of the sun's aureole based on atmospheric aerosol models, *Appl. Opt. 10,* 1263 (1971).

2. A. E. S. Green, et al., Light scattering and the size-altitude distribution of atmospheric aerosols, *J. Colloid Interface Sci. 39,* 520 (1972).

3. A. E. S. Green, et al., Remote sensing of atmospheric aerosols, *Proc. of the VIIth International Symposium on Remote Sensing of Environment, Ann Arbor, Michigan,* Vol. III, p. 1749, 1971.

4. A. Deepak, Double Scattering Corrections to the Theory of the Sun's Aureole, NASA TM X-64842 (1973).

5. R. D. McPeters and A. E. S. Green, Photographic aureole measurements and the validity of aerosol single scattering, *Appl. Opt. 15,* 2457 (1976).

6. A. Deepak and R. R. Adams, Equidensity and Isophote Contour Mapping of Solar Aureole for Determining Physical and Spatial Characteristics of Atmospheric Aerosols, Technical Digest, *Remote Sensing of the Atmosphere, Optical Society of America Spring Meeting, Anaheim,* March 17-20, 1975. (Paper No. WC-2.)

7. A. Deepak, A Photographic Solar Aureole Isophote Technique for Measurement of Atmospheric Aerosol Properties, *Proc. of the International Radiation Conference, Garmisch-Partenkirchen, West Germany,* August 18-28, 1976, Science Press, Princeton, New Jersey, 1977.

8. D. Deirmendjian, Theory of the Solar aureole, II, *Ann. Geophys. 15,* 218 (1957).

9. D. Deirmendjian, Use of Scattering Techniques in Cloud Microphysics Research: 1. The Aureole Method, Rand Report No. R-590-PR, the RAND Corporation, Santa Monica, California 90406 1970.

10. K. Murai, Spectral measurements of direct solar radiation and of the Sun's aureole (II), *Meteor. and Geophys. 19,* 447 (1968).

11. K. Sekihara and K. Murai, On the measurement of atmospheric extinction of solar radiation and the Sun's aureole, *Meteor. and Geophys. 19,* 57 (1961).

12. B. Rydgren, A photometric study of solar aureole under various weather conditions, *Tellus. 20,* 55 (1968).

13. R. Eiden, Calculations and measurements of the spectral radiance of the solar aureole, *Tellus. 20,* 380 (1968).

14. H. C. van de Hulst, in "The Sun" (G. P. Kuiper, Ed.), p. 257. U. of Chicago Press, Chicago, 1963.

15. G. E. Shaw, Radiative and Physical Properties of Atmosphere Aerosols at the South Pole, *Abstracts of papers presented at the Interdisciplinary Symposia, International Union of Geodesy and Geophysics, XVI General Assembly (Grenoble, France),* 25 August - 6 September 1975, p. 194.

16. J. T. Twitty, J. J. Parent, J. A. Weinman, and E. L. Eloranta, Aerosol size distributions: Remote determination from airborne measurements of the solar aureole, *Appl. Opt. 15,* 980 (1976).

17. A. E. S. Green and J. D. Martin, in "The Middle Ultraviolet: Its Science and Technology" (A. E. S. Green, Ed.), p. 140. John Wiley & Sons, Inc., New York, 1966.

18. H. C. van de Hulst, "Light Scattering by Small Particles," John Wiley and Sons, Inc., New York, 1962.

19. Harvey L. Malchow and Cynthia K. Whitney, Inversion of Scattered Radiance Horizon Profiles for Gaseous Concentrations and Aerosol Parameters. Inversion Methods in Atmospheric Remote Sounding, NASA CP-004, 1977 (Ed. Adarsh Deepak).

DISCUSSION

Fymat: You have mentioned that you are using the solar aureole, but it seems to me that you are also using lidar and multi-wavelength extinction. I am, therefore, a little bit lost on what measurements you are employing and what parameters you are attempting to recover.

Deepak: Normally, you use a radiometer to perform an extinction measurement of the direct solar radiation. What I am suggesting is that the extinction data are not sufficient to obtain the size distribution accurately. But, if in addition one has measurements of the angular distribution of scattered radiance, one gets a much better idea of the size distribution. That is one point. Monostatic lidar measurements, on the other hand, give you a good vertical resolution of aerosol backscattering coefficient from which information about the altitude distribution of aerosol concentration can be obtained. What I have shown is that whereas the lidar backscattering ratio data one can fit with any of the size distributions[1] obtained with ν = 4.0 to 5.2, the solar aureole isophote data one could fit with only one of those size distributions. Only one of them reproduced all these isophote curves including the bumps seen in the curves.

Fymat: I have two very brief questions. Are you doing all the work assuming the Junge distribution? And, are you fitting for one Junge parameter?

Deepak: Let me show the size distribution function that I used, but did not have the time to show earlier.[1] It is a function with four constants and those curves of size distribution and isophotes that you saw were obtained from this function by varying the parameter ν. For large particles, the size distribution behaves as a Junge distribution. We have one scaling parameter and three adjustable constants here. The aim is to obtain the set of parameters which gives the best fit to the experimental values of the multispectral phase function, the isophotes, the extinction coefficient, as well as the lidar backscattering ratio. So one tries to best fit as many different measurements obtained by different methods by adjusting the parameters in order to get at the so-called "unique" size distribution.

(In response to Dr. Deirmendjian's show of hand): Well, I must also mention that Professor Deirmendjian suggested a similar approach to aureole measurements a long time ago and tried to show the same thing.

---

[1] See Eq. (34) in text.

Deirmendjian: Well, I have no comment then! In 1956, that is what I wanted to show and I did take Junge-type distribution with different slopes and I demonstrated that the aureole would change.

Irvine: You assumed an index of refraction?

Deepak: Yes, I did.

Irvine: In some problems that could make a big difference, but I guess here, since you are mostly concerned with the diffraction peak, it may not be so sensitive.

Deepak: Yes, that is indeed another advantage of the forward scattering, that it is not as sensitive to that effect. However, it does have *some* sensitivity. I have done a parametric study with about 15 different indices of refraction--varying the values of both the real part as well as the imaginary part. Only one index of refraction, with a value of 1.55-i(0), gave a good fit to the data. When I used a small value for the imaginary part, the curves tended to flatten out. So, there is that small ballpark of error in the index of refraction value, depending on the coarseness of grid of values chosen.

Irvine: It might have been useful to have some chemists at the conference who consider these sorts of inversion problems using laboratory data; that is, deducing hydrosol properties from scattered intensities and polarizations. You also assume single scattering?

Deepak: As a first approximation, yes. That is why this method is valid for relatively clear days. But I think perhaps Jerry Twitty and definitely Dr. Green have done some work with multiple scattering in the solar aureole. Dr. Green might elucidate further upon this point in his talk later.

Irvine: It would be relatively easy to take these models you deduce and see how much second-order scattering there is to see how good an approximation single scattering is?

Deepak: Yes, I did some calculations on that. Second order contribution was not that much. It was within 5 to 6% of the single scattering contribution.

Twitty: With regard to multiple scattering, that is obviously a function of how much aerosol is actually in the atmosphere. So to estimate it you have got to say something about your optical depth. What Adarsh has done is essentially independent of that because he has assumed single scattering. So up to whatever point multiple scattering becomes important, his results hold.

Deepak: Up to about optical depths of 0.5 or so, the method is good.

Twitty: Now in the work I did for my doctoral thesis, radiance measurements along the solar almucantar in the same geometry that Adarsh showed, I did not actually include any multiple scattering but it can be shown to what extent that was important for the specific data that I used in the analysis, which had quite low optical depths, about 0.1 for the aerosol part.

Herman: 0.01 or 0.1?

Twitty: 0.1.

Herman: What did you find for results on multiple scattering?

Twitty: I used our data from about $3°$ to about $25°$, which is very similar to what Adarsh shows, and the multiple scattering is about 5% of the radiance at $25°$. So, I think it is similar to other people's results, including yours.

P. Russell: I wonder about the sensitivity of this method to the stratospheric aerosol. In particular, what happens to your simulated isophotes if you just eliminate the stratospheric aerosol from the simulation?

Deepak: I have done a number of numerical parametric studies on that, but did not have time to show the results. We added aerosol layers, first two and then three layers, and moved them up and down in altitude. The shift in the isophote pattern was seen at about $15°$ away from the Sun. In the near forward direction, no noticeable shift in the shape of the isophotes was detected. But for larger angles there was definitely a shift. One could show that introduction of upper aerosol layers does affect the shape of isophotes at larger angular distances from the Sun. The isophote curves changed a little bit at angles beyond $10°$ or so, even though the aerosol layer introduced into the model was optically very thin.

P. Russell: Does that mean then that the size distribution parameter that you extract from this method is sort of a mean parameter for the tropospheric and stratospheric aerosols?

Deepak: Exactly, all results are averaged over the entire atmosphere--for example, we obtain average refractive index and average size distribution--because we are using the assumption of separability of the altitude size distribution here.

P. Russell: Would the fact that the stratospheric aerosols seem to have their effects in a different angular region help you to separate the stratospheric and tropospheric?

Deepak: Well, one of the ways you can do this is to go up in a plane and take aureole photographs, but it is very expensive. And that is one of the reasons for trying to improve the techniques of the inversion of this ground-based data so that stratospheric aerosol effects can be separated. For this purpose, work is in progress on optimizing a few different inversion schemes. I hope to use some of the techniques that have been developed by Harvey Malchow and Cindy Whitney. The results are not quite complete yet.

P. Russell: It looks very impressive, very low cost.

van de Hulst: On the matter of the influence of multiple scattering, it is possible to make a very easy rough estimate of when it becomes important. If I stand looking at the Sun, there is a drop of a factor of 1000 from the solar disk to where the aureole starts. If I integrate over a solid angle of say 10 times the radius of the Sun, which is a solid angle of 100 times the Sun, then the integrated radiation coming from the aureole is still a factor 10 below that of the Sun. That means that multiple scattering must be somewhere in the order of 10%. And this holds for me standing on the ground. Another factor of two is lost for the average aerosol particle in the layer somewhere up in the atmosphere. By this estimate in the normal clear air situation, the multiple scattering is rarely important beyond some 5%.

Reagan: It was not clear whether you were trying to rationalize and obtain agreement between the aureole data and the extinction and the lidar backscatter measurements. I might add, amplifying on the index of refraction aspect, that the relationship between the extinction coefficient and the backscatter coefficient does indeed change rather markedly as you change either the real or imaginary component. If you are trying simultaneously to make those agree with the aureole data, you would indeed have had some sensitivity to index. Did you go back and check on this and iterate in that sense?

Deepak: The same computer code calculated all the things I have shown. I ran the programs for different m' and m" values. The shapes of the lidar backscattering ratio curves came out to be nearly identical except for the differences in magnitude at peaks of aerosol concentration, which differences could be removed by adjusting the scaling parameter in $n(r)$. From this, I could not clearly distinguish between the various size distributions for the same refractive index. However, I feel that further work needs to be done on this aspect.

ANALYTIC MODEL APPROACH TO THE INVERSION

OF SCATTERING DATA

Alex E. S. Green and Kenneth F. Klenk
University of Florida

We apply an analytic model approach which has been developed in nuclear studies to several simple atmospheric inversion problems. We illustrate by past work on the solar aureole that this method gives a sharp determination of aerosol size distribution parameters. We show that this analytic approach, together with ground level point sampling data measurements, may be used to infer information on the tropospheric ozone profile.

I. ATMOSPHERIC AND NUCLEAR OPTICS

Many of us who are now involved in atmospheric inversion problems were previously involved in analogous problems in other disciplines. As is natural, we try to bring to bear the experience, sense of aesthetics, or prejudices, if you will, which we have acquired in these other fields. The beauty of this conference as it is developing following some of the earlier papers is the sense of open-mindedness which has emerged. It is as if this conference has said, "Let a thousand flowers blossom."

In my own (Green) case, my main prior involvement with inversion problems has been in connection with two nuclear physics endeavors based largely upon scattering data--(1) inferring the nature of the fundamental interaction between neutrons and protons, and (2) inferring the detailed nature of the nuclear potential manifest in the shell and optical models of the nucleus. Let me use the first problem as an illustration of how understanding is advanced in scattering inversion problems.

When the neutron was discovered in 1932, the fundamental problem in nuclear physics became that of inferring the basic force between neutrons and protons. The approach followed has been to perform scattering experiments, i.e., fire neutrons or protons on hydrogen targets, and examine the emerging angular distributions (phase functions) and polarizations at various energies (wavelengths) of the outgoing particles. The hope was to be able to test various proposed two-body potentials which when inserted into the Schrödinger equation or Dirac equation might account for these data within statistical error. This was the main line of approach in nuclear physics until the early 1960s when the only phenomenological models which could fit the 0 to 400 MeV array of scattering data and auxiliary data such as the properties of the bound two-body system (the deuteron) were exceedingly complex, requiring as many as 40 adjustable parameters in their description.

A breakthrough came in the mid-1960s when the discovery of the ω, ρ and η mesons by particle physicists led to the revival of meson theory of nuclear forces initiated by Yukawa in 1935. With the additional physical constraints of meson theory, it suddenly became possible to fit the scattering data with one boson exchange model requiring only five to ten adjustable parameters, rather than the 40 parameters of purely phenomenological models. Although the final story is not yet told, the nuclear physics community, since 1967 (Refs. 1 and 2), has felt a great aesthetic sense of relief that the fundamental law of nuclear physics is not as monstrous as it had appeared to be in the early sixties.

Thus, as some of the earlier speakers have already suggested, it is the additional physics, physical judgment, and physical information which one brings to bear with the scattering data which will often determine the success and utility of an inversion scheme.

THE INVERSION OF SCATTERING DATA

Studies leading to the nuclear shell and optical models (Ref. 3) are even more analogous to the atmospheric inversion problem. For over 40 years, experiments have been performed in which various nuclear particles accelerated from 1 MeV to 100 GeV energy range are scattered from various nuclear targets. Many lead to optical-type angular distributions (phase functions), polarizations, scattering, and absorption cross sections. Many of these data patterns can be accounted for by assuming a complex energy dependent nuclear potential (complex wavelength dependent index of refraction). Even the terminology of this subject, such as "the cloudy crystal ball model," reflects the light scattering analogy. Now nuclear opticians, like atmospheric opticians, divide up into a school concerned with average gross properties and a school concerned with statistical fluctuations. Both groups have greatly enriched the subject, although, as in the light scattering case, the communications between the schools has not always been the best.

My own specialized pursuits of atmospheric optics (apart from a stint in World War II) began in 1959 just after an intensive involvement with nuclear optics (Ref. 4). In these pursuits, I have mostly used the gross structure-nuclear optical modelers approach. The style here has been to use analytic models whose parameters are determined by nonlinear least square adjustment to experimental data. Then we look at the systematics of the parameters with the ultimate objective of relating them to more fundamental physical parameters, e.g., those in the basic nuclear force. I would like now to illustrate this nuclear optical approach with a few simple-minded attacks on some atmospheric optics problems.

II. AUREOLE STUDIES

Deirmendjian and Sekera (Refs. 5 and 6) very early recognized the importance of Mie particles in the theory of the solar aureole. Their work was motivated by an attempt to account for some reported anomalously high transmission of the ultraviolet part of direct sunlight. It is interesting to note that D. S. Saxon, who collaborated with Deirmendjian and Sekera on scattering from dielectric spheres (Ref. 7) during the same time frame, played an important role in the development of the nuclear optical model.

Our work, on determining aerosol size distributions from solar aureole intensities, has used a type of atmospheric modeling and single scattering theory which we first used in a satellite ozone sounding analysis (Ref. 8). That the skylight in the neighborhood of the sun can be used to good advantage may be surmised from the optical theorem of nuclear and atomic scattering theory and bistatic radar analysis (Ref. 9). These works indicate that whereas backscatter cross sections ($180°$ scatter) vary in a complex manner with particle shape and index of refraction, forward scatter cross sections are primarily determined by the volume of the particle. This property was first utilized in the doctoral theses of Adarsh Deepak (Ref. 10) and Barton J. Lipofsky (Ref. 11) which, among other things, involved photographic and photoelectric studies of the solar aureole in the visible region (Ref. 12). More recently, these studies have been extended into the ultraviolet (Refs. 13 and 14).

Dr. Deepak has already described some of the features of this work in his talk on the Aureole Isophote Method and in connection with his Stratospheric Aerosol Photographic Experiment (SAPE) proposal. Let me add some words here on the advantages of this type of experiment.

In Fig. 1, we show the densitometric traces of a measured photographic aureole made with the obscuring disc technique. The traces labeled r and l are through the solar almucantar whereas the traces labeled b and t are below and above the sun in the solar meridian. The beauty of this experiment is the reference intensity provided by the solar disc, so in essence we have on our measuring medium a comparison between diffuse sky intensity and the direct

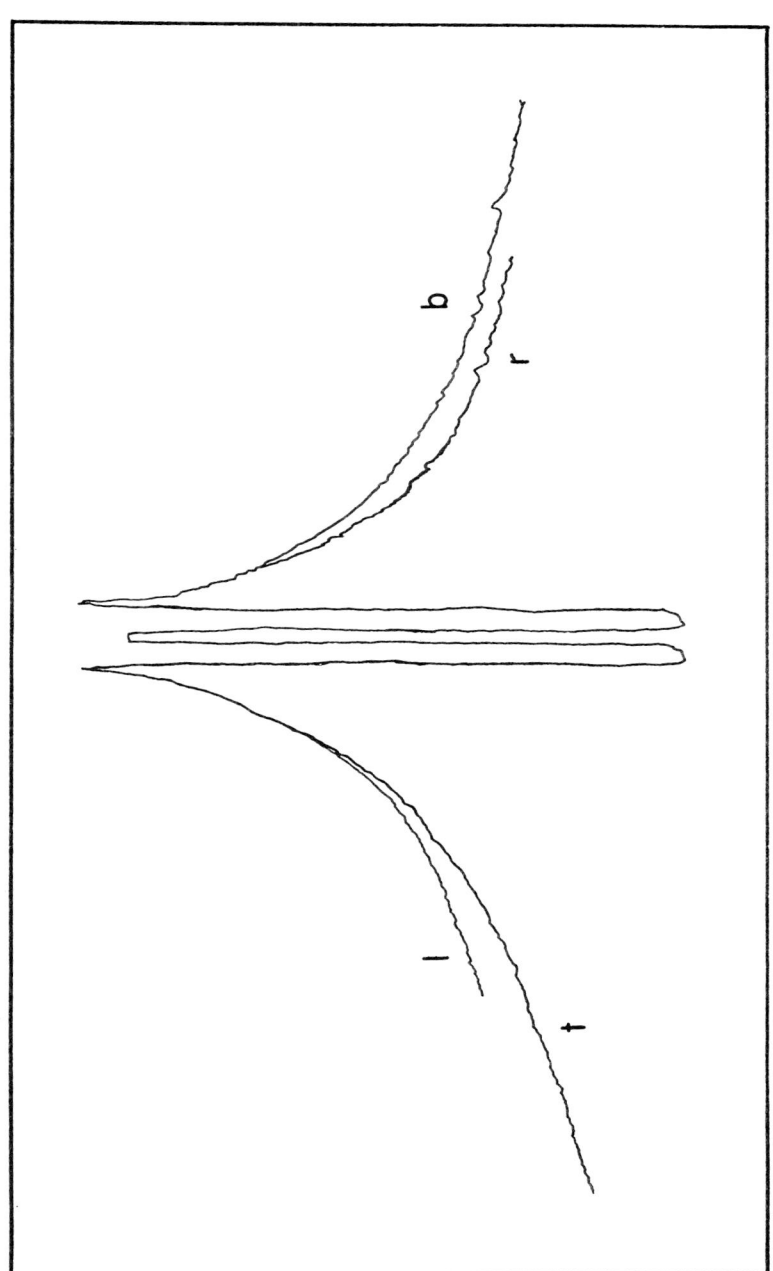

Fig. 1. Densitometric trace of aureole photograph with obscuring disc for horizontal and vertical scans with l, r, t, and b denoting positions of scans on the left, right, top, and bottom sides of the disc, respectively.

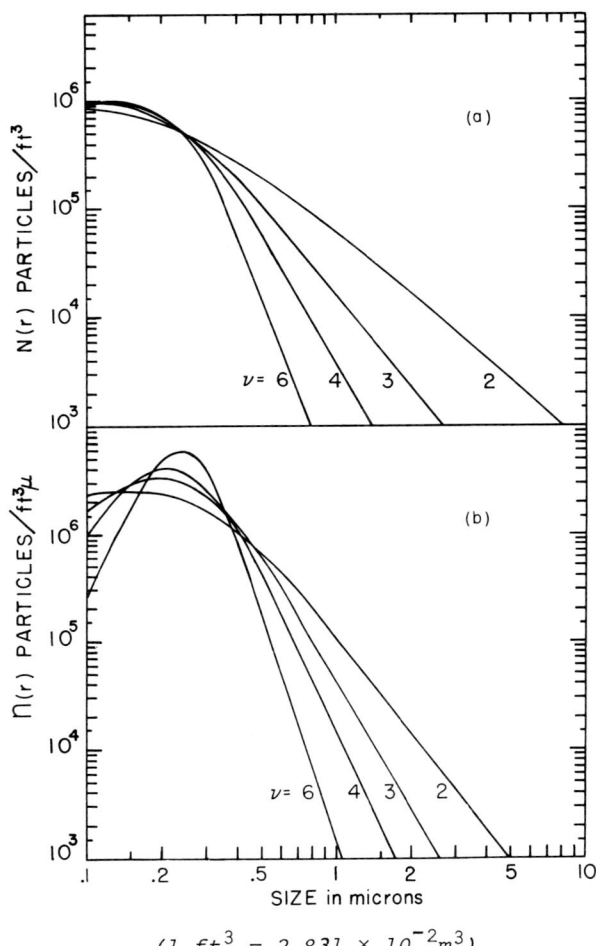

Fig. 2a. Examples of oversize distribution.

Fig. 2b. Corresponding examples of distribution.

solar intensity, attenuated by 10^4 by a Neutral Density 4 (ND4) filter. Thus far, in our work at the University of Florida, we have only exploited the information content in the almucantar trace of the solar aureole.

For the convenience of analysis, we use an analytic size distribution characterized by the cumulative distribution function (see Fig. 2).

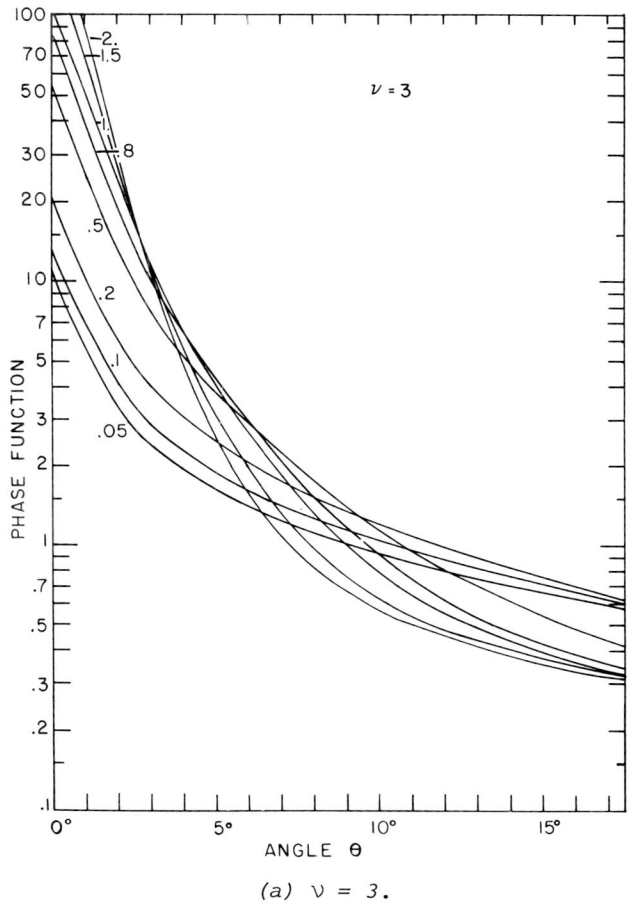

(a) $\nu = 3$.

Fig. 3. Phase functions for model distribution.

$$N(r) = N_o / [1 + (r/a)^\nu] \tag{1}$$

where N_o, ν and a are adjustable parameters. This corresponds to the differential size distribution

$$n(r) = \frac{dN}{dr} = N_o \frac{\nu}{a^\nu} \frac{r^{\nu-1}}{[1 + (r/a)^\nu]^2} \tag{2}$$

This analytic size distribution is a generalization of the Junge power law which is well behaved as $r \to 0$.

We have calculated a library of normalized phase function for this two-parameter size distribution function. Figure 3 illustrates several examples. Figure 4 illustrates some recent work at 640 nm showing the sharpness in the determination of ν obtained by this type of analysis (Ref. 14).

Spectral turbidity measurements, such as determined by a multi-channel Sun photometer, also give information about the particle size distribution (Ref. 15). Green and Sawada (Ref. 16) have determined the relative spectral turbidities associated with

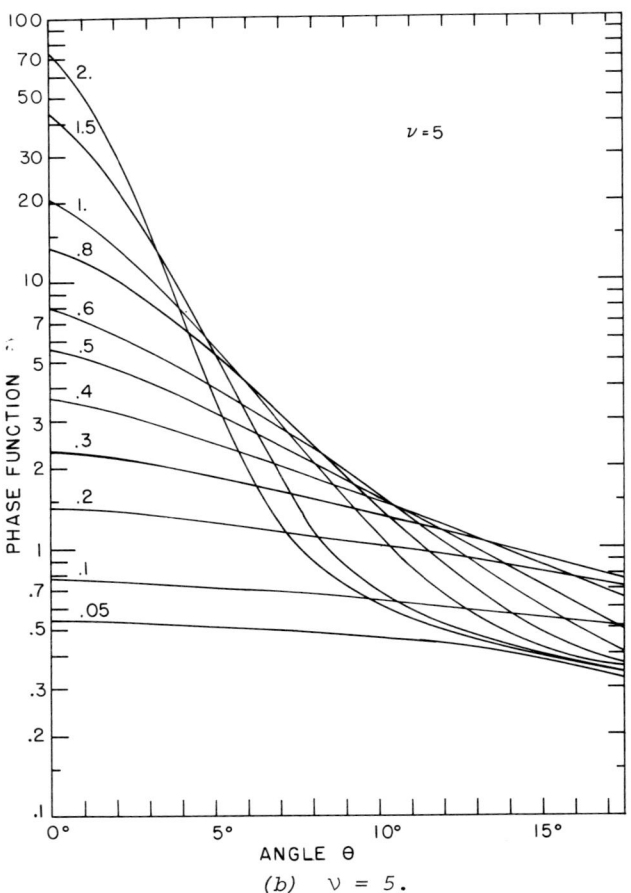

(b) ν = 5.

Fig. 3. Concluded.

THE INVERSION OF SCATTERING DATA

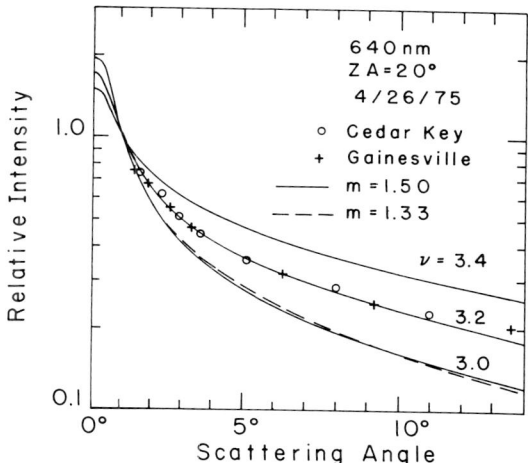

Fig. 4. Measured aureole intensity along the almucantar compared with calculations for three ν variations in the aerosol size distribution. (From Ref. 14.)

aerosol size distributions characterized by Eqs. (1) and (2). Thus, aureole data augmented by spectral turbidity data can lead to a very sharp determination of ν, as well as N_o.

We have considered the problem of multiple scattering in the solar aureole (Ref. 17), particularly in the ultraviolet where Rayleigh scattering becomes so important. Figure 5 shows an intercomparison of three calculations (Ref. 18): (1) a single scattering treatment, (2) a multiple scattering calculation using Monte Carlo techniques, and (3) a multi-channel calculation. The calculation indicates that multiple scattering can be quite significant, particularly in the ultraviolet.

In subsequent work, McPeters and Green (Ref. 14) have found that Rayleigh scattering is the source of most of this multiple scattering in the solar aureole except at high particulate optical depths. Accordingly, they have proposed an analysis technique which uses single aerosol and Rayleigh scattering augmented by

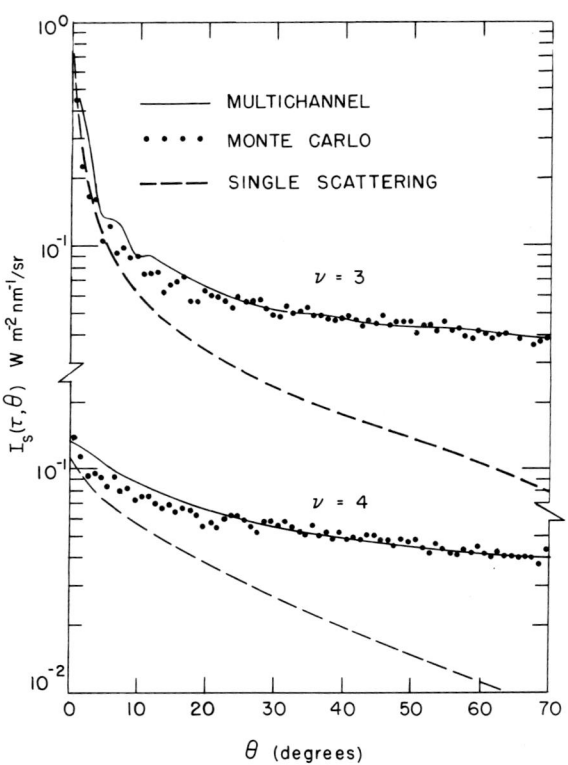

Fig. 5. Absolute intensity as a function of detector zenith angle for two different Mie scattering functions as calculated by three separate methods. The solid lines represent the results of the multi-channel calculations, the dots are the Monte Carlo results, and the dashed curves are the single scattering calculation. Note the break in the scale to avoid superposition of the two sets of results. (From Ref. 18.)

THE INVERSION OF SCATTERING DATA

multiple Rayleigh scattering as determined by the use of the tables of Coulson, Dave, and Sekera (CDS) (Ref. 19). Figure 6 is an illustration of a fit to data based upon such an analysis.

III. DIFFUSE TO DIRECT RATIO MEASUREMENTS

In connection with our ultraviolet studies in support of the Climatic Impact Assessment Program (Ref. 20), we have attempted to realistically characterize the diffuse solar radiation or sky radiation in the ultraviolet. As it turns out, sky radiation in the ultraviolet is often of greater biological consequence than direct sunlight. Green, Sawada and Shettle (Ref. 21) have developed an approximate analytic formula which describes the diffuse spectral irradiance in the ultraviolet region by adapting a single scattering analysis to the systematics of Bener's experiments (Ref. 22) and to theoretical calculations of Shettle and Green (Ref. 23).

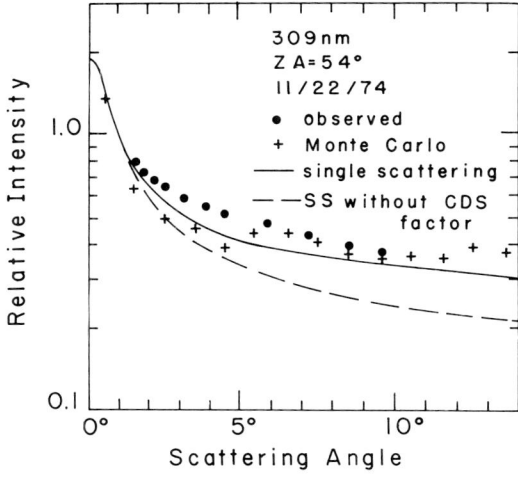

Fig. 6. *Normalized ultraviolet aureole intensities along the almucantar compared with modified single scattering calculation, with and without the CDS Rayleigh multiple scattering correction factor, and compared with a Monte Carlo full multiple scattering calculation. (From Ref. 14.)*

More recently, Chai and Green (Ref. 24) have recognized the merits of measurements of the ratio of diffuse to direct spectral irradiance as a simple indicator of atmospheric optical properties. The ratio method is analogous in some respects to our aureole method except that there we must attenuate the direct solar intensity by four orders of magnitude to compare it to the sky intensity. Figure 7 is an example of measurements of the total sky irradiance, the direct irradiance, and the ratio. The diffuse and direct both vary very markedly, and reflect the fluctuations in the extra-terrestrial solar spectral irradiance and in the ozone extinction coefficients as well as in the wavelength dependence of instrumental sensitivity. However, the diffuse to direct ratio is only slowly varying but still sensitively depends upon such interesting characteristics as atmospheric particulate loading, ground albedo, and sky cover. This ratio method avoids problems associated with the difficulty of absolute spectral irradiance measurements which, at this time, are limited to about 8% in the ultraviolet region.

It should be remarked that Herman, et al. (Ref. 25) have theoretically examined such a ratio method in the visible region in an attempt to estimate the imaginary part of the index of refraction of atmospheric aerosols.

We shall next consider another example of inverting of optical data with the aid of the analytic modeling approach and auxiliary information, such as may be obtained with simple ground-based instruments.

IV. THE INVERSION OF THE LOW ALTITUDE OZONE PROFILE

Tropospheric ozone is a constantly varying atmospheric component which changes with the season of the year, location, and time of day. Both photochemical and stratospheric transport processes are important sources of tropospheric ozone and are responsible for the strong diurnal, seasonal, and spatial variations.

THE INVERSION OF SCATTERING DATA

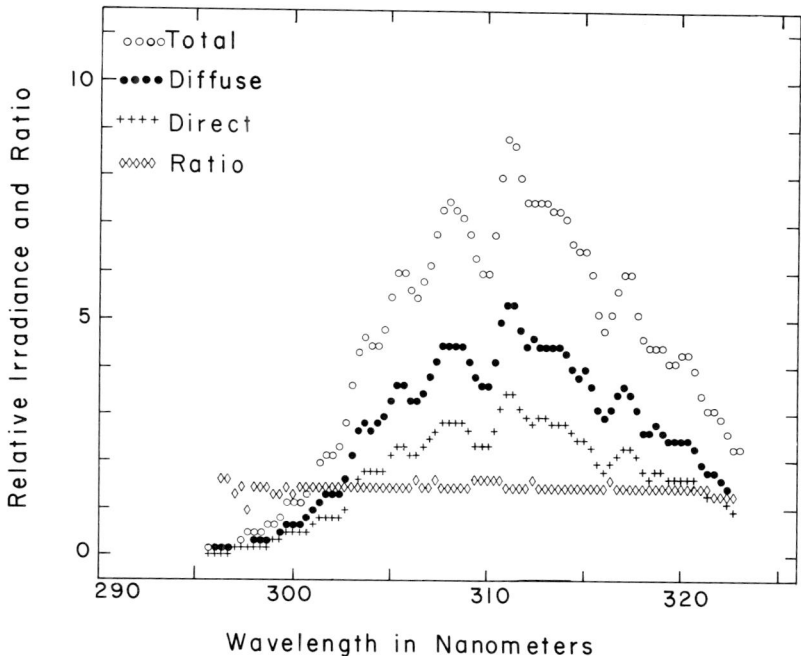

Fig. 7. Total, diffuse, direct, and the ratio of diffuse to direct components of solar irradiances as a function of wavelength as recorded on 31 July 1975 at around 13:50 EST. The magnitudes of the irradiances are in relative scale; they are proportional to the photomultiplier output in 100 μV. (From Ref. 24.)

A systematic program of balloon-borne ozonesonde observations has provided valuable data on the altitude structure of ozone which can be used as a guide in constructing a versatile analytic model (Ref. 26). We consider in what follows the possibility of inferring the tropospheric ozone profile from diffuse to direct ratio measurements in the middle ultraviolet in conjunction with ozone point-sampling at the ground. We model the altitude profiles of the atmospheric components with analytic functions because such a technique simplifies data inversion and provides a convenient way of communicating the resulting profiles.

Green (Ref. 8) has modeled the stratospheric ozone column density as a function of altitude y with a distribution used extensively in the nuclear studies (the so-called Wood-Saxon function)(Refs. 3 and 4)

$$w(y) = \frac{w_o(1 + e^{-y_o/h})}{1 + e^{(y - y_o)/h}} \tag{3}$$

Here w_o is the total ozone thickness and y_o and h are parameters. The density profile $\rho(y)$ is given by

$$\rho(y) = -\frac{dw}{dy} = \frac{w_o}{h} \frac{(1 + e^{-y_o/h}) e^{(y - y_o)/h}}{(1 + e^{(y - y_o)/h})^2} \tag{4}$$

The parameter y_o is the altitude at which the density function peaks and h scales the width of the distribution. Green (Ref. 2) shows how y_o, h and w_o can be approximately inferred from solar backscatter ultraviolet (UV) measurements. In their recent analysis on ground level UV, Shettle and Green (Ref. 23) add an exponential term to this function to allow for the tropospheric ozone component. Here to characterize tropospheric ozone profiles which are concave, i.e., the density decreases with altitude above the ground and then increasing at the tropopause, we add a second term of the form of Eq. (3) or (4) with the parameters y_o' and h' and where w_T is now the sum of $w_o + w_o'$.

In Fig. 8, several profiles corresponding to various values of h' and $\rho(0)$ are shown. Here $y_o' = 0$ and $w_o = 0.29$ atm-cm; and y_o and h are set to 23 km and 4 km, respectively. Extreme concave profiles can be obtained as well as curves of almost constant density. Furthermore, convex profiles can be generated with two distributions by setting y_o' to be a positive number. Convex profiles are observed with greatest frequency in the summer months at latitudes above 40°N.

THE INVERSION OF SCATTERING DATA

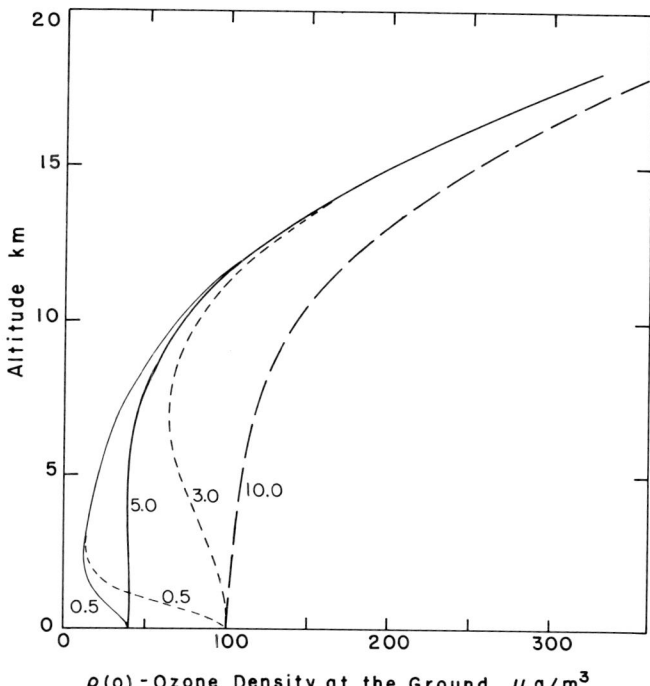

Fig. 8. Ozone density profiles for a number of different values of the parameter h_1' and ground density, with $y_o = 23$ km, $h = 4$ km and $y_o = 0$.

Assuming we know the ground level ozone density and also that the ozone profile can be represented by a two component model with $y' = 0$, we can test the sensitivity of the diffuse to direct ratio to the parameter h'. In Fig. 9, we plot the ratio versus h' for $\rho(0) = 40, 70, 100$ ($\mu g/m^3$) for a wavelength of 300 nm. We take $y_o = 23$ km and $h_o = 4$ km. Furthermore, we have assumed the air and aerosol profiles of Shettle and Green (Ref. 23) which are based on Elterman's 1964 data (Ref. 27). The total aerosol optical depth is 0.411; the aerosol is characterized by the cumulative size distribution given by Eq. (1) with $\nu = 3$ and $a = 0.03$ μm; the

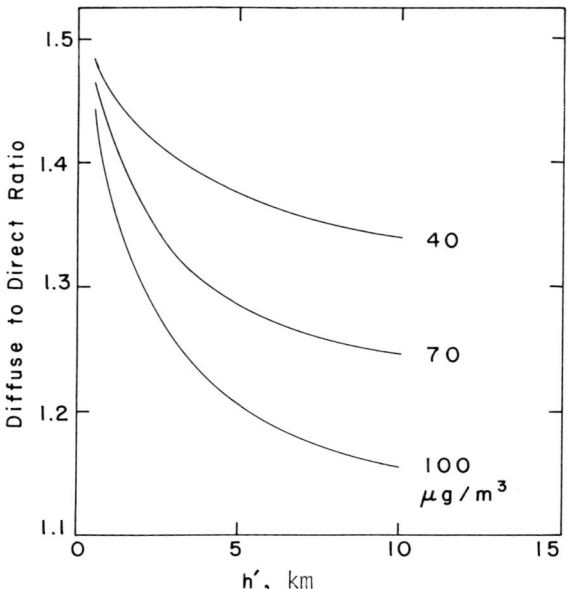

Fig. 9. Diffuse to direct ratio dependence on the scale parameter h' for ground ozone densities of 40, 70, and 100 $\mu g/m^3$.

aerosol index of refraction is 1.5 + .01i; a ground albedo of 0.05 is assumed; and the Sun is directly overhead. The ratios are calculated by using the multiple scattering technique of Shettle and Green (Ref. 23).

The diffuse to direct ratio then can be used to infer a value of h' from a set of curves as in Fig. 9 when used with point sampling measurements which give the ozone concentration at the ground.

Other simple and inexpensive measurements can be made simultaneously with the ratio and point-sampling measurements in order to further delimit the inversion and reduce the uncertainties. For example, aureole photography can be employed to infer the aerosol size distribution parameters (Refs. 10, 12, and 14). Also,

THE INVERSION OF SCATTERING DATA 313

multi-wavelength photometry provides valuable information on the aerosol optical thickness and its wavelength dependence (Refs. 15 and 16).

The two most nebulous parameters are the aerosol single scatter albedo and the ground albedo. It is important to know how uncertainties in these parameters propagate through the inversion. The aerosol single scatter albedo will depend on the aerosol index of refraction and size distribution. By using aureole photography to pin down the size distribution, one can draw on experience from previous investigations to fix the aerosol index of refraction within a range of confidence. For example, bistatic laser and aureole photography methods (Ref. 28) indicate that the index of refraction of a typical Gainesville aerosol is 0.005 ± 0.005 for the imaginary part and 1.50 ± 0.05 for the real part. The diffuse to direct ratio is found to be relatively insensitive to this range of possible error. Similar considerations apply to the ground albedo. In the theoretical model a ground albedo of 0.05 was assumed which is compatible with measurements by Furukawa and Heath (unpublished reports, 1973) of various natural surfaces for the wavelength region 310 to 380 nm. For example, they found that for scrub desert the ground albedo was 0.04 over the 310 to 380 nm region. For farmland, 70% tilled and 30% covered with vegetation, the ground albedo was found to be 0.07 to 0.08 for the 310 to 340 nm region.

Small errors in the ground albedo (≈ 0.02) do not significantly affect the calculations. Furthermore, once the ground albedo is known for a given location, the daily variations of h' can be determined, unless, of course, the surface changes because of snow cover, cultivation, or the like.

If the true optical depth is used in the inversion, then underestimating the ground albedo or the aerosol single-scatter albedo will lead to calculated ratios which are too small. Overestimation will lead to calculated ratios which are too large. Diffuse to direct ratio measurements in the 320 to 340 nm region can be used to infer effective aerosol optical depths here since in this region the ratios are rather insensitive to tropospheric ozone. The effective optical depths can be extrapolated to smaller wavelengths. These effective optical depths will be somewhat different from the true aerosol optical depths and will tend to compensate for errors in the ground albedo and aerosol single-scatter albedo. The diffuse to direct ratio around 300 to 320 nm is insensitive to the altitude distribution of the aerosols so long as the ozone profile near the ground is not changing too rapidly. Also, detailed knowledge of the stratospheric ozone structure or thickness is not required.

V. CONCLUDING REMARKS

Strictly numerical methods of inversion are becoming predominant in remote sensing these days. These are, of course, valuable to infer the irregularities and statistical fluctuations in atmospheric properties. The analytic model method which we have illustrated can be a valuable supplement to such numerical methods. They are particularly useful when used in conjunction with dynamical models of atmosphere structure because of the additional physical input of such models. When the models are joined to ground-based point sampling data, this remote-sensing-analytic model approach gives approximate answers to important questions involved in many public policy decisions on atmospheric pollution.

SYMBOL

a   size distribution parameter, corresponds approximately to the size of particle where n(r) peaks

h   ozone distribution parameter which is proportional to width of ozone density function; prime denotes another value of h

m   real part of aerosol refractive index

I_s	absolute scattered intensity
$n(r)$	differential aerosol size distribution
$N(r)$	cumulative aerosol size distribution
N_o	size distribution parameter which is equal to total number of aerosol particles
r	aerosol particle radius
w_o	total ozone thickness; prime denotes another value of w_o
$w(y)$	the ozone thickness function
y	altitude
y_o	altitude at which ozone density peaks; prime denotes another value of y_o
ZA	zenith angle
θ	scattering angle
ν	size distribution parameter which determines power law dependence of $n(r)$ at large r
$\rho(y)$	ozone density function
τ	optical depth

ACKNOWLEDGMENT

Supported in part by the Atmospheric Sciences Programs of the National Science Foundation and National Aeronautics and Space administration.

REFERENCES

1. A. E. S. Green, M. H. MacGregor and R. Wilson (Eds.), Proceedings of International Conference on N-N interaction, *Rev. Mod. Phys. 39*, 495 (1967).

2. A. E. S. Green, The fundamental nuclear interaction, *Science, 169*, 933 (1970).

3. A. E. S. Green, T. Sawada and D. S. Saxon, "The Nuclear Independent Particle Model, The Shell and Optical Models. Academic Press, Inc., 1968.

4. A. E. S. Green, C. E. Porter and D. Saxon (Eds.), *Proc. of the International Conference on the Nuclear Optical Model, Florida State University Studies, Tallahassee, Florida,* 1959.

5. D. Deirmendjian and Z. Sekera, Theory of the solar aureole, Part I; Scattering and radiative transfer, *The Rand Corporation, Santa Monica, California,* November 1957 [Internal Report.]

6. D. Deirmendjian and Z. Sekera, Atmospheric turbidity and the transmission of ultraviolet sunlight, *J. Opt. Soc. Am. 46,* 565 (1956).

7. D. S. Saxon, Z. Sekera, D. Deirmendjian and R. S. Fraser, On the scattering of plane electromagnetic waves by dielectric spheres, *University of California, Los Angeles,* 1957 [Internal Report.]

8. A. E. S. Green, Attenuation by ozone and the Earth's albedo in the middle ultraviolet, *Appl. Opt. 3,* 203 (1964); see also A. E. S. Green, The middle ultraviolet and its space applications, *Convair General Dynamics, San Diego, California,* ERR-AN-185, July 1962.

9. K. M. Siegal, Bistatic Radar Cross Sections, *Proc. IEEE, 45,* 1137 (1960).

10. Adarsh Deepak, "Second and Higher Order Scattering of Light in a Settling Polydispersed Aerosol," Ph.D. Dissertation, University of Florida, Gainsville, 1969. [Available from Xerox Univ. Microfilms, Ann Arbor, Michigan 48106.]

11. B. J. Lipofsky, "Single Scattering of Light by Polydispersed Aerosols," Ph.D. Dissertation, University of Florida, Gainsville, 1970. [Available from Xerox Univ. Microfilms, Ann Arbor, Michigan 48106.]

12. A. E. S. Green, A. Deepak and B. J. Lipofsky, Interpretation of the sun's aureole based on atmospheric aerosol models, *Appl. Opt. 10*, 1263 (1971).

13. R. D. McPeters, "Scattered Sunlight in the Atmosphere from the Middle Ultraviolet through the Near Infrared," Ph.D. Dissertation, University of Florida, Gainsville, 1975. [Available from Xerox Univ. Microfilms, Ann Arbor, Michigan 48106.]

14. R. D. McPeters and A. E. S. Green, Photographic aureole measurements and the validity of aerosol single scattering, *Appl. Opt. 15*, 2457 (1976).

15. G. E. Shaw, "An Experimental Study of Atmospheric Turbidity Using Radiometric Techniques," Ph.D. Dissertation, University of Arizona, Tucson, 1971. [Available from Xerox Univ. Microfilms, Ann Arbor, Michigan 48106.]; see also G. E. Shaw, J. A. Reagan and B. M. Herman, "Investigations of atmospheric extinction using direct solar radiation measurements made with a multiple wavelength radiometer, *J. Appl. Meteorol. 12*, 374 (1973).

16. A. E. S. Green (Ed.), Report of the coordinated program for the remote sensing of atmospheric aerosols clean air preservation and enhancement research, *University of Florida, Gainesville, Report*, September 10, 1971.

17. A. Deepak, Double scattering corrections to the theory of the sun's aureole, NASA TM X-64800, December 1973; see also Ref. 16.

18. D. R. Furman, A. E. S. Green and T. Mo, Multistream and Monte Carlo calculations of the sun's aureole, *J. Atmos. Sci. 33*, 537 (1976).

19. K. L. Coulson, J. V. Dave and Z. Sekera, "Tables Related to Radiation Emerging from a Planetary Atmosphere with Rayleigh Scattering." University of California Press, Berkeley, California, 1960.

20. Alan J. Grobecker, (Ed.), Impacts of climatic change on the biosphere, *Dept. of Transportation Climatic Impact Assessment Program,* DOT-TST-75-55, September 1975.

21. A. E. S. Green, T. Sawada and E. P. Shettle, The middle ultraviolet reaching the ground, *Photochem. Photobiol. 19,* 251 (1974).

22. P. Bener, *Technical Report, European Research Office, U.S. Army, London,* Contract No. DAJA37-68-C-1017, 1972.

23. E. P. Shettle and A. E. S. Green, Multiple scattering calculation of the middle ultraviolet reaching the ground, *Appl. Opt. 13,* 1567 (1974).

24. A. T. Chai and A. E. S. Green, Measurement of the ratio of diffuse to direct solar irradiances in the middle ultraviolet, *Appl. Opt. 15,* 1182 (1976).

25. B. M. Herman, P. S. Browning and J. J. DeLuisi, Determination of the effective imaginary term of the complex refractive index of atmospheric dust by remote sensing: The diffuse-direct ratiation method, *J. Atmos. Sci. 32,* 918 (1975).

26. W. S. Hering and T. R. Borden, Jr., Ozonesonde observations over North America, AFCRL-64-30II, July, 1964.

27. L. Elterman, *Environmental Research Paper No. 46, Air Force Cambridge Research Laboratories,* AFCRL-64-740.

28. G. Ward, K. M. Cushing, R. D. McPeters and A. E. S. Green, Atmospheric aerosol index of refraction and size-altitude distribution from bistatic laser scattering and solar aureole measurements, *Appl. Opt. 12,* 2585 (1973).

THE INVERSION OF SCATTERING DATA 319

DISCUSSIONS

Deirmendjian: With regard to the importance of multiple scattering in the aureole, I don't want to blow my horn, but I certainly wish to protect Professor Sekera's memory. In my 1956 doctoral thesis, I looked at the problem of the solar aureole at his suggestion. The method used was essentially Sekera's idea and consisted of treating the aureole as a perturbation on the Rayleigh multiple scattered skylight. And that is precisely what I did. Have you seen my original 1957 paper on the aureole?

Green: Yes, we have seen your paper.

Deirmendjian: It was exactly that. You increase the optical thickness of the Rayleigh atmosphere by adding a perturbation optical thickness due to the aerosols. When you do that, it is a kind of hybrid method where single scattering on the aerosols produces the aureole, but multiple scattering, mainly on the Rayleigh particles, produces the rest of the background skylight. I thought you didn't make that clear.

Green: We weren't aware of that aspect of your paper back then. We were aware of your work on the aureole and we have quoted it in our work.

Deirmendjian: I think that was the principal point. I have since reexamined this method in 1970. I did not publish the results in the open literature due to lack of funds. But they are available in a formal 1970 Rand Corporation report, in which I introduced new phase functions. Indeed, the curves that are obtained look very much like some of the measurements I have seen. Subsequently, I intended to compare them with the measurements and look into their use to get information about the aerosol size distribution. At the time, I was unable to do this for lack of support.

Green: Well, there was no intention to slight you or Professor Sekera. We have in our work in this area acknowledged this. The UV problem has a new interest in light of its biological aspects. So part of this work was directed toward answering some particular questions about the radiance of the UV aureole where some serious problems remained. My point about the low altitude ozone distribution is that you can take advantage of the extra scattering associated with aerosols to extend the path and the absorption in the low altitude ozone layer. Thus, you can actually drag out a little information about the high altitude distribution, but little about the low altitude profile.

Weinman: While I agree with the previous speaker that the aureole technique is a powerful one for determining size distributions, I should point out that invisible cirrus clouds can plague this

technique. While there are frequent cases where you can get nice
smooth functions to which you can apply the theory, this hazard
always exists. If one looks at this data, one can pick up cirrus
clouds which are not at all visible to the naked eye. I think
that it takes a certain amount of judicious discrimination to get
cases to which you can apply the radiative transfer theory.

Green: I think Adarsh has shown some results where contours of
isodensity on a film show distortions that are suggestive of a
thin cloud layer. In some of our work, we have noted some problems
of that nature, which one can, perhaps, use as information content.

Deepak: Yes, we have been looking at the shapes of the solar
aureole isophotes rather carefully and we see small bumps. The
isophote curves are not very smooth. That could be due to the
presence of thin clouds. In fact, we have an airport nearby;
when aircraft take off we can easily detect the presence of con-
trails from the systematic distortion in the shape of the isophotes
over the region of the contrail image.

Fymat: I was very interested in your conclusion in determining the
scale height for the troposphere.

Green: For the tropospheric ozone distribution.

Fymat: Yes, but you need, as you say, the ozone distribution at
the lower altitudes. However, it seemed to me, and Dr. Mateer may
care to comment on it, that the actual ground-based Dobson measurements are really being phased out in view of the fact that they
produce different results on inversion using different methods.
So, if we cannot get the distribution at the lower altitudes, how
practical is your conclusion?

Green: Well, insofar as profile, the Umkehr method (not the Dobson
method) relies, and Carl can correct me since he is much more
expert, on the setting of the sun and inverting sky radiances. Our
method could be worked at high noon since it is instantaneous. We
would use wavelength information and we have, in effect, shown by
simulated tests that the diffuse to direct ratio is only sensitive
to the addition of extra ozone below about 10 kilometers. So that
while it is not a complete profile, if you use it in conjunction
with ground level chemical measurement, it gives you information
which relates to the lower atmospheric ozone and it is insensitive
to the stratospheric ozone which is far more abundant. Now am I
correct that the Umkehr method more or less gives you the most
information about the higher atmospheric ozone? So there is no
real contradiction with anything previously known.

THE INVERSION OF SCATTERING DATA

Mateer: The Umkehr measurement is a different kind of measurement. It is a ratio measurement of two wavelengths as opposed to what you are doing--a ratio of diffuse to direct sunlight.

Green: Yes. We also are helped if we use two wavelengths as well. It eliminates a little bit of the absolute radiometry problem. It gives us another check.

OPEN DISCUSSIONS I

King: Alex, what do you think we can learn about solving the atmospheric temperature inversion problem from the similar type inversion problem in inferring nuclear potential in nuclear physics?

Green: Well, I was discussing this type of thing with Dr. Wark, and the analogy would be if we brought to bear to the problem all the physical dynamics data, not only at one location but at adjacent locations. Thus, if we had relatively few questions that we posed to the radiative transfer problem, for example, to get horizontal gradients in temperature profile, and perhaps include such facts that there tends to be a tropopause and we are mainly interested where the temperature break occurs, and ask questions which embody all of our experience as to reasonable temperature profiles. Then when we apply this external information to an inversion problem, it is usually much easier to come to a physically reasonable answer. Now I don't know if that's a good paraphrase of the analogy. We do have one problem which I have to confess compromises the analogy. In the case of the nuclear force problem we think all neutrons and protons are the same, and unfortunately the atmosphere takes on so many different states that we do have that extra complication. Thus, if you tried to unfold many, many details of the instantaneous atmosphere, you may be in trouble with this approach. On the other hand, if you are satisfied with answering the types of physical questions that our dynamical meteorologists, pollution experts, or biologists want answered, I think you get a lot of useful information by these modeling approaches.

Herman: Alex, I would like to ask you, what is the sensitivity of the direct diffuse--insofar as solving for ozone--to variations in aerosol contents? Because as you know we are trying to apply it to learn various parameters of the aerosol. Now you are using it to learn about the ozone but it still has a sensitivity to aerosols even at the ultraviolet wavelengths.

Green: Yes, I should, unless I commit another error of omission, mention that Ben has used the direct-to-diffuse ratio to infer the imaginary part of the index of refraction working at 500 nanometers. We were not aware of this work when we started ours, but were aware of it during its course. I think the trick in our case is to choose two wavelengths--one in the region where the ozone is absorbing and one where it is nonabsorbing. Then we take advantage of whatever we know about our aerosols, including a ground-level aerosol measuring device, plus an aureole measuring device. I would not go to anything expensive like a lidar. But I think if we bound our aerosol model somewhat we find by sensitivity analyses that we do get some information about the low level ozone. This is what we would pick up by using two wavelengths, just as Dobson does. We would go to about 320 nanometers and then to 305 nanometers. We would expect that the aerosols do not change very much, so that we hold the aerosols confusion factor down.

Goulard: I am trying to apply inversion techniques to the field of combustion. I get the profiles of temperature or concentrations following the methods that Dr. Chahine has developed. But, I get them in terms of optical thickness. And when I want to convert it to physical thickness, which is what the combustion people want, I find that I cannot depend on the constant concentration of CO_2 as you do in the atmosphere. Could someone give me some help on how to get around this problem in frequency scanning only?

Chahine: Let me answer Professor Goulard's question on the solution in terms of the optical depth. You know, the independent variable in the radiative transfer equation is τ but we set $d\tau = (\partial\tau/\partial z)dz$ in the equation and obtain a solution as a function of the physical scale z. As you know $\partial\tau/\partial z$ is obtained on the basis of an atmospheric model. Thus, the solution is truly a function of τ although it is usually presented as a function of z. The transformation from the τ-scale to the z-scale is done through a model.

Goulard: You assume Laplace Law of the atmosphere and a constant mixing ratio?

Chahine: Yes, we have to do this.

Goulard: And for a flame?

Chahine: You would have to develop a model for your flame and correlate your optical depths with a physical scale.

Goulard: Is the concentration of CO_2 really constant throughout the atmosphere?

Green: Well, it is, yes, within a small period of time. There are some people who think it is growing.

Gille: The concentration of CO_2 in the troposphere is quite constant although not absolutely so. It varies seasonally by perhaps one part per million out of a background of 330 with a long-term trend of, perhaps, 0.7 ppm per year.

COMPARISON OF LINEAR INVERSION METHODS BY

EXAMINATION OF THE DUALITY BETWEEN

ITERATIVE AND INVERSE

MATRIX METHODS

Henry E. Fleming
National Environmental Satellite Service, NOAA

Linear numerical inversion methods applied to atmospheric remote sounding generally can be categorized in two ways: (1) iterative, and (2) inverse matrix methods. However, these two categories are not unrelated; a duality exists between them. In other words, given an iterative scheme, a corresponding inverse matrix method exists, and conversely. This duality concept is developed for the more familiar linear methods. The iterative duals are compared with the classical linear iterative approaches and their differences analyzed. The importance of the initial profile in all methods is stressed. Calculations using simulated data are made to compare accuracies and to examine the dependence of the solution on the initial profile.

I. INTRODUCTION

When working with different linear inversion methods, one finds that some of the iterative and inverse matrix methods yield similar accuracies. Therefore, one might conjecture that these methods are simply duals of each other, that is, given an iterative method, there exists a corresponding inverse matrix method that yields the same answer, and conversely. It turns out that this is true for a significant class of linear inversion methods.

Furthermore, the iterative methods are not quite the classical ones in that they are not iterated to convergence. Instead, they are iterated only until the residual variances are equal to the variance of the instrumental noise. This leads to solutions that are different from the classical ones. Consequently, the duality between methods is not exact, that is, the solutions to the dual equations agree only to within a certain degree of accuracy. This requires us to speak of "virtual duality" instead of duality in the exact sense.

The general outline of the paper is to first develop the duality principal for the least squares solution. For this case the duality is exact, but this solution is unstable. Hence, we must resort to regularized solutions for which we can have only virtual duality.

The principle of virtual duality first is established for the Twomey-Phillips solution. Then, the results are extended to a more general class of solutions. Finally, the question of the dependence of the solution on the initial profile is addressed. Results of calculations using simulated data are given at appropriate places throughout the paper.

II. THE DUALITY PRINCIPLE

A. The Least Squares Solution

Most atmospheric remote sounding problems are, or can be reduced to, a Fredholm integral equation of the first kind, which can be approximated numerically by a linear system of equations (Ref. 1) of the form

$$\underset{\sim}{A}\, \underset{\sim}{x} = \underset{\sim}{y} \tag{1}$$

The n × m (m > n) matrix $\underset{\sim}{A}$ is the matrix of weighting functions, $\underset{\sim}{x}$ is the m-dimensional source-function vector, and $\underset{\sim}{y}$ is the vector of n measured values.

COMPARISON OF LINEAR INVERSION METHODS

To illustrate the duality principle, we begin with the least squares solution of Eq. (1), namely,

$$\hat{\underline{x}} = \underline{A}^T(\underline{AA}^T)^{-1}\underline{y} \tag{2}$$

where the superscript T denotes the transposed matrix. Of the infinite number of solutions to Eq. (1), the solution in Eq. (2) is the unique solution which has minimum Euclidean norm and exists because the weighting functions are chosen deliberately to be linearly independent. Since A has full rank (i.e., rank n), this solution is also the generalized inverse (or Moore-Penrose pseudo-inverse) solution of Eq. (1).

On the other hand, the classical iterative solution (i.e., solution by successive approximations) to Eq. (1) is given by

$$\underline{x}^k = \underline{x}^{k-1} + \alpha \underline{A}^T(\underline{y} - \underline{A}\underline{x}^{k-1}) \tag{3}$$

where $\underline{x}^k$ is the kth successive approximation to $\underline{x}$ and $\alpha > 0$ is a convergence (relaxation) factor. We say that Eq. (3) is the dual of Eq. (2), and conversely, because in the limit Eq. (3) yields the solution in Eq. (2), and conversely. This, then, is an example of what we mean by the *duality principle* between inverse matrix and iterative inversion methods.

To show that Eq. (3) is the dual of Eq. (2), we write Eq. (3) in terms of the initial approximation vector $\underline{x}^o$ by combining iterations to arrive at the following equations:

$$\begin{aligned}
\underline{x}^k &= \alpha \underline{A}^T \left[\underline{I}_n + (\underline{I}_n - \alpha \underline{AA}^T) + \ldots + (\underline{I}_n - \alpha \underline{AA}^T)^{k-1} \right] \underline{y} \\
&\quad + (\underline{I}_m - \alpha \underline{A}^T \underline{A})^k \underline{x}^o \\
&= \underline{A}^T \left[\underline{I}_n - (\underline{I}_n - \alpha \underline{AA}^T)^k \right] (\underline{AA}^T)^{-1} \underline{y} + (\underline{I}_m - \alpha \underline{A}^T \underline{A})^k \underline{x}^o \\
&= \underline{A}^T (\underline{AA}^T)^{-1} \underline{y} + (\underline{I}_m - \alpha \underline{A}^T \underline{A})^k \left[\underline{x}^o - \underline{A}^T (\underline{AA}^T)^{-1} \underline{y} \right] \tag{4}
\end{aligned}$$

where we used the identity

$$\left[I + (I - B) + \ldots + (I - B)^{k-1} \right] B = I - (I - B)^k \tag{5}$$

and we repeatedly used the identity

$$(I_m - \alpha A^T A)^k A^T = A^T (I_n - \alpha A A^T)^k , \tag{6}$$

and where I_ℓ denotes the ℓ-dimensional identity matrix. Note that a superscript on a vector (a lower case letter) is an iteration index while a superscript on a matrix (a capital letter) represents a power.

Now the iterative scheme in Eq. (4) converges if the Euclidean norm of the matrix difference with the exponent k in the last line is less than unity. Since

$$\left| \alpha \|A^T A\| - \|I_m\| \right| \leq \|I_m - \alpha A^T A\| < 1$$

and since $\|I_m\| = 1$, we have

$$0 < \alpha \|A^T A\| < 2.$$

Under the condition that $\alpha < 2/\|A^T A\|$, Eq. (4) becomes

$$\lim_{k \to \infty} x^k = A^T (A A^T)^{-1} y = \hat{x}.$$

Since this solution agrees with the solution in Eq. (2), we have established the exact duality between Eqs. (2) and (3).

While Eqs. (2) and (3) are an example of the duality principle, in practice, neither solution is of much value because the matrix AA^T is generally ill-conditioned with respect to matrix inversion. The ill-conditioning arises from the fact that typically the weighting functions, which are the rows of A, are very smooth and broad, and, hence, overlapping. Furthermore, the vector y is always contaminated by measurement noise which is

COMPARISON OF LINEAR INVERSION METHODS 329

amplified by $(AA^T)^{-1}$. Consequently, the solutions in Eqs. (2) and (3) are unstable and, hence, of no use in remote sounding applications.

Figure 1 shows just how unstable least squares (generalized inverse) solutions can be. It is based on 139 radiosonde profiles from March 1973 which were almost uniformly distributed between

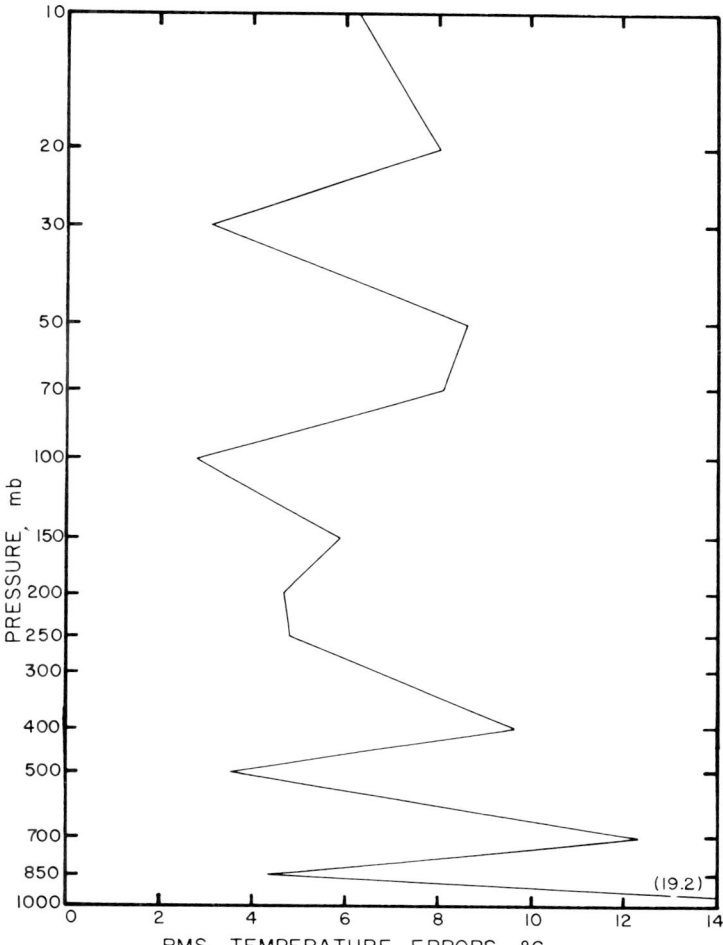

Fig. 1. Profile of the rms errors between the least squares solutions and radiosondes for 139 simulated cases. (1 bar = 100 kPa.)

the equator and 70° N. Clear radiances were calculated for the six CO_2 Vertical Temperature Profile Radiometer (VTPR) channels and were contaminated with noise by a random number generator using a normal distribution with a standard deviation of 0.25 mW/$(m^2 \text{ sr cm}^{-1})$. These radiances were then used in the least squares retrievals. The curve in Fig. 1 represents the 139 rms temperature differences (at the 15 mandatory pressure levels indicated) between the retrieved profiles and the radiosonde profiles (the "truth"). Clearly, these rms errors are unacceptably large--they vary from around 3°C at three levels to 19.2°C at 1000 mb. (1 bar = 100 kPa.) To obtain stable and physically realistic solutions to Eq. (1), one must resort to regularization methods which stabilize the ill-conditioned matrix that is to be inverted.

B. The Twomey-Phillips Solution

The first regularized solution of Eq. (1) that we will consider is that of Twomey and Phillips (Ref. 2), which is also known as the "minimum information solution" (Ref. 1). Regularization methods are designed specifically to solve Eq. (1) in the form

$$A\underset{\sim}{x} = \underset{\sim}{y} + \underset{\sim}{\varepsilon} \tag{7}$$

where $\underset{\sim}{y}$ is still the n-dimensional vector of measured quantities, but $\underset{\sim}{\varepsilon}$ is the n-dimensional vector of measurement errors, so that $\underset{\sim}{y} + \underset{\sim}{\varepsilon}$ would be the errorless vector. The Twomey-Phillips solution of Eq. (7) is

$$\hat{\underset{\sim}{x}} = \underset{\sim}{x}^o + A^T(AA^T + \gamma I_n)^{-1}(\underset{\sim}{y} - A\underset{\sim}{x}^o) \tag{8}$$

where $\underset{\sim}{x}^o$ is an initial approximation to the solution and γ is the smoothing (or regularization) parameter.

We must now establish the iterative dual to Eq. (8), which requires the matrix identity

$$A^T(AA^T + \gamma I_n)^{-1} = (A^TA + \gamma I_m)^{-1}A^T \tag{9}$$

COMPARISON OF LINEAR INVERSION METHODS

This identity permits us to write Eq. (8) in the form

$$\hat{x} = x^o + 1/\gamma A^T(y - A\hat{x}) \tag{10}$$

Proper interpretation of this equation yields the required iteration formula. We start the iterative process by setting $\hat{x} = x^o$ on the right-hand side of Eq. (10), that is, we introduce a one-step lag between the $\hat{x}$'s on the left- and right-hand sides of the equation. Then x^o and $\hat{x}$ on the right-hand side are updated with the last approximation of $\hat{x}$ on the left. If α is used as the convergence factor, the following iterative equation results:

$$x^k = x^{k-1} + \alpha/\gamma\, A^T(y - A x^{k-1}) \tag{11}$$

This interpretation of Eq. (10) is purely heuristic, but it will be justified in the next section.

Now, the convergent solution of Eq. (11) is not the dual of the solution in Eq. (8). To see this, use the relationships in Eq. (4) with A^T replaced everywhere by $1/\gamma A^T$. Then for $\alpha < 2\gamma/\|A^T A\|$, we have

$$\lim_{k \to \infty} x^k = A^T(AA^T)^{-1} y, \tag{12}$$

which again is the least squares solution of Eq. (1) and, therefore, not the dual of the solution in Eq. (8). In the next section the procedure for finding the virtual dual of the solution in Eq. (8) is described. Recall from the introduction that we cannot have exact duality for regularized solutions.

C. The Virtual Iterative Dual

In Ref. 3, it is shown that the generalized inverse solution of Eq. (1), even when AA^T is singular, can be obtained from the limiting process

$$x = \lim_{\gamma \to 0} A^T(AA^T + \gamma I_n)^{-1} y \tag{13}$$

Furthermore, it is shown that by aborting the passage to the limit zero at that value of γ for which the residual $\| A\hat{x} - y \|$ is equal to the norm of the random error vector ε of y, we obtain the Twomey-Phillips solution.

The virtual iterative dual to the solution, in Eq. (8), is found in an analogous way. Instead of passing to the limit as in Eq. (12), one stops the iteration process at that step for which the residual $\| Ax^k - y \|$ is equal to the norm of ε in Eq. (7). Just as the amount of smoothing of the solution in Eq. (8) is determined by how soon the approach of γ to zero in Eq. (13) is aborted, the amount of smoothing in the virtual iterative dual in Eq. (11) is determined by how soon the iteration process is stopped.

The equation that ensues when the iteration of Eq. (11) is terminated after k steps is given by Eq. (4) with A^T replaced everywhere by $1/\gamma A^T$, namely,

$$x^k = A^T(AA^T)^{-1} y + (I_m - \alpha/\gamma A^T A)^k \left[x^o - A^T(AA^T)^{-1} y \right] \quad (14)$$

To prove that this solution is the virtual dual of the solution in Eq. (8), we must show that for some finite k_o, the solution x^{k_o} of Eq. (14) is such that $x^{k_o} \simeq \hat{x}$ from Eq. (8), and that $\| Ax^{k_o} - y \| \simeq \| \varepsilon \|$. The symbol "$\simeq$" is to be read as "is virtually equal to", whose meaning is made more precise in the subsequent discussion.

III. PROOF OF VIRTUAL DUALITY

A. Eigenvalue Decomposition

We begin the proof of virtual duality by first reducing the dimensionality of matrices from m to the smaller dimension n where necessary. The first such case is the power matrix in Eq. (14). Subtract x^o from both sides of Eq. (14), apply the

COMPARISON OF LINEAR INVERSION METHODS

identity in Eq. (5), then apply the identity in Eq. (6), and finally use the identity in Eq. (5) again to obtain the forms

$$x^k - x^o = \left[I_m - (I_m - \alpha/\gamma A^T A)^k\right]\left[A^T(AA^T)^{-1} y - x^o\right]$$

$$= A^T\left[I_n - (I_n - \alpha/\gamma AA^T)^k\right](AA^T)^{-1}(y - Ax^o) \quad (15)$$

Now to prove that $x^{k_o} \simeq x$, we make the following substitution in Eq. (15):

$$(I_n + 1/\gamma AA^T)^{-1} \quad (16)$$

for $(I_n - \alpha/\gamma AA^T)^k$ and show that this does indeed transform Eq. (14) into Eq. (8). Then we show that the expressions in Eq. (16) are virtually equal for the iteration index k_o. Finally, we show that the index k_o is such that $\|Ax^{k_o} - y\| \simeq \|\epsilon\|$. Consequently, the two solutions in Eqs. (14) and (8) must be virtually the same and the virtual duality between them is established.

We start by making the substitution, of Eq. (16), in Eq. (15) and apply the identity

$$I - (I + B)^{-1} = (I + B)^{-1}B = B(I + B)^{-1} \quad (17)$$

to arrive at the Eq. (8). This proves that if the substitution of Eq. (16) is made in Eq. (15), then Eq. (14) is indeed the virtual dual of the solution in Eq. (8) for $k = k_o$.

Next, we show that in the expression in Eq. (16), the power matrix is virtually equal to the inverse matrix. To do this, we must consider the diagonalization of the symmetric matrix AA^T. Associated with the $n \times n$ matrix AA^T are the $n \times n$ orthogonal matrix of eigenvectors U and eigenvalues $\lambda_1 \geq \lambda_2 \geq \ldots \geq \lambda_n > 0$ such that

$$AA^T = U\Lambda U^T \quad (18a)$$

$$U^T U = UU^T = I_n \quad (18b)$$

$$\underset{\sim}{\Lambda} = \text{diag}(\lambda_1,\ldots,\lambda_n) \tag{18c}$$

where diag( ) denotes a diagonal matrix.

Now apply Eqs. (18a) and (18b) to the matrices in Eq. (16). This yields the decompositions

$$(\underset{\sim}{I}_n + 1/\gamma \underset{\sim}{A}\underset{\sim}{A}^T)^{-1} = \underset{\sim}{U}(\underset{\sim}{I}_n + 1/\gamma \underset{\sim}{\Lambda})^{-1}\underset{\sim}{U}^T \tag{19a}$$

and

$$(\underset{\sim}{I}_n - \alpha/\gamma \underset{\sim}{A}\underset{\sim}{A}^T)^k = \underset{\sim}{U}(\underset{\sim}{I}_n - \alpha/\gamma \underset{\sim}{\Lambda})^k \underset{\sim}{U}^T \tag{19b}$$

Thus, the whole question of virtual duality between Eqs. (8) and (11) comes down to comparing a function of eigenvalues of the form

$$f_k(z) = (1 - \alpha z)^k \tag{20a}$$

with one of the form

$$h(z) = (1 + z)^{-1} \tag{20b}$$

where $z = \lambda/\gamma > 0$.

B. Proof that the Eigenvalues are Virtually Equal

Recall that there are several constraints on the problem. For convergence we need $|1 - \alpha z| < 1$, or $0 < \alpha z < 2$, but to prevent alternating signs in the iteration process, we use the further restriction $0 < \alpha z \leq 1$. This gives us the inequalities $0 \leq f_k(z) < 1$ and $0 < h(z) < 1$.

To see that $h(z) \simeq f_{k_0}(z)$, consider Fig. 2 in which the curves for $f_k(z)$ with $k = 1,2,4,8,16,32,$ and 128 are plotted for $\alpha = 0.01$. This value for α is typical. The dashed curve is that of $h(z)$ for the domain $0 \leq \alpha z \leq 1$, that is, for $0 \leq z \leq 100$. Note that the variable x in Fig. 2 corresponds to z in the text.

Figure 2 is to be interpreted in the following manner. Given the eigenvalues of the operator $1/\gamma \underset{\sim}{A}\underset{\sim}{A}^T$, which are located on the

COMPARISON OF LINEAR INVERSION METHODS

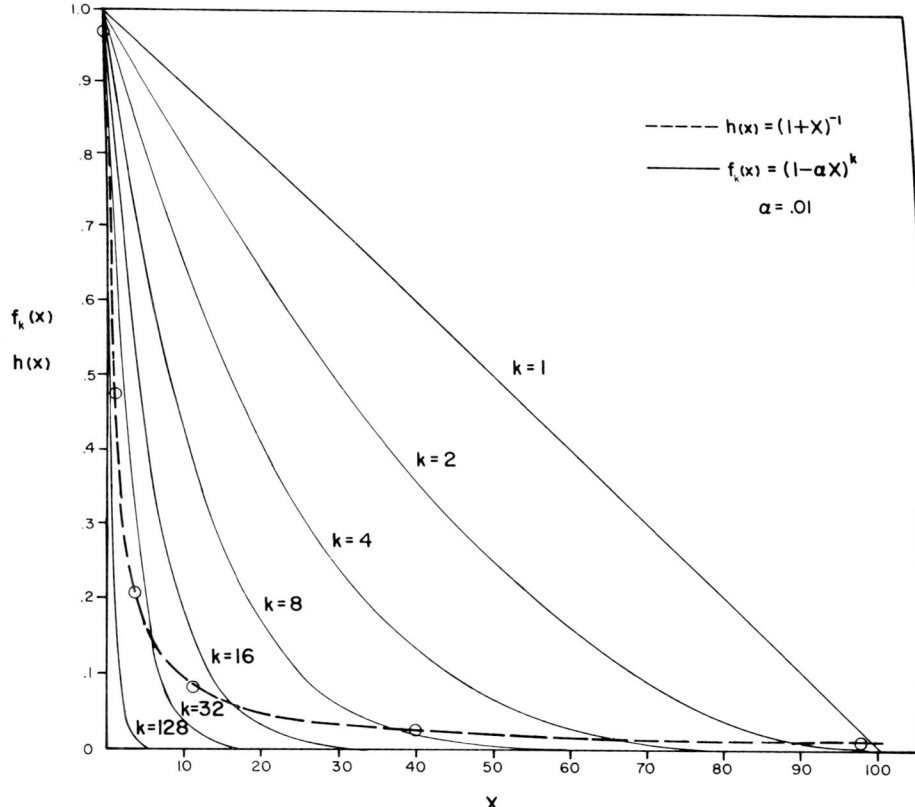

Figure 2. The functions of eigenvalues $h(x)$ (the dashed curve) and $f_k(x)$ (the family of solid curves) show how close the eigenvalues are for the case $\alpha = 0.01$. The circles are the six eigenvalues of the inverse matrix corresponding to the six eigenvectors of $1/\gamma \underset{\sim\sim}{AA}^T$. (Note that the variable x in the figure corresponds to the variable z in the text.) (1 bar = 100 kPa.)

x-axis, the eigenvalues of the inverse operator, in Eq. (19a), are the corresponding points on the dashed curve. Similarly, corresponding to these same eigenvalues on the x-axis are the eigenvalues of the approximating operator in Eq. (19b) on the family of solid curves for the various iteration indices k. Thus, if one of the k-curves in Fig. 2 were to lie everywhere on the dashed curve, that would be the curve $f_{k_o}(z)$ which we are seeking, because then any eigenvalue of the one operator would be an eigenvalue of the other, and the two operators, in Eqs. (19a) and (19b) would be identical--giving us the duality we are looking for. However, it is obvious from the figure that we cannot have perfect agreement between any of the k-curves and the dashed curve. Instead, there is a curve k_o which is in best "virtual" agreement with the dashed curve in the sense that the maximum departure between the curves is minimized.

An example will clarify the situation. A typical set of eigenvalues for $1/\gamma AA^T$ based on the six CO_2 weighting functions of the VTPR sounder is given by {0.03, 1.10, 3.84, 11.01, 39.87, 97.88}. The corresponding eigenvalues of the inverse operator, in Eq. (19a), are shown in Fig. 2 as circles on the dashed curve. Notice that the eigenvalues are distributed over the entire extent of the x-axis. This can always be made to occur because the convergence factor α is chosen in such a way that $(\alpha/\gamma) \| AA^T \| < 1$. Since $\| AA^T \|$ is equal to the largest eigenvalue of AA^T, picking α to be close to $\gamma / \| AA^T \|$ guarantees that the largest eigenvalue of $1/\gamma AA^T$ is near $f_1(x) = 0$, that is, near $x = 1/\alpha$ in Fig. 2.

On the other hand, the smallest eigenvalue will always be close to zero, because the matrix AA^T is ill-conditioned with respect to matrix inversion, that is, the ratio between the largest and smallest eigenvalue (the P-condition number) is in considerable excess of the value 100. Of course, it was this ill-conditioning that forced us to use regularization methods in the first place.

The smallest eigenvalue of $1/\gamma AA^T$ is the important one, because it corresponds to the largest one in the inverse operator in Eq. (19a). In fact, since the smallest eigenvalue is almost zero, the corresponding one in the inverse matrix will be almost one. It is clear from Fig. 2 that the approximate operator

Eq. (19b) will also have its largest eigenvalue near unity, and, in fact, the curves h(x) and $f_k(x)$ have their best agreement there. This is very encouraging because the norm of the operators in Eqs. (19a) and (19b) is equal to their largest eigenvalues. Therefore, the two operators will agree in the most important sense--their norms will be virtually the same.

We know from the theory of empirical orthogonal functions, see Ref. 4, that the structure of any positive definite matrix, such as in Eqs. (19a) or (19b), is determined primarily by its largest eigenvalue, and decreasingly so with decreasing eigenvalues; the smallest one contributes the least. It is clear from Fig. 2 that the larger eigenvalues of the matrices, in Eqs. (19a) and (19b), match up the best and that the intermediate ones match up to a lesser extent. The smaller ones compare favorably again, but we just noted that they are not very important.

It is now clear that, graphically at least, a case can be made for the existence of virtual duality between solutions, in Eqs. (11) and (8). (A more precise argument is developed in a subsequent section.) In other words, there exists a k_o such that the curve $f_{k_o}(z)$ is very similar to the curve h(z). That says that the eigenvalues of the matrices, in Eqs. (19a) and (19b), are almost the same; consequently, the matrices themselves are almost the same. This, in turn, says that Eqs. (15) and (8) give virtually the same solutions which, finally, establishes the duality between solutions in Eqs. (11) and (8).

C. Convergence Properties

Figure 2 is useful also in analyzing convergence properties, but first we simplify the interpretation of the figure by letting $t = \alpha z$. Then, t, h, and all the f_k's vary between 0 and 1, and Eqs. (20a) and (20b) become, respectively,

$$f_k(t) = (1 - t)^k \tag{21a}$$

and

$$h(t) = (1 + t/\alpha)^{-1} \tag{21b}$$

Now imagine that the x-axis in Fig. 2 has been replaced by a t-axis that varies between 0 and 1. This allows us to answer convergence questions very easily, because now the family of f_k-curves remains fixed for variable α and the single dashed curve moves instead.

The first question is: What happens when one chooses a smaller α? From Eq. (21b), it is clear that a smaller α produces a smaller value of h. Thus, the elbow of the dashed curve in Fig. 2 will move even further into the lower left-hand corner of the figure and become more L-shaped. This results in better agreement with one of the f_k-curves, but a greater number of iterations are required for convergence. Therefore, the smaller the choice of α, the truer the duality, but the larger the number of iterations that are required.

The converse of the foregoing is that as α increases, the number of iterations decreases, but the duality between solutions, in Eqs. (11) and (8), degrades. If the value of α is made too large, it may exceed $\gamma/\|AA^T\|$, in which case Eq. (11) diverges. What is the graphical interpretation of a divergent solution in Fig. 2? In this case, the dashed curve will lie entirely above the curve labeled "k = 1," except at the point (0,1). Under these circumstances, no amount of iteration will yield a solution--not even the first iteration. In fact, the more the solution is iterated, the further the solution, in Eq. (11), will move from the correct solution in Eq. (8).

If we have two initial approximations of different accuracy, will the more accurate approximation require fewer iterations than the less accurate one? A better initial approximation means that

γ in Eqs. (8) and (11) must be increased. This implies that α can be increased, since $\alpha < \gamma/\|AA^T\|$. Hence, our question is answered in the affirmative. More is said about the initial approximation in a subsequent section.

Finally, Fig. 2 clearly illustrates the important fact that the iterative analogue of regularization is to terminate the iterations after a finite number of steps. The dashed curve in the figure lies completely to the right of the solid curve labeled "k = 128". Therefore, if one were to iterate to convergence, the agreement between $h(z)$ and $f_\infty(z)$ would be quite poor. Clearly, the optimum number of successive approximations is finite--being fewer than 128 when $\alpha = 0.01$.

D. Proof that the Error Norm is Satisfied

All that remains to be proved is that index k_o is such that $\|Ax^{k_o} - y\| \simeq \|\varepsilon\|$, where x^{k_o} is given by Eq. (11), or equivalently, Eq. (14). To prove this, multiply Eq. (14) by the matrix A and apply the identity, in Eq. (6), to the resulting equation, that is,

$$Ax^{k_o} - y = (I_n - \alpha/\gamma AA^T)^{k_o} (Ax^o - y)$$

$$\simeq (I_n + 1/\gamma AA^T)^{-1} (Ax^o - y) \qquad (22)$$

where the relationship, in Eq. (16), was used for the second line.

On the other hand, multiplication of Eq. (8) by the matrix A and application of the identity in Eq. (17) yield

$$A\hat{x} = Ax^o + \left[I_n - \gamma(AA^T + \gamma I_n)^{-1} \right] (y - Ax^o)$$

which implies that

$$A\hat{x} - y = (I_n + 1/\gamma AA^T)^{-1} (Ax^o - y) = \varepsilon \qquad (23)$$

where the last equality follows from the fact that $\hat{x}$ satisfies Eq. (7). By comparing Eqs. (22) and (23), we see that $\| Ax^{k_o} - y \| \simeq \| \varepsilon \|$. Thus, the proof of virtual duality between the solutions in Eqs. (8) and (11), for $k = k_o$, is complete.

IV. CLOSENESS OF THE DUAL SOLUTIONS

A. Closeness of the Eigenvalues

In the previous section, we investigated the closeness of the curves $h(z)$ and $f_{k_o}(z)$ geometrically. Now we study the closeness of these two curves analytically to get a numerical estimate of their difference. The function $h(z)$ is fixed, but the functions $f_k(z)$ vary with k. Thus, the first task is to find that $k = k_o$ for which $f_k(z)$ is closest to $h(z)$. It is impossible to know k_o *a priori*, but an acceptable choice for k_o is the one that minimizes the squared differences between corresponding eigenvalues of the matrices in Eq. (16). However, the closed form solution of this minimization problem for arbitrary eigenvalues and arbitrary α is intractable. As an alternative, we will do the problem analytically, and with the aid of several approximations, arrive at a reasonable (but not optimal) *a priori* estimate of k_o.

Let $t = \alpha z$, then, since $0 \leq t \leq 1$, an analytic measure of the closeness between $h(z)$ and $f_k(z)$ of Eqs. (20a) and (20b) is given by the integral

$$F(k) = \int_0^1 \left[1/(1 + t/\alpha) - (1 - t)^k \right]^2 dt \qquad (24)$$

To find the k that minimizes this difference, we solve the equation $dF/dk = 0$, which implies that

$$\int_0^1 \frac{(1 - t)^k \ln(1 - t)}{1 + t/\alpha} dt = -(2k + 1)^{-2} \qquad (25)$$

If we could evaluate this integral in closed form for α and k, we would know k as a function of α. This cannot be done, but the approximation $(1 - t)^k \ln(1 - t) \simeq te^{-kt}$, valid for sufficiently large k, puts the integral into the form

$$-\int_0^1 \left[te^{-kt}/(1 + t/\alpha) \right] dt \simeq -\alpha/(2k) \tag{26}$$

In evaluating the integral in Eq. (26), it was assumed that k was sufficiently large so that $e^{-k} \simeq 0$, and the mean-value theorem was used to obtain the estimate

$$\int_\alpha^{\alpha+1} \frac{e^{-kt}}{t} dt = \frac{1}{\alpha + \theta} \int_\alpha^{\alpha+1} e^{-kt} dt \simeq \frac{e^{-\alpha k}}{2\alpha k}$$

for $0 < \theta < 1$. It was found empirically that a reasonable value for θ is α. This choice is based on the observation that normally $\alpha << 1$, and so in the interval $[\alpha, \alpha + 1]$ the function $1/t$ contributes much more to the integrand at α than at $\alpha + 1$.

Equations (25) and (26) can now be combined to yield

$$k = \frac{1}{4\alpha} \left[(1 - 2\alpha) + (1 - 4\alpha)^{1/2} \right]$$

But since α is small, the approximation $(1 - x)^n \simeq 1 - nx$ is valid, and we obtain the following explicit expression for a k that makes the difference, in Eq. (24), small, but does not necessarily minimize it:

$$k_o = \frac{1}{2\alpha} - 1 \tag{27}$$

The formula in Eq. (27) is an acceptable approximation for k_o whenever $\alpha < 0.1$.

Recall that in Fig. 2, we chose $\alpha = 0.01$, in which case $k_o = 49$ by the formula in Eq. (27). Examination of Fig. 2 confirms that the curve $k = 49$ is a reasonable estimate, particularly for matching the larger eigenvalues of the inverse operators.

We can now go back to Eq. (24) and estimate $F(k_o)$. Using

the approximation

$$F(k_o) \leq \int_0^1 \frac{dt}{(1 + t/\alpha)^2} + \int_0^1 (1 - t)^{2k_o} dt,$$

we find by Eq. (27) that

$$F(k_o) \leq 2\alpha/(1 - \alpha^2) \tag{28}$$

The importance of the relationship, in Eq. (28), is that it says that the integrated difference of Eq. (24) approaches 0 as $\alpha \to 0$. In other words, $h(z)$ and $f_k(z)$ can be made arbitrarily close by making α sufficiencly small. Thus, the smaller α is chosen, the closer the virtual duality is to true duality.

B. Closeness of the Twomey-Phillips Solution to Its Dual

In the previous section, we studied the closeness of the eigenvalues of the dual matrices; now, we examine the closeness of the solutions themselves. Shown in Fig. 3 are the rms differences between the Twomey-Phillips inverse matrix solution and its iterative dual for two different initial approximations. The dashed curve represents the rms differences when forecasts are used as the initial profile, and the solid curve represents the rms differences when average climatological initial profiles are used.

The structure of Fig. 3 is the same as that of Fig. 1, except that the abscissa is in tenths of degrees instead of degrees. The data used is also the same as that described for Fig. 1, that is, the same 139 profiles and corresponding simulated radiances are used.

The results in Fig. 3 are as expected; the difference between the two curves is negligible. This indicates that the dual solutions agree with one another irrespective of the initial profile used, as long as the same one is used by both inversion methods.

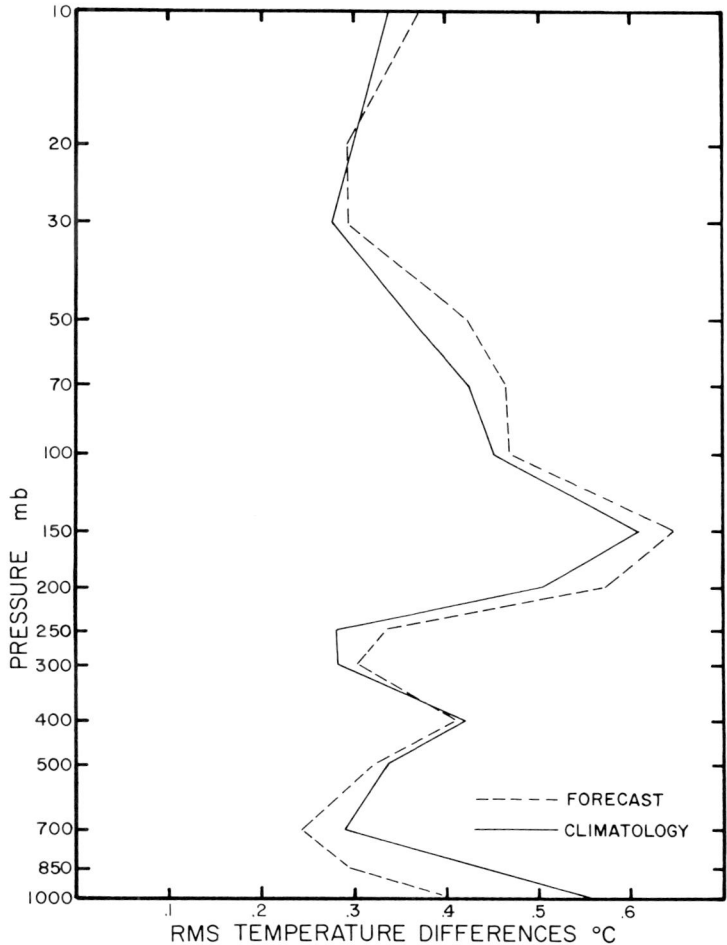

Fig 3. Profiles of the rms differences between the dual solutions of the Twomey-Phillips method using forecast (dashed curve) and climatological (solid curve) initial profiles. (1 bar = 100 kPa.)

On the other hand, the accuracy of each of the two curves, although reasonable, is not quite as good as expected. There are three reasons for this.

First, the γ used in the inverse matrix Twomey-Phillips solution was the same value for all 139 cases. This prevented some of the solutions from satisfying the requirement that $\| A\hat{x} - y \| \simeq \| \varepsilon \|$. The scalor γ should have been adjusted in each solution relative to the magnitude of the difference $\| Ax^o - y \|$. However, we do not have a reasonable *a priori* method for making this adjustment, and so a constant average value for γ was used.

Second, the number of iterations for the iterative dual was limited to 100 which, according to Eq. (27), is twice the number needed for our choice of $\alpha = .01$. Nevertheless, a few of the cases did not converge in the sense that $\| Ax^{k_o} - y \| \simeq \| \varepsilon \|$. Again, the problem stems directly from the choice of γ, which, in turn, affects the convergence factor α, since $\alpha < \gamma / \| AA^T \|$.

Third, the α used was always about the largest value permissible in order to keep the number of iterations low. However, we saw in the previous section that by decreasing the value of α and increasing the number of iterations, we could have improved the accuracy.

The overall accuracy in Fig. 3 is roughly between 0.3 and $0.4^\circ C$, except at the tropopause and at the surface. At these two levels, there is less agreement between the dual solutions, apparently due to the usual problem of obtaining accurate solutions there by any method.

V. EQUATIONS OF A MORE GENERAL FORM

A. The Dual Equations

Details of the principle of virtual duality were worked out in the previous sections for the Phillips-Twomey inversion method

COMPARISON OF LINEAR INVERSION METHODS

as an outgrowth of the duality principle for the least squares solutions. However, we need not stop there. The principle of virtual duality can be generalized to include most of the better known regularized linear inversion methods.

The most general of the regularized inverse matrix solutions of Eq. (7) that we will consider has the form

$$\hat{x} = x^o + VA^T(AVA^T + W)^{-1} (y - Ax^o) \quad (29)$$

where V and W are arbitrary symmetric positive definite matrices of dimensions $m \times m$ and $n \times n$, respectively, subject to the constraint $\|A\hat{x} - y\| \simeq \|\varepsilon\|$. Hence, V and W have inverses.

The iterative dual of Eq. (29) is of the form

$$x^k = x^{k-1} + \alpha VA^T W^{-1} (y - Ax^{k-1}) \quad (30)$$

which is obtained in the same way that Eq. (11) was derived from Eq. (8), except that the identity, in Eq. (9), must be replaced by the more general identity

$$VA^T(AVA^T + W)^{-1} = (A^T W^{-1} A + V^{-1})^{-1} A^T W^{-1} \quad (31)$$

The dual Eq. (30) converges provided $\alpha < 2/\|VA^T W^{-1} A\|$. As before, we limit the number of iterations to $k = k_o$, where k is such that $\|Ax^{k_o} - y\| \simeq \|\varepsilon\|$. If we were to iterate to convergence, we would have

$$\lim_{k \to \infty} x^k = VA^T W^{-1} (AVA^T W^{-1})^{-1} y = VA^T (AVA^T)^{-1} y \quad (32)$$

which is the weighted least squares solution with weighting matrix V.

Note that Eqs. (29) and (30) are simply generalizations of Eqs. (8) and (11), respectively, in which $1/\gamma A^T$ has been replaced by $VA^T W^{-1}$. Consequently, all the previous discussion relating to Eqs. (12) through (28) holds for the more general situation in which $1/\gamma A^T$ is replaced everywhere by $VA^T W^{-1}$. In particular, the key substitution, in Eq. (16), is replaced by the substitution

$$(I_n + \underset{\sim}{A}\underset{\sim}{V}\underset{\sim}{A}^T\underset{\sim}{W}^{-1})^{-1} \tag{33}$$

for $(I_n - \alpha \underset{\sim}{A}\underset{\sim}{V}\underset{\sim}{A}^T\underset{\sim}{W}^{-1})^k$, and the eigenvalue analysis is applied to $\underset{\sim}{A}\underset{\sim}{V}\underset{\sim}{A}^T\underset{\sim}{W}^{-1}$ instead of $\underset{\sim}{A}\underset{\sim}{A}^T$. Therefore, the eigenvalues λ_i are different as is the definition of z in Eqs. (20a) and (20b). However, none of the analyses associated with Fig. 2 change. Only the locations of the circles on the dashed curve in Fig. 2 and the scale of the abscissa must be altered. Thus, we see that virtual duality holds between Eqs. (29) and (30) as well.

We are now ready to reap the benefits of all our work to this point. Establishment of the principle of virtual duality for the general Eqs. (29) and (30) permits us to examine the various linear inverse matrix and iterative inversion methods and immediately write down their respective duals. Furthermore, knowledge of one of the duals implies immediate knowledge of the other without further investigation.

B. Special Cases

In this section, we look at the more familiar linear inversion methods in the context of the duality principle. Included in the matrix inverse methods are those by Twomey and Phillips (Refs. 1 and 2), Twomey H-matrix types (Refs. 2 and 5), Crone (Ref. 6), and Rodgers, Strand and Westwater (Refs. 1, 7, 8 and 9). Table 1 indicates the particular forms that $\underset{\sim}{V}$ and $\underset{\sim}{W}$ have for the special cases just cited. (Note that the identity, in Eq. (31), is required to reconcile Table 1 with the expressions in the original publications.)

Since the methods listed in Table 1 are all inverse matrix methods, their iterative duals are of particular interest. For example, the Rodgers-Strand-Westwater case by Eq. (30) becomes

TABLE 1

*Identification of Inverse Matrix Methods
in Terms of the V and W Matrices*

Method	V	W
Twomey-Phillips	I_m	γI_n
Twomey H-matrix	$1/\gamma H^{-1}$	I_n
Crone	$1/\gamma I_m$	$(AA^T)^{-1}$
Rodgers-Strand-Westwater[1]	$\left\{\begin{array}{c} \underset{\sim}{R} \\ \underset{\sim}{S}_f \end{array}\right.$	$\left.\begin{array}{c} \underset{\sim}{E} \\ \underset{\sim}{S}_\varepsilon \end{array}\right\}$

[1] $\underset{\sim}{R} \equiv \underset{\sim}{S}_f$ is the profile covariance matrix, and
$\underset{\sim}{E} \equiv \underset{\sim}{S}_\varepsilon$ is the noise covariance matrix.

$$\underset{\sim}{x}^k = \underset{\sim}{x}^{k-1} + \alpha R A^T \underset{\sim}{E}^{-1} (\underset{\sim}{y} - \underset{\sim}{A}\underset{\sim}{x}^{k-1}) \tag{34}$$

which is a form that apparently has never appeared in the literature. This is true as well of the iterative Twomey H-matrix and iterative Crone methods.

Among the iterative methods, we have those by Landweber (Ref. 10), W. L. Smith (Refs. 1 and 11), Conrath-Revah (Ref. 12), and Strand (Ref. 10). These are summarized in Table 2 in the same format as Table 1.

Of particular interest are the inverse matrix duals of the iterative methods of Table 2. For example, by Eq. (29) and the identity, in Eq. (31), the Smith and Conrath-Revah cases can be written in the apparently new form

$$\hat{\underset{\sim}{x}} = \underset{\sim}{x}^o + (\underset{\sim}{A}^T\underset{\sim}{A} + \underset{\sim}{D})^{-1}\underset{\sim}{A}^T(\underset{\sim}{y} - \underset{\sim}{A}\underset{\sim}{x}^o) \tag{35}$$

which is a Twomey H-matrix type solution with $\underset{\sim}{H}$ equal to a diagonal matrix.

Listings and categorizations similar to those of Tables 1 and 2 could be made for square inversion systems, such as the linearized Chahine (Ref. 13), Chiu (Ref. 14), expansion in bases functions (Refs. 8, 9, and 15), and empirical orthogonal functions (Ref. 4). However, the theory and duality procedures for square

TABLE 2

Identification of Iterative Methods in Terms of the V and W Matrices

Method	V	W
Landweber	I_m	I_n
W. L. Smith (linearized)[1]	D^{-1}	I_n
Conrath-Revah[2]	G^{-1}	I_n
Strand	$(A^T A + \gamma I_m)^{-1}$	I_n

$$\left.\begin{array}{l} {}^1 D = \text{diag } (A^T \vec{I}_n) \\ {}^2 G = \text{diag } (A^T \vec{I}_m) \end{array}\right\} \quad \vec{I}_\ell = \underbrace{(1,1,\ldots,1)}_{\ell}{}^T$$

systems are essentially special cases of the rectangular systems just described; therefore, there is no need for further illustration.

VI. THE INITIAL APPROXIMATION

A. Theory

At the beginning of this paper, where the iterative dual of the least squares solution was discussed (Eq. (4)), we saw that in the limit, as $k \to \infty$, the term containing the initial approximation $\underset{\sim}{x}^o$ vanished. This is a case in which the solution is independent of the initial approximation. Unfortunately, this is not true for any of the regularized solutions discussed in this paper.

To see just how the solutions depend upon the initial approximation, let $\bar{\underset{\sim}{x}}$ represent the solution of any of the inversion methods of this paper when no initial approximation is used, that is, when $\underset{\sim}{x}^o \equiv \underset{\sim}{0}$. Then, Eq. (29) can be written

$$\hat{\underset{\sim}{x}} = \bar{\underset{\sim}{x}} + \left[I_m - V A^T (A V A^T + W)^{-1} A \right] \underset{\sim}{x}^o \qquad (36)$$

and Eq. (14), with the substitution $V A^T W^{-1}$ for $1/\gamma A^T$, can be written

$$\underset{\sim}{x}{}^{k_0} = \underset{\sim}{\bar{x}} + (\underset{\sim}{I}_m - \alpha \underset{\sim\sim}{VA}{}^T \underset{\sim}{W}{}^{-1} \underset{\sim}{A})^{k_0} \underset{\sim}{x}{}^o \qquad (37)$$

Application of the identity

$$\underset{\sim}{I}_m - \underset{\sim\sim}{VA}{}^T (\underset{\sim\sim\sim}{AVA}{}^T + \underset{\sim}{W})^{-1} \underset{\sim}{A} = (\underset{\sim}{I}_m + \underset{\sim\sim}{VA}{}^T \underset{\sim}{W}{}^{-1} \underset{\sim}{A})^{-1} \qquad (38)$$

to Eq. (36) permits us to write it in a form similar to Eq. (37), namely

$$\underset{\sim}{\hat{x}} = \underset{\sim}{\bar{x}} + (\underset{\sim}{I}_m + \underset{\sim\sim}{VA}{}^T \underset{\sim}{W}{}^{-1} \underset{\sim}{A})^{-1} \underset{\sim}{x}{}^o \qquad (39)$$

Since the n × m (n < m) rectangular matrix $\underset{\sim}{A}$ has rank n, the rank of $\underset{\sim\sim}{VA}{}^T\underset{\sim}{W}{}^{-1}\underset{\sim}{A}$ is also n. Consequently, $\underset{\sim\sim}{VA}{}^T\underset{\sim}{W}{}^{-1}\underset{\sim}{A} \neq \pm \underset{\sim}{I}_m$, and the terms in Eqs. (37) and (39) containing the factor $\underset{\sim}{x}{}^o$ cannot vanish. Hence, the dependence of the solutions, in Eqs. (37) and (39) on $\underset{\sim}{x}{}^o$ is a permanent and significant feature of regularized solutions for rectangular systems. This is the price one must pay for stability.

If the matrix $\underset{\sim\sim}{VA}{}^T\underset{\sim}{W}{}^{-1}\underset{\sim}{A}$ is diagonalized by the m × m orthogonal matrix $\underset{\sim}{P}$, then the matrices in Eqs. (37) and (39) also can be diagonalized by $\underset{\sim}{P}$, that is

$$(\underset{\sim}{I}_m - \alpha \underset{\sim\sim}{VA}{}^T \underset{\sim}{W}{}^{-1} \underset{\sim}{A})^{k_0} = \underset{\sim}{P}(\underset{\sim}{I}_m - \alpha \underset{\sim}{\Lambda})^{k_0} \underset{\sim}{P}{}^T \qquad (40a)$$

and

$$(\underset{\sim}{I}_m + \underset{\sim\sim}{VA}{}^T \underset{\sim}{W}{}^{-1} \underset{\sim}{A})^{-1} = \underset{\sim}{P}(\underset{\sim}{I}_m + \underset{\sim}{\Lambda})^{-1} \underset{\sim}{P}{}^T \qquad (40b)$$

The functions of eigenvalues in Eqs. (40a) and (40b) are again of the form in Eqs. (20a) and (20b), respectively. Thus, not only do $\underset{\sim}{x}{}^{k_0}$ and $\underset{\sim}{\hat{x}}$ of Eqs. (37) and (39) depend on $\underset{\sim}{x}{}^o$, they depend on $\underset{\sim}{x}{}^o$ in virtually the same way. This result is certainly not unexpected, considering the origins of Eqs. (37) and (39).

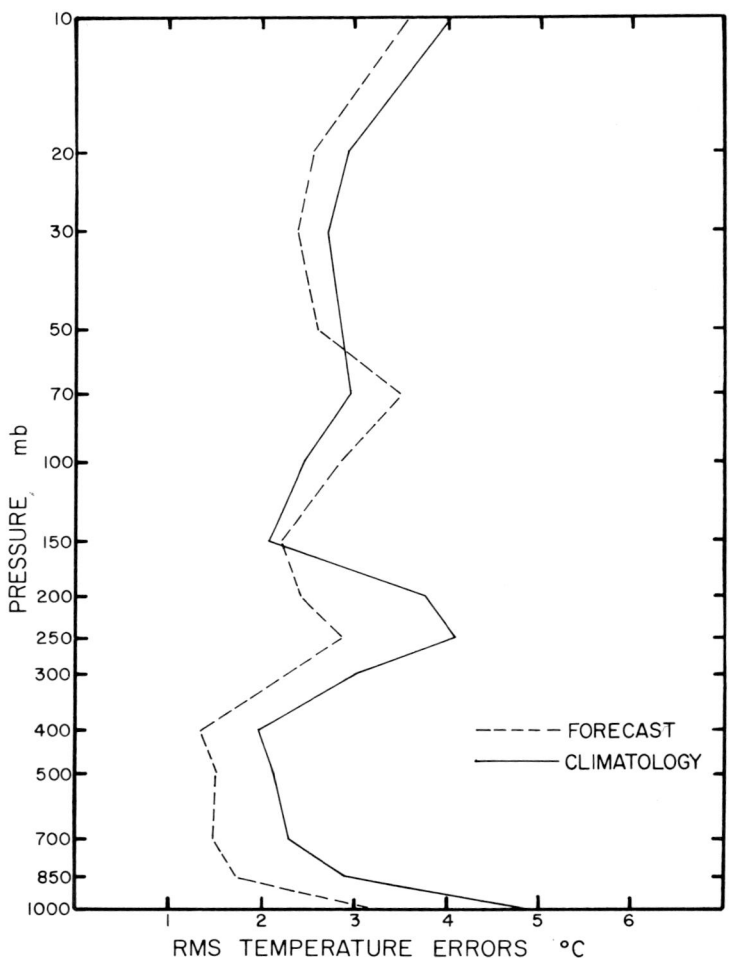

Fig. 4. Profiles of the rms errors between the 139 radiosondes and the Twomey-Phillips inverse matrix solutions when forecasts (dashed curve) and climatology (solid curve) are used as the initial profiles. (1 bar = 100 kPa.)

B. Results

Examples of the dependence of the solution on the initial approximation are afforded by the Twomey-Phillips inverse matrix solutions as shown in Fig. 4. As in Fig. 1, the curves are rms differences between 139 retrieved temperature profiles using simulated radiances, and the corresponding radiosonde profiles (used as truth). The dashed curve in Fig. 4 represents the rms errors, at the 15 mandatory pressure levels, when forecasts are used for the initial profile, while the solid curve represents the rms errors when climatology is used for the initial profile.

As expected, the retrievals based on the forecasts are more accurate than those based on climatology, except at 70, 100, and 150 mb. The pattern reversal at these three levels is due to 23 profiles between $0°$ and $18°N$ that have much better climatological initial profiles than forecast initial profiles in the vicinity of the tropopause. This anomaly occurred because the numerical forecast model applies only north of $18°$ latitude. Below that latitude a single average initial profile is used in lieu of a forecast, and this profile just happens to be worse than the climatological one in the tropical tropopause region.

VII. CONCLUDING REMARKS

True duality exists for the least squares solution, but the solution is unstable. Therefore, regularization methods must be used to achieve stability. The price one pays for this stability is the dependence of the solution on the initial approximation.

Virtual duality has been established for a very general class of linear regularized solutions. The regularization process for the iterative solutions is accomplished by terminating the iteration process when the norm of the residual errors is equal to the norm of the errors in the radiances.

The convergence process for the iterative duals can be demonstrated graphically. On the other hand, an estimate was derived for the terminal iteration index k_o associated with any given convergence factor α. For a given k_o the closeness of the eigenvalues of the dual solutions can be determined from the pertinent equations. The concept of "virtual duality" is meaningful because by reducing the size of α, the eigenvalues of the dual operators can be made arbitrarily close.

Explicit formulas for the inverse matrix solutions and their iterative duals are given for the more common of the regularized linear inversion methods. These include the following methods: Twomey-Phillips, Twomey H-matrix, Crone, Rodgers-Strand-Westwater, Landweber, W. L. Smith, Conrath-Revah, and Strand. This list is by no means complete. In addition, apparently new, inversion methods have arisen as duals of some of the known methods.

Finally, the whole point in developing the principle of vitual duality is that knowledge of one of the duals immediately implies knowledge of the other. Thus, the study of inversion methods is more unified through the link of duality.

SYMBOLS

$\underset{\sim}{A}$	$n \times m$ matrix of weighting functions (Eqs. (1) and (7))
$\underset{\sim}{B}$	arbitrary square matrix
$\|\underset{\sim}{B}\|$	Euclidean norm of square matrix B
D	$\mathrm{diag}(A^T \vec{1}_n)$
$\underset{\sim}{E}$	$\equiv \underset{\sim}{S}_\varepsilon$
G	$\mathrm{diag}(A^T A \vec{1}_n)$
$\underset{\sim}{H}$	Twomey H-matrix
F(k)	measure of closeness of eigenvalue functions f_k and h for arbitrary k (Eq. (24))
$f_k(z)$	$(1 - \alpha z)^k$ (Eq.(20a))
h(z)	$(1 + z)^{-1}$ (Eq. (20b))

COMPARISON OF LINEAR INVERSION METHODS 353

$\underset{\sim}{I}$	identity matrix; subscript ℓ, m and n denote the dimension of the square matrix
k	iteration index
k_o	terminal iteration index
$\underset{\sim}{P}$	m × m orthogonal matrix of eigenvectors
$\underset{\sim}{R}$	$\equiv \underset{\sim}{S}_f$
$\underset{\sim}{S}_f$	profile covariance matrix
$\underset{\sim}{S}_\varepsilon$	noise covariance matrix
T	transpose of a matrix
t	αZ
$\underset{\sim}{U}$	n × n orthogonal matrix of eigenvectors
$\underset{\sim}{V}$	arbitrary m × m symmetrix positive definite matrix (Eqs. (20) and (30))
$\underset{\sim}{W}$	arbitrary n × n symmetric positive definite matrix (Eqs. (29) and (30))
$\underset{\sim}{x}$	m-dimensional source function vector (Eqs. (1) and (7))
$\underset{\sim}{x}^o$	m-dimensional initial approximation vector
$\underset{\sim}{x}^k$	kth successive approximation of the m-dimensional iterative solution vector (Eqs. (3), (11), and (30))
$\underset{\sim}{\hat{x}}$	m-dimensional inverse matrix solution vector (Eqs. (2), (8), and (29))
$\underset{\sim}{\bar{x}}$	m-dimensional solution vector when no initial approximation is used (i.e., $x^o = 0$) (Eqs. (36), (37), and (39))
$\underset{\sim}{y}$	n-dimensional vector of measurements (Eqs. (1) and (7))
z	eigenvalue
α	scalar convergence (relaxation) factor (Eqs. (3), (11), and (30))
γ	smoothing (regularization) parameter (Eqs. (8) and (11))
$\underset{\sim}{\varepsilon}$	n-dimensional error vector (Eq. (7))
θ	parameter ($0 < \theta < 1$)
$\underset{\sim}{\Lambda}$	diagonal matrix of eigenvalues λ_i
λ	ith largest eigenvalue
$\approx$	"virtually equal to"

REFERENCES

1. H. E. Fleming and W. L. Smith, Inversion techniques for remote sensing of atmospheric temperature profiles in "Temperature: Its Measurement and Control in Science and Industry" (Harmon H. Plumb, Ed.), Vol. 4, pp. 2239-2250. Instrument Society of America, Pittsburgh, 1972.
2. S. Twomey, On the numerical solution of Fredholm integral equations of the first kind by the inversion of the linear system produced by quadrature, *J. Assoc. Comput. Mach.* 10 97 (1963).
3. H. E. Fleming, Status of mathematical inversion techniques, *Technical Digest of the Optical Society of America Spring Conference on Remote Sensing of the Atmosphere, Anaheim, Calif.* March 19-21, 1975, pp. ThA1-1 to ThA1-4.
4. J. C. Alishouse, L. J. Crone, H. E. Fleming, F. L. Van Cleef and D. Q. Wark, A discussion of empirical orthogonal functions and their application to vertical temperature profiles, *Tellus, 19*, 477 (1967).
5. S. Twomey, The application of numerical filtering to the solution of integral equations encountered in indirect sensing measurements, *J. Franklin Inst. 279,* 95 (1965).
6. L. Crone, The singular value decomposition of matrices and cheap numerical filtering of systems of linear equations, *J. Franklin Inst. 294,* 133 (1972).
7. C. D. Rodgers, Remote sounding of the atmospheric temperature profile in the presence of cloud, *Q. J. R. Meteorol. Soc. 96,* 654 (1970).
8. O. N. Strand and E. R. Westwater, Statistical estimation of the numerical solution of a Fredholm integral equation of the first kind, *J. Assoc. Comput. Mach. 15*, 100 (1968).
9. O. N. Strand and E. R. Westwater, Minimum-RMS estimation of the numerical solution of a Fredholm integral equation of the first kind, *SIAM J. Numer. Anal. 5,* 287 (1968).

10. O. N. Strand, Theory and methods related to the singular-function expansion and Landweber's iteration for integral equations of the first kind, *SIAM J. Numer. Anal. 11*, 798 (1974).

11. W. L. Smith, Iterative solution of the radiative transfer equation for the temperature and absorbing gas profile of an atmosphere, *Appl. Opt. 9*, 1993 (1970).

12. B. J. Conrath and I. Revah, A review of nonstatistical techniques for the estimation of vertical atmospheric structure from remote infrared measurements, *Proc. of a Workshop on the Mathematics of Profile Inversion*, Ames Research Center, Moffett Field, Calif. 1972. [NASA TM X-62,150, pp. 1-36.]

13. M. T. Chahine, Determination of the temperature profile in an atmosphere from its outgoing radiance, *J. Opt. Soc. Am. 58*, 1634 (1968).

14. J. S. H. Chiu, A self-contained iterative algorithm for a numerical solution to the radiative transfer equation, *Beitr. Phys. Atmos. [Contrib. to Atmos. Phys.] 48*, 185 (1975).

15. D. Q. Wark and H. E. Fleming, Indirect measurements of atmospheric temperature profiles from satellites: I. Introduction, *Mon. Weather Rev. 94*, 351 (1966).

DISCUSSIONS

Staelin: Have you considered whether you can convert from the virtual duality principle to a true duality principle by altering the way the matrix is regularized? You employed a constant times the identity matrix.

Fleming: Yes.

Staelin: Suppose one regularized, for example, by using different elements on the diagonal. In other words, is there some way to make it a true duality rather than only a virtual duality?

Fleming: I couldn't find any. The difficulty is this: in any iterative scheme when you telescope together the various iterates, you always end up with a matrix that is raised to a power. Of course, its dual always involves an inverse matrix. I just don't see any way in which a power matrix can ever be made exactly equal to an inverse for a finite number of iterations. That is the real heart of the difficulty.

Strand: I wanted to say that I notice the version that you considered is a scalar α, at least at first.

Fleming: Yes.

Strand: And this, I believe, you can obtain without using different alphas by dividing both sides of the equation by suitable factors in the first place. I was in contact with Landweber one time and he told me that I could get the effect of my D matrix in this way. Well, the D that I had was something slightly different in that my α was a matrix. So, instead of using your scalar matrix, one fudges the thing part way by getting a matrix that does part of the job and then iterates with that for the rest of it. I would be surprised if there wasn't the same kind of duality for that situation. In fact, in the case when D equals Twomey's matrix, we are able to find a very close known relationship between the γ that you use, say, for ten iterations and the γ used for one iteration, which is essentially Twomey's process to start with. So, although you didn't demonstrate this, I would certainly think that the same kind of duality would hold for a more general assignment of what I call the D matrix.

Fleming: Yes, intuitively I feel it is true for the linear problem. That might be another way to approach the problem.

Cerni: You said that your solution must always depend on the initial guess, and there exists no alternative. Certainly you can entirely replace, if you choose, the constraint to an initial guess with just a smoothing constraint. Then there will be no initial guess dependence. What sort of results do you get with purely a smooth constraint in the inversion?

Fleming: Getting rid of the initial guess is equivalent to using a zero vector in the x_o vector. You can smooth all you like, but the larger you make your smoothing parameter in any of these formulations, the closer you drive the solution toward the initial approximation. So, if you started with a zero vector, you are just going to force the answer to look more and more like that constant value, which is certainly smooth. But it is still dependent on the initial guess. Do you disagree? Did I answer your question?

Cerni: No, I mean in certain instances we get fine results with a smoothing constraint alone by reducing the high frequency components of the solution. With no initial guess, we get a very representative solution.

Fleming: I don't know in what context you are talking about this. First of all, do you have a linear problem?

Cerni: Yes, it has been linearized.

Fleming: Normally, what kind of vectors are you looking for? Are you looking for something that is close to zero?

Cerni: No.

Fleming: The chances are that you have a stable system and then the regularized solutions are not necessary. Did you look at the conditioning of the matrix you are inverting? If it is well-conditioned you don't need a regularization method. In that case, you should use the least squares solution. We saw that the least squares solution is independent of initial approximation. So I think what we are saying is that your problem is in a different context than mine is. [Added in proof: In retrospect, it appears that Dr. Cerni always uses the same first guess (the implied zero vector); therefore, the question of the dependence of the solution on the initial guess is not germane.]

Herman: When you say that the final solution is dependent on the initial guess, what degree of dependence do you mean? I will be showing this afternoon, and earlier in the week, some of the Arizona group have also shown that inversions of real data starting with widely varying first guesses almost overlap each other. Now this was done using an H matrix and not the other technique you were talking about, which you said does not fit in this category, but, nevertheless, it is relatively insensitive to the initial guess and is an ill-conditioned matrix.

Fleming: I am simply saying that the better the initial approximation, the better reliance you can have on the solution, resulting in narrower error bars around your solution. I am not saying it is dependent in a particular way; I am saying there is a strong dependence which differs with each different kind of initial approximation and with each different method of solution.

Rodgers: One would hope that the solution would depend on the initial guess because the initial guess is a virtual measurement and the solution also depends on the measurements. That wasn't what I really wanted to ask. I feel sure that some kind of linear transformation can turn the more general linear method (where you have two covariance matrices) into a form with a dual by suitable eigenvector transformation.

Fleming: I feel it can, but I just don't know how.

Rodgers: Well, you can transform the covariance matrix to a diagonal form or a unit form.

Fleming: Yes, but even if they are diagonal, they don't commute. You can't move the diagonal from one side of the matrix to the other side. This is where the whole thing broke down on me.

Rodgers: You can turn a covariance matrix into a unit matrix by a suitable scale transformation and rotation. That's what you need, isn't it?

Fleming: But then will it be the same matrix? Will it hold for A^TA as well as the transformed matrix?

Rodgers: Well, it will give you a different A. If you apply a suitable transformation to X, such that what is equivalent to the covariance of X turns into unit matrix, then A is going to change to something else. And then I think you end up with a form that has a dual.

Fleming: If you can do that, my theory applies and one has duality. [Added in proof: In the final text of the paper, this difficulty has been overcome, and the virtual duality principle is shown to hold for all of the better known linear inversion methods.]

Twitty: This is just a comment. You told Dr. Cerni that it sounded like his solution might fall in the category of having a stable solution, because the inverse of the matrix really exists. In most of the cases we have, that is not true. We have an unstable matrix. There must be something that goes continuously from the case where you have to regularize it and where you don't.

COMPARISON OF LINEAR INVERSION METHODS

There must be some measure, or scale, in this analysis that could measure how stable a situation you have.

Fleming: Yes, and that is usually the so-called P-condition number of A, which is the ratio of the largest to the smallest eigenvalue of A^TA. That is one way of looking at it, if you want to do an eigenvalue analysis on the matrix.

Twitty: Yes, but perhaps a measure that would relate to something more physical, because we are talking about whether we have virtual data or not virtual data in the solution.

Fleming: One quick thing that can be done is to solve the problem in the best way you know. Then perturb the input vector of measured quantities in a systematic and realistic way and solve the problem again. If the new solution differs from the original solution in a proportionate and physically acceptable way, the problem is well-conditioned. Otherwise, some kind of regularization and/or additional virtual data are needed.

Twitty: Yes, I understand you can take a specific problem like that and determine the regime you are in. You seem to have been attacking the problem for getting more physical insight into things.

Westwater: You mentioned that the duality did not exist when you considered the second or first derivative constraint. Have you used this type of constraint in temperature retrievals? And if you have, how does that compare with the standard Twomey minimization around an initial guess?

Fleming: They are generally very close. The nice thing about the smoothing techniques is that the answer is relatively insensitive to the kind of smoother you use. The type of constraint used is critical only in those regions where the weighting functions are close to zero. Let me give you an example. We were looking at some stratospheric sounding problems at one time which involved the so-called "sudden warming." If you use a sounder such as the VTPR for which most of the weighting functions peak the troposphere, there is little information about the stratosphere. So if you have a sudden warming situation, the radiances corresponding to the stratosphere increase drastically. Consequently, when you retrieve a temperature profile, all that excess energy is dumped into the lower part of the atmosphere where the weighting functions peak, and you lose all the information on the stratospheric warming. After experimenting with the various kinds of smoothers, we found that we could restore the sudden warming portion of the temperature profile, which is above the weighting function peaks by using the right kind of smoother. In other words, the smoothing constraints provide the information where the weighting functions trail off to zero, but have little effect on the rest of the solution. This is completely in the vein of Clive Rodgers' talk. You have to know *a priori* the kind of information you are working with and use it in the correct way. All this is as much an art as it is a science.

Twomey: Henry, I never thought I'd hear you say that!

INVERSION OF PASSIVE MICROWAVE REMOTE

SENSING DATA FROM SATELLITES

David H. Staelin
Massachusetts Institute of Technology

Global passive microwave observations from Earth-orbiting satellites have mapped humidity and liquid water over ocean, temperature profiles, ice and snow, and other geophysical parameters. Future satellites will extend the altitude range above 100 km and will expand the list of trace constituents that can be monitored.

In most applications, the inversion problem is adequately approximated as linear with jointly Gaussian statistics, and, thus, a linear retrieval performs well. In some cases, the problem is typically factored into a decision process followed by appropriate linear or quasi-linear processes. Certain problems, however, require more powerful nonlinear or nonstationary procedures, such as Kalman filtering.

I. INTRODUCTION

Passive microwave sensing techniques are employed in geophysics, radioastronomy, biomedical and other areas. The special value of these techniques results primarily from the facts that

1. microwave receivers can sense thermal radiation with one-second sensitivities as high as 10^{-2}K.

2. microwave radiation penetrates most matter much more readily than does infrared radiation, which is the usual alternative technology; and

3. many gases have microwave resonances which can readily be resolved, and which permit observation even in some cases where

other spectral regions are inadequate. Geophysical applications have been reviewed by Staelin (Ref. 1), Tomiyasu (Ref. 2), Waters (Ref. 3), and others.

One important property of microwave thermal radiation results because $h\nu \ll kT$ for most cases of interest (where h is Planck's constant, ν is frequency, k is Boltzmann's constant, and T is temperature). For example, if $T \gtrsim 100$ K, then ν must be much less than 2000 GHz (2×10^{12} Hz). This constraint is usually met, and then Planck's radiation law passes to the Rayleigh-Jeans limit, where thermal radiation intensity I is

$$I - \frac{2kT}{\lambda^2} \quad (W\ m^{-2}\ Hz^{-1}\ ster^{-1}) \tag{1}$$

(λ is wavelength) and the power in a single-mode transmission line is

$$P = kT \quad (W\ Hz^{-1}) \tag{2}$$

The linear relation between temperature and power generally results in more nearly linear retrieval problems when data are to be interpreted.

It also follows that the responsiveness ($\Delta P/\Delta T$) of receivers is not degraded for low-temperature sources. Receiver sensitivities ΔT_{rms} (K) typically depend on radio-frequency bandwidth B (Hz), integration time τ (sec), receiver noise temperature T_R (K), and the antenna or source temperature T_A (K), as follows:

$$\Delta T_{rms}(K) \approx 2\ \frac{T_A + T_R}{\sqrt{B\tau}} \tag{3}$$

The source temperature, viewed from a terrestrial satellite, might be ~250 K, and T_R (below 200 GHz) could range from 10 to 5,000 K, with 1000 K being typical. B might range from 10^6 to 10^9 Hz, with 10^8 being typical; therefore, ΔT_{rms} for 1-sec averaging might be 0.01-10 K, with 0.24 being typical. Note that the receiver sensitivity actually improves for colder targets. Receiver and antenna

MICROWAVE REMOTE SENSING DATA

technology for remote sensing is similar to that for radioastronomy, which has been reviewed by Kraus (Ref. 4), and others.

The equation of radiative transfer appropriate for microwaves ($10^8 - 10^{12}$ Hz) in the terrestrial atmosphere is usually approximated as that for a slightly lossy medium in local thermodynamic equilibrium. Scattering from tropospheric inhomogeneities is very weak and, like refractive and ionospheric effects, is usually neglected for passive microwave remote sensing purposes. Exceptions occur primarily for

1. tropospheric rays parallel ($\pm \approx 1°$) to the geoid, when refractive effects become important; and

2. scattering from cloud droplets, which becomes important only when $\lambda \gtrsim \frac{d}{\pi n}$, where n and d are the refractive index and diameter of the drops, respectively.

Below 60 GHz, this situation is important primarily for heavy precipitation. Even then the absorption of those same clouds can overwhelm the scattering effects, which generally remain a modest perturbation in computations of cloud brightness temperatures. Microwave scattering from clouds has been discsused in many places, such as Refs. 5 and 6. Scattering is most often discussed in the context of radar, because then even a weak scattered signal from a cloud is essentially the only signal.

The usual simplicity of microwave radiative transfer also simplifies data interpretation. The equation of radiation transfer can usually be approximated as

$$T_B = RT_{sky} e^{-2\tau_o} + (1-R) T_{surf} e^{-\tau_o}$$

$$+ Re^{-\tau_o} \int_o^\infty T(\ell') \alpha(\ell') e^{-\int_o^{\ell'} \alpha(\ell') d\ell} d\ell'$$

$$\int_o^\infty T(\ell) \alpha(\ell) e^{-\int_\ell^\infty \alpha(\ell) d\ell} d\ell \qquad (4)$$

where T_B is the brightness temperature viewed from a satellite. ℓ is distance along a ray emitted upward from the terrestrial surface, and ℓ' is distance along a ray incident upon the surface; $\ell = \ell' = 0$ refers to the surface, and $\ell = \infty$ refers to some point above the atmosphere. R is the surface reflectivity, T_{sky} and T_{surf} are the brightness and kinetic temperatures of space and the terrestrial surface, respectively. $\alpha(\ell)$ is the atmospheric absorption coefficient and τ_o is the integral of α along ℓ from 0 to ∞.

The equation of radiative transfer is thus generally dominated by atmospheric absorption, thermal emission processes, and surface reflection characteristics. Surface characteristics vary considerably, some surfaces being largely specular, such as the ocean (Ref. 7), and some scattering or absorbing strongly, such as snow or ice (Ref. 8) or forested land (Ref. 9). The surface reflectivity of the ocean or fresh water might be ≈0.4 to 0.6, whereas that of ice and snow might be ≈0.7 to 1.0, and that of land might be ≈0.8 to 1.0.

The dominant microwave absorbers in the atmosphere are H_2O, O_2, O_3, clouds, and precipitation. Figure 1, prepared by P. W. Rosenkranz,[1] illustrates the typical opacity of the terrestrial atmosphere due to water vapor, oxygen, and clouds. Many typical trace constituents also have weak but quite observable spectra. The microwave spectral properties of the atmosphere have been reviewed by Staelin (Ref. 1), Waters (Ref. 3), Rosenkranz (Ref. 10), Liebe (Ref. 11), and others. With the recently available sensitive receivers in the 100-300 GHz region, many additional microwave spectral lines are being detected in the atmosphere for the first time, such as those of O_3 and CO near 100-200 GHz.

Below altitudes of ≈75 km pressure broaderning dominates the width of most microwave molecular resonances and, thus, provides a means for determining the pressure altitude at which

---

[1] Personal communication.

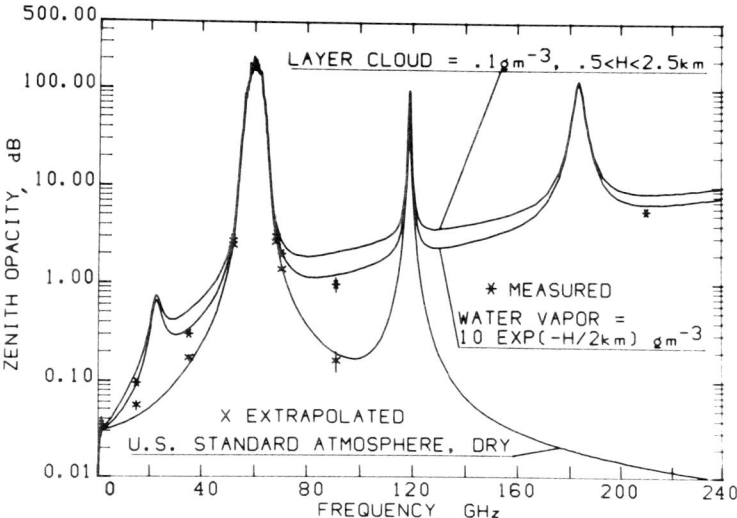

Fig. 1. Sea level zenith opacity for oxygen in the U.S. Standard Atmosphere and for 2 cm of precipitable water vapor distributed exponentially with 2 km scale height. The effects of a 2-km thick stratus cloud are also shown. Typical measurements for O_2 alone and O_2 plus H_2O are indicated by asterisks. (Figure supplied by Rosenkranz.)

contributions to an observed spectral line originated. Typical microwave receivers achieve adequate sensitivities with bandwidths of ≈1-100 MHz, which are generally much less than the observed linewidths of 2-2000 MHz. This ability to measure easily the precise shapes of most spectral lines contrasts with the situation for most infrared detectors, which typically observe broad bands that overlap many resonances. Even gaseous selective absorption infrared filters only partially limit this smoothing of spectra and consequent degradation of altitude information.

The altitude dependence of microwave emission differs in another way from that of infrared sensors. The linear relation between temperature and radiated microwave power differs from the strongly nonlinear temperature dependence of the Planck function in the thermal infrared. This results in altitude sensitivities of microwave spectra that are only modestly dependent on atmospheric temperature profiles, whereas these temperatures can have a profound effect on spectra in the thermal infrared.

II. SOUNDING OF ATMOSPHERIC TEMPERATURE PROFILES

Two basic geometries are appropriate for satellite sensing of atmospheric temperature profiles: the "nadir mode" wherein the satellite sensor views the atmosphere at any desired angle such that the ray intercepts the terrestrial surface, and the "limb-scanning mode" for which the ray does not intercept the earth but passes through the atmosphere at the limb of the planet. Only nadir-mode microwave systems have been built to date.

At present only two satellites carry passive microwave sounders for temperature profile measurement: Nimbus-5 carries Nimbus E Microwave Spectrometer (NEMS), launched December 12, 1972 (Ref. 12); and Nimbus-6 carries Scanning Microwave Spectrometer (SCAMS), launched June 11, 1975 (Ref. 13). They each have two channels near 1-cm wavelength for observing the surface and atmospheric water, and three channels near 0.6-cm wavelength which are on the opaque wings of the 0.5-cm oxygen absorption complex; they measure thermal radiation originating from different atmospheric levels (0-18 km)

Temperature profiles may be determined because each channel $T_B(\nu)$ responds differently to the temperature profile $T(\ell)$ as characterized by standard weighting functions $W(\nu,\ell)$:

$$T_B(\nu) \simeq T_{const} + \int_0^\infty T(\ell)W(\nu,\ell)d\ell \qquad (5)$$

The microwave temperature weighting functions presented in the literature to date are formed by simply factoring $T(\ell)$ from the appropriate terms of Eq. (4) to produce Eq. (5). Typical weighting functions are presented in Fig. 2. In fact, more meaningful

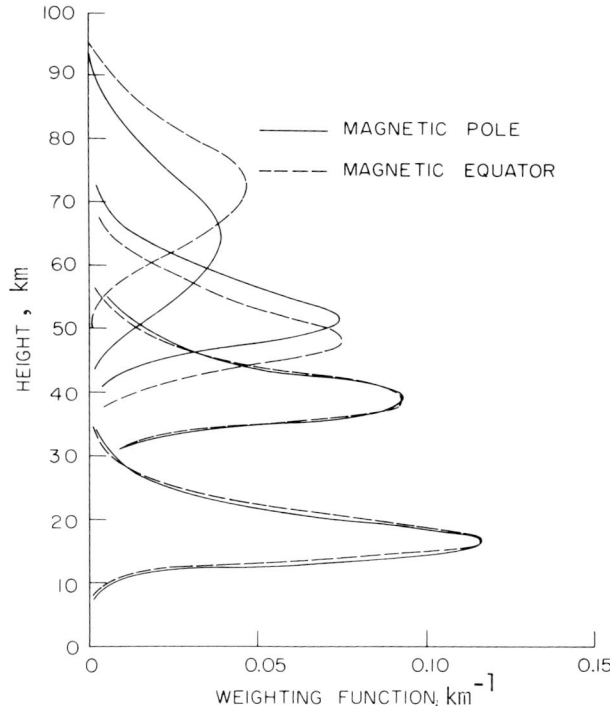

Fig. 2. Representative atmospheric temperature weighting functions for frequencies near the 60-GHz oxygen band. A nadir-viewing satellite is assumed. A continuous distribution of such weighting functions can be obtained between 0 and 75 km.

weighting functions could be found that are less temperature dependent. If one expands the integrand of Eq. (5) in a Taylor series about a presumed temperature profile and keeps only the first order term $T(\ell)W'(\nu,\ell)$ in the integral, then the resulting temperature weighting function $W'(\nu,\ell)$ is usually not greatly different from the customarily presented $W(\nu,\ell)$ because of the linearity of the microwave spectrum.

These weighting functions are very similar to those in the infrared region. The microwave functions are generally broader in the lower troposphere where the lapse rate is significant and the Planck function, therefore, sharpens the infrared temperature weighting functions based on the CO_2 bands near 15 µm and especially 4.3 µm. Near the tropopause and in the lower stratosphere, however, the Planck function severely broadens the infrared functions, and the low spectral resolution of present infrared sounders broadens them further at still higher altitudes.

The different behavior of the microwave and infrared weighting functions suggests optimum satellite systems will incorporate both spectral regions, e.g., the narrow 4.3 µm weighting functions for the 0-8 km region, and the narrower microwave functions for altitudes above ≈8 km. One present compensating virtue of broadband infrared sensors is that their greater sensitivities permit shorter integration times, and, therefore, they can yield higher spatial resolution. Thus, there can be complex system design trade-offs between the ranges of vertical, horizontal, and temperature resolution that various microwave and infrared sensors can provide.

A second important property of microwave temperature sounders is their ability to penetrate all cirrus clouds and almost all nonprecipitating clouds. A study of the degradation of NEMS temperature retrievals by clouds (Ref. 14) shows that only intense precipitation bands, subh as the Inter-Tropical Convergence Ozone (ITCZ) or major extratropical storms, have any measurable effect on temperature retrievals. This effect arises from perturbations in the brightness temperature of ≈ -5 to +2 K for channels sensing ≈3-8 km altitude. These effects are normally negligible because most heavy clouds have temperatures close to those of the observed brightness temperatures, and because most heavy clouds do not fill the entire ≈300-km viewing zone of NEMS. Approximately 0.4% of NEMS observations for the 4-km weighting function were perturbed more than 1 K, and these were primarily over the ITCZ (Ref. 14).

Because the radiative transfer problem in the opaque O_2 bands is nearly linear in temperature, and since the temperature variations are nearly a jointly Gaussian process, a linear least-square-error retrieval process for temperature profiles is nearly optimum; conventional linear regression procedures can be used. Analyses of the accuracy of such temperature retrievals for NEMS were performed by Waters, et al. (Ref. 15) and Wilcox and Sanders (Ref. 16). Typical accuracies are presented in Fig. 3 (From Ref. 15).

The reported inaccuracies in the retrievals arise from (1) receiver noise and calibration errors, amplified by the retrieval process; (2) errors in the presumed atmospheric propagation characteristics; (3) retrieval errors resulting from the inability of coarse weighting functions to respond to fine structure in the true temperature profiles, which is most pronounced near the surface and the tropopause; and (4) errors in the comparison sources, which are typically single or smoothed sets of radiosondes. Receiver noise for NEMS was ≈ 0.15 K per sounding. Calibration errors are believed to be less than 2 K and essentially constant; therefore, they contribute primarily to mean retrieval errors of $\approx \pm 2$ K, but very little to root-mean-square (rms) errors about the mean. Errors in the propagation equations are believed to be small and to produce systematic errors largely indistinguishable from calibration errors. Most of the inferred retrieval errors are believed to result from coarse weighting functions and radiosonde errors. Evidence for this hypothesis follows from comparison of the rms and mean errors between the observed T_B values and those computed theoretically for interpolations of the corresponding 0^ℓ and 12^ℓ National Meteorological Center (NMC) temperature field analyses; these rms errors about the mean regression line are ≈ 1.1 K, 0.7 K, and 0.7 K for channels peaking near 4 km, 11 km, and 18 km, respectively. Removing the ≈ 0.15 K contribution of receiver noise implies rms NMC errors of only ≈ 0.7 K, which is not implausible, even if all propagation and calibration errors were zero.

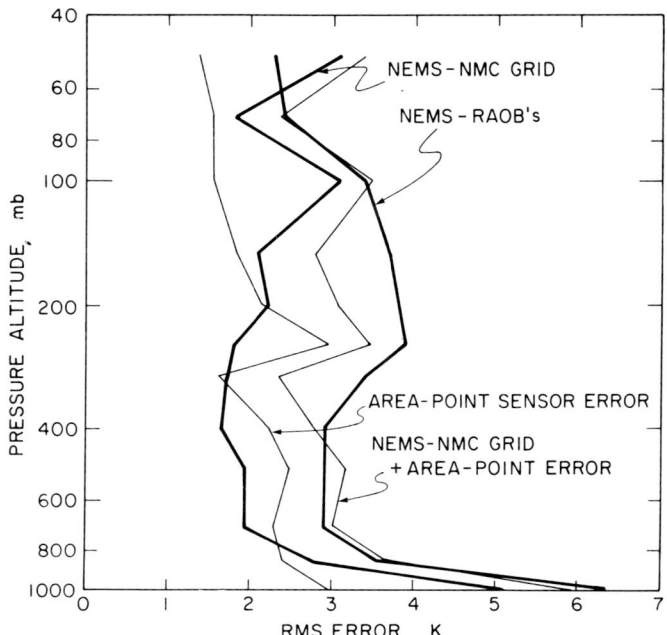

Fig. 3. Discrepancies (rms) between (1) NEMS temperature retrievals (mean error deleted) and NMC analyses interpolated in time and space in data-rich regions, (2) NEMS retrievals and 82 coincident Radiosonde Observations (RAOB's), (3) a statistically modeled point-temperature sensor and a hypothetical area sensor sampling $\sim 4 \times 10^4$ km^2 for statistics above 54 N latitude, and (4) the geometric sum of (1) and (3). (1 bar = 100 kPa.)

The data of Fig. 3 reveals that much more error was deduced when comparisons were made with individual radiosondes instead of the NMC grid. There are two hypotheses to explain this difference. The first explanation is that the excessive errors for radiosondes are due to aliasing errors; the large area averaged by NEMS is better represented by the NMC analyses, which are derived by averaging nearby radiosondes. This hypothesis has been tested by Ledsham,[2] who computed the auto-correlation functions of the

---

[2] Personal communication.

atmospheric temperature field at several altitudes and latitudes, and then predicted the aliasing errors that could be expected. These errors are plotted in Fig. 3 as the area-point-sensor error. This result is consistent with the observed rms discrepancies between the NMC analyses and individual radiosondes, and consistent with the discrepancies between the NEMS comparisons with NMC analyses and radiosondes, all shown in Fig. 3.

The second hypothesis (Ref. 17) is that the radiosondes are more representative than the NMC analyses, which smooth the fine vertical structure in the comparison profiles and, thus, yield artificially smaller rms errors. If this second hypothesis is correct, then there may be an unidentified extra source of error. Although this appears unlikely, a more definitive study remains to be done.

SCAMS is similar to NEMS but scans left to right in 13 1-sec steps with $7.5°$ beamwidths. This yields maps of temperature profiles ≈ 2500 km wide and with $\approx 150-300$ km resolution. Examples are shown in Fig. 4.

The 60-GHz band of oxygen is not the only opaque region that can be used for temperature sounding; the isolated 118.75 GHz resonance of O_2 is even more opaque than the 60-GHz complex. Although cloud effects can be perhaps twice as great near 118 GHz, the effects are still important primarily in the ITCZ and large extratropical storms. The weighting functions are slightly broader at 118 GHz than near 60 GHz, and can reach to altitudes near 80 km at the very center of the resonance.

Limb-scanning observations of the atmospheric temperature profile require large antennas in order to obtain useful angular resolution. The excellent altitude resolution for limb-scanning systems arises because the majority of the emission originates within ≈ 2 km of the ray tangent height if the atmosphere is not opaque. If the satellite is 1000 km from the tangent point for a

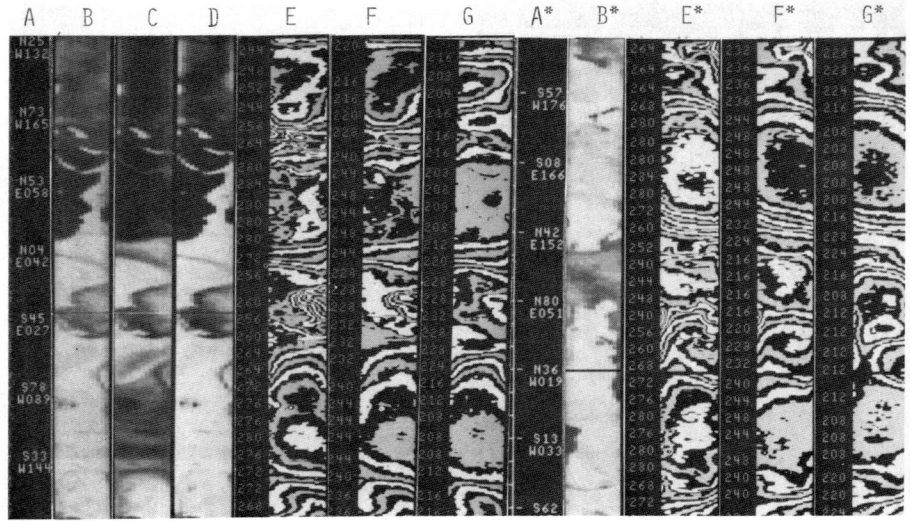

Fig. 4. Representative maps produced by SCAMS, on 3 and 5 February 1976 * indicates 5 February data. (A) latitude and longitude at nadir (image center), (B) brightness temperature at 31.6 GHz (black is hot; note land masses), (C) retrieved water vapor over ocean (black is humid), (D) retrieved liquid water over ocean (black is heavy clouds or precipitation), (E) retrieved average air temperatures "K" 1000-500 mb (note uncompensated effects of high elevation land over Antarctica); contours are each 2K wide, (F) average temperature 500-250 mb, and (G) average temperature 250-100 mb.

given ray, then an antenna beamwidth of ≈ 7 arc minutes is required to achieve 2-km vertical resolution. This requires an antenna ≈ 3 m long in the vertical direction at 60 GHz, or ≈ 1.5 m at 118 GHz. These dimensions are within the capability of the space shuttle and modern antennas.

The techniques one can employ at microwave frequencies (Ref. 18) are again analogous to those available in the infrared (Ref. 19). Weighting functions available near 125 GHz are illustrated in Fig. 5, and brightness temperatures as a function of tangent height are illustrated in Fig. 6 for two different model atmospheres.

One of the problems in limb-scanning is determination of the absolute pressure altitude of the scan. Accuracies better than ≈200 m are desirable. Note that the warmer atmosphere in Fig. 6 appears hotter at those view angles (or tangent heights) corresponding to greater optical depths and appears cooler at others. This effect arises because the hotter gas is more nearly transparent and actually emits less per meter of path length. For each observing frequency there is a "neutral point," a view angle or ray-tangent pressure for which the brightness is largely independent of the atmospheric temperature profile. Thus, this constant pressure altitude can conceptually be located by determining the view angle which exhibits the known neutral-point brightness temperature; in practice a more mathematical estimation procedure would be used.

Figure 6 suggests that the brightness temperature varies about 0.2 K in 10 meters, so a radiometer sensitivity of 0.2 K can enable the reference pressure altitude to be inferred with an accuracy several times worse than this, say 50 meters, which is more than adequate. Corrections must also be made for refractive effects and the effects of trace constituents; these also introduce errors of varying significance. A variety of standard linear and nonlinear retrieval procedures can be adapted to this problem.

There are two other independent ways to determine atmospheric temperatures from limb-scanning spectra. Above those altitudes where pressure broadening dominates, i.e., above ≈ 60 km, Doppler broadening becomes important. Measurements of Doppler width can yield estimates of the translational temperature of the oxygen. The second technique has long been used for optical spectra; the relative populations of a series of energy levels is determined, and these yield, in the case of O_2, a rotational temperature.

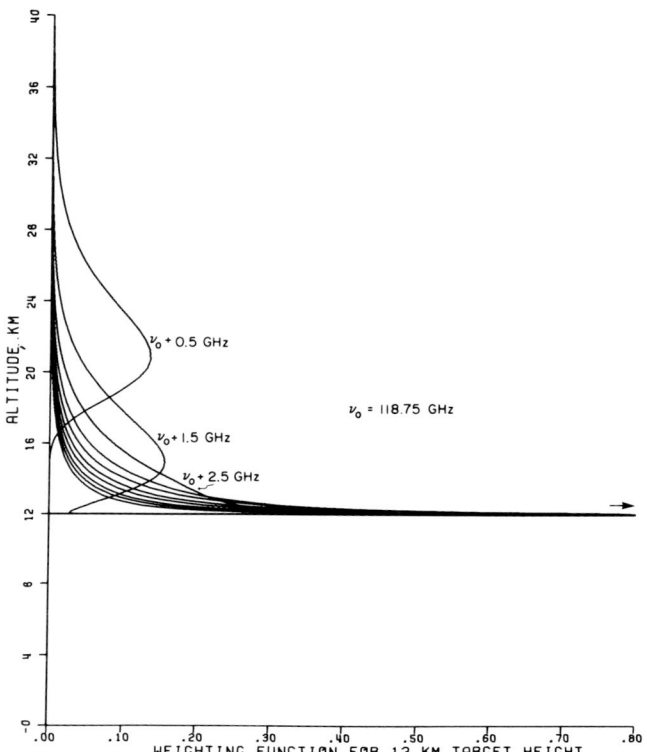

Fig. 5. Limb-scan-mode temperature weighting functions obtained near 118.75 GHz for a ray tangent at 12-km altitude. The peaks at 12 km are truncated. These functions can be translated continuously in altitude by scanning the limb.

These techniques have not yet been employed for atmospheric probing, but have been proposed for a limb-scanning experiment on the space shuttle.

Several avenues for improvement of microwave retrievals exist. The present linear and quasilinear techniques are nearly optimum for single position retrievals, but SCAMS and future temperature sounders will generally map the temperature field in three dimensions. Some appropriate smoothing could be employed to improve

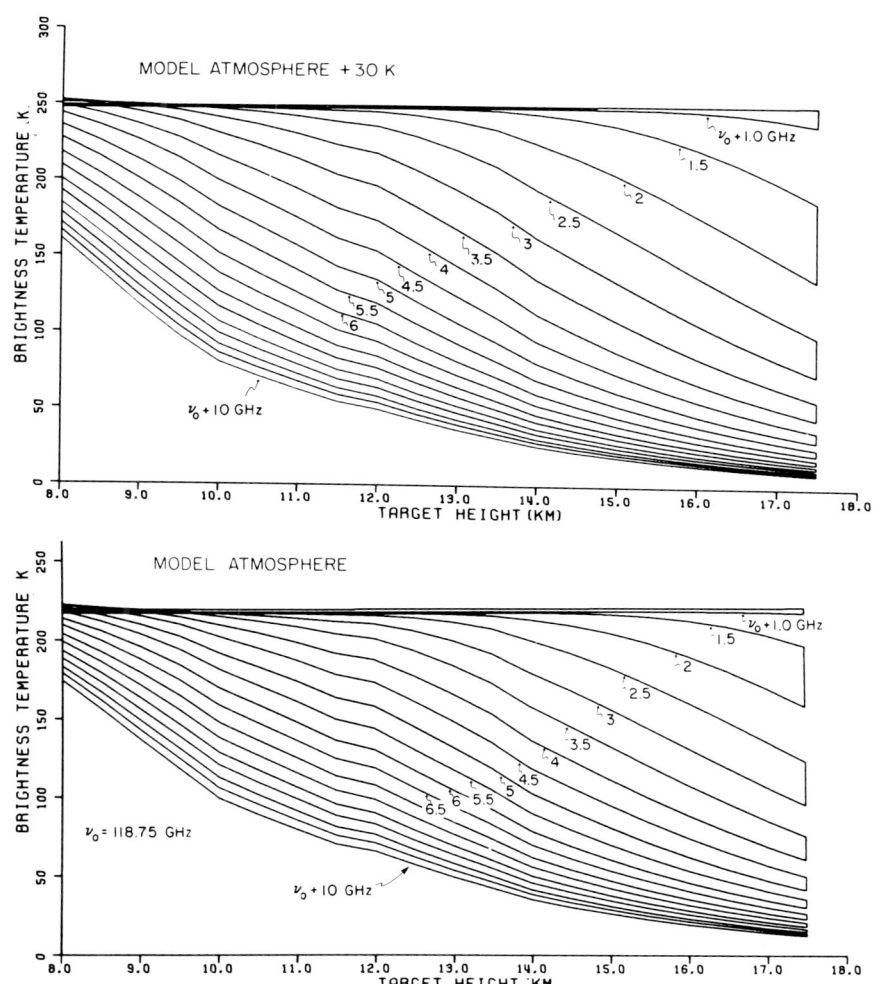

Fig. 6. Predicted limb brightness temperatures near 118.75 GHz as a function of ray tangent height for two model atmospheres: the U.S. Standard Atmosphere and one uniformly 30 K warmer. Refractive effects were incorporated.

temperature retrievals because the temperature field is correlated over ≈ 1000 km in the troposphere, and ≈ 2000 km in the stratosphere. Similar correlations exist in time, and they also could be utilized, as could independent data sets, such as the current NMC analysis, etc. These correlations can be extrapolated to greater distances in time and space by means of atmospheric prediction models; thus, enabling each retrieval to incorporate still larger sets of independent observational data.

Furthermore, those heavy precipitation cells that sometimes introduce errors into retrievals are generally much smaller than the correlation length of the temperature field, and should, therefore, be detectable and partially correctable.

One formulation for such problems is that of Kalman-Bucy filters (Ref. 20). In this construct, the process to be estimated is assumed to be drawn from the class of first-order continuous time infinite-state Markov processes. Its action upon input data is to optimally estimate the truly random elements controlling transitions between states. These random elements provide information which is then used along with deterministic elements, to update the filter's estimate of the current state of the process, i.e., the observables. This model permits estimation over a finite observation window and estimation of nonstationary processes, neither of which can be treated by conventional Wiener theory. Furthermore, nonlinear relationships between the random elements and the estimated observables can be introduced in a natural manner. This and the Markov process assumption produce a very convenient approach for retrieval of temperatures or any other meteorological parameters. In fact, such a retrieval process could even be combined with the numerical weather prediction process, with the present numerical prediction models being incorporated as part of the processor that estimates the observables.

III. SOUNDING OF ATMOSPHERIC COMPOSITION

Figure 1 illustrated resonances of oxygen and water vapor. Other important molecules include O_3, N_2O, and CO. Stronger resonances below 200 GHz are included in Table 1.

TABLE 1

Strongest Microwave Resonances below 200 GHz of Trace Constituents

Species	Resonance (GHz)
H_2O	22.235
	183.310
O_3	101.737
	110.836
	125.087
	142.175
	165.784
	184.378
	195.430
$N_2^{14}O^{16}$	100.492
	125.614
	150.735
	175.856
$C^{12}O^{16}$	115.271

The spectra of these and other molecules was recently reviewed by Waters (Ref. 3).

Such spectral lines can be observed passively against three different backgrounds: (1) cold space reflected by the ocean; (2) an opaque atmosphere at a different temperature that provides contrast; and (3) cold space, as in the limb-scanning mode.

Passive microwave satellite measurements of atmospheric composition have been made only of water vapor and liquid water over ocean; the high reflectivity of ocean ($\approx$ 60 percent) near the 1.35-cm resonance provides a cold background for thermal emission from both water vapor and liquid water. Figure 7 illustrates the

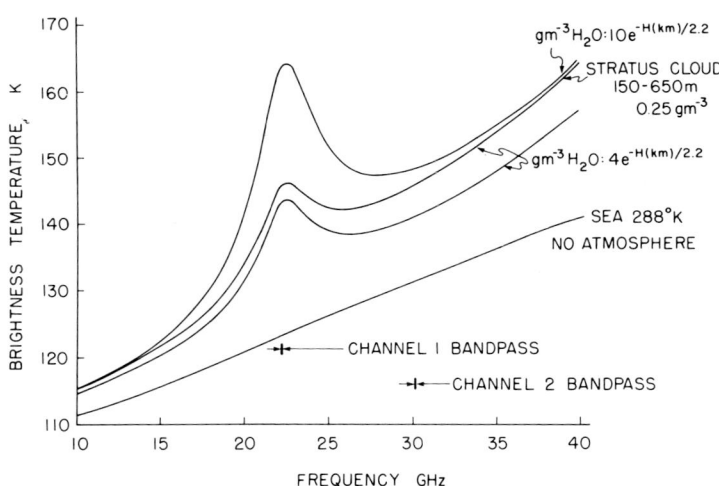

Fig. 7. Computed brightness temperature of smooth ocean at 288 K overlaid by (1) no atmosphere; (2) a nominal atmosphere, including a modest amount of water vapor; (3) the same plus a heavy stratus cloud; and (4) a more humid but cloud-free atmosphere.

brightness temperatures that would be observed at nadir over a smooth ocean surface.

The spectrum in the absence of any atmosphere would not be flat (see Fig. 7) because the dielectric constant of sea water depends on frequency as well as on temperature, and salinity. This temperature dependence has the curious property that the sea brightness temperature is nearly independent of sea temperature at wavelengths near 1 cm; it increases and decreases slightly with temperature for longer and shorter wavelengths, respectively. This fact largely removes sea temperature as a major source of error in atmospheric water retrievals; climatic values for sea temperature are generally sufficiently accurate.

Atmospheric water vapor appears in emission against the cold ocean background with a line shape and amplitude that is nearly proportional to the absorption coefficient of the atmosphere. This almost linear relationship results from the Rayleigh-Jeans approximation of Eq. (1) and because the atmosphere is nearly transparent, as seen in Fig. 1, even at the center of the resonance and over the tropics. The H_2O line width for each atmospheric layer is proportional to its pressure, and, thus, the total line shape contains information about the water vapor profile which can be retrieved. Linear and quasi-linear techniques have been used successfully for estimating water vapor abundances and profiles; the problem is sufficiently linear that they generally suffice (Refs. 21, 22, and 23).

Water vapor weighting functions $W(\nu,\ell)$ may be defined as

$$B_B(\nu) \approx T_{const}(\nu) + \int \rho_{H_2O}(\ell) W(\nu,\ell) d\ell \qquad (5)$$

where $W(\nu,\ell)$ ($K\ g^{-1} m^{+2}$) is independent of the water vapor profile $\rho H_2O(\ell)$ ($g\ m^{-3}$) for small perturbations of $\rho_{H_2O}(\ell)$ about a standard atmosphere. Because water vapor weighting functions near 1-cm wavelength have widths with a half-amplitude of ≈ 2 pressure scale heights or ≈ 16 km, water vapor retrievals can extract little more than the total tropospheric abundance and perhaps a water vapor scale height ($\rho_{H_2O}(\ell)$ usually is exponential with a scale height near 2.2 km).

The presence of clouds and precipitation complicates water vapor retrievals only slightly because the nonresonant character of cloud absorption can be distinguished from the resonant character of water vapor (see Figs. 1 and 7), as noted by Staelin (Ref. 21) and others (Refs. 22 and 23). Rosenkranz, et al. (Ref. 22) concluded theoretically that rms accuracies of 1.3 mm precipitable water vapor and 0.06 mm liquid water could be obtained from observations over the ocean at only two frequencies, such as 22.2 and

31.4 GHz. Comparisons of such two-frequency water vapor estimates with radiosondes yield rms errors of ≈ 1.2 and 4.4 mm (Ref. 23); the poor agreement is very likely due to

1. the point-sampling character of radiosondes as contrasted with the areas of $\approx 10^5$ km^2 averaged by NEMS or SCAMS, and

2. the inability of radiosondes to respond well to low humidity.

Theoretical errors can be reduced through use of more than two sounding frequencies.

Heavy precipitation can be partially opaque ($\gtrsim 4$ dB attenuation) near 1-cm wavelength for precipitation rates above ≈ 3 mm hr^{-1}. Because such precipitation is obten confined to convective regions with an areal extent much smaller than the $\approx 10^5$ km^2 sampled by NEMS or SCAMS, the water vapor retrievals are not significantly affected, as suggested in Fig. 8 which presents water vapor and liquid water retrievals from NEMS over an Indian Ocean typhoon (Ref. 23). If high angular resolution sounders were used, then the water vapor retrievals over the opaque storm cells would naturally be less accurate.

The sounding of water vapor profiles near the core of the opaque 183.3 GHz resonance requires a different retrieval procedure, similar to that used for interpreting satellite observations of the opaque infrared bands of water vapor. If the temperature profile is known, then each value of brightness temperature on the observed resonance corresponds to the altitude which is being reached at that frequency; this altitude level is usually near an optical depth of unity. We now define the contribution function, $C(\nu,\ell)$, which describes the degree to which the observed brightness temperature is determined by the temperature at various altitudes; it is naturally peaked.

$$T_B(\nu) \approx \int_0^\infty T(\ell) C(\nu,\ell) d\ell \qquad (6)$$

$C(\nu,\ell)$ goes to zero above the atmosphere and at large optical

MICROWAVE REMOTE SENSING DATA

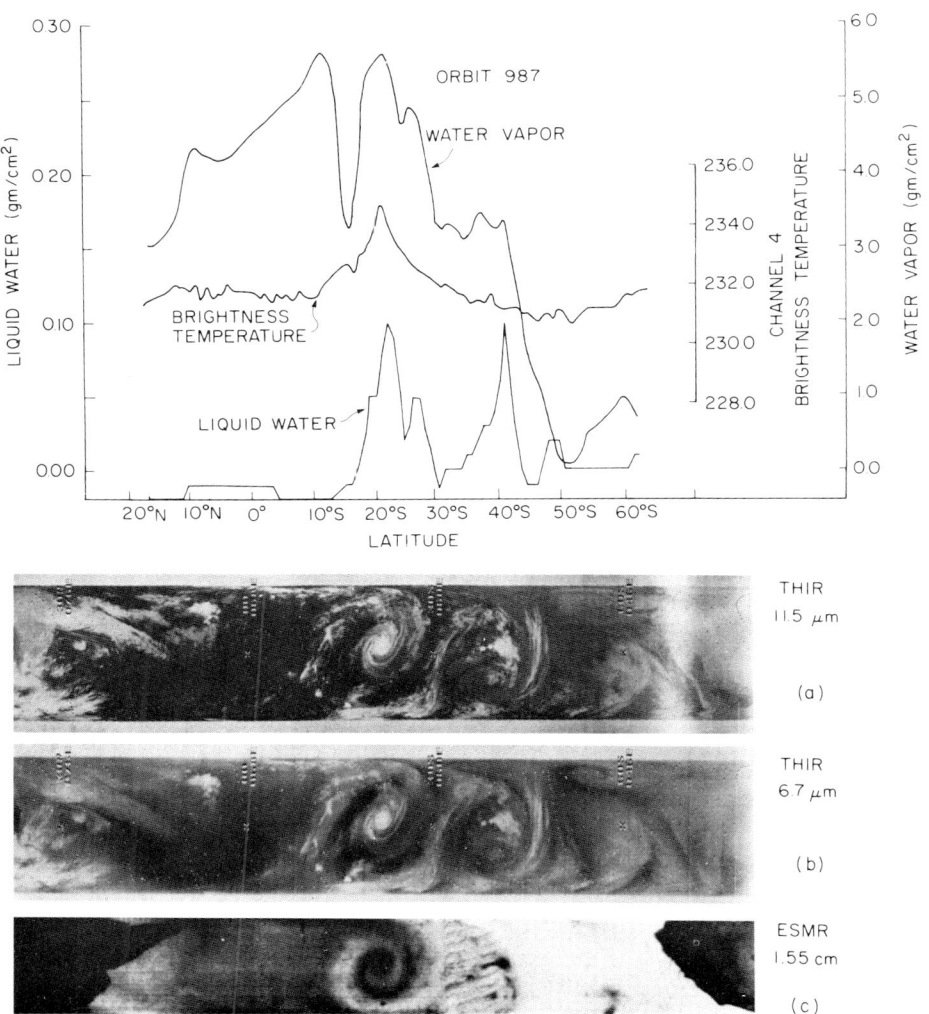

Fig. 8. Typhoon 22 February, 1973. Images were made near 19 GHz by Electrically Scanned Microwave Radiometer (ESMR) (dark is warm and humid), Temperature Humidity Infrared Radiometer (THIR) 6.7 μm (dark is dry), and THIR 11.5 μm (dark is warm or cloud-free). NEMS retrievals along the center of these images are presented for liquid water, water vapor, and 54.9 GHz brightness temperatures. Note the lack of interference between the water vapor and liquid water retrievals.

depths. The width of this function typically would approximate the water vapor scale height in the altitude zone being sounded. Because the contribution function depends strongly on the water vapor profile, it is quite different from the usual weighting function, a concept which is inappropriate here. The peak of the contribution function for the 183 GHz H_2O resonance typically would fall between 0 and ≈ 8 km for frequencies between the wing and the line center, respectively. Still more opaque resonances at shorter wavelengths enable water vapor profiles to be sensed to still higher altitudes. Little can be learned in regions with zero lapse rate, however, except perhaps the total water vapor content of that layer. Most other resonances, such as those of N_2O, CO, and some of O_3 are generally not opaque and can be observed over ocean or against the background of an opaque troposphere; the contrast depends upon the temperature difference between the atmosphere at the peak of the resonant contribution function and that of the background. Although no such microwave measurements have yet been made from space, they are quite feasible.

The most sensitive and precise way to measure atmospheric composition is by means of limb scanning. The path length of the radiation is ≈ 300 km near the tangent point (where the ray is parallel to the terrestrial surface or geoid), as contrasted with ≈ 10 km for nadir observations. This factor of ≈ 30 is augmented another factor of ≈ 10 by the greater contrast provided by cold space, as opposed to that supplied by the underlying atmosphere for nadir observations. Waters (Ref. 3) has analyzed this possibility for several trace constituents and it appears that CO might be sensed up to ≈ 100 km or more using the 230 GHz resonance, H_2O to ≈ 100 km using 183 GHz, O_3 to ≈ 100 km at 184 GHz, etc. Such constituents might be sensed down to altitudes of ≈ 8 km, depending on the altitudes of water clouds. The abundances can be measured with accuracies of a few percent with 2 km vertical resolution for constituents that are sufficiently abundant. Exploitation of these techniques will require somewhat more precise antennas than those used heretofore, but they are within the present state of the art and compatible with the constraints imposed by the space shuttle program.

IV. SOUNDING OF ATMOSPHERIC PRESSURE

There are at least two approaches to the sounding of atmospheric pressure using microwave techniques and a single satellite. These include measurement of

1. the pressure-dependent differential atmospheric absorption near a strong resonance of a constituent like O_2, which has a nearly constant mixing ratio; this can be done in principle using radar techniques from a satellite viewing nadir over ocean; and

2. brightness temperatures at the limb at one or more frequencies, as a function of altitude of the ray tangent point, which must be known to a precision greater than that desired for the measurements of pressure altitude.

The successful operation of any of these techniques would require a level of precision in one or more particulars beyond the present state of the art; several of these necessary advances appear to be feasible.

Pressure measurements have been made on other planets for decades by observing the widths of spectral lines formed by absorption of sunlight reflected from the planetary surface or from cloud layers. Radar measurements of Venus near 3- and 10-cm wavelength were used in a similar manner by Barrett and Staelin (Ref. 24) to support the contention that the surface pressure on Venus was very high; CO_2 was the presumed microwave absorber. The same technique can be employed from Earth-orbiting satellites, for example, using radar at two frequencies ≈ 0.5-1 GHz apart, centered near 52-52.5 GHz. Such radar echoes might suffer attenuations differing by 5-10 dB due to the nearby O_2 absorption band; the absorption is a function of the mass, temperature, and pressure of the atmosphere, particularly in that portion below 3 km where most absorption would occur. Since such O_2 absorption is approximately proportional to the square of pressure, differential attenuation should be measured with an accuracy of $\approx (10 \text{ dB}/1000) \times 2 = 0.02$ dB if errors of ≈ 1 mb are to be achievable. This is an unusually stringent requirement and would require very precise calibration. The measurement and interpretation problem is compounded by the facts that (1) the radar echo strength varies randomly at the two operating frequencies, (2) sea state reflectivity and absorption by clouds and water vapor are slightly different at the two frequencies, and (3) the absorption is somewhat temperature dependent (fortunately, the temperature dependence is near a minimum at 52 GHz). These difficulties notwithstanding, pressure accuracies of a few mb appear to be achievable in principle.

The use of limb-scanning pressure sensors has the advantage that such systems could map the entire Earth rather than only nadir points over ocean. The technique involves use of the neutral-point concept described earlier, wherein a particular brightness temperature at a particular frequency tends to occur for a ray tangent height at a particular known pressure altitude. Because this neutral-point is slightly dependent on the temperature profile, and since the ray is refracted so as to displace the apparent tangent height by perhaps several hundred meters, a slightly non-linear inversion procedure would appear to be appropriate. Although no inversion experiments have been performed, it appears likely that pressure altitudes could be determined with an accuracy of ≈ 25-100 m by use of O_2 absorption above ≈ 10 km altitude. This altitude must, of course, be referenced to the Earth, which would require the satellite position, altitude, and angular orientation of the instrument to be known with accuracies on the order of kilometers, meters, and 1 arc sec, respectively; this appears to be feasible but difficult.

MICROWAVE REMOTE SENSING DATA

Similar procedures could be employed in the infrared region, but even the slightest trace of haze or clouds could become a serious source of error here. An advantage of infrared is that the antenna would be smaller and easier to point precisely.

Neither of these two pressure measuring techniques provides a complete pressure profile; the radar measures an average pressure over the lowest few km of the atmosphere, and the passive limb-sounding technique can determine pressure altitudes $\approx 10\text{-}60$ km (i.e., that range over which pressure broadening dominates the O_2 line widths, and water vapor and refractive effects are small). Either technique must be supplemented by temperature profile measurements to yield a complete pressure profile, and the accuracy of these derived profiles degrades for altitudes far from the pressure reference. Therefore, a good combined system, if feasible, might use (1) limb-scan measurements to map globally the pressures near the tropopause, (2) nadir radar measurements to provide more accurate but more sparse surface pressures over ocean, and (3) microwave or infrared temperature profile spectrometers for enabling the pressure measurements to be extrapolated to all altitudes.

SYMBOLS

B	receiver bandwidth, Hz
$C(\nu,\ell)$	water vapor contribution function
d	diameter of droplets
h	Planck's constant
I	intensity, $W\ m^{-2}\ Hz^{-1}\ ster^{-1}$
k	Boltzmann's constant
ℓ, ℓ'	distance along ray path
n	refractive index of droplets
P	power density, $W\ Hz^{-1}$
$\Delta P/\Delta T$	ratio of power change to temperature change
R	surface reflectivity

T	temperature, K
T_A	antenna temperature, K
T_{const}	an appropriate constant temperature, K
T_R	receiver noise temperature, K
T_{sky}	sky brightness temperature above atmosphere, K
T_{surf}	surface temperature, K
ΔT_{rms}	receiver sensitivity, K
$W(\nu, \ell)$	temperature or water vapor weighting function
$W'(\nu, \ell)$	first order temperature weighting function
λ	wavelength, m
ν	frequency, Hz
ρ_{H_2O}	water vapor density, g m^{-3}
τ	receiver integration time, sec
τ_o	atmospheric opacity, nepers

ACKNOWLEDGMENT

This work was supported under NASA Contract NAS 5-21980 and Air Force Contract F 19628-75-C-0122.

REFERENCES

1. D. H. Staelin, Passive remote sensing at microwave wavelengths, Proc. IEEE, 57, 427 (1969).

2. K. Tomiyasu, Remote sensing of the Earth by microwaves, Proc. IEEE, 62, 86 (1974).

3. J. W. Waters, Absorption and emission by atmospheric gases, in "Methods of Experimental Physics" (M. L. Meeks, Ed.), Vol. 12, Part B. Academic Press, New York, 1976.

4. J. D. Kraus, "Radio Astronomy." McGraw-Hill Book Co., New York, 1966.

5. D. E. Kerr (Ed.), "Propagation of Short Radio Waves." McGraw-Hill Book Co., New York, 1951.

6. L. Tsang, J. A. Kong, E. Njoku, D. H. Staelin, and J. W. Waters, Theory for microwave thermal emission from a layer of cloud or rain, *IEEE Trans. Antennas Propag.*, September 1977.

7. W. J. Webster, Jr., T. T. Wilheit, D. M. Ross, and P. Gloersen, Spectral characteristics of the microwave emission from a wind-driven foam-covered sea, *J. Geophys. Res. 81,* 3095 (1976).

8. P. Gloersen, W. Nordberg, T. J. Schmugge, T. T. Wilheit, and W. J. Campbell, Microwave signatures of first year and multi-year sea ice, *J. Geophys. Res. 78,* 3564 (1973).

9. T. Schmugge, P. Gloersen, T. Wilheit, and F. Geiger, Remote sensing of soil moisture with microwave radiometers, *J. Geophys. Res. 79,* 317 (1974).

10. P. W. Rosenkranz, Shape of the 5 mm oxygen band in the atmosphere, *IEEE Trans. Antennas Propag. AP-23,* 498 (1975).

11. H. J. Liebe, G. G. Gimmestad, and J. D. Hopponen, Atmospheric oxygen microwave spectrum--experiment versus theory, *IEEE Trans. Antennas Propag. AP-25, 3,* 327 (1977).

12. D. H. Staelin, A. H. Barrett, J. W. Waters, F. T. Barath, E. J. Johnston, P. W. Rosenkranz, N. E. Gaut, and W. B. Lenoir, Microwave spectrometer on the Nimbus-5 satellite: meteorol- and geophysical data, *Science, 182,* 1339 (1973).

13. D. H. Staelin, P. W. Rosenkranz, F. T. Barath, E. J. Johnston, and J. W. Waters, Microwave spectroscopic imagery of the Earth, *Science,* August, 1977.

14. D. H. Staelin, A. L. Cassel, K. F. Kunzi, R. L. Pettyjohn, R. K. L. Poon, and P. W. Rosenkranz, Microwave atmospheric temperature sounding: effects of clouds on the Nimbus 5 satellite data, *J. Atmos. Sci. 32,* 1970 (1975).

15. J. W. Waters, K. F. Kunzi, R. L. Pettyjohn, R. K. L. Poon, and D. H. Staelin, Remote sensing of atmospheric temperature profiles with the Nimbus 5 microwave spectrometer, *J. Atmos. Sci. 32*, 1953 (1975).

16. R. W. Wilcox and F. Sanders, Comparison of layer thickness as observed by Nimbus E microwave spectrometer and by radiosonde, *J. Appl. Meteorol. 15*, 956 (1976).

17. K. F. Kunzi, A. G. Piaget, and C. B. Ruchti, The accuracy of the terrestrial atmospheric temperature profile derived from Nimbus 5 microwave spectrometer (NEMS) data, *Proc. of Symposium on Meteorological Observations from Space, COSPAR, Philadelphia Pa.*, June 8-10, 1976. [National Center.]

18. J. W. Waters, Prospects for microwave limb-sounding of the atmosphere, *Remote Sensing of the Atmosphere, Opt. Soc. Am.*, Anaheim, Calif., March 19-21, 1975, Conference Digest, WA4:1-4.

19. J. C. Gille and F. B. House, On the inversion of limb radiance measurements I: Temperature and thickness, *J. Atmos. Sci. 28*, 1427 (1971).

20. H. L. Van Trees, "Detection, Estimation, and Modulation Theory, Part I. John Wiley and Sons, Inc., New York, 1968.

21. D. H. Staelin, Measurements and interpretation of the microwave spectrum of the terrestrial atmosphere near 1-centimeter wavelength, *J. Geophys. Res. 71*, 2875 (1966).

22. P. W. Rosenkranz, F. T. Barath, J. C. Blinn III, E. J. Johnston, W. B. Lenoir, D. H. Staelin, and J. W. Waters, Microwave radiometric measurements of atmospheric temperature and water from an aircraft, *J. Geophys. Res. 77*, 5833 (1972).

23. D. H. Staelin, K. F. Kunzi, R. L. Pettyjohn, R. K. L. Poon, R. W. Wilcox, and J. W. Waters, Remote sensing of atmospheric water vapor and liquid water with the Nimbus-5 microwave spectrometer, *J. Appl. Met. 15,* 1204 (1976).

24. A. H. Barrett and D. H. Staelin, Radio observations of Venus and the interpretations, *Space Sci. Rev. 3,* 109 (1964).

DISCUSSION

Twomey: We seem to have uncovered another duality here apart from the one Henry Fleming spoke about. When the infrared people compare their retrievals with the radiosondes they say we don't seem to have done quite as well as we hoped for with this. When the microwave people compare their results with the radiosondes they say the radiosondes don't seem to be doing very well.

Westwater: You mentioned, I believe, that your H_2O retrievals differed from the theoretical results by about a factor of three and I assume that is in standard deviation. Did the deviations show any systematic component? Were they greater for the larger water content or were they more or less uniform throughout the entire distribution of water vapor in the atmosphere?

Staelin: Essentially uniform. That is an interesting question. One would expect nonuniformity in the errors because we were using a linear estimator in this initial procedure, even though the problem is nonlinear Here I think a partial explanation lies in the statistics of the atmosphere. I think the water vapor scale heights may be systematically larger where we have larger amounts of water vapor, and this alters our retrievals because our weighting function is nonuniform with altitude. These two effects may tend to cancel; the underestimate due to high H_2O opacity balances the overestimate that results for large-scale heights.

Westwater: And the second question, was there any correlation between the error and the amount of liquid water content that was also inferred? In other words, were the clouds affecting the retrievals somewhat more than what your theory would predict?

Staelin: Ground truth data is poor near intense storms. However, satellite passes over the ITCZ have yielded water vapor retrievals that vary smoothly and continuously across regions where the liquid water retrievals vary abruptly from zero to maximum and back. Such insensitivity to liquid water suggests this is not a problem at this spatial resolution of the order of 200 km.

Planet: You pointed out the nicety of microwave soundings in a cloudy atmosphere. Could you make some comments on a comparison between a microwave-only temperature sounding and an infrared-only temperature sounding as you go from a cloudy atmosphere to partially cloudy atmosphere to clear atmosphere?

Staelin: That's a loaded question. I think it depends on the details of the systems. In a clear atmosphere, the principal difference between infrared and microwaves is that the temperature weighting functions in the microwave region are decidedly narrower above the tropopause. Below the tropopause, where the temperature

MICROWAVE REMOTE SENSING DATA

gradient is significant, the infrared channels yield comparable widths and the four-micron functions, in particular, become narrower as they approach the surface. In partial cloudiness, it depends on the infrared scientist's agility in retrieving through partially cloudy conditions. Infrared cannot penetrate unbroken cirrus, whereas cirrus is totally transparent at the microwave frequencies we are using. In heavy precipitating clouds, the microwave channels sounding below ≈ 10 km will be affected, and it again depends on one's agility in handling that problem. The percentage of the time that these sensor types have problems for various weather conditions has been analyzed to some extent by W. Smith and, perhaps, others. I don't recall the quantitiative results, but there clearly is an advantage to the microwave system.

J. Russell: In the weighting functions you showed for the limb mode, I think you stated it was 2 kilometers width at the half-height point. Was that for an infinitesimal beam width?

Staelin: Yes. And half-height in this case is not really the right term because the peak amplitudes of those weighting functions are much greater than shown in the figure which was clipped at the right margin.

J. Russell: How much does the width change with a finite beam width?

Staelin: One would convolve the two functions. To first order, I would take the geometric sum of the two widths. If we have a 2-kilometer beam and a 2-kilometer weighting function, that would be roughly 3 kilometers.

Chahine: How small can the field of view be in your 12-channel sounder, reasonably?

Staelin: The 12-channel sounder that presently exists has a $7.5°$ field of view, which is the same as present microwave sounder unit on the Tiros-N Satellite.

Chahine: Which is how many kilometers at the surface?

Staelin: I forget the altitude of Tiros-N. I think the footprint is on the order of 150 kilometers.

Kaplan: How much worse off are you with clouds if you use the 2 millimeter oxygen band?

Staelin: We have been studying that question. The 2-millimeter oxygen band has an advantage that a smaller antenna will give you comparable resolution. Roughly speaking, the 118-GHz opacity of a given cloud or cloud layer is roughly a little more than twice what it would be at 60 GHz. But because of the statistics of cloud opacity as a function of position on the Earth, it appears we will not be affected much more than we are at 60 GHz. To evaluate that exactly will be difficult because there are no good statistics that are relevant.

Unidentified Speaker: Do you see any advantage in going to shorter wavelengths, higher frequencies, say, millimeter and sub-millimeter waves in future systems?

Staelin: I think it is in the study of trace constituents that we will obtain the greatest advantage, because their line strengths typically increase at shorter wavelengths. There is another advantage, particularly for the initial synchronous satellite systems in that smaller antennas will yield better resolution. I think that not too far into the future we will have antennas of almost any arbitrary size available. So *that* advantage will disappear after some decades. Although the insensitivity to clouds perseveres into the short millimeter wavelengths, it does degrade and at hundreds of gigahertz even ice clouds become much more significant.

Wark: There's one thing that seems to be swept under the rug a lot of times and that is all the inverse problems performed near the surface. There is clearly a difference between the sort of thing we saw Henry Fleming present and what you have presented in the 1000 to 800 millibar range. This is extremely important from a meteorological standpoint because the thing becomes very bad there. The rest of the sounding is correspondingly bad. Could you comment on how microwaves affect this result near the surface versus your method of retrieval?

Staelin: If the atmosphere is clear and the surface is known, then performance depends on the shape of the kernel. As I mentioned the kernels can be narrower in the 4 µm band, and even somewhat narrower in the 15 µm band, depending on the temperature profile. In the presence of clouds, infrared retrievals will degrade. In terms of the retrieval technique, we have used primarily linear regressions. The temperature retrieval errors increase near the surface because we are comparing our 3-channel retrievals to rather detailed profiles obtained by radiosondes or from the NMC analyses, both of which exhibit inversions and other sharp discontinuities near the surface, which no broad kernel could recover.

Wark: But your results of the thousand millibars were considerably worse than anything we have seen here.

Staelin: That is largely because we did not put in the surface temperature. Had we used the NMC forecast surface temperature or an infrared surface temperature estimate, our results would

MICROWAVE REMOTE SENSING DATA 393

have been better.

Wark: All right, then it is the method; not the consequence of the employment of microwaves. It is the method of retrieval where you do not employ a piece of information?

Staelin: That's right. We definitely perform better if we know the surface temperature either from an infrared surface temperature measurement or from *a priori* information, etc.

Susskind: I will make a comment about some of the previous things. We have been analyzing some of the information contained in the HIRS IR data and SCAMS data, both on Nimbus 6, and with regard to some of the questions that have been answered. The quality of the sounding between SCAMS alone and HIRS alone is very comparable one to another. In the presence of clouds, if it is completely overcast you are out of the game in the infrared. But, say, when it is 80-85 percent cloudy, the quality of the soundings is comparable. And, likewise, with surface temperatures. The quality of the soundings in the lower troposphere is, in fact, comparable. We got results of, say, $3°$ RMS error at the surface in the microwave. The errors are largest at the surface and at 850 mb and are a little larger in the microwave than in the infrared, but we interpret this to be due to some problems in calculating the surface emissivity exactly. But differences are very small--maybe 1/10 or 2/10 degrees in the RMS sense between the microwave and infrared. And we have come to the conclusion that the sounders are quite comparable one to the other--slightly worse errors down low, but nothing drastic. I might add that our results are obtained with no additional surface information but the observations and a climatology guess.

Staelin: I think one has to be careful in distinguishing between comparisons of particular instruments and comparisons in principle. In other words, these particular satellites have both infrared and microwave channels which might be chosen differently. Certainly with a larger budget one would use more than three microwave temperature-sounding channels. Then the performance limits are ultimately set by the widths of the kernels, and the effects of surfaces and clouds. Surface emissivity effects are observable only for weighting functions that peak within $\approx$ 4 km of the surface. Extreme variations for the 4 km channel of NEMS and the 2-km channel of SCAMS were approximately 1.5 K and 10 K, respectively, in brightness temperature as the satellite passed from ocean to land; this can largely be compensated by use of window channels.

Fraser: Meteorologists, of course, would like to have the pressure measured to 10/10 of a percent of the surface pressure. What development efforts are underway to do that with microwave?

Staelin: No major efforts to my knowledge.

Fraser: Because it is so important and you indicated that it might be feasible. Is there some reason for this?

Staelin: I should say none that are funded. There has been effort and interest in this in a number of quarters. Some of these techniques have been used in astronomy. Radar pressure sounding data, for example, was used on Venus over ten years ago by Barrett and myself to support the contention the atmosphere was dense. Limb sounding measurements, of course, have also been used in astronomy. The task of applying these techniques to actual working instruments involves profound engineering problems. Desmond Smith in Scotland has been interested in this problem for many years. I don't know the present status of that work. I had one student write his thesis on this subject. But there has been no significant funding. Proposals have been submitted, but other things have priority.

Chahine: One of your former students, Dr. Joe Waters, is working on the development of pressure sounder at JPL.

Staelin: I think it is an interesting problem and obviously of great importance if one can be successful. More effort seems to be warranted.

APPLICATION OF STATISTICAL INVERSION TO

GROUND-BASED MICROWAVE REMOTE SENSING

OF TEMPERATURE AND WATER

VAPOR PROFILES

E. R. Westwater and M. T. Decker
NOAA/ERL/Wave Propagation Laboratory

Surface-based observations of downwelling microwave thermal emission are related to temperature and humidity profiles via a standard integral equation of radiative transfer. Both in clear and in cloudy atmospheres, statistical inversion techniques can be used to retrieve profiles from a data vector of brightness observations and surface meteorological constraints. For the clear case, we illustrate accuracy predictions and profile retrievals for (a) single-frequency angular-scanned data, (b) multi-frequency angular scanned data, and (c) multi-frequency zenith data. For case (c) we compare predicted and achieved accuracies in a recently conducted joint NOAA-JPL radiometric experiment. Finally, we present retrievals of cloud-contaminated radiometric data.

I. INTRODUCTION

The continuous measurement of temperature and humidity in the Earth's boundary layer is an important requirement in some areas of meteorological research. Ground-based microwave radiometric measurements of temperature structure show promise of meeting this need and the technique has been investigated by several groups (Refs. 1, 2, 3, and 4). Limited information on the moisture profile is also radiometrically available (Refs. 5 and 6).

In this review, we outline the application of statistical

inversion methods to a few of the increasingly complex measuring techniques that have evolved over the years. For each method, statistical inversion appears capable of extracting maximum information from the measurements.

The extent to which inversion theory applies to a problem depends strongly on the solution of the direct problem; i.e., given the profile, can we calculate the measurements to within the experimental accuracy? We therefore spend some time discussing the accuracies of microwave thermal emission calculations.

We then review temperature retrieval results from single- and multiple-frequency angular scanning radiometers. Finally, we conclude with recent results in multi-frequency sensing of temperature and moisture profiles.

II. MICROWAVE ATTENUATION AND EMISSION IN CLEAR AND CLOUDY ATMOSPHERES

Measurements of microwave radiant power are commonly expressed as an equivalent black body temperature, or brightness temperature, T_b. Except during rain, atmospheric scattering is small relative to absorption. For a nonscattering atmosphere in local thermodynamic equilibrium, the downward brightness temperature at frequency ν is given by

$$T_b(\nu) = T_b^{(ext)} \tau_\nu + \int_0^\infty T \alpha_\nu \exp\left(-\int_0^s \alpha_\nu(s')ds'\right) ds \qquad (1)$$

where T is absolute temperature (K), α_ν is absorption coefficient (km^{-1}), $T_b^{(ext)}$ is brightness temperature external to the Earth's atmosphere (K), τ_ν is transmission through the atmosphere, and s is path length from the receiver to emitting volume (km). In the troposphere, microwave absorption is due principally to molecular resonances of O_2 (60 GHz) and H_2O (22 GHz) and to clouds and rain. In general, the absorption is a strong function of composition and a weak function of temperature. Thus, around 22 GHz, the emission varies principally because of variations in water concentration,

APPLICATION OF STATISTICAL INVERSION

whereas at 60 GHz, the emission from the well-mixed constituent O_2 depends mainly on temperature. For ground-based applications, the term $T_b^{(ext)}$ describes discrete external sources, such as the Sun or Moon, and the continuum "big bang" contribution of 2.9 K. Calculations of the microwave absorption coefficient are shown in Fig. 1.

A. Water Vapor Attenuation

Water vapor attenuation arises from the rotational transition at 22.235 GHz and the nonresonant contribution of submillimeter and infrared lines. At frequencies below 100 GHz, the spectrum is the sum of a resonant term and a contribution from the higher frequency lines that varies as ν^2. An excellent summary and discussion of the theoretical experimental basis of water vapor attenuation calculations are given by Waters (Ref. 7).

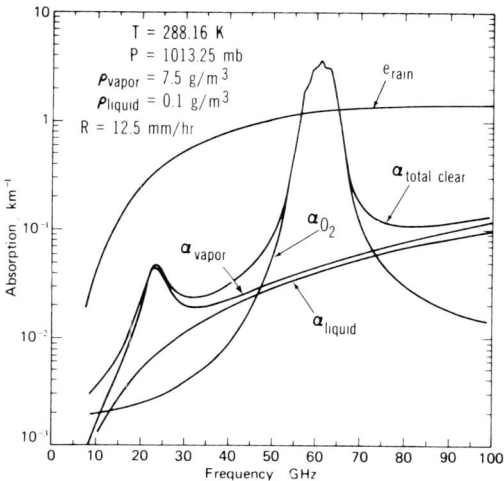

Fig. 1. Microwave atmospheric absorption in clear air, clouds, and rain. α = absorption coefficient, e = attenuation coefficient. (1 bar = 100 kPa)

For remote sensing of water and temperature, we must accurately calculate water vapor emission and attenuation from profiles of meteorological variables. Since several equations to calculate vapor attenuation exist in the literature (Ref. 7), we thought that a comparison of more recent models in brightness calculations would be at least suggestive of the accuracy to which the clear air direct problem is solved in the 20-35 GHz band. Thus, we present, in Table 1, comparisons of calculations of brightness temperature for five selected radiosonde profiles. The calculations labeled BA used constants derived by least squares from the absorption data of Becker and Autler (Ref. 8); L labels calculations using constants derived by Liebe (Ref. 9) using dispersion spectroscopy; finally W labels the results using parameters given by Waters (Ref. 7). BA and L assume the Van Vleck-Weisskopf line shape; W uses that of Gross-Zhevakin-Naumov. When compared with radiometric accuracies that approach 1 K, the agreement is not completely satisfactory, especially for the profiles with larger water content. If one accepts the more recent results (L and W), the agreement is within 5%.

B. Oxygen Attenuation

Beginning with the classic theoretical paper by Van Vleck (Ref. 10) on microwave absorption by molecular O_2, a large amount of theoretical, laboratory, and atmospheric research has been devoted to the understanding of attenuation from this constituent. As a consequence, the knowledge of O_2 absorption parameters and of related pressure broadening theory has steadily increased. Two recent advances are of note. The first is the development of the dispersion spectroscopy technique by Liebe and its application to the determination of the spectroscopic parameters of O_2 (Ref. 11). The second is Rosenkranz's (Ref. 12) theoretical attenuation model which accounts for overlapping lines in the oxygen complex.

As in the previous case of water vapor, we wanted to

TABLE 1

Comparison of Calculated Zenith Brightness
Temperatures at 22 and 31 GHz Resulting
from Various Choices of Spectral
Line Parameters

Profile	Liquid water content	22.235 GHz			31.65 GHz		
		BA	L	W	BA	L	W
1	0.8	23.4	22.3	21.4	14.7	15.0	14.5
2	1.1	28.6	27.2	26.1	16.5	16.9	16.3
3	1.3	31.1	31.5	30.1	18.1	18.5	17.7
4	2.8	68.2	64.5	61.5	29.0	30.1	28.3
5	3.6	74.9	70.8	67.5	31.2	32.4	30.4

determine the degree to which contemporary absorption models agreed with each other. To this end, we performed calculations of zenith brightness using three absorption models: the first (RMC) used the Van Vleck-Weisskopf line shape with constants given by Reber, Mitchell and Carter (Ref. 13); the second (R) used Rosenkranz's (Ref. 12) line shape and the constants given in his paper; the third (L) used Liebe's measurements (Ref. 11), the Rosenkranz line shape, and Rosenkranz's value of the non-resonant line width. The results are shown in Table 2.

There is close agreement between all three models at 53.8 and 55.5 GHz and R and L agree well also at 52.8 GHz. The difference between RMC and the other two at the most transparent channel, 52.85 GHz, is almost constant (≈ 4.2 K) and occurs primarily because of the difference in absorption prediction in the pressure range 100-500 mb.

C. Comparison of Clear Air Emission Measurement and Calculations

Five-channel microwave observations were taken at Pt. Mugu, California and were kindly provided by B. Gary and N. Yamane of

TABLE 2

Comparison of Calculated Zenith Brightness Temperatures (K) in the Oxygen Band Resulting from Various Absorption Models

Profile	52.85 GHz			53.85 GHz			55.45 GHz		
	RMC	R	L	RMC	R	L	RMC	R	L
1	194.8	190.6	191.3	251.2	250.8	251.2	281.5	281.6	281.7
2	191.5	187.2	187.7	250.7	250.3	250.6	280.8	281.0	281.0
3	193.8	189.6	190.0	257.0	256.8	256.9	289.7	289.8	289.9
4	190.6	186.3	187.0	248.3	248.0	248.4	278.7	278.9	278.9
5	189.9	185.5	186.0	252.4	252.1	252.4	287.4	287.6	287.6

Jet Propulsion Laboratory (JPL). Their radiometer was similar to the SCAMS system used on the Nimbus 6 satellite (Ref. 14). During the three-week period of observations, thrice-a-day radiosondes obtained standard meteorological soundings of temperature, pressure, and humidity. Calculations of brightness temperature were made by using the constants of Becker-Autler for water vapor, and those of Liebe for O_2. The comparison of measurements and calculations are shown in Table 3. Considering the difficulties in making absolute radiometric measurements and in observing the same volume of air with radiosonde and radiometer, the agreement in the O_2 band is quite good. Note, however, the relatively large variance at the 22 and 31 GHz channels. Because of the difficulty in obtaining reliable direct measurements of humidity profiles by radiosondes (Refs. 15 and 16), these differences may not be caused by incorrect absorption coefficients.

D. Attenuation by Clouds and Rain

Attenuation from a distribution of spherical particles of known dielectric properties can be calculated by classical electromagnetic theory. Depending on the ratio of particle size to

TABLE 3

Average and Root-Mean-Square (rms) Differences (K) between Measured and Calculated Brightness Temperatures (N = 24 Radiosonde Observations)

Frequency (GHz)

Difference	22.234	31.65	52.85	53.85	55.45
Average	2.5	1.7	0.2	0.1	-1.3
Root-mean-square	3.4	1.8	1.8	0.8	1.4
Percent	12.6	11.2	0.9	0.3	0.5

wavelength, simplicity or complexity prevails. In the domain here this ratio is small, called the Rayleigh region, attenuation is independent of size distribution and is directly proportional to total mass of droplets. In addition, scattering is negligible relative to absorption. For large particles, Mie theory must be used, and attenuation depends on size distribution in both absorption and scattering. For our purposes, we will consider water clouds with modal radii less than 50 μm to be in the Rayleigh region for frequencies less than 100 GHz. Calculations of water absorption for a cloud liquid water content of $\rho_{liquid} = 0.1$ g/m^3 are shown in Fig. 1. Depending on the frequency, spherical ice particles absorb from one to two orders of magnitude less than an equivalent amount of water.

In contrast to nonprecipitating clouds, rain (and hail) both scatters and absorbs microwave energy. The attenuation coefficient e must be calculated from Mie theory for both absorption and scattering coefficients. Calculations of e for a moderate rain of 12.5 mm/hr are also shown in Fig. 1. We assumed a Laws and Parsons (Ref. 17) size distribution for this rain rate (liquid) water content = 0.6 g/m^3). It is clear that rain dominates all other sources of attenuation except in the vicinity of the oxygen complex.

The atmospheric thermal emission spectrum in the presence of a cloud can differ considerably from that obtained under clear conditions. In detail, the amount of contrast depends on the height profiles of temperature, humidity, and liquid content. However, the largest contribution to the difference is the liquid thickness with the emission being relatively insensitive to the height and geometric thickness of the cloud (Ref. 18). Table 4 shows estimated rms differences between clear and cloudy brightness temperatures for the July climatology of Pt. Mugu, California. These differences clearly show the need for cloud correction. Similar calculations for ocean climatologies indicated rms differences about twice as large as those of Table 4.

III. STATISTICAL INVERSION OF GROUND-BASED MICROWAVE RADIOMETER DATA TO RECOVER VERTICAL TEMPERATURE PROFILES

Minimum-variance statistical inversion is used to estimate a parameter vector $\underset{\sim}{p}$ from a data vector $\underset{\sim}{d}$ according to the well-known prescription

$$\hat{\underset{\sim}{p}} = <\underset{\sim}{p}> + <\underset{\sim}{p}'\ \underset{\sim}{d}'^T> <\underset{\sim}{d}'\ \underset{\sim}{d}'^T>^{-1} \underset{\sim}{d}' \qquad (2)$$

In Eq. (2), $\hat{\underset{\sim}{p}}$ is the estimator of $\underset{\sim}{p}$, $<\cdot>$ refers to ensemble averages over joint distributions of $\underset{\sim}{p}$ and $\underset{\sim}{d}$, and primes denote

TABLE 4

Rms Differences between Clear and Cloudy Zenith Brightness Temperatures

Frequency, GHz	Rms difference, K
22.235	3.6
31.65	7.6
52.85	6.4
53.85	2.0
55.45	0.1

departures from mean values. The assumptions and derivations leading to this equation are given by Rodgers (Ref. 19). The covariance matrix of this estimate, $S_{p-\hat{p}}$, is given by

$$S_{p-\hat{p}} \equiv \langle (p - \hat{p})(p - \hat{p})^T \rangle = \langle p' p'^T \rangle - \langle p' d'^T \rangle \langle d' d'^T \rangle^{-1} \langle d' p'^T \rangle, \qquad (3)$$

The ith diagonal element of $S_{p-\hat{p}}$ represents the residual variance of the estimate of the ith parameter and is a direct measure of solution quality. If p represents a discretized profile of some variable, then the matrix trace of $S_{p-\hat{p}}$ is a useful measure of overall solution quality. Here p usually represents vertical temperature or humidity profiles at a sufficiently dense set of height coordinates, and d is some function of observed brightness temperatures.

The application of these equations to ground-based (g.b.) sensing of meteorological profiles differs in several respects from the corresponding satellite retrieval problem:

(a) Satellite retrievals use *a priori* statistics appropriate to a latitudinal region; a ground-based application requires only single-station statistics. Consequently, the *a priori* variance of the g.b. ensemble is usually much less than that used for satellites.

(b) For the g.b. problem, direct observations of the desired profile at the surface can usually be obtained. This can be imposed as an exact constraint on the inferred profile, which, in addition, further reduces the *a priori* variance at all levels, since we are now averaging over an ensemble with fixed surface conditions.

(c) A practical consequence of (a) and (b) is that linear methods are frequently appropriate, because of the relatively small variation about the initial guess profile.

A. Temperature Sensing by Inversion of Single-Channel Angular-Scanned Radiometric Data

Initially, ground-based temperature sensing was attempted by fixed-frequency angular scanning methods (Refs. 1,20, and 21). A typical set of weighting functions is shown in Fig. 2. Although these weighting functions attain their maxima at the surface, linear combinations of them can be made to peak at various altitudes yielding a spatial resolution that degrades with altitude (Ref. 3).

Weighting functions give the system response to a delta-function input; another meaningful system characterization is the total variance of the observations, and its partitioning into the contributions from all relevant meteorological variables. An example of this analysis is shown in Fig. 3, in which partial variances in T_b from fluctuations in temperature, relative humidity, and pressure are shown as a function of elevation angle. Although this channel is in the O_2 band, the fluctuations due to humidity exceed those from temperature for angles greater than $20°$. The reduction in "noise" due to surface constraints is shown in Fig. 4, where, in particular, the variance in T_b due to pressure fluctuations is reduced to almost zero. However, the contamination of the temperature "signal" due to the humidity "noise" is still extensive and really requires a water channel for its removal.

As mentioned earlier, the diagonal elements of $S_{\underset{\sim}{T}-\underset{\sim}{\hat{T}}}$ are a measure of solution quality. Plots of the square roots of these diagonal elements as a function of altitude indicate the expected standard deviation of the solution over an ensemble of profiles. An example of this type of plot is Fig. 5, which shows theoretical retrieval accuracy for several choices of operating frequency for the August climatology of Salt Lake City, Utah. These calculations predict an accuracy of somewhat better than 1 K up to 3 km in altitude. Retrieval of profiles from radiometric data (clear

APPLICATION OF STATISTICAL INVERSION

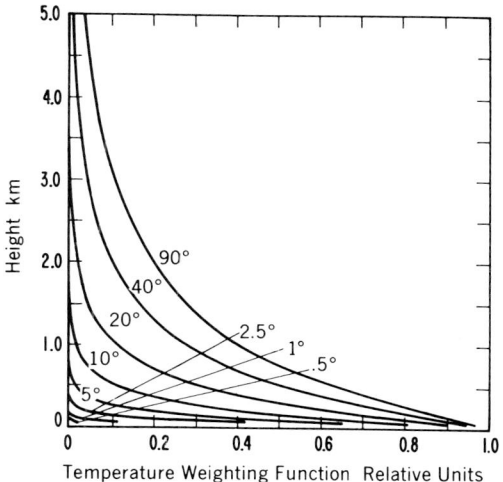

Fig. 2. Temperature weighting functions for ground-based angular scanning at 54.5 GHz.

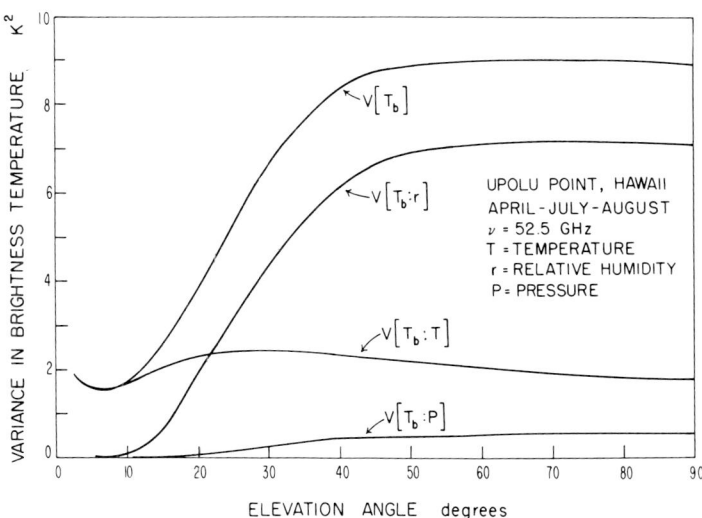

Fig. 3. Contribution to fluctuations in brightness temperature at 52.5 GHz from temperature, humidity, and pressure.

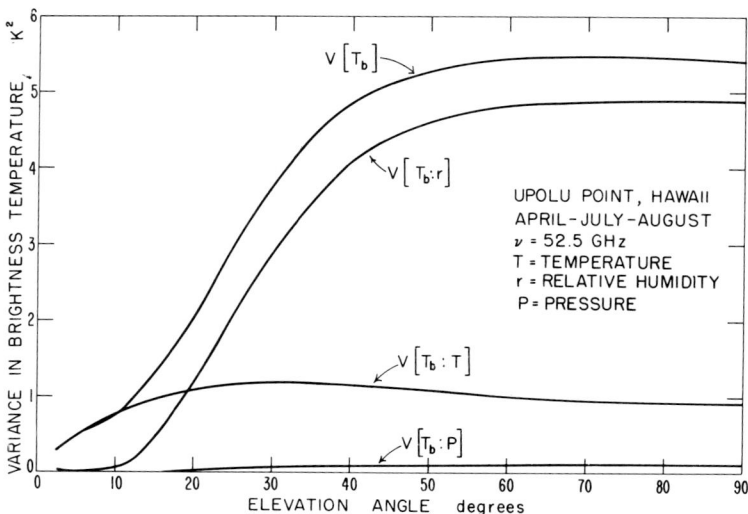

Fig. 4. Contribution to fluctuations in brightness temperature at 52.5 GHz from temperature, humidity, and pressure. Constrained surface conditions.

skies only) gave accuracies very close to these predictions (Ref. 21). Examples of profile recoveries from single-channel angular scan data are given in Figs. 6 and 7.

B. Temperature Sensing by Inversion of Multi-Spectral Angular-Scanned Radiometric Data

Although the simplicity of a single-frequency angular-scanning radiometer is attractive, scanning with a multi-channel system can significantly improve retrieval accuracy (Ref. 22). An example of the amount of improvement can be seen in Fig. 8 in which theoretical retrieval accuracy for several systems is shown as a function of altitude. Table 5 gives an explanation of the spectral combinations of Fig. 8; combination 1A refers to experimental noise level of 0.1 K and represents somewhat of a practical

APPLICATION OF STATISTICAL INVERSION

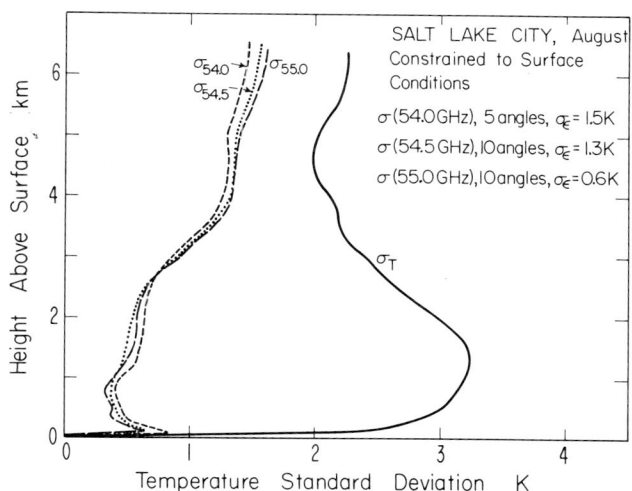

Fig. 5. Theoretical accuracy in retrieving vertical temperature profiles by inversion of single-frequency angular scanned radiometer data. σ_ε = assumed instrument noise level. σ_T = a priori standard deviation for fixed surface conditions.

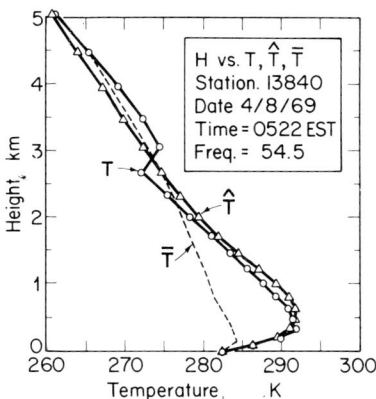

Fig. 6. Temperature profile derived from single-frequency angular scanned radiometer data at Cincinnati, Ohio. A priori statistics from Dayton, Ohio. T = radiosonde profile. $\bar{T}$ = mean profile for constrained surface conditions. $\hat{T}$ = inferred profile.

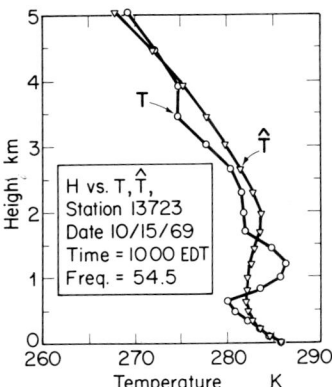

Fig. 7. Temperature profile derived from single-frequency angular scanned radiometer data at Raleigh-Durham, North Carolina. A priori statistics from Greensborough, North Carolina - radiosonde profile. $\hat{T}$ = inferred profile.

TABLE 5

Explanation of Spectral Combinations Employed to Estimate Temperature Profiles

Combination	Input data	
	Elevation angle (deg)	Frequency (GHz)
1	5	54.5, 55.5
	10	54.5, 55.5
	15	54.5, 55.5
	30	53.5, 54.5, 55.5
	60	52.5, 53.5, 54.5
	90	52.5, 53.5, 54.5
2	90	52.5, 53.5, 54.5, 55.5

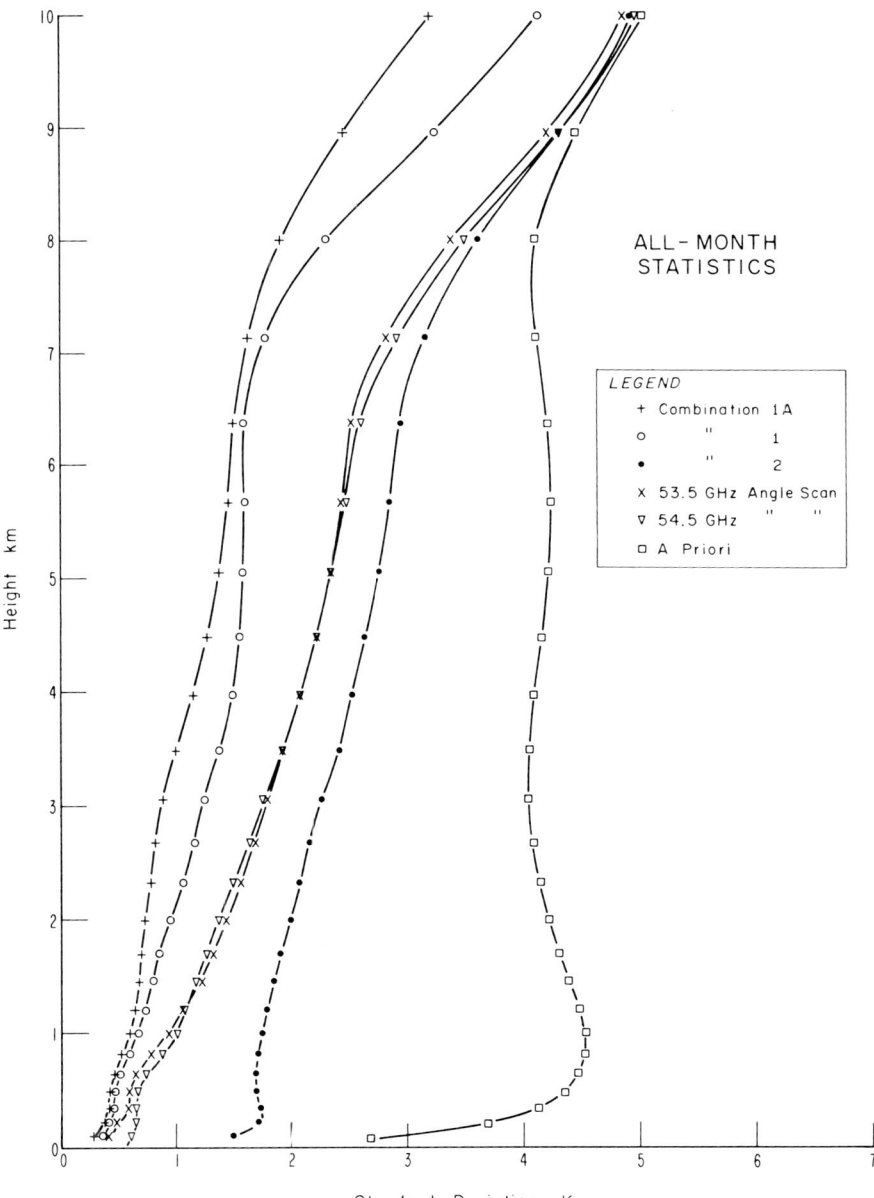

Fig. 8. Theoretical accuracy in retrieving vertical temperature profiles from various combinations of multi-spectral multi-angle radiometer data, Denver, Colorado (after Ref. 20).

limit in measurement accuracy; the noise levels in other combinations were obtained from comparison of measurements and calculations of brightness temperature. The multi-spectral angular-scan combination 1 improves the retrieval accuracy over the single-frequency scans by a margin of about two to one.

Typical examples of the low-altitude temperature structure that can be recovered is shown in Figs. 9, 10, and 11, which show retrievals of a low altitude elevated inversion, a ground-based inversion, and a low-level super-adiabatic profile. Note expecially the improvement in the recovery of the elevated inversion over the single-frequency result of Fig. 7.

IV. RECENT RESULTS IN INVERSION OF GROUND-BASED MULTI-SPECTRAL RADIOMETRIC DATA TO INFER TEMPERATURE AND HUMIDITY PROFILES

We are currently investigating the feasibility of sensing of temperature profiles from an ocean data buoy-mounted radiometer. Because of buoy motion, the previously discussed angular scanning techniques are not practical; in addition, the cloud problem dictates the need for two moisture channels to sense and to correct for clouds. Cloud correction algorithms appropriate for microwave passing sensing have been published by Westwater, et al., Rosenkranz, et al., and Fowler, et al. (Refs. 18, 23, and 24).

The National Weather Service desires that the temperature of layers, 100 mb in thickness, be determined with accuracy ± 1 K. Although temperatures of the lowest two layers could be useful, retrievals to 500 mb are wanted. Predictions for ocean climatologies, using Eq. (3), showed that under clear conditions, a three-frequency zenith looking radiometer could meet the lower altitude requirements and could be reasonably close to those at the higher altitudes (Ref. 18). Accuracy predictions are shown in Fig. 12 for several systems under consideration; the temperature weighting functions for system C are shown in Fig. 13.

APPLICATION OF STATISTICAL INVERSION 411

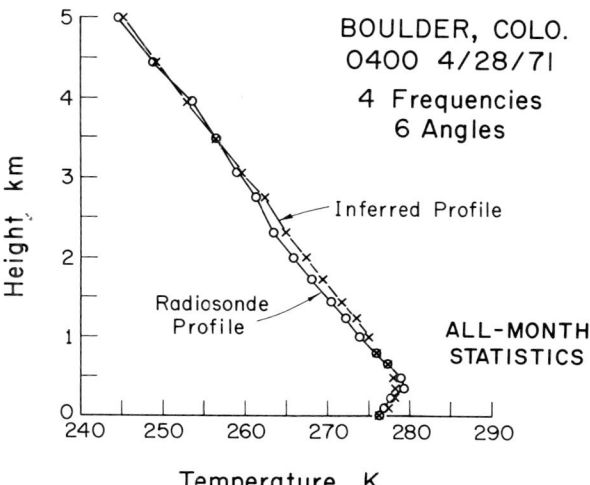

Fig. 9. Inferred and radiosonde temperature profiles from multi-spectral multi-angle radiometer data for 0400, April 28, 1971. A priori statistics from Denver, Colorado. (After Ref. 20.)

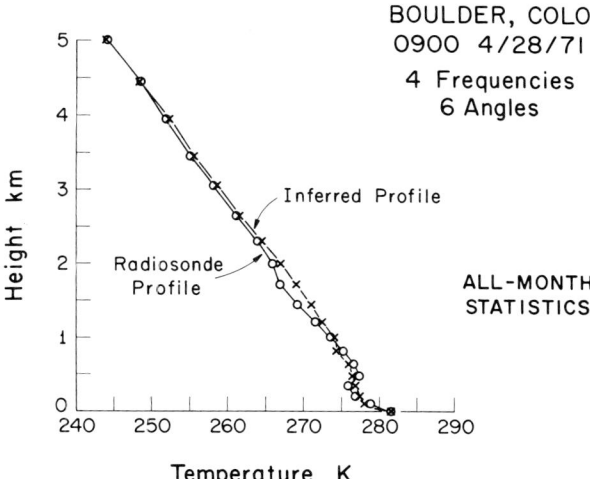

Fig. 10. Inferred and radiosonde temperature profiles from multi-spectral multi-angle radiometer data for 0900, April 28, 1971.

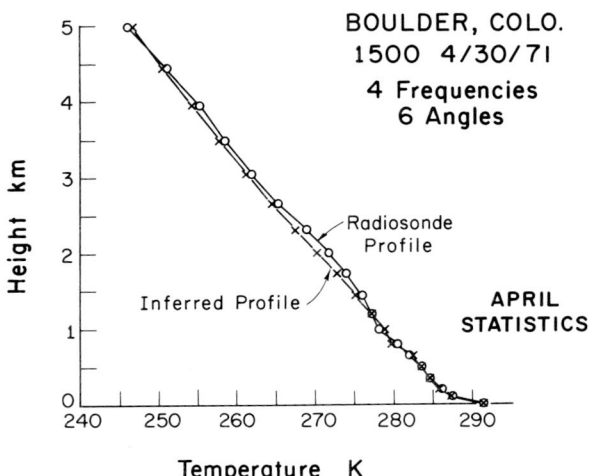

Fig. 11. Inferred and radiosonde temperature profiles from multi-spectral multi-angle radiometer data for 1500, April 30, 1971.

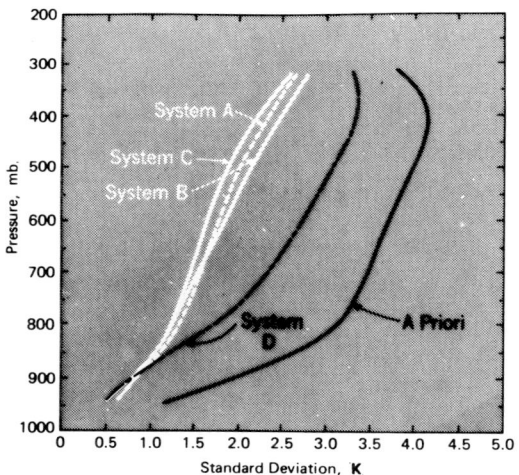

Fig. 12. Comparison of several radiometric systems in retrieving temperature layers of 100-mb thickness during clear conditions at ocean climatologies. Five-station rms average. Instrumental error = 0.5K. Systems: A(52.8, 55.4, 58.8, 20.6, 31.65 GHz); B(54.0, 55.4, 58.8, 20.6, 31.65 GHz); C(52.8, 54.0, 55.4, 20.6, 31.65 GHz); D(55.4, 58.8 GHz).

APPLICATION OF STATISTICAL INVERSION

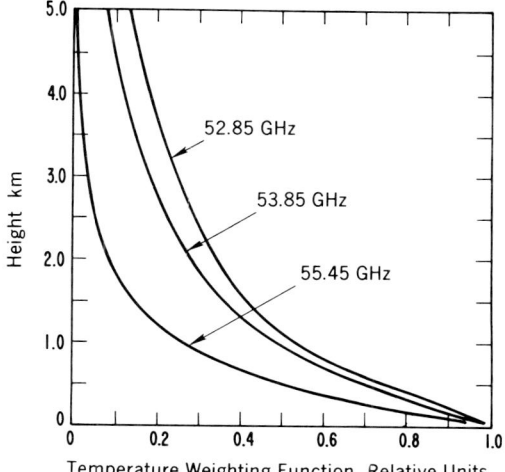

Fig. 13. Temperature weighting functions for ground-based sensing by SCAMS radiometer.

As mentioned in Section II.D, both the 52.85 GHz and the 53.85 GHz channels require cloud correction. The results shown in Fig. 14 indicate that with the addition of the two water correcting channels, and with a careful choice of frequency in the O_2 band, retrieval accuracies approaching those in clear air can be obtained. On the basis of the calculations shown in Figs. 12 and 14, the radiometric system C (SCAMS) was chosen for experimental verification of the temperature sensing capability of the buoy-based system.

The experimental program to confirm the theoretical predictions is being conducted jointly by National Oceanic and Atmospheric Administration (NOAA) and the Jet Propulsion Laboratory. During a three-week measurement period in March 1976, at Point Mugu, California, 21 suitable radiometer-radiosonde observations were obtained. Although some of the data were taken when there were visual observations of clouds, the radiometrically estimated liquid content was less than 0.02 mm; hence, the clear air retrieval

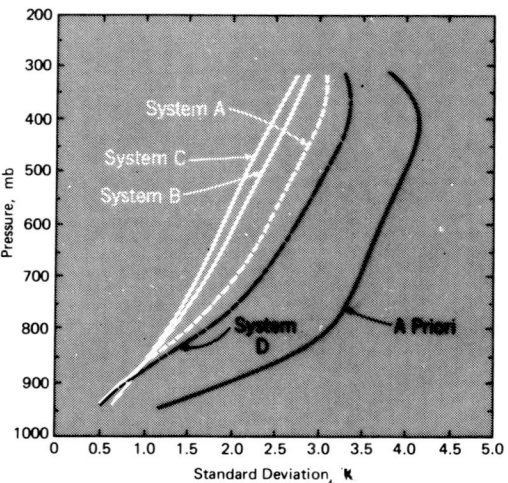

Fig. 14. Comparison of several radiometric systems in retrieving temperature layers of 100-mb thickness during cloudy conditions (liquid thickness < 4 mm H_2O). Five-station rms average. Instrumental error - 0.5 K. Systems: A(52.8, 55.4, 58.8, 20.6, 31.65 GHz); B(54.0, 55.4, 58.8, 20.6, 31.65 GHz); C(52.8, 54.0, 55.4, 20.6, 31.65 GHz); D(55.4, 58.8 GHz).

algorithms were used exclusively. The a priori data base was two years of twice daily soundings during February, March and April, taken in 1973-1974. The statistical summary of experiment results are compared with a priori predictions in Fig. 15. Although the observed variability of the 21 profiles about the three-month a priori is somewhat greater than the theoretical average, the achieved retrieval accuracies are in close agreement with predictions. It is somewhat unusual that theory predicts (a) no reduction in variance at about 750 mb above the surface, and (b) a modest reduction in variance above this level. The experiment results confirm this prediction.

Low resolution information on the water vapor distribution is also contained in the five-channel measurements. Typical water

APPLICATION OF STATISTICAL INVERSION

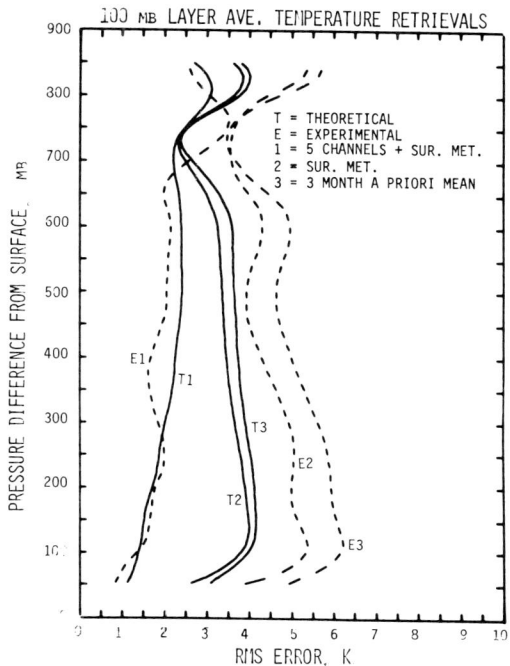

Fig. 15. RMS accuracy in retrieval of 100-mb layer averaged temperature, Pt. Mugu, California, March 1976, 21 profiles.

vapor weighting functions for the SCAMS system are shown in Fig. 16. Statistical retrieval algorithms were again applied to the 21 sets of radiometer observations to infer water vapor profiles. The summary of results is shown in Fig. 17, and typical temperature and humidity retrievals are given in Figs. 18 and 19.

Another set of observations was taken by JPL at Pt. Mugu in July 1976. In all, 22 concurrent radiometric and radiosonde observations were obtained. The statistical retrieval algorithm, in Eq. (2), was used to estimate the cloud liquid water content (LWC) of each of the profiles. As mentioned earlier, we classify a profile as cloudy if the estimated LWC is greater than 0.02 mm (a LWC of 0.02 mm will give a 1 K change in brightness temperature

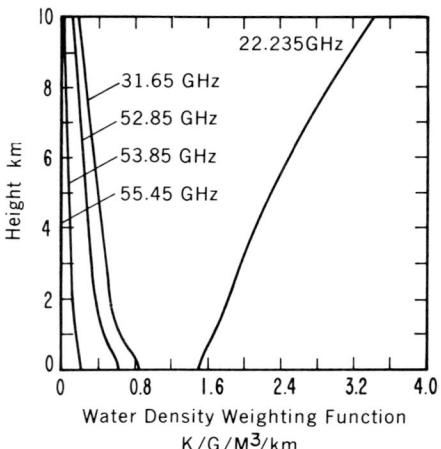

Fig. 16. Water vapor density weighting functions for ground-based SCAMS radiometer.

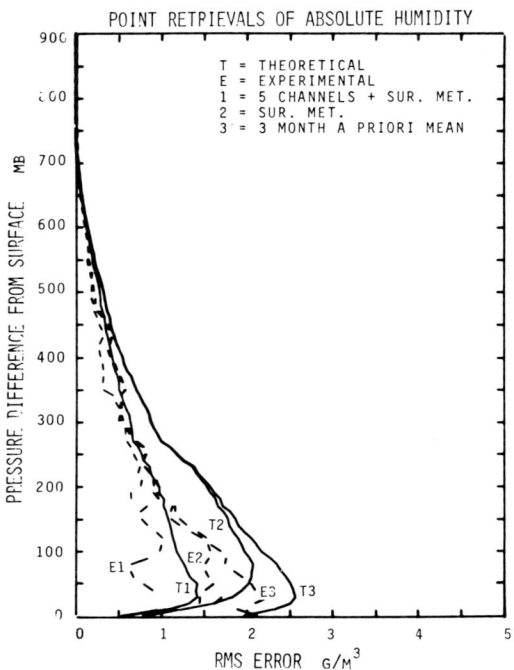

Fig. 17. Rms accuracy in retrievals of absolute humidity, Pt. Mugu, California, March 1976, 21 profiles.

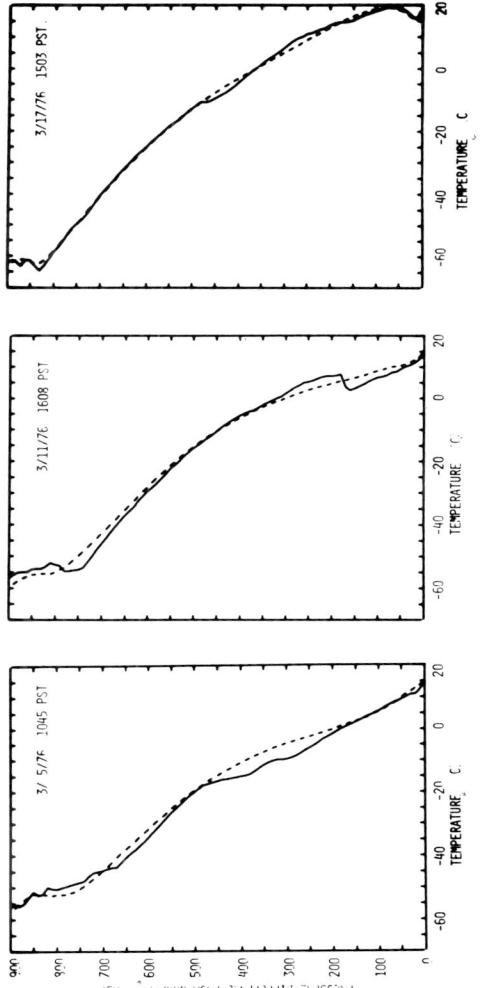

Fig. 18. Examples of temperature profiles retrieved from SCAMS radiometric data (dashed lines) compared with concurrent radiosondes (solid lines).

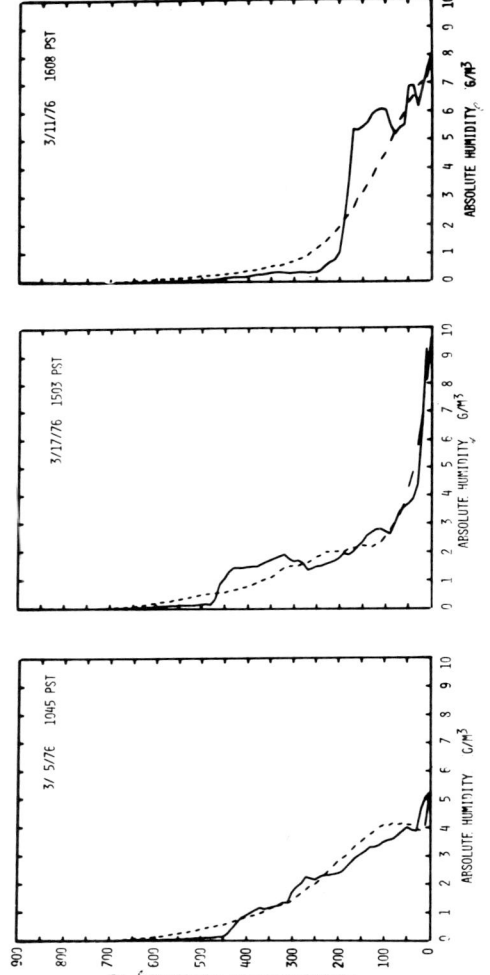

Fig. 19. Examples of water vapor profiles retrieved from SCAMS radiometric data (dashed lines) compared with concurrent radiosonde profiles (solid lines).

in the window channel). With this classification, 17 of the 22 profiles were radiometrically cloudy. We are currently investigating the application of various cloud correction algorithms to these data. However, with a very simple correction method, namely, no correction at all, the upper two channels can yield useful retrieval. Examples of retrievals obtained during cloudy conditions are shown in Figs. 20 and 21, and the statistical comparison of all 22 profiles is given in Fig. 22. Although there is a substantial reduction in variance in the region from 50 to 150 mb above the surface, it is likely that cloud correction can reduce this variance still further.

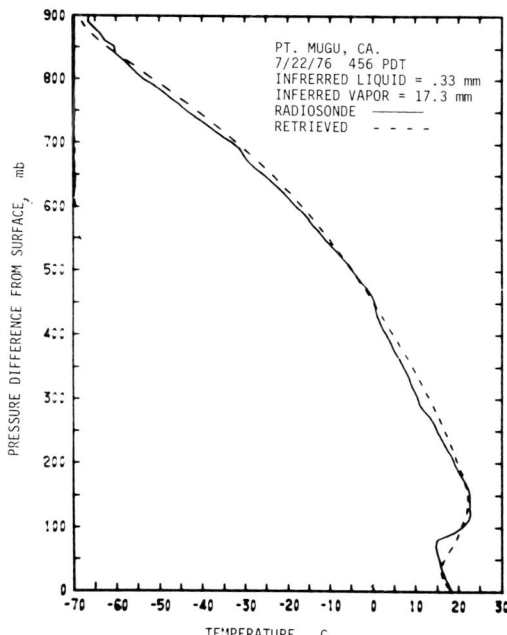

Fig. 20. Example of temperature profile retrieved from two oxygen channels during cloudy conditions.

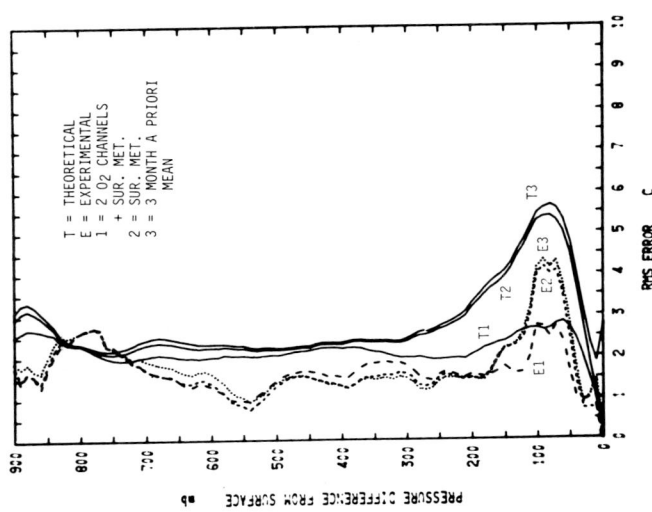

Fig. 22. Rms differences between radiometrically retrieved and radiosonde temperature profiles, 22 profiles: 5 clear + 17 cloudy.

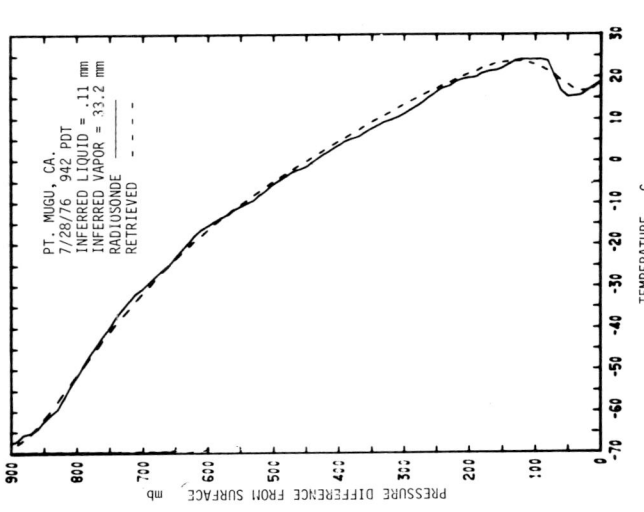

Fig. 21. Example of temperature profile retrieved from two oxygen channels during cloudy conditions.

APPLICATION OF STATISTICAL INVERSION

V. CONCLUSIONS

Statistical retrieval algorithms have proven to be quite useful in ground-based radiometric sensing, not only in profile recovery, but in making a reasonable *a priori* prediction of measurement achievability. Thus, we feel we can confidently answer questions relevant to system design, such as

    a. what is the climatological variation of retrieval accuracy?

    b. for a given number of channels, what is the optimum location of measurement ordinates?

    c. what noise levels are required to give specified retrieval accuracies?

We are currently investigating extensions of the statistical technique to the microwave cloud problem.

SYMBOLS

$\underset{\sim}{d}$	generalized data vector (Eqs. (2) and (3))
e_{rain}	extinction coefficient of rain, km^{-1} (Eq. (1))
$\underset{\sim}{p}$	generalized parameter vector (Eqs. (2) and (3))
$\underset{\sim}{\hat{p}}$	statistical estimator of p (Eqs. (2) and (3))
P	atmospheric pressure, mb (Figs. 1, 3 and 4)
r	relative humidity (Figs. 3 and 4)
R	rain rate, mm/hr (Fig. 1)
s, s'	path length, km from receiver to emitting volume (Eq. (2))
$S_{\underset{\sim}{q}}$	covariance matrix of random vector q (Eq. (3))
T	absolute temperature, K (Eq. (1), Figs. 1, 3 to 7)
$\bar{T}$	average absolute temperature (Fig. 6)
$T_b(\nu)$	brightness temperature, K (Eq. (1), Figs. 3 and 4)
$T_b^{(ext)}$	brightness temperature, K external to the Earth's atmosphere (Eq. (1))
$V[T_b]$	variance in brightness temperature, K^2 (Figs. 3 and 4)

$V[T_b:P]$ variance in brightness temperature, K^2 due to fluctuations in atmospheric profile parameter p (Figs. 3 and 4)

α_ν absorption coefficient, km^{-1} (Eq. (1), Fig. 1)

ν frequency, GHz (Eq. (1))

ρ_{liquid} density of liquid water, g/m^3 (Fig. 1)

ρ_{vapor} absolute humidity, g/m^3 (Fig. 1)

σ standard deviation (Fig. 5)

σ_ε standard deviation of instrumental noise, K (Fig. 5)

REFERENCES

1. C. R. Hosler and T. J. Lemmons, Radiometric measurements of temperature profiles in the planetary boundary layer, *J. Appl. Meteorol. 11*, 341 (1972).

2. G. F. Miner, D. D. Thornton, and W. J. Welch, The inference of atmospheric temperature profiles from ground-based measurements of microwave emission from atmospheric oxygen, *J. Geophys. Res. 77*, 975 (1972).

3. E. R. Westwater, J. B. Snider and A. V. Carlson, Experimental determination of temperature profiles by ground-based microwave radiometry, *J. Appl. Meteorol. 14*, 524 (1975).

4. A. T. Yershov, Y. V. Lebskiy, A. P. Naumov, and V. M. Plechkov, Determination of the vertical temperature profile from ground-based measurements of the atmospheric radiation at λ = 5 mm, *Izv. Atmos. Oceanic Phys. 11*, 1220 (1975).

5. J. W. Waters and D. H. Staelin, Statistical inversion of radiometric data, *Q. Prog. Rep. No. 89, Res. Lab. Elect., Massachusetts Institute of Technology*, 1968.

6. A. T. Yershov, A. P. Naumov, and V. M. Plechkov, Determination of the vertical humidity profile from ground-based thermal-radar measurements of atmospheric absorption, *Izv. Vuz. Radiofizika*, 15 (1972).

7. J. W. Waters, Absorption and emission by atmospheric gases, in "Methods of Experimental Physics" (M. L. Meeks, Ed.), Vol. 12, Part B, Ch. 2.3. Academic Press, New York, 1976.

8. G. E. Becker and S. H. Autler, Water vapor absorption of electromagnetic radiation in the centimeter wavelength range, *Phys. Rev.* 70, 300 (1946).

9. H. J. Liebe, Calculated tropospheric dispersion and absorption due to the 22-GHz water vapor line, *IEEE Trans. Antennas Propag.* AP-17, 621 (1969).

10. J. W. Van Vleck, The absorption of microwaves by oxygen, *Phys. Rev.* 71, 413 (1947).

11. H. J. Liebe, et al., Atmospheric oxygen microwave spectrum-experiment versus theory, *IEEE Trans. Antennas Propag.* AP-25 3, 327 (1977).

12. P. W. Rosenkranz, Shape of the 5-mm oxygen band in the atmosphere, *IEEE Trans. Antennas Propag.* AP-23, 498 (1975).

13. E. E. Reber, R. L. Mitchell, and C. J. Carter, Attenuation of the 5-mm wavelength band in a variable atmosphere, *IEEE Trans. Antennas Propag.* AP-18, 472 (1970).

14. D. H. Staelin, A. H. Barrett, P. W. Rosenkranz, F. T. Barath, E. J. Johnson, J. W. Waters, A. Wouters, and W. B. Lenoir, The scanning microwave spectrometer (SCAMS) experiment, *The Nimbus 6 Users Guide*, NASA, Goddard Space Flight Center, Greenbelt, Maryland, 1975.

15. F. Ostapoff, W. W. Shinners, and E. Augstein, Some tests on the radiosonde humidity error, *NOAA Tech. Rep. ERL 194-AOML 4, U.S. Dept. of Commerce*, 1970 [Superintendent of Documents: C55:13:ERL 194-AOML-4].

16. J. F. Morrissey and F. J. Brousaides, Temperature-induced errors in the ML-476 humidity data, *J. Appl. Meteorol. 9*, 805 (1970).

17. J. O. Laws and D. A. Parsons, The relation of raindrop size to intensity, *Trans. Am. Geophys. Union, 24*, 432 (1943).

18. E. R. Westwater, M. T. Decker, and F. O. Guiraud, Feasibility of atmospheric temperature sensing from ocean data buoys by microwave radiometry, *NOAA Tech. Rep. ERL 375-WPL 48*, 1976.

19. C. D. Rodgers, Retrieval of atmospheric temperature and composition from remote measurements of thermal radiation, *Rev. Geophys. Space Phys. 14*, 609 (1975).

20. W. D. Mount, A. C. Anway, C. V. Wick, and C. M. Maloy, Capabilities of millimeter wave radiometers for remotely measuring temperature profiles pertinent to air pollution, *Final Report, Contract No. PH. 86-67-76, Sperry Rand Research Center, Sudbury, Massachusetts*, February 1968.

21. E. R. Westwater, Ground-based determination of low altitude profiles by microwaves, *Mon. Weather Rev. 100*, 15 (1972).

22. J. B. Snider, Ground-based sensing of temperature profiles from angular and multi-spectral microwave emission measurements, *J. Appl. Meteorol. 11*, 958 (1972).

23. P. W. Rosenkranz, F. T. Barath, J. C. Blinn III, E. J. Johnston, W. B. Lenoir, D. H. Staelin, and J. W. Waters, Microwave radiometric measurements of atmospheric temperature and water from an aircraft, *J. Geophys. Res. 77*, 5833 (1972).

24. M. G. Fowler, N. D. Sze, and N. E. Gaut, The estimation of clear sky emission values from cloudy radiometric data, *Air Force Cambridge Research Laboratories-TR-75-0440*, 1975.

DISCUSSION

Susskind: When you do your analysis of the SCAMS data, do you subtract or do you find any bias in the brightness temperatures? I know Dave Staelin mentioned there was the possibility of a 2° bias in the measurements. And when we do our analysis we have to subtract something from the brightness temperatures.

Westwater: On the first or the second slide that I showed, we did have a small bias in the upper channel--a 54.5 channel and this was a bias of about 1.3°. The biases in the other two channels were essentially entirely below the noise levels. These were in the order of about 2/10 of a degree. Now with resepct to that question, we also did retrieval with subtracting a bias and retrievals with just the raw data. And for the temperature retrievals it made very little difference in the RMS errors. But there is somewhat of a bias in this in the data.

Susskind: I wonder if I could address this quickly to Dave Staelin. You mentioned that there was a bias of approximately 2° with the SCAMS measurement. Is that still there? Namely, that due to instrument calibration.

Staelin: There are two origins for bias in the satellite instruments. One is the instrument calibration. We recalibrated in orbit using NMC data yielding corrections on the order of a degree or so. There is another correction in the case of our data from both the Nimbus 5 and 6 satellites, which I did not discuss, and that is the transmittances. Either they have systematic errors which are larger than anything that has been indicated in the laboratory spectroscope data, to date, or the radiosondes have systematic errors which are larger than their manufacturer would probably like to accept. This is still a residual effect on the order of one or two degrees. But we have yet to track down its origin. It can be removed empirically and is not a problem now.

Susskind: We do find roughly a systematic difference between the observations and our ability to calculate them of approximately 2°. It's hard to tell if it is in the measurement or in our ability to calculate them.

Westwater: There is one other parameter in the Rosencrantz theory that is quite difficult to measure by Liebe's laboratory dispersion technique. This does not affect the calculated radiances at the strongly attenuating frequencies. However, at a frequency of around 50 GHz or 52 GHz, they are much more sensitive to this parameter that was not determined well from Liebe's measurements. In our calculations, we used the value of the nonresonant parameter that was given by Rosencrantz himself and it seemed to give a much better agreement with our measurements.

Fleming: Would you care to comment on just how you adjust for clouds when you sense them?

Westwater: The technique that we originally tried was to estimate the equivalent clear air brightness from the entire set of radiance observations and then use the equivalent clear brightness in retrieval. However, we ran into difficulties in applying this technique to the 52.8 GHz channel, so that the retrievals you saw were uncorrected brightness measurements at 53.8 and 55.45 GHz. We are still trying to unscramble the eggs again and what has happened to the cloud correction at the lower attenuating oxygen channel.

Fraser: Perhaps I misunderstood your last slide. I understood you to say that you were quite pleased with the water vapor retrieval. But, as I read the last slide, the theoretical and the experimental standard deviations, which I interpret as deviations in the radiosondes observations, except for 50 to 70 millibars above the ground, were essentially the same. If they are, then why are you so pleased with experimental data?

Westwater: The slide I showed at the end was the clear air data that was taken in March. And at least the total integrated water contents agreed quite well with that estimated by the radiosondes. However, in the period in July we did not obtain adequate agreement, primarily because of the cloud correction difficulties that I mentioned before. One of the reasons I am really pleased with the data is we really expected to get one parameter and one parameter alone. That would just be the total integrated water content. The amount of structure shown in the retrievals was an order of magnitude greater than anything I would have expected or that anyone would have expected from a weighting function which is essentially constant with altitude. There is very little structure in the weighting function itself.

INVERSION METHODS IN TEMPERATURE AND AEROSOL

REMOTE SOUNDING: THEIR COMMONALITY AND

DIFFERENCES, AND SOME UNEXPLORED

APPROACHES

Alain L. Fymat
Jet Propulsion Laboratory
California Institute of Technology

Departing from conventional research lines, this paper considers in parallel the two remote sensing problems of temperature profiling and aerosol characterization (complex refractive index, size distribution). These problems are formally identical and differ only in the explicit form of the source function which, for aerosols, includes contributions from both single and multiple scattering processes. The functional dependence on the desired atmospheric parameters, however, is considerably more complex in the case of aerosols. Both problems are essentially nonlinear. However, when the observables are the spectral extinction or the single scattering of the source radiation, the associated problem is completely analogous to the linearized temperature inversion problem; viz. the solution of a first-kind Fredholm integral equation must be obtained. Four questions attach to such an equation: existence, uniqueness, stability and construction of the solution, which are all analyzed. Methods for obtaining the solution of the linear problem are classified following three main categories (i) derivation of properties that all solutions satisfy, which must then be properties of the actual solution; (ii) regularization of the ill-posed problem; and (iii) data changes within their domain of uncertainty in order to avoid the basic instability. A number of unexplored methods (e.g., reduction to a second-kind equation, singular value and other decompositions, invariant imbedding, etc.) are indicated. Solutions of the nonlinear problem are also classified. Lastly, a two-step strategy is proposed for retrieving first the complex refractive index of aerosols, in a manner that is essentially independent of their size distribution, and then the size distribution itself. The first step involves the inversion of spectral extinction ratios using the author's Minimization Search Method, while the second step rests on an analytical integral transform of light scattered within a narrow forward cone.

I. INTRODUCTION

Vertical temperature profiling and aerosol microstructure reconstruction have generally formed two separate branches of research in remote sensing theory. This separation also exists in the retrievals of the constant concentration of thermally active gases and the complex refractive index of aerosols. While in each case of these two situations, the underlying physical processes are certainly different, the corresponding mathematical inversion problems are nevertheless very similar. Thus, a greater cross-fertilization between these two traditionally divided activities can only result in advances in the solutions to these problems. In fact, one should proceed further and even explore whether the physical basis in one area can be used advantageously in the other area. For example, measurement of the in- and out-of-band components of Rayleigh scattering by an atmospheric absorber of known density such as O_2, CO_2, can provide the altitude distribution of molecular absorption from which the temperature profile can subsequently be recovered. Likewise, Mie scattering by aerosol stratifications can locate the altitudes of temperature inversions trapping these particulates. Information derived from these and similar scattering techniques can be used to improve temperature retrievals in the same way as temperature and gaseous composition determinations can help in the investigation of aerosols.

The aim of this article is to set forth the commonality and differences between the methodologies that have been developed, or that can be employed, in the remote sounding of temperature and aerosols. The following section provides the equations of remote sensing, appropriate for a variety of viewing geometries, in both their differential and integral formulations, and expressions for the internal and external field sources. Section III formulates the temperature sounding problem in the infrared and microwave regions of the spectrum with analyses of the weighting functions and of the nature of the associated mathematical problem. Section IV considers the particulate sounding problem for reflected, directly and diffusely transmitted radiation at low and large scattering angles, and provides a tabular analysis of the nature of the associated inverse problems. Section V analyzes the nature of the problem of finding a solution to linear first-kind Fredholm integral equations in terms of existence, uniqueness, stability, and construction of the solution. It also provides a preliminary classification of methods of solution of such equations. The corresponding classification for nonlinear methods is given in Section VI. The final section proposes a two-step strategy developed by the author for solving the particulate sounding problem. There, the complex refractive index would be retrieved from spectral extinction ratios while the particle size distribution would be reconstructed from angular forwardly scattered radiances at a given wavelength.

II. EQUATIONS OF REMOTE SENSING

Consider the situation represented by Fig. 1: a slab atmosphere bounded between altitudes z_1 and z_2, or optical depths, τ_1 and 0, respectively. Any point P in the atmosphere is located in a cartesian frame of reference by its spherical coordinates: $\vec{r}$ = radius vector, θ = zenith angle measured from the z-axis, and ϕ = azimuth angle measured from the x-axis. Upward directions are denoted $\vec{\Omega}^+ \equiv (\alpha, \beta, \gamma)$; likewise downward directions are represented by $\vec{\Omega}^- \equiv (\alpha, \beta, -\gamma)$, where $\alpha = \sin\theta \cos\phi$, $\beta = \sin\theta \sin\phi$ and $\gamma = \cos\theta$ are direction cosines. The equations of remote sensing are derived from the radiative transfer equation (RTE) expressing the principle of energy conservation within an elemental atmospheric volume. They can be obtained in either a differential or an integral form, as follows.

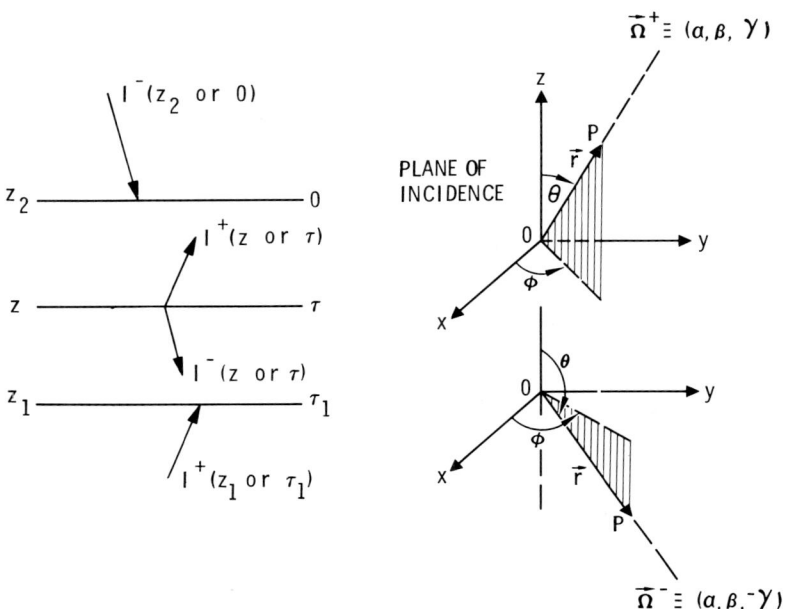

Fig. 1. Illustrating the geometry for remote sensing.

A. Differential Formulation

In its most general form, the monochromatic RTE at frequency ν for either upwelling or downwelling radiation is written as

$$(\vec{\Omega}^{\pm} \cdot \vec{\nabla}) I_\nu^{\pm} = -K_\nu (I_\nu^{\pm} - J_\nu) \tag{1}$$

where the specific intensity (or radiance) I, the source-function J, and the volume extinction coefficient K are all functions of location and direction: $F \equiv F(\vec{r}; \vec{\Omega})$, $F \equiv I, J, K$. The differential operator $(\vec{\Omega} \cdot \vec{\nabla})$ is specified by the geometry of the atmosphere; various relevant expressions of this operator are summarized in Table 1. It is usually assumed that K is either directionally averaged or independent of direction. The same assumption is separately valid for the absorption coefficient, k_ν, and the scattering coefficient, σ_ν, the sum of which equals K.

TABLE 1

Various Forms of the Differential Transfer Operator ($\vec{\Omega}^{\pm} \cdot \vec{\nabla}$) for Geometries of Interest to Remote Sensing

ATMOSPHERIC GEOMETRY	DIFFERENTIAL TRANSFER OPERATOR
1a. SPHERICAL (TWILIGHT, LIMB, TERMINATOR)	$\alpha \frac{\partial}{\partial x} + \beta \frac{\partial}{\partial y} + \gamma \frac{\partial}{\partial z}$ $= (1-\mu^2)^{1/2} \left[\cos\phi \frac{\partial}{\partial x} + \sin\phi \frac{\partial}{\partial y} \right] \pm \mu \frac{\partial}{\partial z}, \quad \mu = \cos\theta$
1b. (ORIGIN AT EARTH'S CENTER AND Z PARALLEL TO LOCAL VERTICAL)	$\pm \mu \frac{\partial}{\partial r} + \frac{1-\mu^2}{r} \frac{\partial}{\partial \mu}$ $+ \frac{(1-\mu^2)^{1/2} (1-\mu_0^2)^{1/2}}{r} \left[\cos\phi \frac{\partial}{\partial \mu_0} + \frac{\mu_0}{1-\mu_0^2} \sin\phi \frac{\partial}{\partial \phi} \right]$
2. SPHERICAL SYMMETRY ($\partial/\partial\mu_0 = \partial/\partial\phi \equiv 0$)	$\pm \mu \frac{\partial}{\partial r} + \frac{1-\mu^2}{r} \frac{\partial}{\partial \mu}$
3. AXIAL SYMMETRY ($\mu = 1$)	$\pm \frac{\partial}{\partial r}$
4. PLANE-PARALLEL (e.g., $\partial/\partial x = \partial/\partial y \equiv 0$)	$\pm \mu \frac{\partial}{\partial z}$

Equation (1) is to be solved subject to appropriate boundary conditions on the radiation field at the top I^- ($z = z_2$ or $\tau = 0$) and the bottom I^+ ($z = z_1$ or $\tau = \tau_1$) of the atmospheric slab. Depending on the wavelength region utilized and on assumptions on the problem, Eq. (1) can either be a scalar or a four-vector equation. The latter situation involves Stokes' representation of a light vector, but it may be noted that other representations, such as a four-coherency vector or a 2 × 2 Jones' matrix, can be used as well. In general, k, σ and K are taken to be scalars. In the following sections, we shall limit ourselves to the case of a plane-parallel stratified atmosphere for which the RTE is simply

$$\pm \mu \frac{\partial}{\partial z} I^{\pm}(z; \mu, \phi) = -K(z)[I^{\pm}(z; \mu, \phi) - J(z; \mu, \phi)] \qquad (1a)$$

Alternatively, introducing the normal optical thickness

$$\tau_\nu(z) = \int_z K_\nu(z)dz \qquad (1b)$$

we also have

$$\pm \mu \frac{\partial}{\partial \tau} I^{\pm}(\tau; \mu, \phi) = J(\tau; \mu, \phi) - I^{\pm}(\tau; \mu, \phi) \qquad (2)$$

(In these equations and in the following, the subscript ν is dropped for convenience. It will, however, be indicated when it occurs for the first time.)

B. Integral Formulation

The integral formulation is simply obtained by formal integration of the RTE. For example, by integration of Eq. (2),

$$I^{+}(\tau; \vec{\Omega}) = I^{+}(\tau_1; \vec{\Omega})e^{-(\tau_1 - \tau)/\mu} + \int_\tau^{\tau_1} J(s; \vec{\Omega})e^{-(s - \tau)/\mu} \frac{ds}{\mu} \qquad (3a)$$

$$I^{-}(\tau; \vec{\Omega}) = I^{-}(0; \vec{\Omega})e^{-\tau/\mu} + \int_0^\tau J(s; \vec{\Omega})e^{-(\tau - s)/\mu} \frac{ds}{\mu} \qquad (3b)$$

which provide the upward and downward radiation fields at any arbitrary level τ in the atmosphere. The first term on the right side of either of these last equations represents the boundary contribution while the integral term provides the atmospheric contribution proper. From these equations, one obtains immediately the field emerging from the top, $I^{+}(0; \vec{\Omega})$, or reaching the bottom, $I^{-}(\tau_1; \vec{\Omega})$, of the atmosphere. In particular, for an infinitely thick medium ($\tau_1 = \infty$),

$$I^{+}(0; \vec{\Omega}) = \int_0^\infty J(s; \vec{\Omega})e^{-s/\mu} \frac{ds}{\mu} \qquad (3c)$$

C. Internal and External Field Sources

The explicit solution of the RTE, in either its integral or its differential form, depends, of course, on assumptions on the nature of the source fields either impinging on the atmosphere or originating within the atmosphere itself. Expressions for the source function are summarized in Table 2. In particular, the scattering source function for a clear atmosphere is

$$J_{NC,S} \equiv J(SS) + J(MS)$$

$$= \frac{1}{4} P_\nu(\vec{r};\, \vec{\Omega},\, \vec{\Omega}_o^-) I_o(\vec{\Omega}_o^-) t_\nu(\vec{r},\, \vec{r}_o;\, \vec{\Omega})$$

$$+ \frac{1}{4\pi} \int P_\nu(\vec{r};\, \vec{\Omega},\, \vec{\Omega}') I(\vec{r};\, \vec{\Omega}')\, d\omega' \qquad (4)$$

where I_o is the incident radiation, t the transmission function and P the scattering phase function (or phase matrix for a light vector). It receives contributions from both single (SS) and multiple (MS) scattering processes in the atmosphere (first and second term, respectively, in the right-hand side of the equation). The integral equation satisfied by $J_{NC,S}$, the so-called auxiliary equation of radiative transfer for the source function, is obtained by substituting Eqs. (3a) and (3b) in the second term in the right-hand side of Eq. (4). The kernel of these equations is a linear combination of the scattering phase functions (or phase matrices) for gaseous (superscript g) and particulate (superscript p) contributions.

$$P_\nu = a_\nu P_\nu^g + (1 - a_\nu) P_\nu^p \qquad (5)$$

where the weight, called turbidity factor, is given by

$$a_\nu = \frac{\sigma_\nu^p}{\sigma_\nu^g + \sigma_\nu^p} \qquad (6)$$

For some gases, it may be necessary to take into account their

TABLE 2

Source-Function Expressions Corresponding to Some Typical Atmospheric States**

ATMOSPHERIC STATE	SOURCE-FUNCTION EXPRESSION
1. LOCAL THERMODYNAMIC EQUILIBRIUM: SCATTERING NEGLECTED	$J_{NC,NS} = B_\nu [T(p)]$
2. LOCAL THERMODYNAMIC EQUILIBRIUM: SCATTERING INCLUDED	$J_{NC} = \tilde{\omega}_\nu J_{NC,S} - (1-\tilde{\omega}_\nu) J_{NC,NS}$
3. PARTLY OBSCURED BY HAZE OR CLOUD	$J^{PO} = N J^O + (1-N) J^{NO}$ (IN INFRARED)
4. ANY ARBITRARY COMBINATION OF PREVIOUS CASES	LINEAR COMBINATION OF PREVIOUS SOURCE FUNCTION EXPRESSIONS

*NC = NOT CLOUDY, NS = NOT SCATTERING, NO = NOT OBSCURED
S = SCATTERING, O = OBSCURED, PO = PARTLY OBSCURED
$\tilde{\omega}_\nu = \sigma_\nu / K_\nu$ = SINGLE SCATTERING ALBEDO, $\sigma_\nu = \sigma_\nu^g + \sigma_\nu^p$, $K_\nu = K_\nu^g + K_\nu^p$
N = FRACTION OF CLOUDINESS OF THE SKY, T = TEMPERATURE, P = PRESSURE

nonsphericity; in this case, the phase function is that corresponding to Rayleigh-Cabannes scattering

$$P_\nu^g \equiv P_\nu^{RC} = b_\nu P_\nu^R + (1 - b_\nu) P_\nu^I \tag{7}$$

itself a linear combination of the Rayleigh phase function (superscript R), applicable to scattering by spherical molecules, and the isotropic (superscript I) phase function. For continuum frequencies, $\nu = f$, the weight $b_f = 2(1 - \delta)/(2 + \delta)$, where δ is the so-called gaseous depolarization factor (e.g., $b_f = 0$ for isotropic scatterers; $b_f = 1.0$ for Rayleigh (spherical) particles or molecules; $b_f = 0.4$ for rod-like scatterers; and $0.4 < b_f < 1.0$ for all shapes between a rod and a sphere) while for spectral line frequencies, $\nu = \ell$, it is the well-known Hamilton's coefficient for

resonantly fluorescent line scattering (e.g., $b_\ell = 0.5$ for H-Ly α).
If the particulates are spherical, P_ν^p is provided by the Mie
expressions. If P_{ij} denote the elements of matrix $\underset{\sim}{P}$, i, j = 1 to
4, we have the following normalization condition

$$\frac{1}{4\pi} \int P_{11}(\vec{r}; \vec{\Omega}, \vec{\Omega}') d\omega' = \frac{1}{4\pi} \int P_{11}(\vec{r}; \vec{\Omega}', \vec{\Omega}) d\omega = 1 \qquad (8)$$

(P_{11} is none other than the phase function.) The phase function
(or phase matrix) also satisfies a number of reciprocity and symmetry relations (not given here).

The equations summarized so far provide the necessary formalism for remote sensing from above the atmosphere or from the surface and for looking at arbitrary targets along arbitrary directions. They include (or can be used to provide) equations appropriate for nadir or zenith viewing, twilight, limb, and terminator scans.

III. TEMPERATURE SOUNDING PROBLEM

The basic assumptions are here: plane-parallel atmosphere, in Local Thermodynamic Equilibrium (LTE), free of clouds and aerosols; negligible gaseous scattering; negligible surface reflection of downward atmospheric radiation; and known active gases. The following paragraphs provide the relevant equations for the thermal and the microwave sounding of the temperature structure. For simplicity, and with the exception of some remarks on the limb geometry, we shall essentially limit these derivations to nadir and zenith viewing geometries.

A. Thermal Sounding

The relevant equations, in differential form, are obtained simply from the derivations of the previous section

$$\pm \frac{\mu}{k'(z)\rho(z)} \frac{\partial I^{\pm}(z; \mu)}{\partial z} = B[T(z)] - I^{\pm}(z; \mu) \qquad (9)$$

where $k'(z) = k(z)/\rho(z)$ is the mass absorption coefficient, and ρ is the absorbing gas density. It is easy to convert this equation from an altitude, z, to a pressure variable, p, by use of the hydrostatic equation

$$-\rho(z)dz = \frac{w(p)}{g(p)} dp$$

where w is the mass mixing ratio of the absorbing gas, and g is the acceleration of gravity. Equation (9) must be solved subject to the boundary conditions:

$$\left.\begin{array}{l} I^+(0; \mu) = B[T(0)] \quad \text{(Nadir viewing)} \\ I^-(\bar{z}; \mu) \equiv 0 \quad \text{(Zenith viewing)} \end{array}\right\} \quad (10)$$

where $\bar{z}$ is the altitude of the atmospheric top (or of the sensing platform). Integrations of Eqs. (9) and (10), or use of Eqs. (3) and (10), yields straightforwardly

$$I^+(\bar{y}; \mu) = B[T(y_s)] t[\bar{y}, y_s; w_\ell(y); T(y); \mu]$$

$$+ \int_{y_s}^{\bar{y}} B[T(y)] \frac{\partial}{\partial y} t[\bar{y}, y; w_\ell(y); T(y); \mu] dy \quad (11a)$$

$$I^-(y_s; \mu) = -\int_{y_s}^{\bar{y}} B[T(y)] \frac{\partial}{\partial y} t[y, y_s; w_\ell(y); T(y); \mu] dy \quad (11b)$$

where $y \equiv p, \log p, \ldots$ is a single-valued function of the pressure, and

$$t(y_2, y_1; \mu) = \exp\left[\frac{1}{\mu} \int_{y_1}^{y_2} \frac{w(y)}{g(y)} k'(y) dy\right] \quad (12)$$

is the slant transmission function between levels y_1 and y_2; its derivative is now commonly referred to as the weighting function.

TEMPERATURE AND AEROSOL REMOTE SOUNDING

B. Microwave Sounding

Equations (9) and (11) are still valid in this case, but Planck's function can be replaced by its approximation provided by the Rayleigh-Jeans formula.

C. Properties of the Weighting Functions

The CO_2 transmittances and weighting functions corresponding to the National Oceanic and Atmospheric Administration's NOAA 2-Vertical Temperature Profile Radiometer are illustrated in Fig. 2. The following general properties of the weighting functions can be listed:

1. For both infrared (IR) and microwave (MW) radiation, they are weak functions of the temperature, mainly through line broadening.

2. Each function covers a finite, although extensive, portion of the atmosphere and exhibits a generally broad peak. This results in low vertical resolution of the retrieved temperature profile, i.e., the fine scale structure cannot be recovered. It may be noted, however, that the weighting functions corresponding to limb observations can be much narrower, thus permitting a finer resolution than with other geometries.

3. Some functions can exhibit several maxima which result in inversion ambiguities. To avoid this situation, it is best to discard such functions.

4. The functions are overlapping, i.e., they are mutually correlated, the degree of correlation increasing with the number of channels (frequency or tangent height). Methods for retrieving the temperature profile will later be described. In those methods where the integral remote sensing equation is approximated by a linear algebraic system, this correlation manifests itself by the linear dependence (ill-conditioning) of the column vectors of the operator matrix. As the number of channels is increased, so does the number of column vectors and their linear dependence.

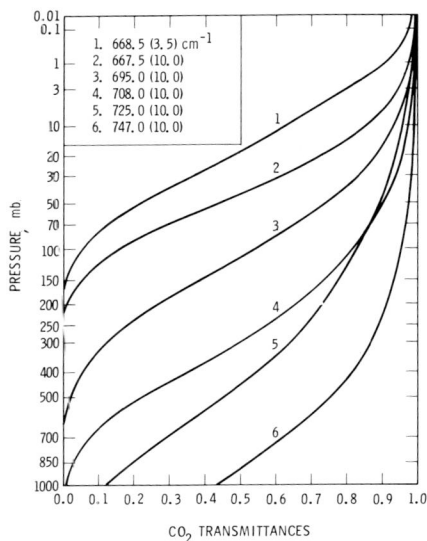 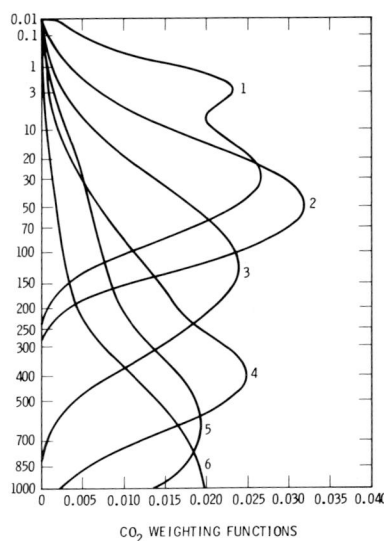

Fig. 2. NOAA 2 - Vertical Temperature Profile Radiometer (from Ref. 1). (Note: Only center frequencies are listed. Filter bandpasses at half-widths are provided within parentheses in the insert.) (1 bar = 100 kPa.)

D. Nature of the Mathematical Problem

The remote sensing equations, Eqs. (9) or (11), are essentially nonlinear in the temperature profile. The sources of nonlinearities are summarized as follows:

1. Temperature dependence of the transmission function: weak in all wavelength regions of interest.
2. Frequency dependence of Planck's function: weak within a single IR band, stronger when several IR bands are considered together; none in the MW region.
3. Scattering by gases and particulates: weak (usually not considered).
4. Surface reflection effects: weak (usually not considered).

TEMPERATURE AND AEROSOL REMOTE SOUNDING

5. Presence of clouds and/or hazes: can be large.
6. Mathematical nonlinearities (constraints, etc.): can be large.

However, when utilizing a single IR band or in the MW region, it is possible to separate the frequency and temperature dependences of Planck's function by using a variety of approximations. In this case, the integral remote sensing equation can be written as the first-kind Fredholm integral equation:

$$I(\nu_j) = \int_Y K(\nu_j; y) B(y) dy, \quad (j = 1, 2, \ldots n) \tag{13}$$

where ν_j are the n sounding frequencies, the kernel K is the transmission function (since it exhibits a much weaker variation with T than does B), B stands for the T-term in Planck's function, and

$$I(\nu_j) = \frac{\text{Measurement } (- \text{ surface term})}{\nu\text{-term in Planck's function}}$$

Prior to reviewing the methods of solution of the temperature sounding problem, we shall first analyze the particulate sounding problem in order to bring into more evidence the commonality and differences between these two problems.

IV. PARTICULATE SOUNDING PROBLEM

The basic assumptions for this problem are: plane-parallel atmosphere, completely hazy or cloudy; known contributions from gases; no thermal radiation effects; and known surface reflection effects.

A. Reflected and Transmitted Radiation

The form of the remote sensing equations used in this case is

$$\pm \mu \frac{\partial I^{\pm}(\tau; \vec{\Omega})}{\partial \tau} = I^{\pm}(\tau; \vec{\Omega}) - J(\tau; \vec{\Omega}) \tag{14}$$

where

$$J = \tilde{\omega}[J(SS) + J(MS)]$$

is provided by Eq. (4), and the boundary conditions are

$$\left. \begin{array}{ll} I^+(\tau_1; \vec{\Omega}) = \text{Prescribed surface reflection} & \text{Reflection} \\ \\ I^-(0; \vec{\Omega}) = \text{Radiation source strength} & \text{Transmission} \end{array} \right\} \quad (15)$$

The corresponding integral formulations are as given in Eqs. (3a)-(3c)

For transmitted radiation, it is interesting to note that it is possible to separate the direct from the diffuse component. In the former case, the corresponding spectral extinction measurements are performed by pointing in the source direction. In this direction, direct radiation is predominant, and the extinction process conserves any incident polarization. Thus, while these measurements are restricted to a single direction and polarization information is lost, nevertheless, the complexity of the corresponding mathematical problem is considerably reduced. The equation for this case is obtained by letting $J = 0$ since there is no self-illumination of the medium by scattering, and

$$I^-(\tau_1; \vec{\Omega}) = I^-(0; \vec{\Omega}) e^{-\tau_1/\mu} \quad (16a)$$

a strictly scalar problem irrespective of the initial polarization. Away from this direction, direct radiation vanishes, and Eq. (3b), or Eq. (14) with a null boundary condition, yields

$$I^-(\tau_1; \vec{\Omega}) = \int_0^{\tau_1} J(s; \vec{\Omega}) e^{-(\tau_1 - s)/\mu} \frac{ds}{\mu} \quad (16b)$$

a vector problem if polarization is considered.

The problem unknowns are, in the macroscale: the volume absorption and scattering coefficients and their vertical profiles (from which the scattering, absorption, and total optical depths at any level in the atmosphere, the corresponding optical thicknesses of the atmosphere, and the single scattering albedo and its vertical profile can be obtained), and the scattering phase-function (- matrix). If the absorption and scattering coefficients are known, then it is sufficient to determine the number density profile in place of the corresponding volume coefficients. In the microscale, these parameters involve the unknown complex refractive index, the particle size distribution, and their vertical profiles.

B. Spectral Extinction

Assuming that the particulates are spheres, Eq. (16a) becomes

$$E[(\lambda);m(\lambda); n(x)] \equiv -\tau_1^g - \mu \ln \left| \frac{\bar{I}^-(\tau_1^p; \vec{\Omega})}{\bar{I}^-(0; \vec{\Omega})} \right|$$

$$= \pi k^{-3} \int_0^\infty x^2 Q_{ext}[x; m_r(\lambda), m_i(\lambda)] n(x) dx \quad (16c)$$

where $k = 2\pi/\lambda$ is the wavenumber, $x = 2\pi r/\lambda$ is the size parameter, r = radius, m_r and m_i are, respectively, the real and imaginary components of the refractive index, $n(x)$ is the size distribution, and Q_{ext} is the Mie efficiency factor for extinction

$$Q_{ext} = \frac{4}{x^2} Re\{S(0)\} = \frac{2}{x^2} \sum_{n=1}^{\infty} (2n+1) Re(a_n + b_n) \quad (17)$$

In this last expression, $S(0)$ is the Mie complex scattering amplitude function in the forward scattering direction, and a_n and b_n are the well-known Mie coefficients. For $|m - 1| \ll 1$, van de Hulst has provided the following approximation to Q_{ext}:

$$Q_H = 4 Re\{K(i\rho + \rho \tan \beta)\} \quad (18a)$$

where

$$K(w) = \frac{1}{2} + \frac{e^{-w}}{w} + \frac{e^{-w} - 1}{w^2}$$

$\rho = 2x|m - 1|$, and $\tan \beta = m_i/(m_r - 1)$. The approximation in Eq. (18a) is found in practice to apply well for $1.33 < m_r < 1.5$. For $1.0 < m_r < 1.5$ and $0.05 \leq m_i \leq 0.25$, Deirmendjian has provided the following *ad hoc* extension of Q_H:

$$Q_D = (1 + D)Q_H \tag{18b}$$

where D is a known constant.

If m_r and m_i are known, or constant over the wavelength range of the measurements, or else slowly varying within this range, Eq. (16c) can be written as a linear first-kind Fredholm integral equation in the size distribution:

$$E(\lambda_j) = \int_X K(\lambda_j; x) N(x) dx \quad (j = 1, 2, \ldots, n) \tag{19}$$

where $K \equiv Q_{ext}$ (or $x^2 Q_{ext}$) and $N = x^2 n(x)$ [or $n(x)$].

C. Single Scattering in a Homogeneous Slab

The corresponding vector equations, obtained by canceling $J(MS)$ are, respectively, on reflection and transmission:

$$I^+(0; \mu, \mu_0) = \frac{\tilde{\omega} f}{4} \underset{\sim}{P}[\Theta; m; n(x)] \vec{I}^-(0; \mu_0) \tag{20a}$$

$$I^-(\tau_1; \mu, \mu_0) = \frac{\tilde{\omega} g}{4} \underset{\sim}{P}[\Theta; m; n(x)] \vec{I}^-(0; \mu_0) \tag{20b}$$

where Θ is the scattering angle, and

$$\left. \begin{array}{l} f \equiv \dfrac{\mu_0}{\mu_0 + \mu} \left[1 - e^{-\tau_1 \left(\frac{1}{\mu} + \frac{1}{\mu_0} \right)} \right] \\[2em] g \equiv \dfrac{\mu_0}{\mu_0 - \mu} \left[e^{-\tau_1/\mu_0} - e^{-\tau_1/\mu} \right] \end{array} \right\} \tag{20c}$$

TEMPERATURE AND AEROSOL REMOTE SOUNDING

Further, for $\tau_1 \ll 1$, $f = g \simeq \tau_1/\mu$, $\tilde{\omega}\tau_1 = \tau_{sc}$ (= scattering optical thickness), and Eqs. (20a) and (20b) degenerate to the single expression

$$\vec{I}^+(0; \mu, \mu_0) = \vec{I}^-(0; \mu, \mu_0) = \frac{\tau_{sc}}{4\mu} \underset{\sim}{P} \vec{I}^-(0; \mu_0) \qquad (21)$$

For incident light that is either natural or fully plane-polarized, Eq. (21) can be reduced to a scalar one. If the phase function (- matrix) is known, or modeled, this equation can be used for retrieving τ_{sc}. Alternatively, if τ_{sc} is known independently, and if m is also known, the same equation can be written as a first kind Fredholm integral equation in the size distribution

$$\frac{4\mu I^+(0; \mu, \mu_0)}{\tau_{sc} I^-(0; \mu_0)} \equiv P(\Theta) = \int_X P(\Theta; x) n(x) dx \qquad (22a)$$

where P has different expressions depending on the incident polarization.

D. Near-Forward Scattering

Let ℓ be the length of the scattering medium. Since this medium is homogeneous, $\tau_{sc} = \sigma\ell$. By introducing the scattering diagram $F = \sigma P$, Eq. (22a) can be written as

$$I_\Theta = \frac{4\mu}{\ell} \left[\frac{I^+(0; \mu, \mu_0)}{I^-(0; \mu_0)} \right] = \int_X F(\Theta; x) n(x) dx \qquad (22b)$$

For incident light that is plane-polarized either along the vertical or the horizontal, the two equations (22b) for this case degenerate to a single equation for all scattering angles. On the other hand, for natural light, this is the case only at forward scattering angles. In this latter case, in the Kirchhoff approximation to Fraunhoffer diffraction,

$$F(\Theta; x) = \frac{x^2 J_1^2(x \sin \Theta)}{k^3 \sin^2 \Theta} \qquad (23a)$$

A considerably better approximation, provided by the works of Penner, and Shifrin and Punina, is

$$F^*(\Theta; x) = \frac{Q_{ext}}{4} F(\Theta; x) \qquad (23b)$$

This approximation is valid for $\Theta_{max} \approx 10^o$ and $x \geq 5$. A detailed analysis of its accuracy (which is actually quite high) for a set of refractive indices is not available. With Eq. (23b), Eq. (22b) becomes

$$I_\Theta = \frac{1}{k^3 \sin^2\Theta} \int_X J_1^2(x \sin \Theta) x^2 n^*(x) dx \qquad (24)$$

where $n^*(x) = (1/4) Q_{ext}^2 n(x)$, and J_1 is Bessel function of the first kind and order unity.

E. Observables and Corresponding Mathematical Problems for Reconstruction of Particle Size Distribution

The situation is summarized in Fig. 3, which is self-explanatory. It may be noted, however, that under certain approximations, the inverse problem for the individual spectral extinctions and for the near-forward radiances arising from single scattering, can be solved analytically. For multiple scattering, for spectral extinction ratios, and for single scattering radiance and polarization ratios, the problem is amenable to a minimization search. In all other cases, the problem is that of solving a first-kind Fredholm integral equation. Since the latter equation is of fundamental importance in both the temperature and the particulate sounding problem, we shall now turn to its analysis.

V. ON THE INVERSION OF FIRST-KIND FREDHOLM INTEGRAL EQUATIONS

Consider the first-kind equation:

$$f(x) = \int_Y A(x, y) g(y) dy \qquad (x \in X) \qquad (25)$$

TEMPERATURE AND AEROSOL REMOTE SOUNDING

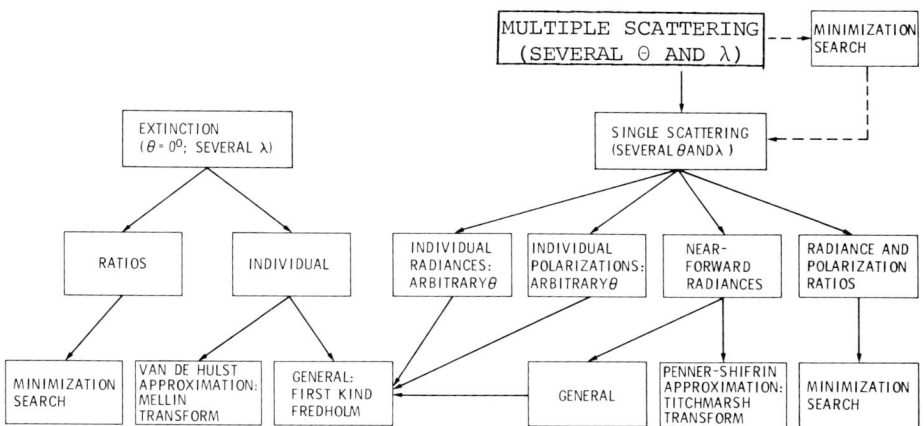

Fig. 3. Observations and corresponding mathematical problems for reconstruction of particle size distribution (from Ref. 2).

There are four important questions bearing on the solution of such an equation (Ref. 3).

A. Questions of Existence, Uniqueness, Stability and Construction

Existence. What conditions must be placed upon f to ensure that there is a corresponding g?

Uniqueness. Given that a solution g_1 exists, are there any others, g_2, g_3, etc. also satisfying the basic equation? Alternatively, are there any nontrivial solutions h(y) to the equation

$$\int_Y A(x, y) h(y) \, dy = 0$$

If the answer is negative, then g(y) is unique. Otherwise, $h \in H$ (annihilator of A), and the knowledge of f(x) can tell nothing whatsoever about those parts of g(y) that belong to H; these must be deduced from information other than that contained in f(x).

Stability. Granted existence and uniqueness, does g(y) depend continuously on f(x)? Alternatively, does the closeness of the computed $\tilde{f}(x)$ to the measured $\bar{f}(x)$ entail a similar closeness of the inferred $\tilde{g}(y)$ to the actual $g(y)$?

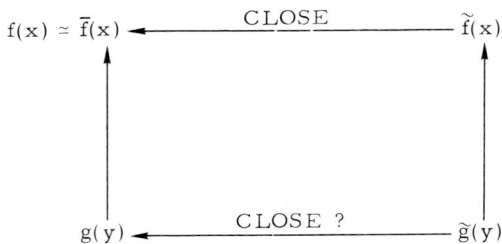

If the answer is in the affirmative, then $g(y)$ is stable; otherwise $g(y)$ is unstable. Physically, instability is the case every time the kernel tends to smooth the behavior of the required function for all values of the parameter x. Under certain circumstances the linearized temperature sounding problem for a thick atmosphere can be reduced to the convolution problem (Ref. 4)

$$f(x) = \int_{-\infty}^{\infty} a(x - y)g(y)dy \qquad (26a)$$

The universal kernel in this equation is plotted in Fig. 4(a). The Fourier transform of Eq. (26) is

$$F(k) = A(k) * G(k) \qquad (26b)$$

$$a(-y) = \exp\left[-(-y) - \exp -(-y)\right]$$

Fig. 4a. *The universal weighting function (from Ref. 4).*

The filtering function A(k) is drawn in Fig. 4(b). It corresponds to a low-pass filter. It is seen that A(k) drops very sharply with increasing k, e.g., for k = 7 it drops by four orders of magnitude. After a certain k-value, the value of A(k) may drop below the experiment noise. In this case, the inversion would correspond to the noise rather than to the signal. This illustrates the necessity of applying a constraint in order to limit the number of k values that can be admitted. This number is also intimately connected with the information content of the measurements under consideration.

Construction. Granted existence and uniqueness, can a procedure be found which will generate g(y) from f(x) to any requested finite accuracy in a finite number of steps?

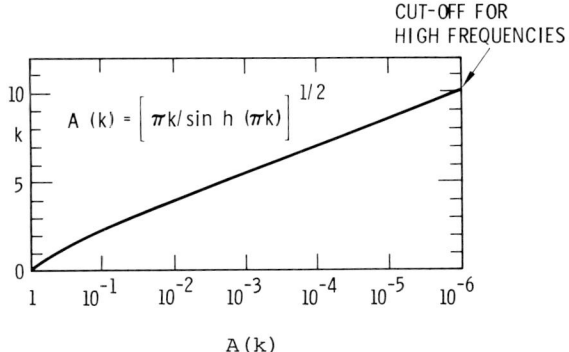

Fig. 4b. Amplitude of the Fourier spectrum A(k) of the weighting function (from Ref. 4).

B. A Preliminary Classification of Methods of Solution

Since the actual solution cannot be recovered from the measurements, there are three courses of action that are left open:

1. Derivation of properties that all solutions share, which must then be properties of the true solution.

2. Introduction of assumptions (physical, mathematical)

about the solution to restrict the class of admissible solutions.

3. Investigation of alternative concepts or conditions.

This is the basis of the classification provided in Fig. 5 for retrieval of the temperature profile, and in Fig. 6 for the reconstruction of the particle size distribution. A review of the various methods there listed has been provided elsewhere (Ref. 5). A number of spectral expansion and least-squares techniques (e.g., truncated singular value decomposition, stepwise regression, Cholesky decomposition, and modified Gram-Schmidt orthogonalization), the reduction to a second-kind Fredholm integral equation, invariant imbedding, etc., remain to be explored, or have not been sufficiently explored. Likewise, alternative or complementary concepts, such as the scattering processes discussed in the Introduction section, need to be further investigated. In the particulate sounding problem, the climatological methods are not applicable, and the analytical methods are integral transforms.

These tabulations of methods will be updated as new concepts and methods are developed for the solution of the two problems of temperature profile and particle size distribution reconstructions.

VI. A PRELIMINARY CLASSIFICATION OF NONLINEAR SOLUTION METHODS

The methods of solution (Fig. 7) are here necessarily iterative. For the temperature sounding problem, Newtonian iteration, relaxation, and Marquardt algorithm have been used. A number of other nonlinear least-squares and search methods are available for study. The same conclusion applies to the size distribution problem where only the minimization search technique and some of its variants have been used.

TEMPERATURE AND AEROSOL REMOTE SOUNDING

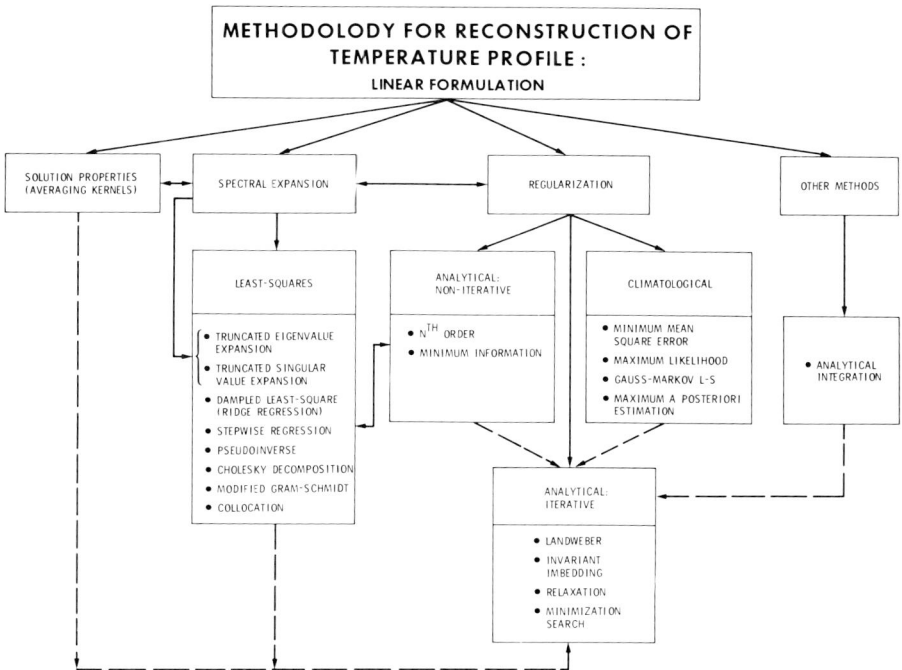

Fig. 5. Classification for retrieval of temperature profile (From Ref. 5).

VII. A STRATEGY FOR THE PARTICULATE SOUNDING PROBLEM

This is a two-step approach in which the complex refractive index is first determined, independently of the particle size distribution, from the inversion of spectral transmission ratios in a narrow wavelength range. This parameter known, the particle size distribution is then reconstructed from an analytical inversion of single scattering radiances in a narrow forward cone. Whenever the conditions attending single scattering are no longer valid, the single scattering solution can be iterated using a number of schemes to provide the multiple scattering solution. The approach is well adapted to a limb scanning experiment in order to also retrieve the vertical profiles of the unknown parameters.

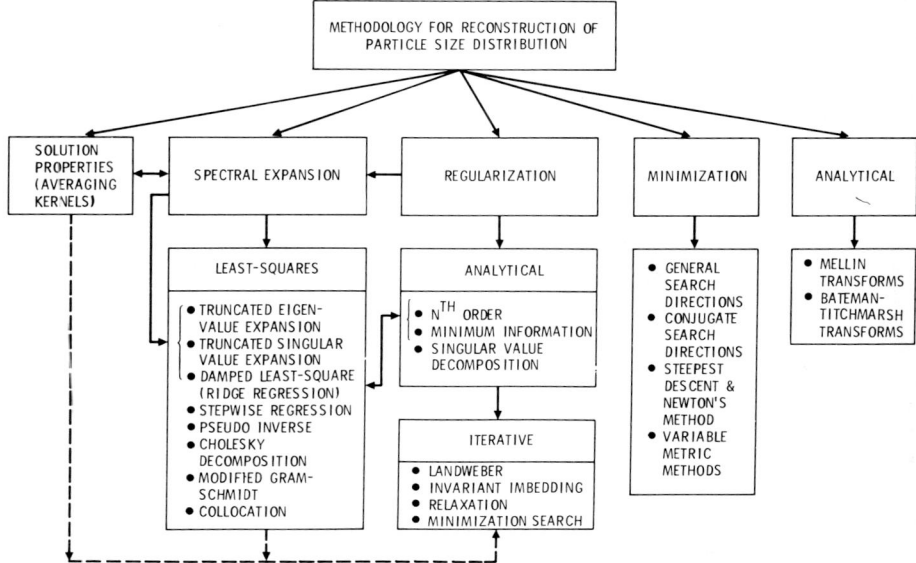

Fig. 6. Classification for reconstruction of particle size distribution (From Ref. 5).

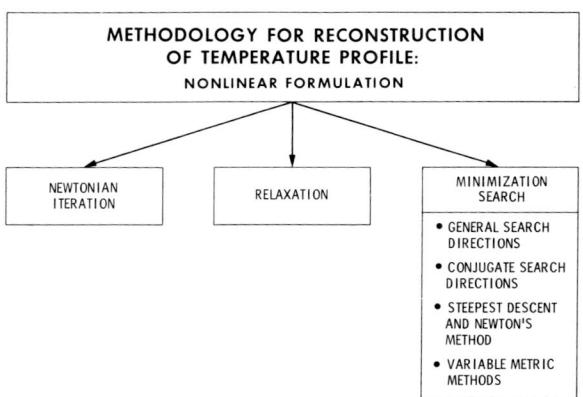

Fig. 7. Methodology for reconstruction of temperature profile.

A. Retrieval of the Complex Refractive Index: Spectral Extinction Ratio Approach

Return to Eq. (16c) providing the expression of the spectral extinction, again under the assumption that the gaseous contribution (τ_1^g) is known. Figure 8 illustrates the variations of the extinction efficiency factor as a function of particle radius for several wavelengths. At any wavelength, the main contribution to the measurement $E(\lambda)$ (see Eq. (16c) comes from the first strongly marked peak of the corresponding curve. If the wavelength interval considered is sufficiently narrow, then Q_{ext} is essentially independent of particle size. For example, for wavelengths extending from 0.4 to 0.9 μm, the contribution to solar extinction measurements originates from particles in the size range 0.3 to 0.6 μm. This latter range would be further reduced if the wavelength interval were smaller. This observation was used for developing a method of retrieving the complex refractive index of aerosols in a manner that is essentially independent of size distribution. The same procedure can be applied to a set of wavelength intervals in order to obtain the spectrum of this parameter. A complete description of the method has been provided elsewhere (Ref. 6). Briefly, we consider the set of spectral ratios:

$$R_{uv} = \frac{E(\lambda_u)}{E(\lambda_v)} \qquad (27)$$

where λ_u and λ_v are any two wavelengths of the ensemble of N wavelengths recorded, λ_v being fixed for any particular set. Obviously, there are N such sets. For simplicity, we shall model the size distribution by the gamma function

$$n(x) = \text{Constant } x^\alpha e^{-\beta x/k} \qquad (28)$$

where the two distribution parameters α and β are expressed in terms of moments of $n(x)$. Such a model, however, is neither required nor necessary for one can generate the actual distribution from some initial guess, $n^{(o)}(x)$, by using the relation

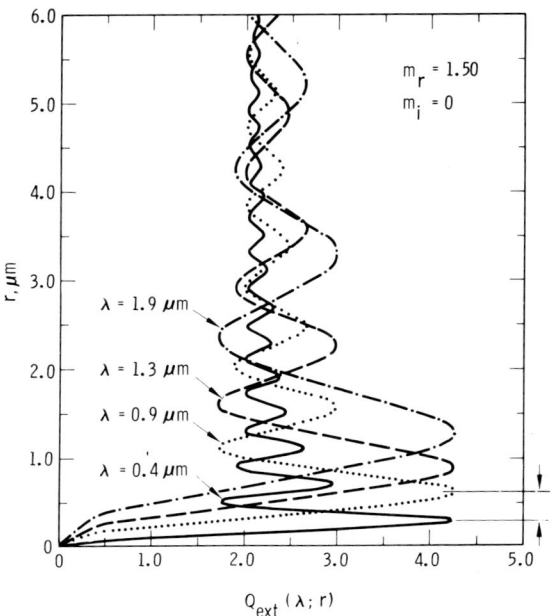

Fig. 8. Extinction efficiency factor as a function of particle radius for several wavelengths (from Ref. 7).

$$n(x_\ell) = \delta_\ell n^{(o)}(x_\ell) \quad (\ell = 0, 1, 2, \ldots,)$$

where the parameters δ_ℓ are unknowns of the problem. The inverse problem then consists in determining m_r, m_i, and α and β (or the δ_ℓ's) from the measured values of R_{uv} at a set of values of the independent variable λ_u. We first reformulate the inversion problem as a problem in minimization theory. Denote the measured ratios by $\tilde{R} \equiv \tilde{R}_{uv}\{m_r, m_i; \alpha, \beta \text{ (or } \delta_\ell)\}$ and the ratios computed from Eqs. (28) and (16c) by $R_{uv} \equiv R_{uv}\{m_r, m_i; \alpha, \beta \text{ (or } \delta_\ell)\}$. Then, define the absolute (or relative) deviations $D_{jn} = \tilde{R}_{jn} - R_{jn}$ [or $(\tilde{R}_{jn} - R_{jn})/\tilde{R}_{jn}$], $n = 1, 2, \ldots, N - 1$, and the deviation vector $\underset{\sim}{D}_j^T = (D_{j1}, D_{j2}, \ldots D_{jN-1})$, where the superscript T denotes matrix transposition. In these definitions, $j = 1, 2, \ldots N$ refers to the fixed wavelength λ_v of the particular set of ratios studied.

The problem is to minimize the objective function:

$$S_j \{m_r, m_i; \alpha, \beta \text{ (or } \delta_\ell)\} = \underset{\sim}{D}_j^T \underset{\sim}{D}_j \tag{29}$$

representing the sum of squares of deviations. It may be noted that while in Eq. (29) we use the Euclidean norm, any other norm could be considered. One could also define the objective function by

$$S_j \{m_r, m_i; \alpha, \beta \text{ (or } \delta_\ell)\} = \underset{\sim}{D}_j^T \underset{\sim}{W} \underset{\sim}{D}_j \tag{30}$$

where $\underset{\sim}{W}$ is a weighting matrix reflecting the degree of confidence in the individual measurements, their spectral resolution, accuracy, etc. It may be noted that the function S_j is the equation of a hypersurface in the parameter space of dimensions: m_r, m_i, α and β or $m_r, m_i, \delta_1, \delta_2, \ldots \delta_\ell$. Equivalently, then, the inverse problem consists in finding the minimum of the surface S_j. This minimum is located by the author's Minimization Search Method (MSM) (Ref. 6).

Figure 9 provides the contour curves of the surface $S_j(m_r, m_i)$ for $\lambda_j = 0.4 \mu m$ and size distribution parameters $\beta = 9$, 15 and 23. The true values of the unknown parameters were taken to be $m_r = 1.44$, $m_i = 0.03$, $\alpha = 2$ and $\beta = 15$. The data ("measurements") were generated in the computer for the nine wavelengths: $\lambda_u = 0.4$, 0.45, 0.49, 0.525, 0.575, 0.61, 0.64, 0.67 and 0.7 μm. The figure shows, irrespective of the β-value, that the unique minimum of the surface S is always located at $m_r = 1.44$, $m_i = 0.03$, i.e., at the true value of the complex refractive index. This result is not surprising in light of our earlier remarks in connection with Fig. 8. It serves to demonstrate that the extinction ratios are insensitive to the size distribution. As a further illustration of this

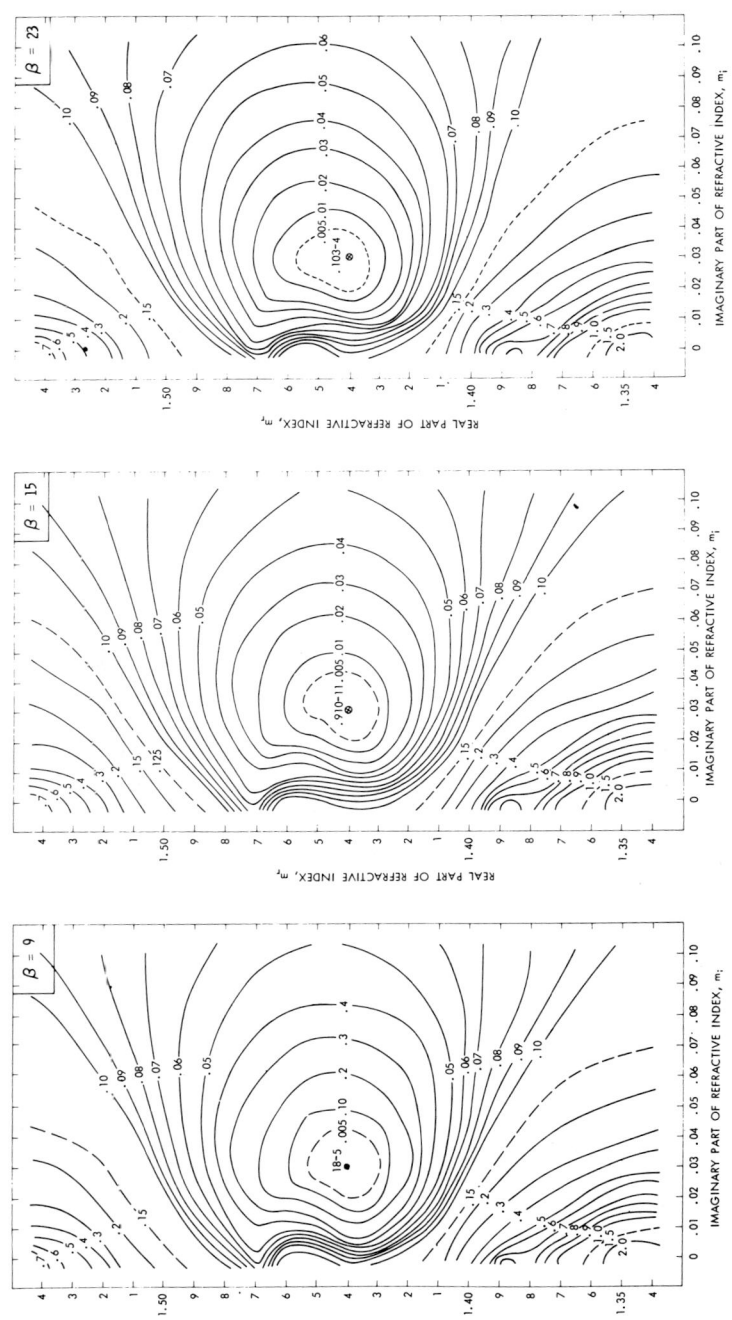

Fig. 9. Contour curves of the surface $S(m_r, m_i)$ for size distribution parameters $\beta = 9$, 15 and 23. Note that the minimum does not move as β increases (from Ref. 6).

insensitivity, Table 3 is provided. This table shows that even for highly accurate measurements within 0.1% accuracy, the insensitivity to β ranges from β = 8 to β = 23 approximately. Figure 9, together with Table 3, demonstrate conclusively the insensitivity of spectral extinction ratios within a narrow wavelength interval to the particle size distributuion. They are, however, extremely sensitive to the refractive index, an interesting circumstance.

An error analysis of this method is summarized in Table 4 for various wavelength sets and experimental accuracies. For example,

TABLE 3

Illustrating the Insensitivity of a Set of Spectral Extinction Ratios to the Particle Size Distribution Parameter (from Ref. 6)

λ \ β	0.45	0.49	0.525	0.575	0.61	0.64	0.67	0.70
8	0.0264	0.0298	0.0445	0.0832	0.1264	0.1696	0.1908	0.1989
9	0.0097	0.0110	0.0164	0.0306	0.0466	0.0625	0.0702	0.0735
10	0.0035	0.0040	0.0060	0.0112	0.0171	0.0229	0.0257	0.0273
11	0.0013	0.0015	0.0022	0.0042	0.0062	0.0084	0.0089	0.0094
12	0.0004	0.0006	0.0008	0.0015	0.0023	0.0030	0.0030	0.0038
13	0.0001	0.0002	0.0002	0.0006	0.0008	0.0010	0.0010	0.0009
14	0.0000	0.0001	0.0001	0.0002	0.0002	0.0003	0.0000	0.0000
15	0.0000	0.0000	0.0000	0.0000	0.0000	0.0000	0.0000	0.0000
16	0.0001	0.0001	0.0001	0.0001	0.0001	0.0002	0.0000	0.0000
17	0.0002	0.0002	0.0004	0.0003	0.0003	0.0004	0.0010	0.0000
18	0.0004	0.0007	0.0008	0.0009	0.0010	0.0010	0.0010	0.0009
19	0.0012	0.0019	0.0023	0.0024	0.0025	0.0025	0.0030	0.0028
20	0.0033	0.0052	0.0061	0.0067	0.0066	0.0066	0.0069	0.0066
21	0.0088	0.0141	0.0167	0.0182	0.0177	0.0177	0.0188	0.0179
22	0.0238	0.0384	0.0453	0.0493	0.0482	0.0478	0.0494	0.0490
23	0.0645	0.1041	0.1229	0.1336	0.1305	0.1304	0.1325	0.1320
24	0.1742	0.2817	0.3325	0.3617	0.3532	0.3503	0.3588	0.3563
25	0.4675	0.7570	0.8943	0.9733	0.9502	0.9432	0.9658	0.9595
26	3.1770	2.9293	1.8157	0.3927	0.0210	0.7765	0.9262	0.6098
27	1.7884	0.0926	0.8605	1.7033	0.1176	0.5073	2.7075	3.5535
28	3.3692	1.5580	4.0406	6.0713	2.9787	4.6707	2.3339	2.4582
29	6.9479	5.2720	4.1998	2.9636	5.7156	4.0361	10.8785	10.9422
30	16.3302	4.0105	11.6555	10.8407	6.6112	7.8650	1.3394	2.5600
31	16.7168	4.6318	11.0039	10.1242	5.8426	7.0861	2.2034	1.7098

Error upper bounds (last digit rounded off):
—— = 0.001%; - - - - - = 0.005%; - - . - - . = 0.01%; x - -x - - = 0.05%; ·········· = 0.1%.

TABLE 4

Illustrating the Sensitivity of the Inverse Solution for the Complex Refractive Index and Size Distribution Parameter to the Number of Extinction Ratios and Their Accuracy (from Ref. 6)

Wavelength ratios	Number of significant figures	Method	m_r	$|\Delta m_r|$ (%)	$-m_i$	$|\Delta m_i|$ (%)	β
$R_{00}, R_{01},$		true values	1.44		0.03		15.0
$R_{02}, R_{03},$		guess	1.38		0.025		13.5
$R_{04}, R_{05},$							
$R_{06}, R_{07},$	4	MSM	1.43998	0.001	0.02999	0.033	13.1909
R_{08}	3	MSM	1.44015	0.010	0.03007	0.233	13.1874
	2	MSM	1.43840	0.111	0.02776	7.467	13.3377
$R_{00}, R_{01},$	3	MSM	1.44009	0.006	0.03009	0.285	13.2460
$R_{03}, R_{04},$	2	MSM	1.43716	0.197	0.02686	10.483	13.3280
R_{06}, R_{08}							
$R_{11}, R_{15},$	3	MSM	1.44093	0.065	0.03009	0.303	13.3913
R_{17}	2	MSM	1.44570	0.396	0.02656	11.455	13.5131

using only three channels at 0.45 μm, 0.61 μm, and 0.67 μm, data with 0.1% accuracy can be used to obtain the real and imaginary parts of the refractive index to within 0.07% and 0.3%, respectively.

B. Reconstruction of the Particle Size Distributuion: Angular Forward Scattering Approach

This approach rests on an analytical inversion[1] cf Eq.(24) using the Bateman-Titchmarsh formula. The result is

---

[1] K. S. Shifrin, Calculation of a certain class of definite integrals containing the square of a first-order Bessel function, *Trudy Vsesoyuznogo Zaochnogo Lesotekhnicheskogo Instituta*, 2 (1956). (Periodical title unofficially translated as "Proc. All-Union Institute of Correspondence on Forestry (or "Wood Technology").) (Note: Copy of this article could not be obtained from the Library of Congress, the periodical could not be verified in serial lists or the article in abstract journals.)

$$n^*(x) = -\frac{2\pi k^3}{x^2} \int_{\Delta\Theta \approx 10°} J_1(y) Y_1(y) y \frac{d(\sin^3\Theta \, I_\Theta)}{d(\sin\Theta)} d(\sin\Theta) \quad (31)$$

where $y = x \sin\Theta$, and Y_1 is Bessel function of the second kind and order unity. This result is extremely interesting for it shows that $n^*(x)$ can be obtained from the data by a simple integration. And, if the refractive index is known, such as by the approach described previously, the size distribution $n(x)$ can also be obtained. In so doing, no *a priori* knowledge of the size

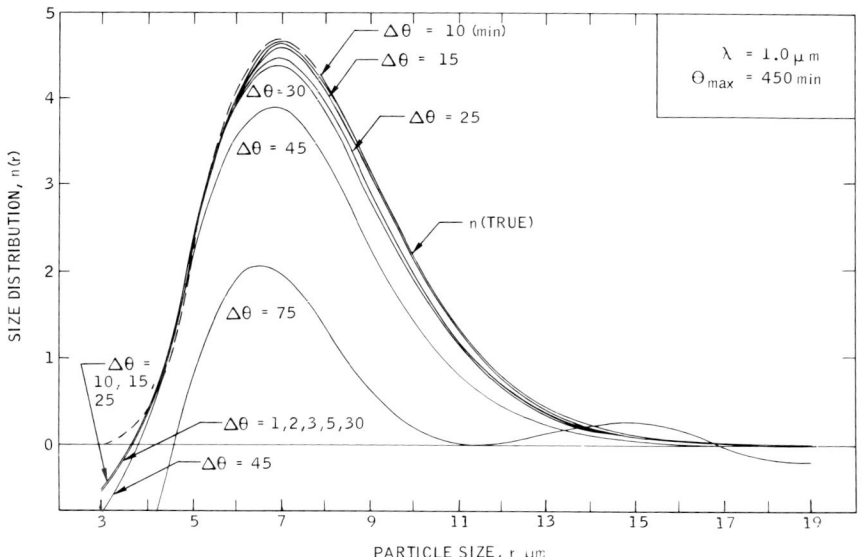

Fig. 10. *Sample reconstructions of the particle size distribution with the angular forward scattering method (from Ref. 3).*

distribution, or its analytical model, is required. Although not critical, the knowledge of the refractive index allows a more accurate reconstruction at the smaller size parameter values. The accuracy obtained is estimated to be better than a few percent.

The present method of determining the size distribution appears preferable to that utilizing spectral extinction data (Ref. 9) in that (i) it does not need an extended spectral interval (0.35 μm - 2.27 μm) over which the aerosol refractive index must be assumed known and constant, (ii) the latter assumption is more readily met in the narrow spectral interval used in our ratio technique for retrieving the complex refractive index; (iii) it is not restricted to an upper size limit of 5 μm with systematic overestimations at larger values.

Figure 10 provides some sample reconstructions for various angular resolutions, $\Delta\Theta$, in a forward scattering cone of half width, Θ_{max} = 450 min. It is seen that $\Delta\Theta \approx 15$ min provides indeed excellent reconstructions.

SYMBOLS

a	turbidity factor
a_n, b_n	Mie coefficients
A	kernel function
A(k)	Fourier transform in Eq. (26b)
b	weighting factor in Eq. (7)
B	Planck's function
D	known constant in Eq. (18b). Also deviation vector in Eqs. (29) and (30)
E	extinction function defined in Eq. (16c)
f(x)	known function (measurements)
f, g	auxiliary functions in Eqs. (20) and (21)
F	scattering diagram
F(k)	Fourier transform in Eq. (26b)
g	gravitational acceleration
g(y) or h(y)	sought function
G(k)	Fourier transform in Eq. (26b)
I	specific intensity (I_o = incident value)

J	source function
J_1	Bessel function
k	wavenumber
$k_\nu (k'_\nu)$	monochromatic volume (mass) absorption coefficient
$K(K')$	volume (mass) extinction coefficient
$K(x)$	kernel function in Eq. (19)
ℓ	length of scattering medium
$m = m_r - im_i$	complex refractive index
$n(x)$	particle size distribution [initial guess $n^{(0)}(x)$]
$n^*(x)$	$Q_{ext}^2 / 4n(x)$
$N(x)$	$n(x)$ or $x^2 n(x)$ in Eq. (19)
p	pressure
$\underset{\sim}{P}$	phase-matrix (P_{11} = phase-function). Superscripts: I = isotropic, R = Rayleigh, RC = Rayleigh-Cabannes, g = gases, p = particulates
Q_{ext}	Mie efficiency factor for extinction
r	particle radius
$\vec{r}$	radius vector
R_{uv}	spectral extinction ratio
S	objective function
$S(0)$	Mie complex scattering amplitude in forward direction
t	transmission function
T	temperature
w	mass mixing ratio of absorbing gas
W	weighting function
$x \in X$	size parameter
$y \in Y$	single-valued function of pressure
y_s	value of y at the surface
$\bar{y}$	at top of atmosphere or at satellite altitude
y	$x \sin \theta$ in Eq. (31)
Y_1	Bessel function

z	altitude (z_1 = value of z at lower boundary; z_2 = at upper boundary; $\bar{z}$ = at top of atmosphere or at satellite altitude)
α, β	expression in moments of size distribution in Eq. (28)
α, β, γ	direction cosines
δ	gaseous depolarization factor
δ_ℓ	displacement constants
θ	zenith angle ($\theta = \cos^{-1} \mu$). Subscript zero for incidence.
Θ	scattering angle ($\Delta\Theta$ = angular resolution)
λ	wavelength
ν	frequency ($\nu = f$ in continuum, $\nu = \ell$ in spectral line)
ρ	absorbing gas density, also $\rho = 2x\|m-1\|$ in Eq. (18a)
$\sigma(\sigma')$	volume (mass) scattering coefficient
τ	optical depth (τ_1 = optical thickness)
τ_{sc}	scattering optical thickness
ϕ	azimuth angle
ω	solid angle
$\tilde{\omega}$	single scattering albedo (in Table 2)
$\vec{\Omega}$	direction vector. Superscript plus (minus) for upward (downward)

ACKNOWLEDGMENT

This paper presents the results of one phase of research carried out at the Jet Propulsion Laboratory under Contract No. NAS 7-100 with the National Aeronuatics and Space Administration.

REFERENCES

1. L. M. McMillin, et al., Satellite Infrared Soundings from Spacecraft, NOAA TR NESS Report 65, 1973.

2. A. L. Fymat, in "Proceedings of 1976 International Radiation Symposium," Garmisch-Partenkirchen, FRG, August 1976, Science Press, Princeton, New Jersey, July 1977.

3. F. Gilbert and R. L. Parker, Inversion Methods in Atmospheric Radiation Research, International Radiation Commission Report, edited by A. L. Fymat, 1976.

4. D. Gautier and I. Revah, Sounding of planetary atmospheres: A Fourier analysis of the radiative transfer equation, *J. Atmos. Sci. 32,* 881 (1975).

5. A. L. Fymat (ed.), Inversion Methods in Atmospheric Radiation Research, International Radiation Commission Report, 1976.

6. A. L. Fymat, Inverse atmospheric radiative transfer problems: A non-linear minimization search method of solution, *Phys. Earth Planet. Inter. 12,* 273 (1976).

7. H. Quenzel, Determination of size distribution of atmospheric aerosol particles from spectral solar radiation measurements, *J. Geophys. Res. 75,* 2915 (1970).

8. A. L. Fymat, in "Monitoring of the Terrestrial Environment from Space" (Rassegna, ed.), pp. 183-195. Internazionale Elettronica Nucleare ed Aerospaziale, Rome, Italy, 1973.

9. G. Yamamoto and M. Tanaka, Determination of aerosol size distribution from spectral attenuation measurements, *Appl. Opt. 8,* 447 (1969).

DISCUSSIONS

Green: Did I understand your curves were going to negative numbers of particles below three microns?

Fymat: In some of these retrievals, when the size distribution is zero at the origin, we have such negative tails. We have numerical difficulties in retrieving these null values. The difficulties disappear, however, when the distribution has a finite value at the origin or were we to vary the half-width of the forward scattering cone according to the particular radius for which we wish to reconstruct the distribution. The result displayed in Figure 10 of my paper included neither of these two features, and was aimed at illustrating results when both the distribution starts from zero and the scattering cone are fixed for all radius values. In this context, providing we remain within the domain of applicability of the theory, the particular radius value at the origin is irrelevant. It has really nothing to do with exactly three microns. It is just that we are using the wavelength of one micron and the particular distribution we are employing for doing these retrievals started at three microns. (1 micron = 1 micrometer.)

Reagan: I have two questions or comments. One on the size distribution limit. We have talked about this before. It appears that the combination of the kernel and the specific shape of the distribution, particularly when you get into some of the bimodal types, indeed does extend the range over which you may hope to invert the size for us. So we deal with, for example, 0.4 to 1.0 micron wavelength range of measurements for radiometer work. And in the inversions, indeed, for certain types of size distributions we seem to have some success getting out to three to five microns. The second thing is with regard to refractive indexes you were showing there. I was curious what the size range assumption might have been on the Junge calculations. For what I gave earlier in the conference[1] for our radiometer work, dealing again from say 0.4 to 1.0 micron, the refractive index seemed to be quite insensitive in effecting the shape of the size distribution for values running from about 1.33 up bo about 1.55. Those calculations were based on Junge integrals from about 0.02 to 10 microns, again for that 0.4 to 1.0 micron range. Perhaps it was the wavelength range that was giving the variation in index that you were showing there.

Fymat: Indeed, we have discussed this problem several times. If I go back only to the extinction efficiency factor, I have great difficulty in trying to understand how you could retrieve size

---

[1] Topical Meeting on "Atmospheric Aerosols, Their Optical Properties and Effects," sponsored by the Optical Society of America and the NASA-Langley Research Center.

distribution up to five microns or more with the wavelengths you use. The work of Yamamato-Tanaka showed that you have to go to larger wavelengths to be able to sound the larger particle radii. Now, if you combined the extinction with other techniques to compensate in a sense for the other wavelengths you are not using, maybe! But, again, looking strictly at the extinction, I have difficulty understanding on the basis of the physics of the problem how you can go to five micron particles, approximately, by restricting yourself to this very narrow wavelength range.

Reagan: Well, as I was saying, perhaps certain size distributions with peak particle concentrations near 5.0 microns may extend that slightly.

Fymat: Slightly perhaps up to one to two microns, but not to five microns.

Twomey: I think I am a little uncomfortable when I hear phrases like "size distribution" and "refractive index." I think people should think a little more of perhaps changing the wording a little bit and saying "size and refractive index distribution." Because, I think we have no guarantee whatever that the composition remains constant and so I'll say over a decade in size distribution. In fact, when I get down towards 0.1 on the micron, we are in a very dangerous region where there are really no good sources of particles in the atmosphere. We are at the lower tail of effectiveness of processes which produce particles from the bulk, and we are at a rather large size when we are at the upper tail of the processes which produce particles from the vapor. So we must expect, I think, especially around that size region, mixed composition and, furthermore, a mixed composition which is size dependent.

Fymat: I fully agree with your statement.

Herman: I think you made a statement during your talk which I am not sure you meant to say, but you said there is *no* size information in the range of 0.4 to 1.0 micron.

Fymat: No, that was not the statement.

Herman: I thought that was what you said.

Fymat: What I said was that, if you look into narrow wavelength intervals, the corresponding spectral ratio measurements are a little sensitive to the size solution. I am not saying there is no information. As you broaden the interval, the size sensitivity increases.

Herman: I would agree with you that there is no information out beyond one or two microns on the size distribution, but there is information from a few tenths of a micron out to one or two microns.

And I think theoretically it is easy to show this. The fact that the kernels do not peak at those points does not mean that there is no information.

Fymat: No, I never said that there is no information. Also, keep in mind that I am working with extinction ratios and not the individual extinctions.

Kuriyan: In our 1974 paper,[2] we have shown that if a Deirmendjian haze H-type distribution, with a variable parameter b, was assumed, then precise multispectral extinction measurements can be used to infer the size distribution parameter b *as well as* the complex index of refraction. The errors in the measurement introduce corresponding errors in the inferred parameters. If a Junge distribution is used, then the problems associated with the limits of integration prevent the determination of the index of refraction in some cases. It seems as if, and no attempt to prove this statement has been made, the log-normal distribution can also be used to infer the size distribution parameter and the index of refraction. Thus, extinction measurements in the visible range cannot be used to arrive at a unique result but can yield, if the haze H type of size distribution is assumed, both the modal radius and the index of refraction. It must be emphasized that both the log-normal and the haze H-type distributions are well behaved at small and large values of the radius and, thus, the integrals can be computed for the infinite range. It seems reasonable to conclude that the origin of the nonuniqueness is the incomplete information that is available when we restrict the measurements to the visible range. It is, of course, impractical to extend measurements to the entire spectrum since absorption due to nonaerosol matter will interfere with the measurements.

Fymat: I have also employed the Deirmendjian distribution. What I have shown is that, for such a distribution, the extinction ratios taken over a narrow wavelength interval, say 0.45 to 0.70 microns, cannot be used to simultaneously reconstruct the particle size distribution out to several microns and retrieve the complex refractive index.[3] Nevertheless, what is interesting is that these extinction ratios enable us to retrieve the complex refractive index essentially independently of the size distribution, at least when a gamma distribution is assumed.

---

[2] Kuriyan, Chahine and Phillips, *J. Atmos. Sci.* 31, 2233-2236 (1974).

[3] A. Fymat, Inverse atmospheric radiative transfer problems: A non-linear minimization search method of solution, *Phys. Earth Planet. Inter.* 12, 273-282 (1976).

van de Hulst: In one of your earlier slides, you had a number of linear combinations, and one of the linear combinations was between the situation of cloud cover and without cloud cover, I remember.

Fymat: Yes.

van de Hulst: Have you looked at all into the validity of such a linear combination, because it is a very general question. If you have a sky partly covered by clouds, of course, the interplay of the light is completely different from what you would have in a *stratified* layer.

Fymat: What I had in mind in putting that linear combination was really a masking effect of the clouds in the same way as we heard earlier for the temperature sounding problem, and not so much in the interaction which takes place between the clouds and the radiation field. But I will have to look more carefully into this latter aspect. The other thing I would like to point out is in a different subject. I think that this kind of extinction experiment can extremely easily be accommodated by a SAGE-type experiment. It is really a matter of choice of wavelengths.

APPLICATION OF MODIFIED TWOMEY TECHNIQUES TO
INVERT LIDAR ANGULAR SCATTER AND SOLAR
EXTINCTION DATA FOR DETERMINING
AEROSOL SIZE DISTRIBUTIONS

B. M. Herman
University of Arizona

It has been shown theoretically (Ref. 1) that the polarization properties of the angularly scattered laser light from a volume of air may be used to determine the size distribution of the aerosol particles within the volume by the use of appropriate inversion techniques. Similar techniques may be employed to determine a "mean" size distribution of the particulates within a vertical column through the atmosphere from determinations of the aerosol optical depth as a function of wavelength. In both of the examples, primarily because of the nature of the kernel functions involved, a modification of an inversion technique originally described by Twomey (Refs. 2 and 3) has been employed. Details of this method will be presented as well as results from actual measurements employing the University of Arizona bistatic lidar and solar radiometer.

I. INTRODUCTION

With the development in recent years of high-powered, pulsed laser-radar (lidar) systems, considerable effort has been directed toward the problem of utilizing these systems to infer the vertical structure of the atmosphere (e.g., Refs. 4, 5, and 6). These studies were based upon an analysis of the backscattered signals only and, as such, were quite limited in information content. As a result, little progress has been made from such studies toward inferring aerosol size distributions (Ref. 7). However, as has been pointed out by Reagan and others (Refs. 8 and 9), considerable additional information may be obtained from the angular scattering properties of the atmosphere by the use of bistatic lidar. Accordingly, such a system has been constructed at the University of Arizona for the purpose of studying the size as well as the height distribution of atmospheric aerosols. The essential features of this system have been described elsewhere (Refs. 8 and 9).

In another parallel line of research, a series of solar radiometers have been designed and constructed to measure atmospheric turbidity at a series of wavelengths throughout the visible and near Infrared (IR) portions of the spectrum. These instruments have been described by Shaw and others (Refs. 10 and 11). The purposes of the solar radiometry have been many. They provide a continuous monitoring of atmospheric aerosol optical depths throughout the wavelength range covered by the filters. Secondly, since one of the filters is at the laser (ruby) wavelength of 6943 nm, the optical depth of the aerosols at this wavelength provides a means to normalize the laser back-scatter signal, thereby providing a better means to interpret the back-scatter returns in terms of the volume back-scatter coefficient. This procedure has been described by Fernald and others (Ref. 12). An additional bonus of solar radiometry is that, by a proper choice of filters, it may be used to infer vertical loadings of absorbing gases, such as ozone and H_2O. Finally, by virtue of the variation of aerosol turbidity, or optical depth with wavelength, a second, completely independent method of determining the aerosol size distribution is provided. The distribution thus obtained is some type of average through the entire vertical column of the atmosphere but is, nevertheless, quite useful. In the following section, the inversion methods used will be outlined, and finally, results from actual data will be presented.

II. THEORETICAL DEVELOPMENT

A. Bistatic Lidar

Figure 1 shows a schematic diagram of the bistatic system. In this diagram, the transmitter sends out a pulse at an angle γ_1 to the local normal, while the receiver views at an angle γ_2 to the local normal. The shaded area at a height z above the ground represents the instantaneous scattering volume, V, which is that volume from which scattered light, scattered through the angle θ, reaches the receiver at a given instant of time. For the case in which the transmitted pulse is essentially all within the receiver field of view, the lidar equation for the nth Stokes parameter of the scattered flux received by the receiver $F_n^{(s)}(\theta, R_2)$ is

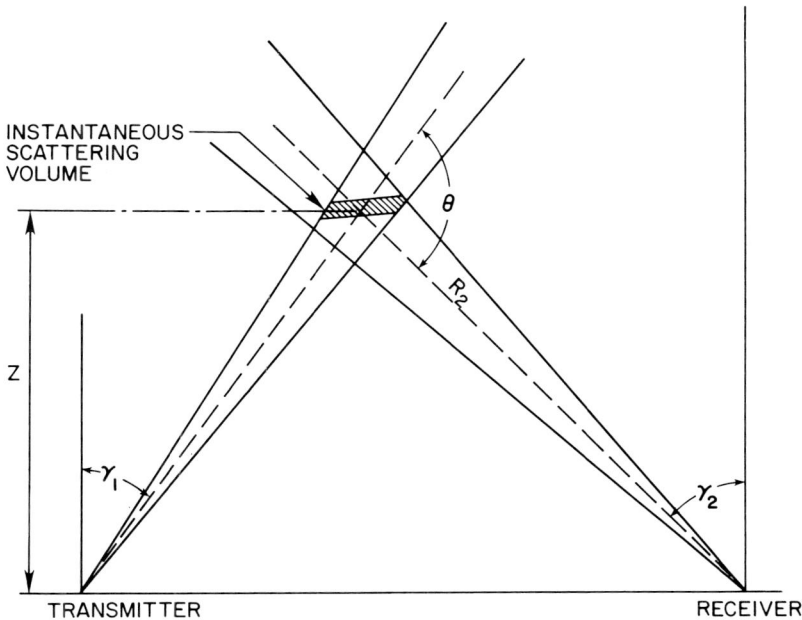

Fig. 1. *Schematic diagram of bistatic system.*

$$F_n^{(s)}(\theta, R_2) = \frac{A_R \ell}{2R_2^2 \sin^2 \frac{\theta}{2}} e^{-\tau(z)(\frac{1}{\cos \gamma_1} + \frac{1}{\cos \gamma_2})} P_{nm}(\theta) F_m^{(t)}$$

$$m, n = 1, 2, 3, 4 \quad (1)$$

where $\tau(z)$ is the optical depth from the ground to height z, ℓ is the pulse length, A_R is the effective receiver aperture, and $F_m^{(t)}$ is the mth Stokes parameter of the transmitted flux. The term $P_{nm}(\theta)$ is the nmth element of the scattering matrix for scattering through the angle θ, with dimensions of cross section per unit volume per steradian. For a more complete discussion of the lidar equation, see Reagan and Herman (Ref. 8). Since $R_2 = z/\cos \gamma_2$, Eq. (1) may be written as

$$F_n^{(s)}(\theta, \gamma_2) = K \frac{\cos \gamma_2}{\sin^2 \frac{\theta}{2}} e^{-\tau(z)(\frac{1}{\cos \gamma_1} + \frac{1}{\cos \gamma_2})} P_{nm}(\theta) F_m^{(t)}$$

$$(m, n = 1, 2, 3, 4) \quad (2)$$

where $K = \frac{A_R \ell}{2z^2}$, a constant for observations made at a constant height, z.

Let us now consider the scattering matrix element, $P_{nm}(\theta)$. As defined, this term has the dimensions of scattering cross-section per unit volume per steradian and is given by

$$P_{nm}(\theta) = \int_0^\infty P_{nm}(\theta, r) n(r) dr + \frac{\rho(z)}{\rho_o} k_{\lambda o} P_{nm}(\theta)_{Ray} \quad (3)$$

where $P_{nm}(\theta, r)$ is the nmth element of the scattering matrix for a particle of radius r, and $n(r)$ is the particle size distribution function. The term $P_{nm}(\theta)_{Ray}$ is the nmth element of the normalized scattering matrix for Rayleigh scattering, $k_{\lambda o}$ is the Rayleigh mass attenuation coefficient at some standard level for the wavelength λ, while ρ_o and $\rho(z)$ are the air densities at the standard level

APPLICATION OF TWOMEY TECHNIQUES   473

and at the height z, respectively. The integral on the right-hand side can be replaced with a sum, such that Eq. (3) becomes

$$P_{nm}(\theta) \approx P_{nm}(\theta, \bar{r}_1) f(r_1) + P_{nm}(\theta, \bar{r}_2) f(r_2) + \ldots$$

$$+ P_{nm}(\theta, \bar{r}_q) f(r_q) + \frac{\rho(z)}{\rho_o} k_{\lambda o} P_{nm}(\theta)_{Ray} \qquad (4)$$

where $P_{nm}(\theta, \bar{r}_j)$ is the nmth element computed at the midpoint radius of the jth interval, and

$$f(r_j) = \int_{r_j}^{r_j + \Delta r_j} n(r) dr \quad (j = 1, 2, \ldots q) \qquad (5)$$

Substituting Eq. (4) into Eq. (2), the lidar equation for the nth Stokes parameter at a scattering angle, θ_i, becomes

$$F_n^{(s)}(\theta_i, \gamma_2) = K \frac{\cos^2 \gamma_2}{\sin^2 \frac{\theta_i}{2}} e^{-\tau(z)\left(\frac{1}{\cos \gamma_1} + \frac{1}{\cos \gamma_2}\right)}$$

$$\times \left[P_{nm}(\theta_i, \bar{r}) f(r) + P_{nm}(\theta_i, \bar{r}_2) f(r_2) + \right.$$

$$\left. P_{nm}(\theta_i \bar{r}_q) f(r_q) + \frac{\rho(z)}{\rho_o} k_{\lambda o} P_{nm}(\theta_i)_{Ray} \right] F_m^{(t)} \qquad (6)$$

There will, thus, be four equations of the form of Eq. (6), one for each of the four Stokes parameters (i.e., for each value of n = 1, 2, 3, 4), for a given angle, θ_i. Thus, the total number of such equations will be 4 × number of scattering angles at which observations are made. In general, we may write

$$A f = g \qquad (7)$$

where $\underset{\sim}{g}$ is a vector of the observations, with components given by

$$F_n^{(s)}(\theta_i, \gamma_2) = K \frac{\cos^2 \gamma_2}{\sin^2 \frac{\theta_i}{2}} e^{-\tau(z)(\frac{1}{\cos \gamma_1} + \frac{1}{\cos \gamma_2})}$$

$$\times \frac{\rho(z)}{\rho_o} k_{\lambda o} P_{nm}(\theta_i) F_{Ray\,m}^{(t)}$$

and of dimensions $4 \times \ell$, where ℓ is the number of scattering angles, θ_i, at which observations are made. We have here assumed that the optical depth, $\tau(z)$, is known so that the Rayleigh contribution term may be computed and subtracted from the observations. The value of $\tau(z)$ may, in fact, be determined to within ± 1% accuracy from other measurements, the technique for which will be described in a later paper. The vector f is the vector of the unknowns to be solved for, with components $f(r_1)$, $f(r_2)$, ... $f(r_q)$, where each component gives the number of particles within the particular increment of radius, as given by Eq. (5). The coefficient matrix, $\underset{\sim}{A}$, is composed of known quantities, the elements of which are given as

$$A_{11} = \left[P_{11}(\theta_1, \bar{r}_1) F_1^{(t)} + P_{12}(\theta_1, \bar{r}_1) F_2^{(t)} + P_{13}(\theta_1, \bar{r}_1) F_3^{(t)} \right.$$

$$\left. P_{14}(\theta_1, \bar{r}_1) F_4^{(t)} \right] \times K \frac{\cos^2 \gamma_2}{\sin^2 \frac{\theta_1}{2}} e^{-\tau(z)(\frac{1}{\cos \gamma_1} + \frac{1}{\cos \gamma_2})}$$

$$\vdots$$

$$A_{1q} = \left[P_{11}(\theta_1, \bar{r}_q) F_1^{(t)} + P_{12}(\theta_1, \bar{r}_q) F_2^{(t)} + P_{13}(\theta_1, \bar{r}_q) F_3^{(t)} \right.$$

$$\left. P_{14}(\theta_1, \bar{r}_q) F_4^{(t)} \right] \times K \frac{\cos^2 \gamma_2}{\sin^2 \frac{\theta_1}{2}} e^{-\tau(z)(\frac{1}{\cos \gamma_1} + \frac{1}{\cos \gamma_2})}$$

Equation continued on next page

$$\begin{aligned}
A_{21} &= \left[P_{21}(\theta_1, \bar{r}_1) F_1^{(t)} + P_{22}(\theta_1, \bar{r}_1) F_2^{(t)} + P_{23}(\theta_1, \bar{r}_1) F_3^{(t)} + \right. \\
&\quad \left. P_{24}(\theta_1, \bar{r}_1) F_4^{(t)} \right] \times K \frac{\cos^2 \gamma_2}{\sin^2 \frac{\theta_1}{2}} e^{-\tau(z) \left(\frac{1}{\cos \gamma_1} + \frac{1}{\cos \gamma_2} \right)} \\
&\vdots \\
A_{51} &= \left[P_{11}(\theta_2, \bar{r}_1) F_1^{(t)} + P_{12}(\theta_2, \bar{r}_1) F_2^{(t)} + P_{13}(\theta_2, \bar{r}_1) F_3^{(t)} + \right. \\
&\quad \left. P_{14}(\theta_2, \bar{r}_1) F_4^{(t)} \right] \times K \frac{\cos^2 \gamma_2}{\sin^2 \frac{\theta_2}{2}} e^{-\tau(z) \left(\frac{1}{\cos \gamma_1} + \frac{1}{\cos \gamma_2} \right)} \\
&\vdots \\
A_{4\ell,q} &= \left[P_{41}(\theta_\ell, \bar{r}_q) F_1^{(t)} + P_{42}(\theta_\ell, \bar{r}_q) F_2^{(t)} + P_{43}(\theta_\ell, \bar{r}_q) F_3^{(t)} + \right. \\
&\quad \left. P_{44}(\theta_\ell, \bar{r}_q) F_4^{(t)} \right] \times K \frac{\cos^2 \gamma_2}{\sin^2 \frac{\theta_\ell}{2}} e^{-\tau(z) \left(\frac{1}{\cos \gamma_1} + \frac{1}{\cos \gamma_2} \right)}
\end{aligned} \quad (8)$$

Thus, the $\underset{\sim}{A}$ matrix will have dimensions $4\ell \times q$, where 4ℓ is the total number of observations, and q is the number of unknowns, the $f(r_j)$. The individual elements of the P scattering matrix for q given value of θ and r, $P_{nm}(\theta_i, \bar{r}_j)$, are determined from Mie theory, while the Stokes parameters of the transmitted light, the $F_n^{(t)}$, are assumed to be [1/2, 1/2, 1, 0] in order that the transmitted light be linearly polarized at 45° to the scattering plane.

In reality, since measurement errors are always present, as are quadrature and roundoff errors in the numerical evaluation of Eq. (3) (i.e., Eq. (4)), Eq. (7) should be written as

$$Af = g + \varepsilon \quad (9)$$

where the components of ε are the errors in each equation. Because of the presence of errors, Eq. (9) no longer possesses a unique solution, and attempts at a direct solution with ε set equal to zero almost always result in poor, highly oscillatory solutions, as shown by Phillips (Ref. 13). Twomey (Refs. 2 and 3) demonstrated possible constraints which are applicable to problems of this sort. One of these constraints is in the form of a smoothing constraint in which the second derivative of the solution points $\partial^2 f(r_j)/\partial r_j^2$ is minimized. This constraint leads to the solution

$$f = (\underset{\sim}{A}^T \underset{\sim}{A} + \gamma \underset{\sim}{H})^{-1} \underset{\sim}{A}^T \underset{\sim}{g} \tag{10}$$

in which $\underset{\sim}{A}^T$ is the transpose of the $\underset{\sim}{A}$ matrix, $\underset{\sim}{H}$ is a smoothing matrix, and γ is a Lagrangian multiplier. The value of γ determines the amount of smoothing, and is so chosen, by experimentation, to give reasonable solutions. The value of γ is not overly critical, in the sense that varying it from the chosen value over a range of an order of magnitude does not seriously affect the solution.

Another possible constraint proposed by Twomey for those cases in which a reasonable estimate of the solution is known consists of minimizing, in a least squares sense, the differences between the actual solution and the initial estimate, or trial solution. This leads to the solution

$$f = (\underset{\sim}{A}^T \underset{\sim}{A} + \gamma \underset{\sim}{I})^{-1} (\underset{\sim}{A}^T \underset{\sim}{g} + \gamma \underset{\sim}{P}) \tag{11}$$

where $\underset{\sim}{I}$ is the identity matrix and $\underset{\sim}{P}$ is the vector of the estimated solution points.

In the present study utilizing angular scattering data, the elements of the A matrix given by Eq. (8) are composed of elements of the scattering matrix, $P_{nm}(\theta, r)$. For values of $r < \lambda$, where λ is the wavelength of the incident light (in this case, $\lambda = 0.6943\ \mu$ representative of a ruby laser), $P_{nm}(\theta, r)$ is, indeed, a smooth function of r. This is demonstrated in Fig. 2, which shows P_{11} for $\theta = 130°$, for radii between $0.01\ \mu m$ and $0.190\ \mu m$. However, as r increases, the matrix elements become an increasingly erratic function of r, as demonstrated in Fig. 3, which shows the same element for radii between $1.20\ \mu m$ and $1.50\ \mu m$. Therefore, in order to get an accurate numerical representation of Eq. (3), it is

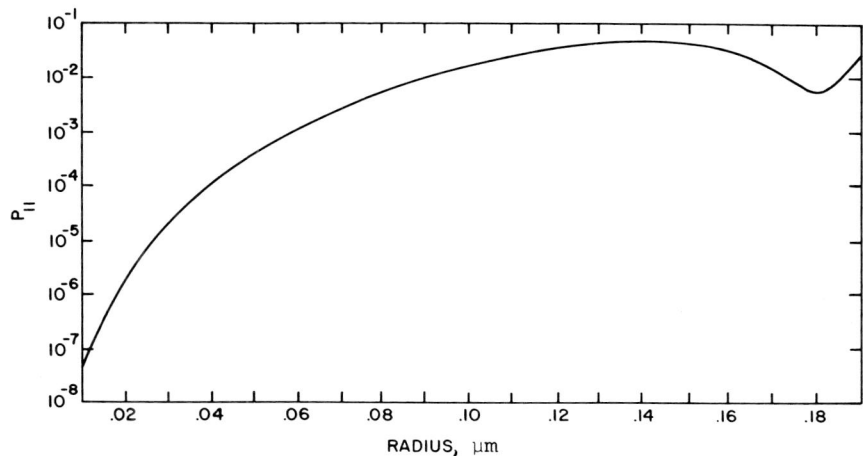

Fig. 2. Matrix element P_{11} for $\theta = 130°$ as a function of particle radius for radii between 0.01 μm and 0.190 μm.

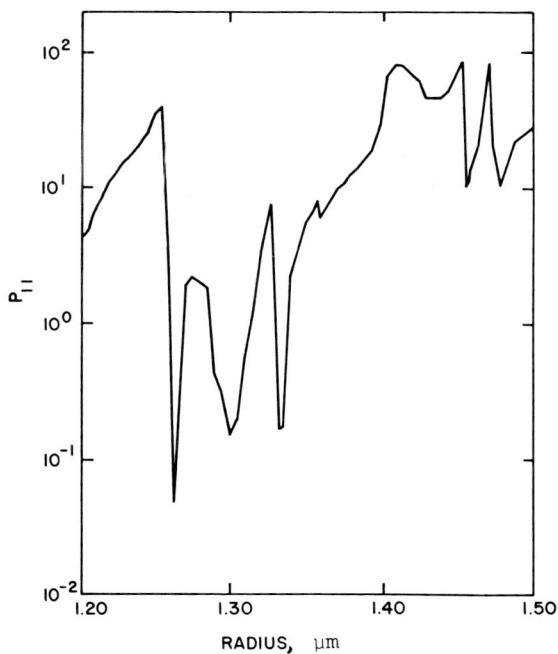

Fig. 3. Matrix element P_{11} for $\theta = 130°$ as a function of particle radius for radii between 1.20 μm and 1.50 μm.

necessary to break the integral up into a large number of very small intervals (Ref. 14). For the range of sizes considered in this work, approximately 1000 intervals in r were considered necessary in order to assure a reasonably accurate representation of the integral in Eq. (3). This, then, would require 1,000 values of the unknown $f(r_j)$, as can be seen from Eqs. (4) and (5), which, in turn, would require the inverse of a matrix, $A^T A$, as in Eqs. (10) and (11) of dimensions 1000 x 1000. In order to simplify this process, the following technique was employed. Each of the relatively large intervals in Eq. (4) was broken down into a series of smaller sub-intervals such that Eq. (4), less the Rayleigh term, becomes

$$P_{nm}(\theta_i)_{aer} \approx f(r_1) \sum_{k_1} P_{nm}(\theta_i, \bar{r}_{k_1}) W_{k_1}(r_{k_1}) +$$

$$f(r_2) \sum_{k_2} P_{nm}(\theta_i, \bar{r}_{k_2}) W_{k_2}(r_{k_2}) + \ldots +$$

$$f(r_q) \sum_{k_q} P_{nm}(\theta_i, \bar{r}_{k_q}) W_{k_q}(r_{k_q}) \qquad (12)$$

where the $W_j(r_j)$ are weighting functions to be described below, and the subscript aer indicates that part of the scattering matrix element due to aerosols only. The coefficients of the unknown $f(\bar{r}_j)$ in Eq. (12) will then appear in the expression for the coefficient matrix, Eq. (8), in place of the $P_{nm}(\theta_i, \bar{r}_j)$. Thus, for example, the expression for A_{11} becomes

$$A_{11} = \left[\sum_{k_1} P_{11}(\theta_1, \bar{r}_{k_1}) W_{k_1}(r_{k_1}) F^{(t)} + \sum_{k_1} P_{12}(\theta_1, \bar{r}_{k_1}) W_{k_1}(r_{k_1}) F_2^{(t)} \right.$$

$$\left. + \sum_{k_1} P_{13}(\theta_1, \bar{r}_{k_1}) W_{k_1}(r_{k_1}) F_3^{(t)} + \sum_{k_1} P_{14}(\theta_1, \bar{r}_{k_1}) F_4^{(t)} \right]$$

$$\times K \frac{\cos^2 \gamma_2}{\sin^2 \frac{\theta_1}{2}} e^{-\tau(z)(\frac{1}{\cos \gamma_1} + \frac{1}{\cos \gamma_1})} \qquad (13)$$

Initially, the weighting functions are unknown, as knowing them would be equivalent to knowing the sought after size distribution function. Thus, an "intelligent" first guess is made, and the problem reduces to one of finding the unknown $f(r_j)$ which, when multiplied by the assumed weighting functions, gives the sought after size distribution function. If the initially assumed weighting functions are exact, the solution vector $\underline{f}$ will be a unit vector (i.e., $f(r_j) = 1$, $j = 1, 2, \ldots q$). Any errors in the initial guess of the weighting function will be mathematically indistinguishable from measurement errors and, therefore, the worse the initial estimate, the worse will be the resulting solution. However, this first solution will still be a better approximation than the first guess, and so the process is repeated. For the second iteration, however, the first solution is used as the weighting function, and a new $f'(r_j)$ is obtained. For the second iteration, the first solution vector components are assumed to be valid at the mid-point of each of the large intervals in r, and are connected by straight lines. Thus, for the second iteration, Eq. (12) becomes

$$P_{nm}(\theta_i)_{aer} = f'(r_1) \sum_{k_1} P_{nm}(\theta_i, \bar{r}_{k_1}) \left\{ f(r_1) + \left[\frac{f(r_2) - f(r_1)}{\Delta r_1}\right] \Delta r_{k_1} \right\} W_{k_1}(r_{k_1}) + \cdots +$$

$$f'(r_{q-1}) \sum_{k_{q-1}} P_{nm}(\theta_i, \bar{r}_{k_{q-1}})$$

$$\times \left\{ f(r_{q-1}) + \left[\frac{f(r_q) - f(r_{q-1})}{\Delta r_{q-1}}\right] \Delta r_{k_{q-1}} \right\} W_{k_{q-1}}(r_{k_{q-1}}) \quad (14)$$

where $\Delta r_1 \ldots \Delta r_{q-1}$ are the increments of Δr for the large intervals, while the Δr_{k_j} are the intervals of Δr in the subintervals, and the $f'(r_j)$ are the components of the new solution vector. In principle, one could continue the iterative process until two successive solution vectors agreed to within any pre-specified amount. In practice, it was found that two such iterations were adequate.

Since most measurements of the size distribution of continental aerosols seem to indicate that these functions follow typically what is known as a Junge distribution (Refs. 15 and 16), given by

$$\frac{dn}{dr} = cr^{-(\nu^* + 1)} \tag{15}$$

a reasonable first guess for the weighting functions would be to assume that the distribution function is of the form given by Eq. (15). Thus, W_{k_j} is given by

$$W_{k_j} = n_{k_j} = \int_{r_{k_j}}^{r_{k_j} + \Delta r_{k_j}} cr^{-(\nu^* + 1)} dr$$

$$= \frac{c}{\nu^*} \left| \frac{1}{r_{k_j}^{\nu^*}} - \frac{1}{(r_{k_j} + \Delta r_{k_j})^{\nu^*}} \right| \tag{16}$$

where c is a normalizing constant determined such that the total integral of Eq. (15) over all sizes yields the proper number density per unit volume, while ν^* is a shaping constant. Typical values of ν^* lie in the range of 2.0 to 4.0.

In the theoretical work to follow, "measurements" of the four Stokes parameters at five scattering angles, θ, at fixed height z, and for various assumed size distributions, were computed from Eq. (2). These twenty "observations" were then used *per se,* and with introduced random errors of 1% and 2%, in order to compute 20 inversion solution points, the $f(r_j)$, by the technique described above. These solution points were taken to apply over each of 20 large intervals in Δr, taken such that log Δr = Constant for all intervals. The volume scattering matrix elements due to aerosols, $P_{nm}(\theta)_{aer}$, were computed from Mie theory with a numerical

APPLICATION OF TWOMEY TECHNIQUES

approximation to Eq. (3), while the Rayleigh component was computed assuming a standard atmospheric density distribution.

Figure 4 shows an inversion solution for a true distribution which is a straight Junge-type with $\nu^* = 2.5$. The initial weighting functions were computed from Eq. (16) assuming $\nu^* = 3.5$ (long dashes). The initial smoothing solution, Eq. (10), is shown as short and long dashes, while the final solution, Eq. (11), is shown as short dashes. For the sake of clarity, the second iteration using the smoothing constraint has been omitted from this figure and from Fig. 5, but in both cases lies roughly intermediate between the first and final solutions. In this case, a nearly perfect inversion is obtained for radii greater than about 0.2. It is important to note here that "exact observations" were

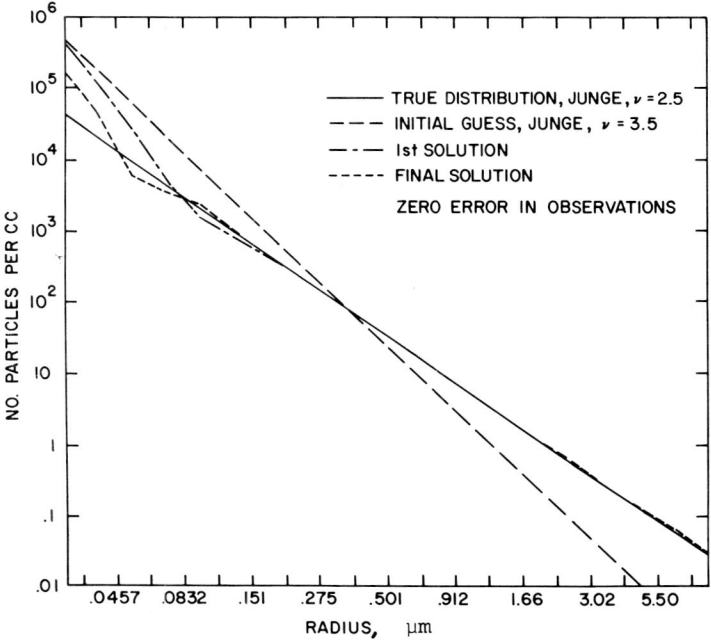

Fig. 4. Inversion results for zero error in observations for Junge-type distribution.

used in this case; that is, observations precisely as computed were used. In the smaller range of sizes, results, while satisfactory, are not quite as good. This is undoubtedly due to the fact that the smaller sizes are in the Rayleigh region for the ruby wavelength, which results in all such particles having the same variation of the scattering matrix elements with scattering angle, θ. Thus, the only information content in the measurements for these small particles, if they were precisely Rayleigh scatterers, is the total scattering cross section, and this can be made up of large numbers of the smallest particles, or fewer numbers of the larger particles of the Rayleigh region, or any intermediate distribution. Since the scattering differs slightly from pure Rayleigh, there is, in fact, a slight amount of information content in the measurements, as can be seen from Fig. 5. In this figure, the true distribution of particles is a modified Junge

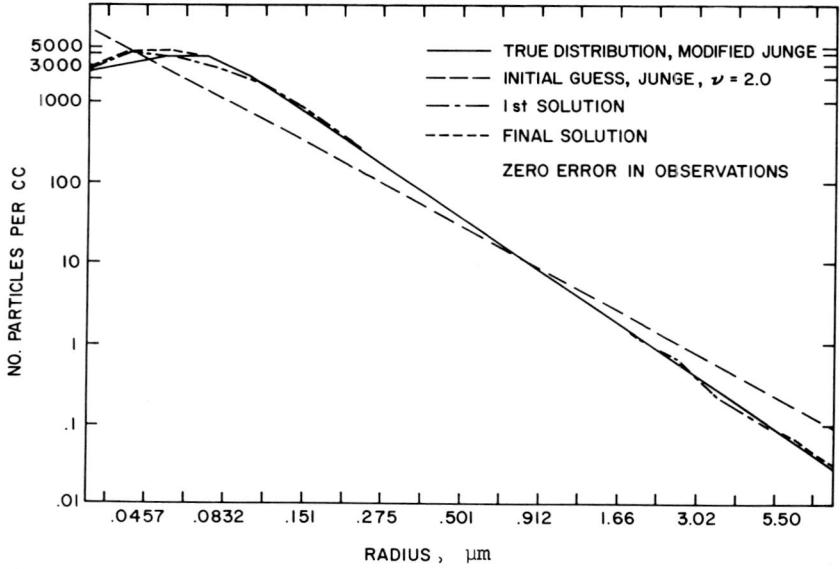

Fig. 5. Inversion results for zero error in observations for modified Junge distribution.

APPLICATION OF TWOMEY TECHNIQUES

distribution, called distribution X, in which a parabola, given by a dn/dr = $a(r_1 - r)^2$ + b was assumed for r ≤ 0.103 and a Junge distribution with ν^* = 2.5 was assumed for r ≥ 0.103 μ. The constants a and b were determined by setting dn/dr and d/dr(dn/dr) equal for both distributions at r = 0.103 μ. As an initial guess for the weighting functions, a straight Junge distribution with ν^* = 2.0 was employed. Again, zero error was placed in the observations. As can be seen from the figure, an excellent solution resulted, even at the smaller sizes where the distribution departed markedly from a Junge type.

Figures 6, 7, and 8 present results in which random errors of 1% and 2% were put into the observations, in addition to the "zero error" results. For the sake of clarity, only the final solutions are shown in these figures. Figure 6 is for a straight

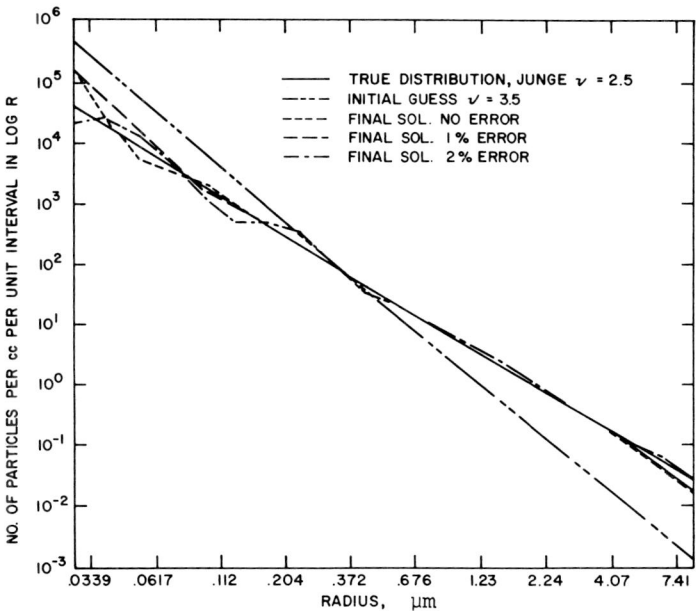

Fig. 6. Inversion solutions for a Junge distribution with ν^* = 2.5, with zero, 1% and 2% errors in the observations.

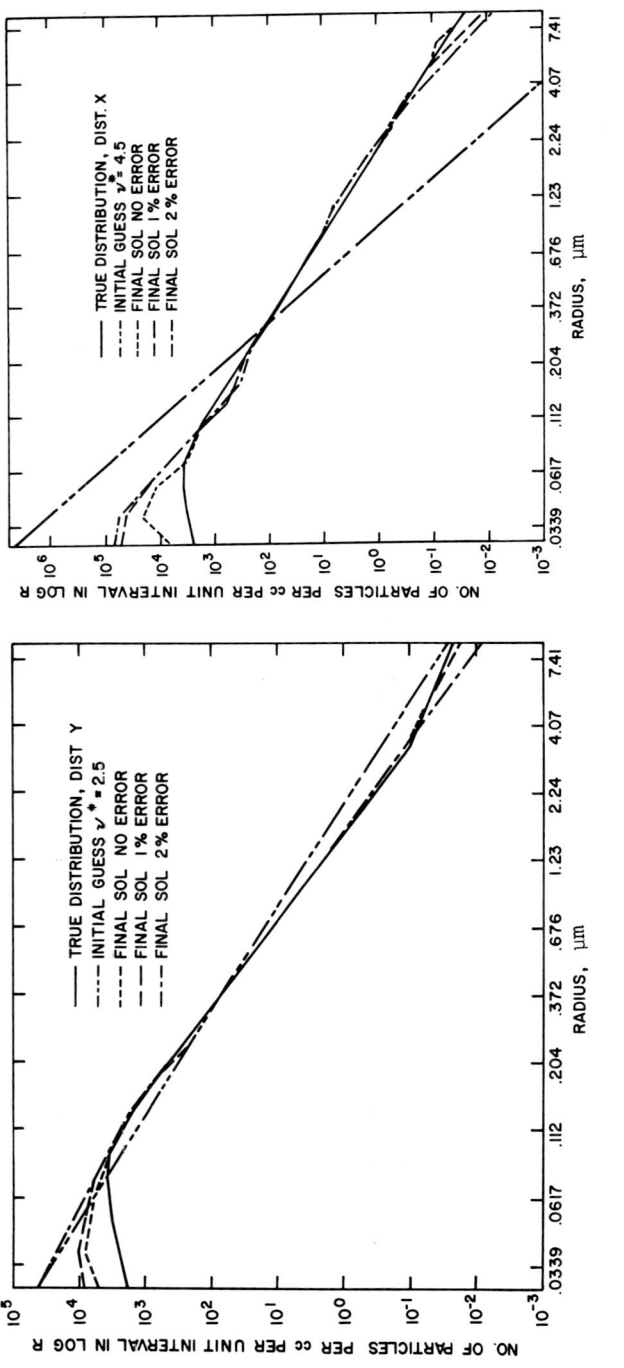

Fig. 7. Inversion solutions for a Junge distribution with distribution X, with zero, 1% and 2% errors in the observation.

Fig. 8. Inversion solutions for a Junge distribution with distribution Y, with zero, 1% and 2% errors in the observation.

APPLICATION OF TWOMEY TECHNIQUES

Junge distribution with $\nu^* = 2.5$. As can be seen, the presence of errors up to 2% degrades the solution only slightly, and, in general, the results show excellent agreement between the true and computed distributions. Figure 7 shows results for distribution X except for these calculations, a much worse initial guess of $\nu^* = 4.5$ was used as compared with the results shown in Fig. 5. By comparison of the final solution with no error in Fig. 7 to the final solution of Fig. 5, it can be seen that this worse initial guess definitely leads to a poorer final solution, particularly at the smallest sizes. Even so, the final solutions must be considered as quite satisfactory, with no appreciable degradation with the introduction of measurement errors, except again at the smaller sizes. Apparently the introduction of even small errors into the observations is enough to essentially eliminate all information content about the smallest sizes, due to the initially low content inherent in these sizes as previously discussed.

Finally, in Fig. 8, results are presented for another distribution, distribution Y. For this distribution X was used for $r \leq 1.3$ μm. This was fitted to a Junge distribution with $\nu^* = 3.0$ for 1.3 μm $\leq r \leq 3.4$ μm, while for $r \geq 3.4$ μm a Junge distribution with $\nu^* = 1.1$ was assumed. The results again are considered quite satisfactory. For zero error and 1% error, it is noted that the solution is able to follow the abrupt change in slope at 3.4 μm, while the 2% error, the constraint required (i.e., the value of γ in Eqs. 10 and 11) is large enough so as to force the solution into a nearly straight line on the log-log plot of the figure. This is true at the small-size end of the distribution also. Here, the zero and 1% error solutions are able to somewhat follow the curvature of the true solution, but with 2% error this feature is completely lost.

B. Solar Radiometer

As mentioned earlier, values of the aerosol optical depth as a function of wavelength also contain considerable information concerning the aerosol size distribution and, therefore, inversion techniques may be employed in order to recover this information. The integral equation which relates aerosol optical depth to an aerosol size distribution can be written as

$$\tau_{aer}(\lambda) = \int_0^\infty \int_0^\infty \pi r^2 Q_t(r, \lambda, m) n(r, z) dz dr \qquad (17)$$

where $n(r, z)dr$ is now the height dependent aerosol size distribution, m the complex refractive index of the aerosols, and $Q_t(r, \lambda, m)$ the extinction efficiency factor from Mie theory. Upon performing the height integration, Eq. (17) can be rewritten as

$$\tau_{aer}(\lambda) = \int_0^\infty \pi r^2 Q_t(r, \lambda, m) n_c(r) dr \qquad (18)$$

where $n_c(r)$ is the columnar size distribution, i.e., the number of aerosols per unit area per unit radius interval in a vertical column through the atmosphere. Thus, $n_c(r)$ is the unknown to be determined through measurements of $\tau_{aer}(\lambda)$, $\pi r^2 Q_t(r, \lambda, m)$ being the known kernel function in this case.

In obtaining $n_c(r)$, the integral in Eq. (10) is again replaced by a summation over coarse intervals in r, each of which is composed of several subintervals as described in the previous section. In order to examine the specific kernel functions which result if that procedure is applied to the present problem, let $n_c(r) = W(r)f(r)$ where $W(r)$ is a rapidly varying function of r while $f(r)$ is varying more slowly. With this substitution, Eq. (18) becomes

$$\tau_{aer}(\lambda) = \int_{r_a}^{r_b} \pi r^2 Q_t(r, \lambda, m) W(r) f(r) dr$$

$$\approx \sum_{j=1}^{} \int_{r_j}^{r_{j+1}} \pi r^2 Q_t(r, \lambda, m) W(r) f(r) dr \qquad (19)$$

where the limits have been made finite with $r_1 = r_a$ and $r_{q+1} = r_b$, and $f(r)$ is assumed constant within each coarse interval. A matrix equation of the form of Eq. (7) thus results where

$$g_i = \tau_{aer}(\lambda_i) \qquad (i = 1, 2, \ldots, p)$$

and

$$A_{ij} = \int_{r_j}^{r_{j+1}} \pi r^2 Q_t(r, \lambda_i, m) W(r) dr \qquad (j = 1, 2, \ldots q) \qquad (20)$$

Writing Eq. (20) as a quadrature over the subinterval from r_j to r_{j+1} results in an expression with weighting functions $W_{k_j}(r_{k_j}) = W(r_{k_j}) \Delta r_{k_j}$ and r_{k_j} are the midpoint radii of each of the k subintervals, and are employed in an identical manner to those previously described. The unknowns to be solved for are the $f(\bar{r}_j)$, one value for each of the larger intervals.

The initial weighting functions were again assumed to be a Junge size distribution given by Eq. (15). In practice, several different values of ν^* are used to calculate the zeroth order weighting function and the final results after successive iterations are intercompared. One test of the procedure is the similarity of the results obtained when different values of ν^* are used for the W_{k_j}.

In both inversion problems described, a proper selection of the magnitude of the constraint, ∂, is of some importance. Although ideally the value of ∂ is closely related to the expected values of the errors, E, in practice it has been found that a selection of ∂ on this basis is not always adequate. The procedure used in the work to follow has been to vary ∂ through a broad range of values about its expected value from an estimate of the errors in order to obtain a minimum value, ∂_{mm}, which yields an acceptable solution. An acceptable solution is one in which all values of the unknown are positive numbers. Values of ∂ too small normally result in highly oscillatory solutions and physically unacceptable negative values for some of the f's, while too large a value of ∂ obviously will cause the solution vector to be dominated by the constraint.

Figure 9 presents kernel functions $\pi r^2 Q_t(r, \lambda, m) W(r)$ for

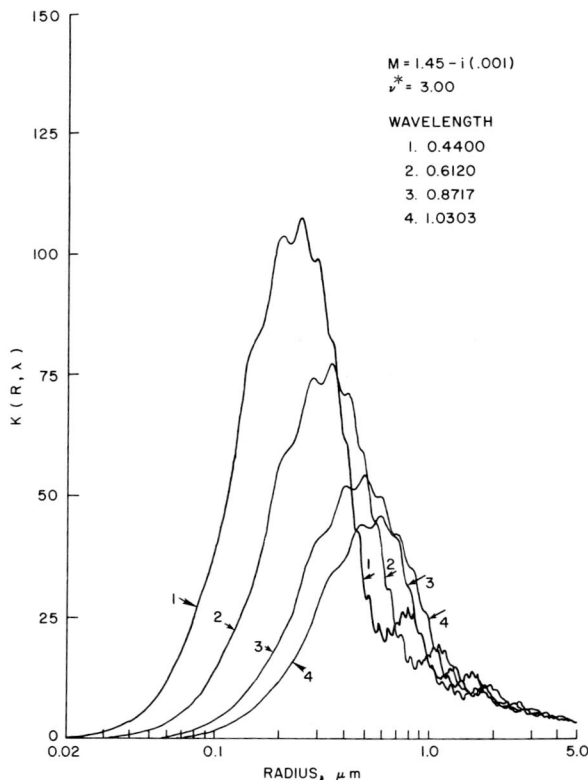

Fig. 9. Kernel functions ($\pi r^2 Q_t$) for four values of λ and for $m = 1.45-00 \; i$.

APPLICATION OF TWOMEY TECHNIQUES

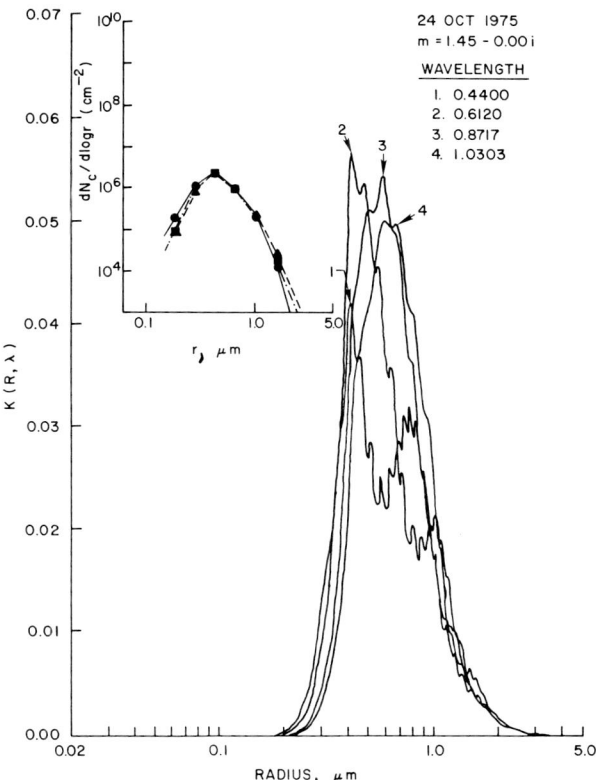

Fig. 10. *Modified kernel functions for the last iteration in the inversion scheme resulting from the size distribution as given in the inset.*

four values of λ and for m = 1.45-00 i. For this case $W(r) = cr^{-(\nu^* + 1)}$ with $\nu^* = 3.0$. As can be seen from these kernels, there apparently is information as to particle number between radii of a few tenths of a micrometer out to approximately 1 micrometer However, with the present technique, successive iterations produce new kernels which are the products of the old kernels and the inverted size distribution from the previous iteration. Figures 10 and 11 present kernels for the last iteration for two actual cases. Figure 10, for October 24, 1975, presents such a kernel

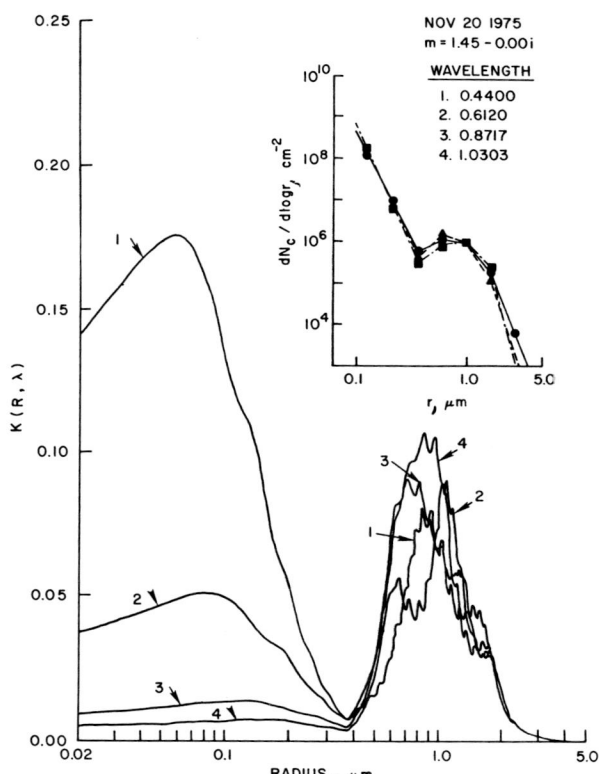

Fig. 11. Modified kernel functions for the last iteration in the inversion scheme resulting from the size distribution as given in the inset.

where the weighting function on the last iteration is as given by the inset in the upper left-hand corner. In this case, the range of information content is actually smaller than the starting kernel function (Fig. 9). However, this kernel is a result of the rather narrow size distribution, log-normal in type, depicted in the insert, and one would not expect information over a broader range of sizes.

Figure 11 shows kernels resulting from a rather common type of size distribution function commonly observed in our work to date. It has the appearance of a Junge type of distribution with

APPLICATION OF TWOMEY TECHNIQUES

a log-normal type superimposed. The kernel functions resulting from this type show a marked contrast to the previous two figures. There appear to be two regions of maximum information content, one for radii somewhat less than 0.1 μm, and another for radii between about 0.7 μm to about 1.5 μm. Under any conditions, it is apparent that with the current selection of wavelengths, the solutions are limited in information to sizes below approximately 1 to 1.5 μm. Inversion results for sizes beyond this (for the case of solar radiometry) are dictated, for the most part, by the constraints that are applied.

III. RESULTS

In this section, results from actual data will be presented. In the case of radiometer results, large numbers of successful inversions have been obtained from which only a few typical ones have been selected. A limited number of bistatic lidar inversions obtained from two cases are selected on days when both radiometer and lidar inversions were obtained.

Figure 12 shows a plot of the measured aerosol optical depth as a function of λ on the left-hand side and the resulting inverted size distribution function on the right. The three different curves are for three different initial weighting functions, $W(r)$, in each case a Junge distribution with exponents $\nu^* = 3.05$, 3.55, and 4.05. The three different solution are presented in order to demonstrate the insensitivity of the final results to the initial guess. Although all cases do not show this excellent agreement, in most cases the agreement is considered to be more than adequate. In general, the solution which produces values of the aerosol optical depth in best agreement with the measurements, in a least squares sense, is the one accepted. Figure 12 demonstrates a size distribution function fairly close to a Junge type. Aerosol optical depths can be seen to be around 0.10 in the mid-visible and demonstrate a nearly linear decrease with λ.

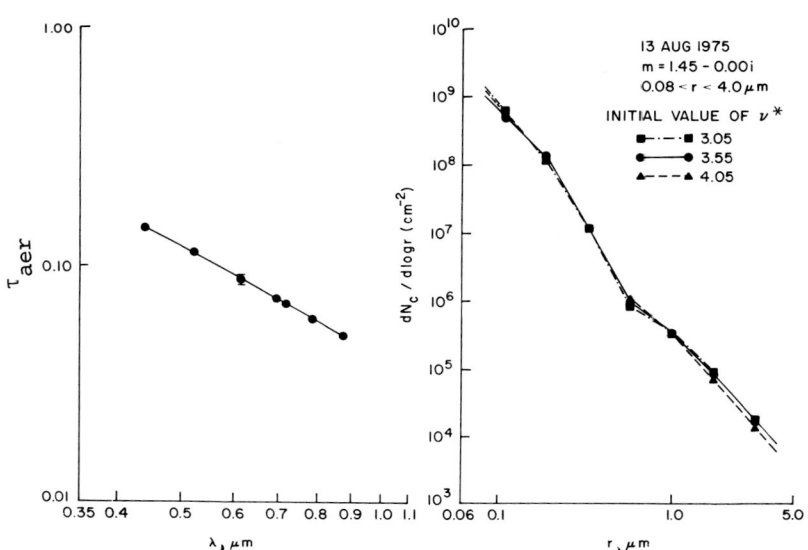

Fig. 12. Left-hand side shows measured aerosol optical depth as a function of λ while the right-hand side shows the resulting inversion for aerosol size distribution for August 13, 1975 data.

Figure 13 shows quite a different situation. For this case, the aerosol optical depth was extremely low, about 0.03 in the red-visible, and shows a nonlinear increase with λ. This type of τ_{aer} vs. λ behavior results in the log-normal type of distribution depicted in the right-hand side of the figure. Again, the three initial first guesses yield solutions in excellent agreement. The sharp peak in this size distribution function at about r = 0.8 μm would give use to a peak in the τ_{aer} vs. λ curve at a value of λ slightly greater than 1.0μm due to the first peak in the Mu values for Q_T. The measured values of τ_{aer} seem to indicate a peak would be reached in this region if measurements were extended out to longer wavelengths.

Figure 14 shows yet another quite common size distribution function commonly observed over Tucson, Arizona. It appears to be a Junge type with a log-normal type superimposed. The τ_{aer} vs.

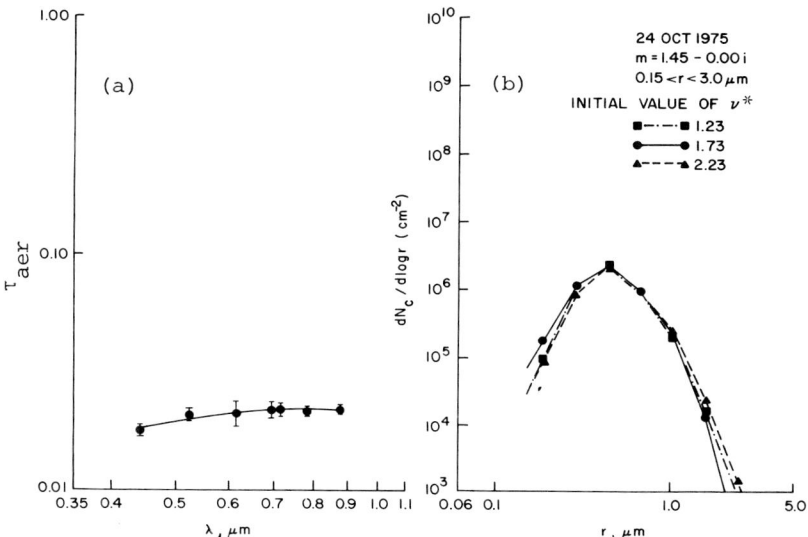

Fig. 13. Left-hand side shows measured aerosol optical depth as a function of λ while the right-hand side shows the resulting inversion for aerosol size distribution for October 24, 1975.

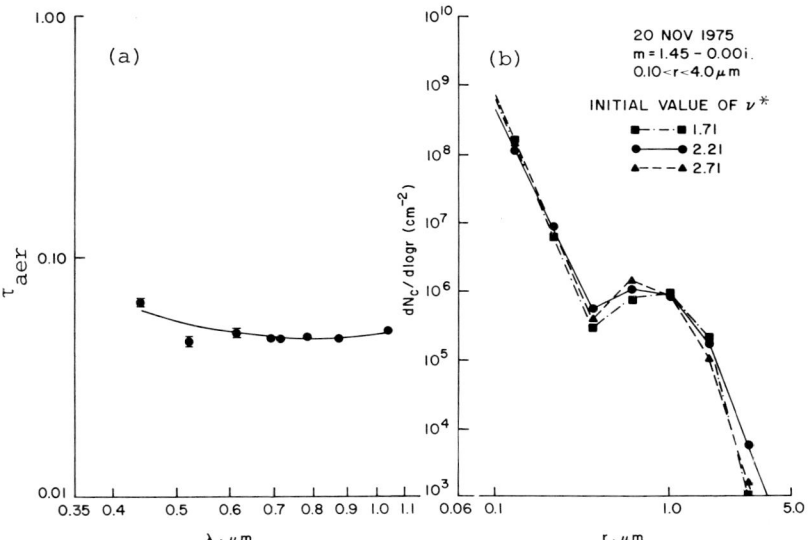

Fig. 14. Left-hand side shows measured aerosol optical depth as a function of λ while the right-hand side shows the resulting inversion for aerosol size distribution for November 20, 1975.

λ measurements which give rise to this type of distribution are curved in an opposite sense to those of Figure 13, as can be seen from the figure. The three initial guesses again yield satisfactory agreement.

Figure 15 shows a series of size distribution functions as determined on four successive days. Although the values of τ_{aer} were quite different on each of the days, they all exhibit a similar shape, consistent with the τ_{aer} vs. λ curves shown on the left-hand side of the figure.

Figure 16 demonstrates another series of three days. The first day of this series was extremely clear with very low aerosol optical depths and was followed by a mild episode of blowing dust on the second day which resulted in aerosol optical depths and was followed by a mild episode of blowing dust on the second day which resulted in aerosol optical depths of about 0.15 to 0.20. On the

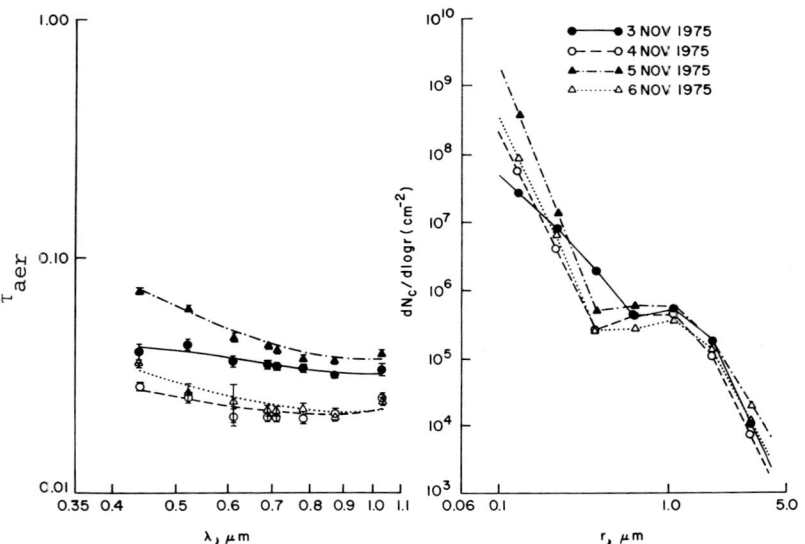

Fig. 15. Measured aerosol optical depths and resulting inversions for aerosol size distribution on four successive days.

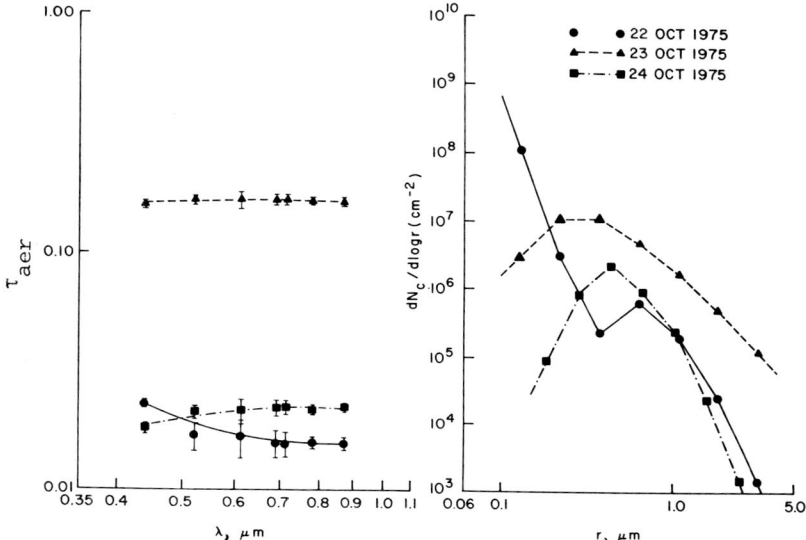

Fig. 16. Measured aerosol optical depths and resulting inversions for aerosol size distribution for a three-day period in which a dust storm occurred on the second day.

third day the optical depth recovered to almost its previous level, but its variation with λ was considerably different. The size distributions resulting from inverting these three days are quite interesting. On the first day, prior to the blowing dust episode, a fairly common size distribution was in evidence, a Junge type with a log-normal superimposed. During the blowing dust episode, this changes to a very broad, log-normal type with a peak at a few tenths of a micrometer. On the third day, the size distribution function is again log-normal in type, but much narrower and with a peak at about 0.6 μm.

There is undoubtedly much information to be gleaned from continued studies of these size distribution functions and their variations with meteorological conditions and their evolution with time. These results are presented here only to indicate the type of information which may be obtained from inversions such as these.

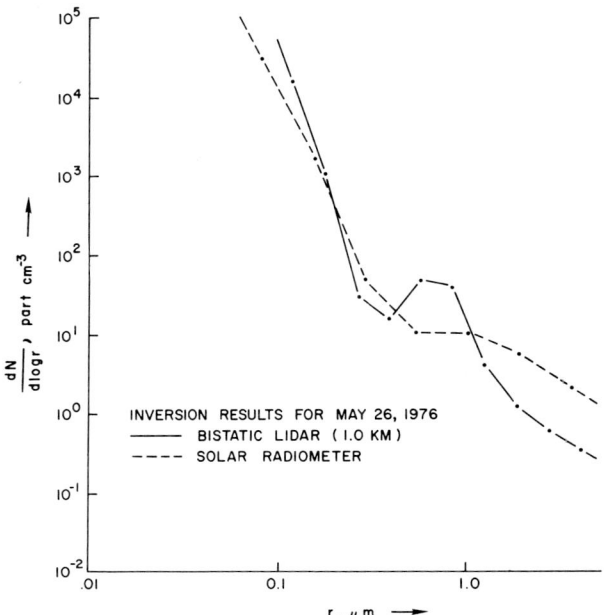

Fig. 17. Inversion for aerosol size distribution on the same day as obtained from bistatic lidar and solar radiometer.

Finally, in Figs. 17 and 18, inverted aerosol size distribution functions are presented for two days comparing results obtained from solar radiometer data with those obtained from bistatic lidar data. In Fig. 17, both inversions resulted in size distribution functions similar to the Junge-log-normal types presented earlier. While the agreement between the two sets of data is not perfect, it is felt that they are similar enough to lend support as to the reliability of this work. In Fig. 18, in addition to the lidar and radiometer inversions, a third distribution is presented. This distribution represents the best fit to the optical depth measurements that could be obtained from a two-slope size distribution function, the details of which need not be given here. In this case, all three methods yield almost identical size distribution functions, which on this particular day is a "modified" Junge type.

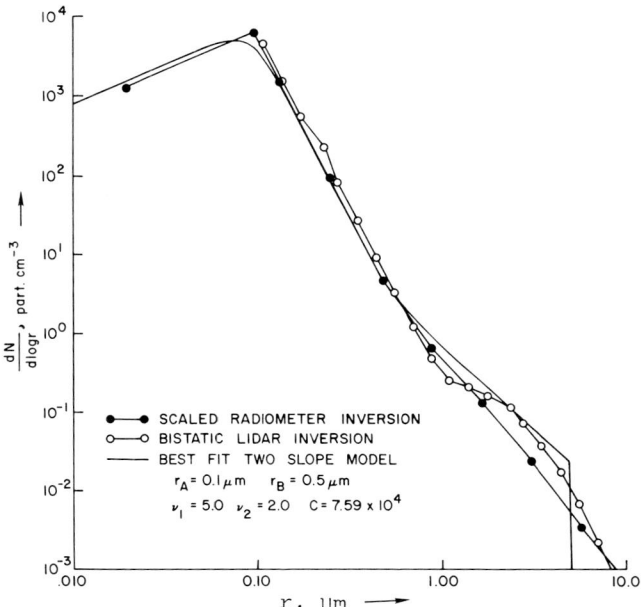

Fig. 18. Inversion for aerosol size distribution on the same day as obtained from bistatic lidar and solar radiometer except for a different day and as a third, best fit, distribution also.

The previous examples are presented in order to demonstrate the potential utility of remote sensing and inversion techniques. Aerosol size distribution functions are, at best, difficult to measure directly. Furthermore, direct measurement methods are generally much more time consuming and costly, factors which render more or less continuous monitoring almost impossible. With the remote sensing techniques described here, continuous monitoring, weather conditions permitting, is almost as easy as, say, monitoring solar radiation. It is felt that with the greater abundance of data which is available through these techniques, it will be possible to learn more about the properties of atmospheric aerosols in a shorter time than has heretofore been possible.

REFERENCES

1. B. M. Herman, S. R. Browning, and J. A. Reagan, Determination of aerosol size distribution from lidar measurements, *J. Atmos. Sci. 28,* 763 (1971).

2. S. Twomey, On the numerical solution of Fredholm integral equations of the first kind by the inversion of the linear system produced by quadrature, *J. Assoc. Comput. Mach. 10,* 97 (1963).

3. S. Twomey, The application of numerical filtering to the solution of integral equations encountered in indirect sensing measurements, *J. Franklin Inst. 279,* 95 (1965).

4. G. Fiocco and L. Smullin, Detection of scattering layers in the upper atmosphere (60-140 km) by optical radar, *Nature, 199* (1963).

5. R. T. H. Collis, Lidar: A new atmospheric probe, *Q. J. R. Meteorol. Soc. 392,* 220 (1966).

6. B. R. Clemesha, G. S. Kent, and R. W. H. Wright, A laser radar for atmospheric studies, *J. Appl. Meteorol. 6,* 386 (1967).

7. E. W. Barrett and O. Ben-Div, Application of lidar to air pollution measurements, *J. Appl. Meteorol. 6,* 500 (1967).

8. J. A. Reagan and B. M. Herman, Bistatic Lidar Investigations of Atmospheric Aerosols, *Proc. 14th Conf. Radar Meteorol., Tucson, Arizona,* November 17-20, 1970.

9. J. A. Reagan, B. M. Herman and R. J. Spiegel, On the Use of Bistatic Lidar in the Study of Atmospheric Aerosols, Proc. 1970 Southwest IEEE Conf., *Dallas, Texas,* April 22-24, 1970, pp. 526-530.

10. G. E. Shaw, J. A. Reagan, and B. M. Herman, Investigations of atmospheric extinction using direct solar radiation measurements made with a multiple wavelength radiometer, *J. Appl. Meteorol. 12,* 374 (1973).

11. G. E. Shaw, R. L. Peck, and G. R. Allen, A filter-wheel solar radiometer for atmospheric transmission studies, *Rev. Sci. Instrum. 44,* 1772 (1973).

12. F. G. Fernald, B. M. Herman, and J. A. Reagan, Determination of aerosol height distributions by lidar, *J. Appl. Meteorol. 11,* 482 (1972).

13. B. L. Phillips, A technique for the numerical solution of certain integral equations of the first kind, *J. Assoc. Comput. Mach. 9,* 84 (1962).

14. J. V. Dave, Effect of coarseness of the integration increment on the calculation of the radiation scattered by polydispersed aerosols, *Appl. Opt. 8,* 1161 (1969).

15. C. Junge, The size distribution and aging of natural aerosols as determined from electrical and optical data on the atmosphere, *J. Meteorol. 12,* 13 (1955).

16. W. Clark and K. Whitby, Concentration and size distribution measurements of atmospheric aerosols and a test of the theory of self-preserving size distributions, *J. Atmos. Sci. 24,* 677 (1967).

DISCUSSION

Pearce: That dust storm graph--I was sort of surprised to see the small particles having been swept away by the storm. Do you have a lot of confidence that that is a real effect?

Herman: This particular dust storm was not locally generated as you can tell by the optical depth. The most common dust storms in Arizona are those generated in the California desert and they blow in from there. We do on occasion get locally-generated dust but on this particular case the optical depth was about 0.15 which indicated that it was from a distant source--relatively distant as opposed to blown up locally. The only explanation that one could give for it is that some of the large particles have fallen out and some of the smaller ones had coagulated. The drop off here I can't explain other than that's what we got. Now, it may be just a poor inversion; I don't know, but we tend to believe it until we find reason not to. On the other side, however, there are just larger number densities but the shape from the peak on over is exactly what we get typically from which we think is locally generated dust. What our feeling is that this is a dust storm from a distant source, but I can't explain why the drop off on the left. But on the right I think it is just simply a case of most of the large particles have fallen out and that is just what is left over.

Twitty: My question is really the same thing, because that really bothers me. Current theories say that the large particles are, in fact, locally generated dust or locally generated surface sources of some kind and the small particles are of a different origin and that the two are more or less independent. So we expect this other bump just to add on to the one that is already there. So the question is, what is your source of that small particle set? How do you account for the Rayleigh scattering?

Herman: We have a calibrated lidar.

Twitty: It is not lidar though.

Herman: Excuse me. Okay, this is not the comparison. The calibration on the radiometer is done...I can't tell you right now. I can't remember how it is done. The radiometer, however, is calibrated to where we know the signal due to the Rayleigh amount. It is calibrated so that we can subtract out the Rayleigh scattering from it. The lidar is calibrated with a pulse to pulse calibration, which gives us a monitor of the power in each pulse and then we calibrate it at the highest level for the Rayleigh calibration.

Twitty: Now what really bothers me, you go back to your weighting function for that inversion. I don't know whether you can find that easily.

APPLICATION OF TWOMEY TECHNIQUES

Herman: I don't think I have one for that day.

Twitty: No but you have one...

Herman: That type of size distribution?

Twitty: That is similar.

Herman: The one with a large amount of particles? (showing graph[1])

Twitty: Yes. And the effect which of course produces the rapidly decreasing distribution causes the number 1, which is the short wavelength, 2, 3, 4, to decrease very rapidly in the weighting function. It could be explained if you are not subtracting sufficient amount of Rayleigh. That's the equivalent to if you have a little bit less Rayleigh at number 1, which is 0.44 μm. then that would bring that whole signal down.

Herman: Oh, yes, improper calibration could cause this to happen.

Twitty: It could have a very great effect on that part.

Herman: I don't argue that point. It is certainly true. We feel we have the best possible calibration that can be done on it. Now I am not saying this is gospel. This is the result of what we have been getting. We are aware of these problems--calibration problems and so forth, and any of these things can throw this inversion or any inversion off. And you are absolutely right. But then this would have been the case on the other days when we didn't get the small particles.

Twitty: Not if on those days you have substantially greater optical depth so that it essentially swamped out that day.

Herman: I don't know what the optical depth was on this day. I don't think it was listed. But we can find it though--at least we can find comparisons. They are not that much different. The worst optical depth case we presented was 0.15; most of them are in the 0.05 to 0.1. Now this case was not a 0.3 or 0.4 optical depth case.

Twitty: That case was a very thin optical depth case, right?

Herman: This was thin, but the dust case was about 0.15. Now when we have locally generated dust storms we get up to 0.3 or 0.4 or higher. The 0.15 is moderately hazy. It is not extreme. I wouldn't call that perturbed.

---

[1] See Fig. 11 in this paper.

Twitty: That is still consistent with my concern about the calibration. And I might make a comment about these kinds of plots, because I think those kinds of plots are really excellent. I looked at those a lot when I was doing inversions, because your weighting functions like that are really your distribution of the signal with respect to the size you are trying to determine. And you can pick right off of that where you clearly have no information with respect to size in the final distribution as you have no signal coming from sizes of that class. It comes right out of the result of the inversion.

Herman: I think that when we get below certainly a few tenths of a micron and above about 2 microns, as I said earlier, I think there are certainly questions here. But I do think we have information between a few tenths of a micrometer and one or two micrometers. It is my feeling that there is some truth to these inversions over that size range. There is no way for anybody to say is it right or wrong. The only way to ever prove this is to make simultaneous measurements by several techniques or go up and sample it directly the same time you make an inversion. We all know the limitations to this.

Twomey: I would just like to comment that under the conditions of blown dust in Arizona when your visibility, say, in the Catalina Mountains and such, is grossly poorer than what is normally sometimes total, the particle concentration as measured by a particle counter invariably in my experience is low. Now that's quite in agreement with what you show up there with reduced number of small particles. And, in aerosol physics this is a very ancient result. There was a gentleman back in 1916 who beat a carpet in a room for half an hour and was very surprised to find that his total particle concentration was less at the end of this episode. His lungs were worse! But the total particle count went down.

Barkstrom: I have a question that is related to the thing that Jerry Twitty asked. First of all, I'd like to know if this is what Dr. Rodgers would classify as a messy nonlinear problem, you know, the worst kind? And, if so, how we can go about calculating an error bar on each of those inverted data?

Herman: I didn't show these. We have done error bars on these. We have error bars on the optical depth determination and then we have error bars, at least a crude estimation, on the size distribution, which I didn't show here.

Barkstrom: Yes, but in light of the nonlinear problem associated with the calibration, is there a problem connected with the error bars that you get out of the linear estimate?

Herman: Who are you asking?

APPLICATION OF TWOMEY TECHNIQUES

Barkstrom: I'd like to sort of ask Dr. Rodgers as well.

Rodgers: I was going to ask you about this one as well! It wasn't very clear to me whether you had a linear problem or a nonlinear problem.

Herman: This is a linear problem but we are iterating because we are trying to bootstrap up our weighting function guess.

Rodgers: Well, if you have to change the weighting function, then it is not a linear problem.

Herman: We are making a first guess to perform that integral of the kernel which we can't do analytically. We have to make some type of weighting function guess to provide a proper integration over the kernel function. Otherwise, if we use just large intervals, we have a very poor representation of the kernel.

Rodgers: The equations are linear?

Herman: The equation itself, if we could do the integral properly, would be linear. We could do it in one iteration.

Rodgers: Yes, so the problem is one of integrating over a peculiar kernel?

Herman: That is the basic problem in particularly the lidar case but also in the radiometer.

Rodgers: I still don't really understand how it comes out nonlinear then?

Herman: It is not a nonlinear problem. We are iterating simply to get a better estimate of the integral over the kernel function. We could get the answer in the first iteration if we wanted, but we are getting a better estimate as we iterate on. I don't think it is equivalent to making it nonlinear. I don't think you can even consider it equivalent to nonlinear.

Rodgers: Well, if it's linear, you don't need to iterate. I just don't understand what the problem is.

Herman: Well, I thought I had explained the technique we are using. As I say, we do not have to iterate. It is not necessary.

THE INVERSION OF STRATOSPHERIC AEROSOL AND

OZONE VERTICAL PROFILES FROM SPACECRAFT

SOLAR EXTINCTION MEASUREMENTS

William P. Chu
Old Dominion University

This paper analyzes the inversions of multi-channel solar extinction measurements in the 0.35-1.0 µm wavelength region to retrieve stratospheric aerosol and ozone vertical profiles using both the constrained linear inversion scheme and the iterative scheme recently developed by Twomey. The inversions of the multi-wavelength solar extinction data obtained from spacecraft have been analyzed based on the inversion of computer simulated data using various atmospheric models with differing amounts of aerosol and ozone in the stratosphere. The sensitivities of the inversion schemes to different experimental errors are discussed in terms of accuracy and resolution of the retrieved profiles.

I. INTRODUCTION

The minor constituents in the stratosphere have received great attention in recent years because of concerns over their roles in affecting the global environment. With mankind's increasing activities in the stratosphere, such as supersonic transport (SST) and Shuttle missions, and ground-based activities, such as those involved with nitrogen fertilizers, an increase in stratospheric aerosol loading together with the removal of stratospheric ozone are possible through various chemical reactions. A realistic assessment on this problem is still not possible due to the lack of information on the distributions of most of the constituents in our atmosphere.

This paper analyzes one of the atmospheric remote sensing methods using spacecraft solar extinction technique to retrieve stratospheric aerosol and ozone vertical profiles. In this remote sensing technique, the spacecraft instrument will direct toward the horizon and measure transmitted solar radiant intensity in a number of spectral intervals with wavelengths from 0.38 to 1.00 μm. The locations of the spectral channels are selected based on considerations of maximum extinction contribution from each individual constituent, interference from other constituents, and available energy for measurements. The radiant intensity data are then inverted to produce vertical profiles for each of the constituents. The inversion procedure can be separated into two steps. The first step involved retrieval of the total vertical extinction profiles at each spectral interval. The second step then separated the contributions from each constituent using the multi-spectral extinction data at each altitude level. Examples of inverted results from simulated data, including experimental errors, are presented and discussed in this paper.

II. EXPERIMENTAL CONCEPT

In the solar extinction experiment, multi-wavelength measurements of the transmitted solar radiant intensity along the line of sight through the atmosphere are made on board the spacecraft. The basic geometry is illustrated in Fig. 1. As the spacecraft emerges from Earth's shadow, the radiometer on board the spacecraft will point to the Sun, measuring solar radiant intensity at several pre-selected wavelength regions. In this paper, we will consider the radiometer with a small field of view in order to achieve higher vertical resolution. The radiometer is assumed to either lock on a fixed position on the solar disk or scan across the solar disk for data acquisition.

INVERSION OF STRATOSPHERIC AEROSOL

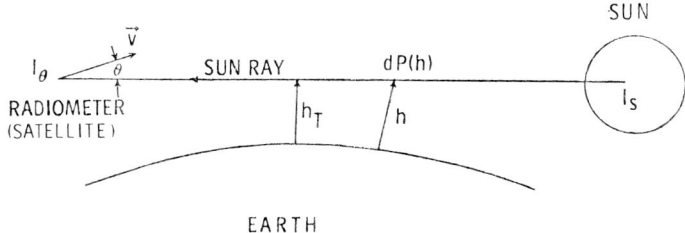

$I_\theta = T|h_T(\theta)| I_S$
LAMBERT - BEER LAW
$T(h_T) = \exp\left\{-2\int_{h_T}^\infty \beta(h)\,dP(h)\right\}$
$\beta(h)$ - EXTINCTION PROFILE
$P(h)$ - REFRACTED OPTICAL PATH

Fig. 1. Geometry of the solar extinction experiment

The monochromatic radiant intensity reaching the radiometer is given by the fundamental equation of radiative transfer (Ref. 1)

$$I_\lambda(h_z) = I_\lambda(h_o)\tau_\lambda(h_z,h_o) + \int_{\tau_\lambda(h_z,h_o)}^{1} J_\lambda(h)\,d\tau_\lambda(h_z,h_o) \qquad (1)$$

Equation (1) relates at wavelength λ, the radiant intensity $I_\lambda(h_z)$ measured at distance h_z, to the radiant intensity $I_\lambda(h_o)$ at boundary h_o, the monochromatic transmittance $\tau_\lambda(h_z,h_o)$ and source function $J_\lambda(h)$ at distance h along path from h_z to h_o.

For solar extinction experiment with spectral channels closed to the visible region, the contribution by atmospheric emission to the radiant intensity is very small and can safely be neglected. The radiative transfer equation is thus greatly simplified to

$$I_\lambda(h_z) = I_\lambda(h_o)\tau_\lambda(h_z,h_o) \qquad (2)$$

where the radiant intensity $I_\lambda(h_o)$ is now the solar radiant intensity and $\tau_\lambda(h_z,h_o)$ is the transmittance of the atmosphere between the Sun and the spacecraft.

In the wavelength range from 0.3 to 1.0 micron, atmospheric attenuation are predominantly caused by aerosol, ozone, and air molecules (Rayleigh component) with minor contributions from NO_2, water vapor and O_2 molecules. Models of the stratospheric extinction as a function of wavelength profile can be constructed from different vertical distributions of aerosol and ozone. Figure 2 shows an extinction coefficient model as a function of wavelength at an altitude of 18 km. Notice that at most of the wavelength shown, the total extinction consists of contributions from aerosol, ozone, and Rayleigh components with different weights. The weightings will also be different for different altitude levels, depending on the distributions of the different constituents.

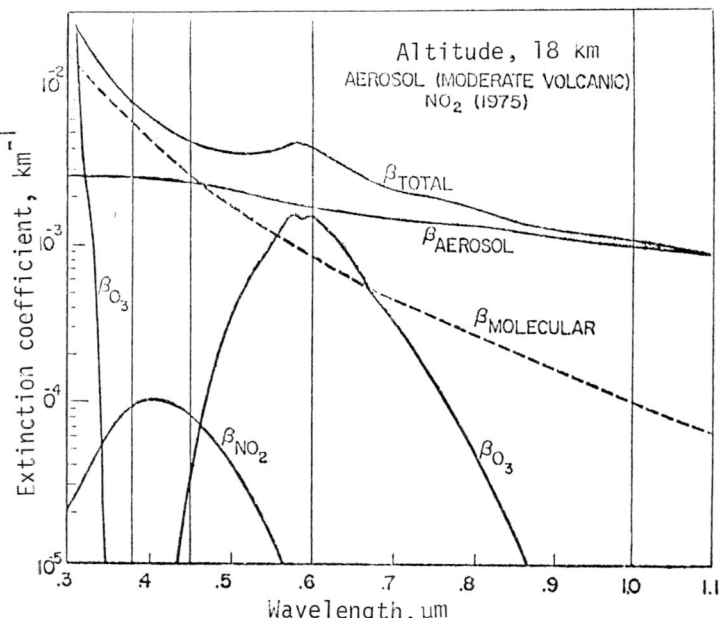

Fig. 2. Extinction as a function of wavelength model with an altitude of 18 km.

If we made the assumptions that the atmosphere is spherically stratified and spherically symmetric, the transmittance function $\tau_\lambda(h_z, h_o)$ will then be given by the Lambert-Beer law as

$$\tau_\lambda(h_z, h_o) = \exp\{-2 \int_{h_T}^{\infty} \beta_\lambda(h) \, dP(h, h_T)\} \quad (3)$$

where $\beta_\lambda(h)$ is the atmospheric extinction coefficient at wavelength λ as a function of vertical height h, and $P(h, h_T)$ is the optical path at height h for line of sight with tangent height h_T. Equivalently, the optical depth $g_\lambda(h_T)$ can be defined as

$$g_\lambda(h_T) = \ln\{1/\tau_\lambda(h_z, h_o)\}$$

$$= 2 \int_{h_T}^{\infty} \beta_\lambda(h) \, dP(h, h_T) \quad (4)$$

The determination of vertical extinction profile at each wavelength λ required the inversion of the integral equation in Eq. (4).

III. INVERSION TECHNIQUES

A. Inversion of Measured Transmittance to Total Extinction Profiles

In Eq. (4), the measured parameter $g_\lambda(h_T)$ is equal to a kind of convolution of the weighting function $\partial P(h, h_T)/\partial h$ with the unknown vertical extinction profile $\beta_\lambda(h)$. The integral can be approximated with a sum over n discrete atmospheric layers with equal thickness and assigning to each layer an averaged extinction coefficient β_j. The integral equation can then be replaced by a system of linear equations

$$g_i = \sum_{j=i}^{n} \beta_j P_{ij} \quad i = 1, 2, \ldots m \quad (5)$$

where P_{ij} is the optical path length in the jth layer with tangent height at the ith layer. Equation (5) can be abbreviated into matrix form

$$g = P\beta \quad (6)$$

where g is the column vector for the measured optical depth, β is the column vector for the unknown extinction profile, and P is the optical path length matrix.

Equation (6) can be solved directly for β either exactly when $n = m$ or in the least square sense when $n < m$

$$\beta = P^{-1} g \qquad n = m \qquad (7a)$$

$$\beta = (P^T P)^{-1} P^T g \qquad n < m \qquad (7b)$$

where P^{-1} is the inverse of P and P^T is the transpose of P. The solutions obtained from the direct inversion are generally unstable due to the presence of noise associated with the measured parameter g. In order to suppress the unphysical oscillations in the inverted solutions, some constraints on the high frequency components of the solutions have to be incorporated in the inversion schemes. We have analyzed two different inversion methods. The first method is the linear constrained inversion as developed by Twomey (Ref. 2) and Phillips (Ref. 3); the second method is the iterative scheme recently developed by Twomey (Ref. 4). We will discuss the two methods separately.

1. Linear Constraint Inversion

The solution to Eq. (6) is given by the following expression

$$\beta = (P^T P + \gamma H)^{-1} P^T g \qquad (8)$$

where normally H is the constrained matrix and γ is a constant smoothing parameter. In our analysis, we used the constrained matrix H which minimizes the second difference of the solution. The smoothing parameter γ whose magnitude is proportional to the noise level of the measurements has been replaced by a diagonal matrix of element γ_{ii}. The reason for using a variable smoothing parameter γ arises from the nonlinear expression for the noise term in Eq. (4). Assuming the radiometer has a noise level e_i associated with each radiant measurement I_i, the deduced

INVERSION OF STRATOSPHERIC AEROSOL

transmittance τ_i will have approximately the same noise level e_i. By computing the optical depth g_i, we have

$$g_i = \ln\{1/(\tau_i + e_i)\} \tag{9}$$

for $e_i \ll \tau_i$, we can expand Eq. (9) and obtain

$$g_i \simeq \ln\{1/\tau_i\} - e_i/\tau_i \tag{10}$$

The second term on the right side is the equivalent noise level for the parameter g_i. Thus, we choose the matrix element γ_{ii} as

$$\gamma_{ii} = \gamma_o \left(\frac{e_i}{\tau_i}\right)/g_i \tag{11}$$

where γ_o is a constant parameter to be adjusted for optimum inverted solutions.

2. Iterative Inversion

The iterative method starts by initially assuming a guessed solution, and the solution is continuously updated through the following iteration:

$$\beta_j^{(k+1)} = \beta_j^{(k)} + (r_i^{(k)} - 1) K_{ij} \beta_j^{(k)} \tag{12}$$

where $\beta_j^{(k)}$ is the solution β_j at k-iteration, $r_i^{(k)}$ is the ratio of measured g_i to the computed g_i from the k-iterated solution, and K_{ij} is the normalized weighting function $\partial P(h, h_T)/\partial h$ where $K_{ij} \leq 1$. The iterative process is terminated when the computed radiant approaches the measured radiant to within the instrument noise level.

B. Inversion of Multi-wavelength Extinction Profiles Into Constituent Profiles

The inversion methods discussed in (A) will generate multi-spectral extinction vertical profiles from the multi-channel extinction measurements. The second step in the inversion process will be to separate the contributions from individual constituent. For the wavelength region of interest, the total extinction coefficient at a fixed altitude level can be written as the sum of contributions from aerosol, ozone, and air molecules as

$$\beta_{total} = \beta_{aerosol}(\lambda) + a_{\lambda i}\beta_{mol}(\lambda_o) + b_{\lambda i}\beta_{o_3}(\lambda_o) \qquad (13)$$

where $\beta_{mol}(\lambda_o)$ is the Rayleigh extinction coefficient at a reference wavelength λ_o, $a_{\lambda i}$ is the wavelength dependence of the Rayleigh extinction, $b_{\lambda i}$ is the wavelength dependence of the ozone absorption profile, and $\beta_{aerosol}(\lambda)$ is the aerosol extinction as a function of wavelength λ. In this study, we have assumed a two-parameter model for the description of the aerosol wavelength dependent extinction coefficient as

$$\beta_{aerosol}(\lambda) = A \lambda^{\alpha} \qquad (14)$$

where A and α are the two parameters. Substituting Eq. (14) into Eq. (13), we then have a system of nonlinear equations with unknown $\beta_{mol}(\lambda_o)$, $\beta_{o_3}(\lambda_o)$, A, and α for each altitude level. The system of nonlinear equations can be solved using the Marquardt (Ref. 5) algorithm. The Marquardt algorithm is a minimum search procedure for a system of nonlinear equations. An initial set of guessed solutions for the unknowns is updated through the following iterative equation

$$\underset{\sim}{x}^{(k+1)} = \underset{\sim}{x}^{(k)} - [\underset{\sim}{J}^T(\underset{\sim}{x}^k)\underset{\sim}{J}(\underset{\sim}{x}^k) + \lambda_k \underset{\sim}{I}]^{-1}\underset{\sim}{J}^T(\underset{\sim}{x}^k)\underset{\sim}{F}(\underset{\sim}{x}^k) \qquad (15)$$

where $\underset{\sim}{x}^k$ is the solution vector for the unknowns at the k-iteration, $\underset{\sim}{F}(\underset{\sim}{x}^k)$ is the residue vector for Eq. (13) at the k-iteration, $\underset{\sim}{J}(\underset{\sim}{x}^k)$ is the Jacobian matrix with element $J_{ij} = \partial f_i/\partial x_j$ where f_i is the ith nonlinear equation. $\underset{\sim}{I}$ is the identity matrix, and λ_k is a control parameter at the k-iteration. The control parameter λ_k is adjusted at each iteration for fast convergence to

INVERSION OF STRATOSPHERIC AEROSOL 513

the minimum. The iterative process is terminated when the difference between two consecutive residue squares is less than 0.1%.

IV. INVERSION RESULTS

Inversion results have been obtained from applying the inversion techniques discussed earlier to simulated spacecraft solar extinction measurements. Several models of the aerosol and ozone profiles were used to generate atmospheric transmittance profiles in different wavelength regions. Wavelengths at 0.38, 0.45, 0.6, and 1.0 micron were chosen to coincide with the four spectral channel locations for the Stratospheric Aerosol and Gas Experiment (SAGE). The simulated radiant intensity data are generated from the atmospheric transmittance profiles by incorporating the spacecraft geometry. The radiometer is assumed to have a field of view of 30 arc seconds, when situated on a satellite at an orbital altitude of 600 km. A total of 80 radiant intensity data points are generated per channel covering tangent altitudes from 10 to 50 km in equal increments of 0.5-km height. Each data point is simulated by assuming that the radiometer is pointing to a different location on the solar disk.

The inversion of these radiant profiles was then performed by introducing experimental errors of various origin and magnitudes to the simulated data. There are three types of erros which directly affect the inversion results. The first type of error is the bias error in the determination of the exact tangent heights for all the data points. The second type of error is the noise associated with each measurement. The third type of error is associated with the uncertainty in knowing the pointing angle of the radiometer for each data point. This pointing error will result in both an erroneous value for the unattenuated solar radiance due to solar limb darkening effect and an erroneous value for the instantaneous tangent height.

The three types of errors mentioned earlier in different magnitudes were added to the simulated radiant data and inversions were performed to determine the sensitivity of these errors on the inversion results. The atmosphere is divided into 45 homogeneous layers of 0.5-km thickness between 10 and 21 km, and increasing thicknesses from 21 to 50 km. The path length matrix is computed by using a ray trace calculation to include atmospheric refraction effect.

Figures 3 to 6 show inversion results for simulated extinction measurements using the linear constrained inversion method. The input aerosol vertical profile used is deduced from an actual lidar backscattered profile with vertical resolution of 0.15 km. The ozone vertical profile used is similar to Elterman's model (Ref. 6) except that artificial layering structures were added. The Rayleigh extinction profile is also deduced from an actual rawinsonde temperature and pressure profiles. In these simulated measurements, experimental errors, including noise levels of 0.1% to full scale radiant intensity and pointing uncertainty of 3 arc seconds, were added to the simulated measurements. The inversion results are presented in terms of aerosol vertical extinction at 1.0 μm (Fig. 3), ozone vertical extinction at 0.6 μm (Fig. 4), Rayleigh vertical extinction at 0.38 μm (Fig. 5), and aerosol optical model parameters (Fig. 6). The aerosol optical model parameter α is assumed to be -0.8 between 17 and 20 km and equal to -1.6 for the rest of the altitude range. Inspections of Figs. 3 to 6 indicate that both aerosol and ozone vertical profiles can be inverted with good accuracy up to an altitude of 40 km. The poor inversion results for the Rayleigh profile below 14 km are caused by the rapid decreasing signal level at the 0.38 μm channel. In Fig. 6, the inverted results for the aerosol optical parameter α within the Junge layer region show some fluctuation and no inversion results can be obtained above 20 km due to the rapid decrease of aerosol concentration in this region.

INVERSION OF STRATOSPHERIC AEROSOL 515

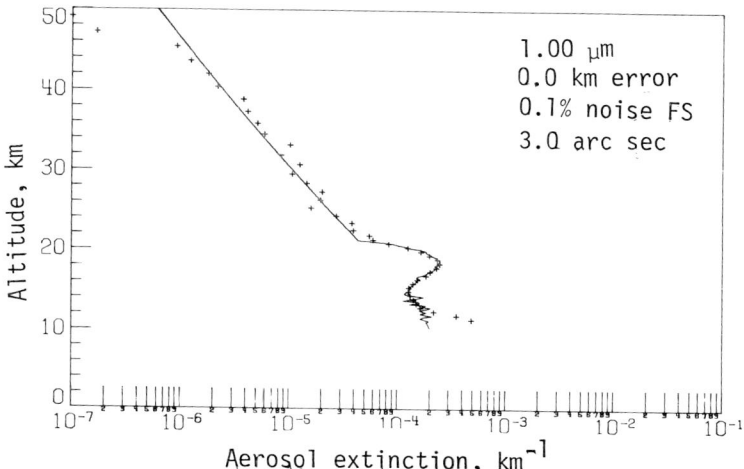

Fig. 3. Inversion results for aerosol vertical extinction profile at 1.00 μm.

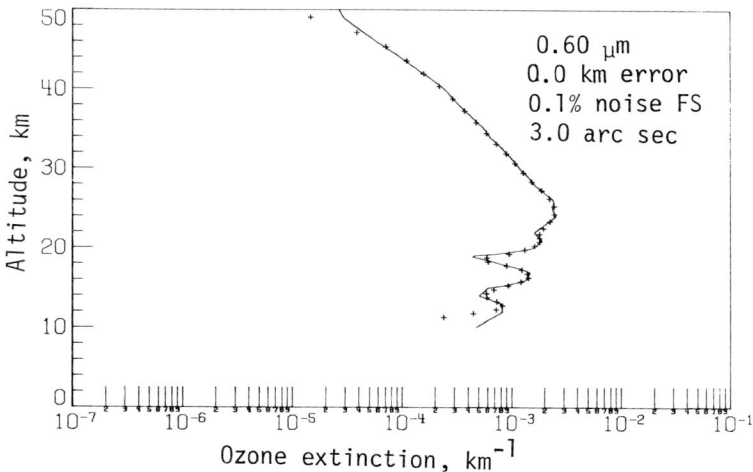

Fig. 4. Inversion results for ozone vertical extinction profile at 0.6 μm.

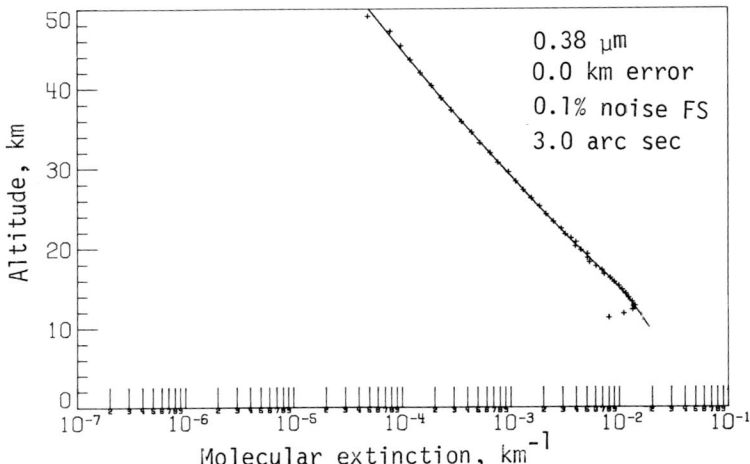

Fig. 5. Inversion results for Rayleigh vertical extinction profile at 0.38 μm.

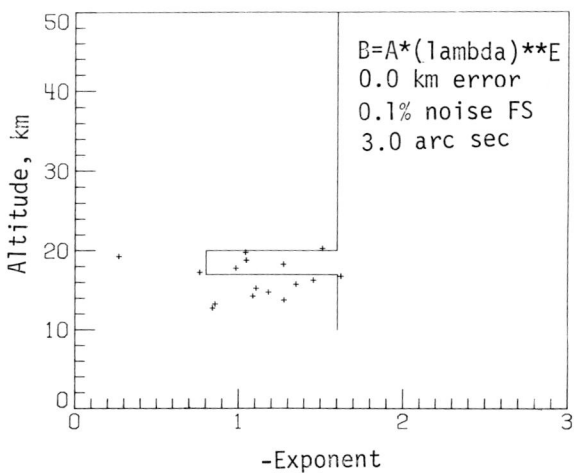

Fig. 6. Inversion results for aerosol optical model parameter α as a function of altitudes.

INVERSION OF STRATOSPHERIC AEROSOL

Figures 7 to 9 show inversion results for the same input models except that the noise level for each measurement has been increased to 1.0% and the pointing uncertainty has been increased to 15 arc seconds. Both the inverted results for aerosol and ozone profiles show considerable decrease in resolution. This behavior is consistent with the inversion technique in which the smoothing parameter α_o in Eq. (11) is increased to accommodate the high noise level in the measurements. The inversion results for the aerosol optical model parameter which are not shown here are similar to Fig. 6 except larger fluctuations were observed.

Figures 10 to 11 show inversion results for a set of different aerosol and ozone vertical profiles. The aerosol profile is deduced from the first published lidar observation of the Volcano de Fuego eruption (Ref. 7) showing sharp layering structures. The peak layer at 19 km is approximately 0.5 km wide. The inversion cannot reproduce this sharp layering structure due to the integrating effect over the radiometer's field of view.

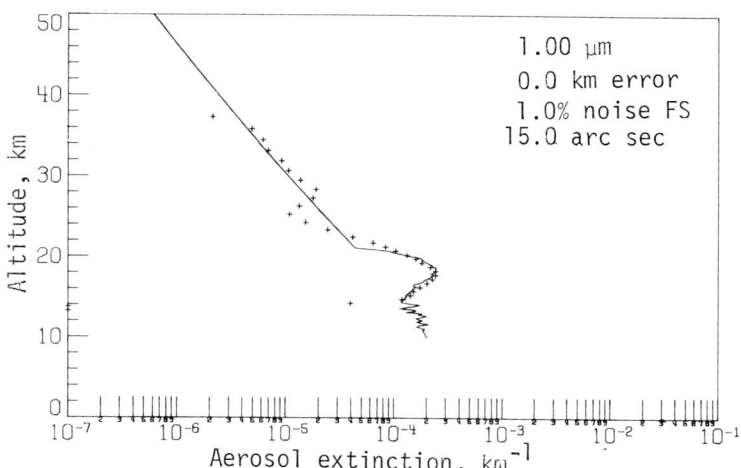

Fig. 7. Inversion results for aerosol vertical extinction profile with increased experimental errors.

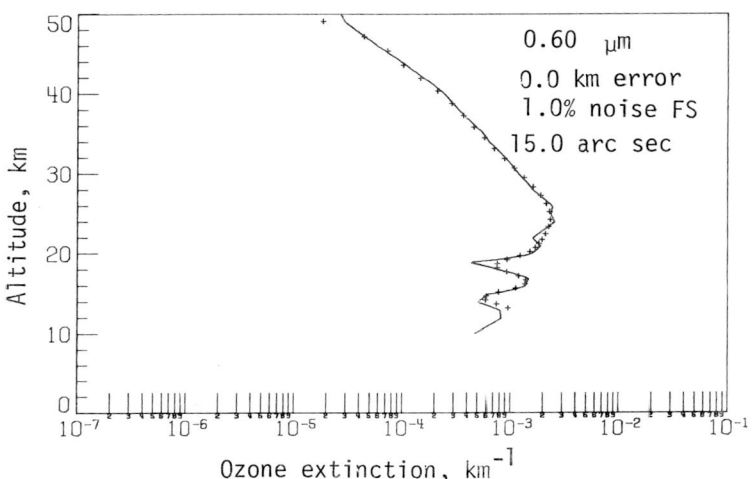

Fig. 8. Inversion results for ozone vertical extinction profile with increased experimental errors.

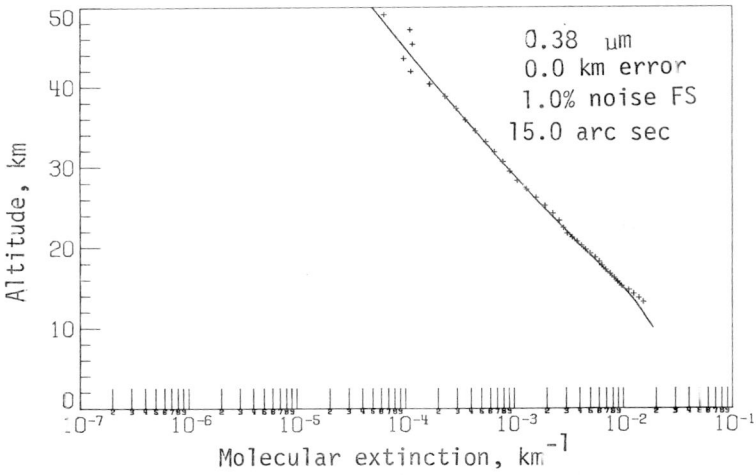

Fig. 9. Inversion results for Rayleigh vertical extinction profile with increased experimental errors.

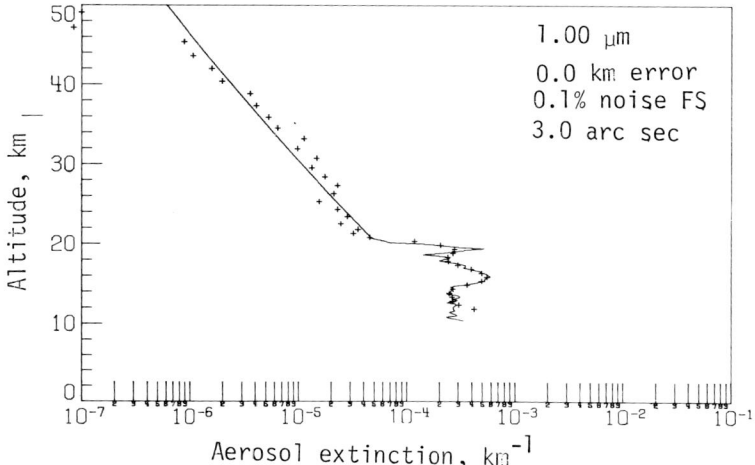

Fig. 10. Inversion results for aerosol vertical extinction profile with different aerosol vertical extinction profile.

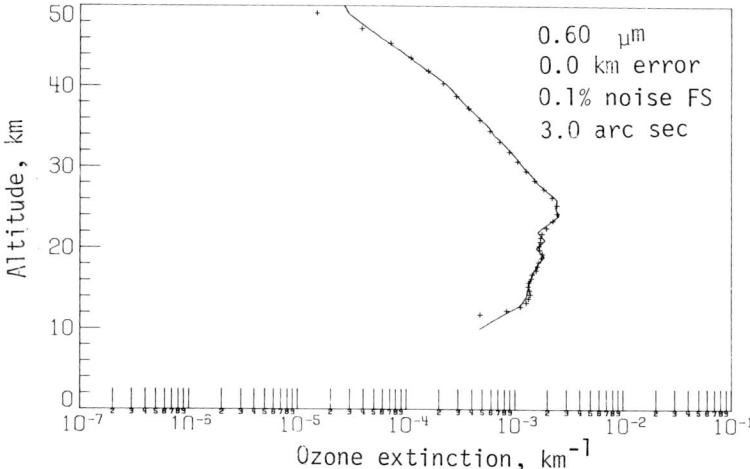

Fig. 11. Inversion results for ozone vertical extinction profile with different ozone vertical extinction profile.

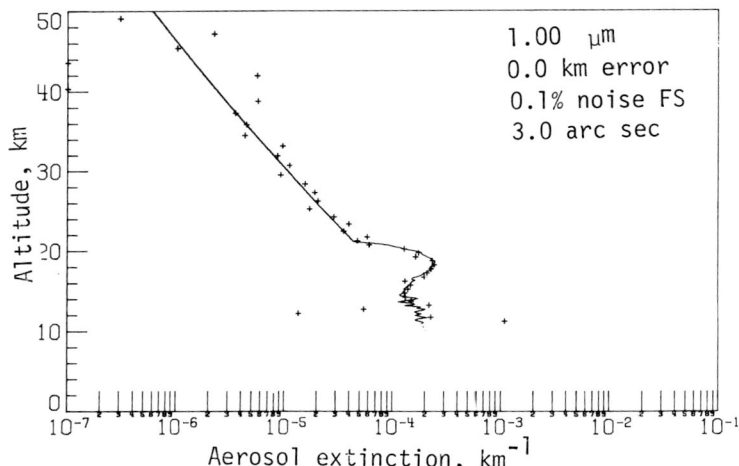

Fig. 12. Inversion results for aerosol vertical extinction profile using the iterative inversion scheme.

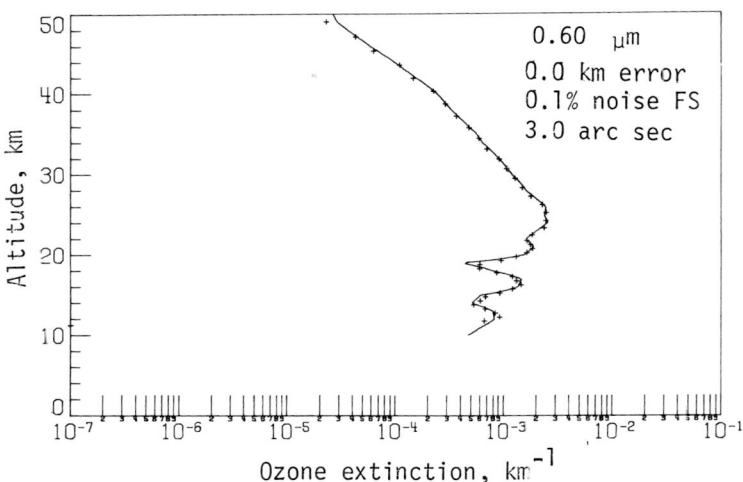

Fig. 13. Inversion results for ozone vertical extinction profile using the iterative inversion scheme.

INVERSION OF STRATOSPHERIC AEROSOL

Figures 12 and 13 show the inversion results using the iterative method as discussed in Eq. (12). The atmospheric model and experimental noise levels are identical to those as shown in Figs. 3 to 6. Comparison of the inversion results in this case to those as shown in Figs. 3 and 4 indicates that they are quite similar. The inverted aerosol profile at altitude above 30 km in this case shows some large amplitude oscillations. This is expected as the noise level in this altitude region is considerably higher than the signal level.

Inversions have also been performed on simulated data including bias errors in the tangent height determination. It is found that bias errors will produce a shift in altitude scale of the inverted profile in direct proportion to the bias magnitude.

V. CONCLUDING REMARKS

This paper has demonstrated that measurements from spacecraft solar extinction experiment can be inverted to produce aerosol and ozone vertical profiles from cloud top up to approximately 50 km altitude. Analysis of the inversion results from simulated measurements including various experimental errors indicated that the resolution of the inverted profiles will be degraded as the errors are increased.

Both the linear constrained inversion method and the iterative method have been used for the inversion of solar extinction measurements. The accuracy of the inverted results from the two different inversion methods are similar.

SYMBOLS

a_λ	coefficient for wavelength dependence of Rayleigh extinction
A	model parameter for aerosol extinction as function of wavelength dependence
b_λ	coefficient for wavelength dependence of ozone absorption
e	noise level
$\underset{\sim}{F}$	residue vector
g_λ	optical depth
h	distance
$\underset{\sim}{H}$	constrained matrix
$\underset{\sim}{I}$	identity matrix
I_λ	monochromatic radiant intensity
$\underset{\sim}{J}$	Jacobian matrix
J_λ	monochromatic source radiant intensity
k	iteration number
K	normalized weighting function
p	optical path length
r	ratio of measured optical depth to computed optical depth
$\underset{\sim}{X}, \underset{\sim}{x}$	solution vector for nonlinear equations
α	model parameter for aerosol extinction as function of wavelength dependence
β_λ	extinction coefficient
γ	smoothing parameter
λ	wavelength
λ_k	control parameter for iterative solution to nonlinear equations
τ_λ	monochromatic transmittance

REFERENCES

1. S. Chandrasekhar, "Radiative Transfer," p. 9. Oxford University Press, New York, 1950.

2. S. Twomey, On the numerical solution of Fredholm integral equations of the first kind by the inversion of the linear system produced by quadrature, *J. Assoc. Comput. Mach. 10*, 99 (1950).

3. D. L. Phillips, A technique for the numerical solutions of certain integral equations of the first kind, *J. Assoc. Comput. Mach. 9*, 97 (1962).

4. S. Twomey, Comparison of constrained linear inversion and an iterative nonlinear algorithm applied to the indirect estimation of particle size distributions, *J. Comput. Phys. 18*, 188 (1975).

5. D. W. Marquardt, An algorithm for least-square estimation of nonlinear parameters, *J. Soc. Ind. Appl. Math. 11*, 431 1963.

6. L. Elterman, Visible and IR attenuation for altitudes to 50 km, *Air Force Cambridge Research Laboratory*, 1968 [AFCRL-68-0153].

7. M. P. McCormick and W. H. Fuller, Lidar measurements of two intense stratospheric dust layers, *Appl. Opt. 14*, 4 (1975).

DISCUSSION

Shettle: On one of your first figures where you show the optical thicknesses, you included nitrogen dioxide, which would add a fifth unknown. I can't recall whether that was a . . .

Chu: No, that was not in the simulation.

Shettle: The question is, would it add a significant effect, particularly when you start trying to get the slope of the aerosol extinction versus wavelength?

Chu: If we assume we don't know anything about nitrogen dioxide, it will contribute some error to the slope on the aerosol curve. It should be pointed out that the peak nitrogen dioxide concentration occurs approximately at 30 km; whereas the aerosol peak is around 18 to 20 km. There is a spatial discrimination as well as a spectral discrimination utilizing this technique.

Shettle: Yes, except you are not looking for it and if you include it, that would give you a fifth unknown. And you still only have four wavelengths.

Chu: In fact, on SAGE we do have five channels.

McCormick: Yes, very recently we have encountered one more channel with lower vertical resolution. Dr. Chu's paper addresses a four-channel instrument only.

Pepin: I would like to comment on that last question. The fact that the NO_2 contributes only a small amount means that one can correct for it quite accurately by not knowing it very well and subtract it out. It's a percent of a small percent.

Shettle: The slope of the aerosol extinction is still sensitive to the NO_2 because it is proportional to the change in the extinction with wavelength, so extra extinction will add to the error.

Pepin: What I am saying is that if it only contributes one percent to the signal and if you only know it to 50 percent, then it will only contribute one-half percent to the inversion problem in terms of signal.

Green: Ben Herman, in his talk, indicated that if the extinction due to aerosols were curved upward, you have one characteristic shape of size distribution and if it were to curve downward, another. What does the effect of assuming a straight line have upon your inversion procedure? Is it then strictly a single Junge power law or broad range or what?

INVERSION OF STRATOSPHERIC AEROSOL

Chu: You mean trying to approximate the aerosol by a straight line?

Green: The power law, the λ^α Law?

Chu: Well, all we can get is two unknown parameters. We are limited by the channel number. We have four measurements and we are looking for four unknowns. This is about the maximum we can do. I don't see any way around the problem besides just trying to describe the aerosol as a two-parameter model. Of course, it might not exactly reflect what nature is doing. But according to the paper that Tom Swissler presented about two days ago, the two-parameter model is quite adequate in describing most of the size distribution. I mean not the bimodal type but the single modal size distribution. Of course, there is some sort of spread in terms of measurmental accuracies, almost quite comparable to what we are seeing here.

Green: Ben had quite a dichotomy. Am I correct in reading your paper, that when you suddenly had an upward curvature you lost a peak? And when you had a downward curvature, I guess you had two humps?

Herman: When it was curved this way, we had the slope with the hump on it--the Junge with a hump on it. When it was curved the other way, in other words, high in the middle and low on either end, it gave kind of a log normal. Now the straightest one that we have, which is almost straight but not perfectly straight, gives not quite Junge, a little bit of deviation from Junge. But don't forget that Junge will give a straight line from zero to infinity in size. Our size limits are finite so you wouldn't expect it to be a perfect relationship.

Pepin: I think that one has to think of the fact that the SAGE experiment is designed to look at the stratosphere. When you are looking at the stratosphere, then you have to ask yourself what kind of aerosols are present in the stratosphere in terms of size distribution. You have to go back to the type of data that was presented in the paper by Harris and Rosen concerning the distribution of sizes that have been measured in the stratosphere; those aerosols look very different than those presented in the paper by Ben Herman, which were tropospheric aerosols. Consequently, the regime that the SAGE experiment has to work in is very different.

Fraser: My question has to do with your assumption of spherical symmetry. With respect to the gases, measurements by Lazarus show that you can get a significant change in the gas concentration in the distance of a few hundred kilometers. Also, one of your primary purposes, as I understand it, in this experiment, is to find out something about volcanic dust put into the stratosphere. I would think that there would be significant gradients in the volcanic dust. Have you examined how that affects your data reduction?

Chu: Well, for the ozone problem we have realized the difficulty. We haven't looked into the details yet about the horizontal homogeneity problem and how they are going to affect our inversion results. What we would probably end up with is sort of a decrease in our resolution if the ozone is really fine structured. As far as the aerosols are concerned, particularly volcanic dust, some of the lidar data taken over Langley recently has shown that we had considerably homogeneous layering structure over quite an extended region.

Fraser: I thought a few days ago someone showed a slide with significant changes in one day?

Chu: One day, yes, but over a single experiment period which is of the order of one minute or so, we are looking over a . . .

Fraser: Yes, but you are looking over a horizontal distance effectively of a few hundred kilometers.

Chu: That's right. A few hundred kilometers--we haven't really seen any large gradients.

McCormick: That is correct. For example, we performed lidar measurements from Kansas City, Missouri during July 1975. Simultaneously, the NCAR lidar in Boulder, Colorado, about 600 miles away, obtained data. A comparison of these data showed very little inhomogeneity in the stratospheric dust. I think you are, perhaps, recalling some of the earlier papers showing data taken soon after the del Fuego volcanic eruption. Our experience, however, is that there is very little change during a data-taking period of a few hours. Even soon after the eruption, we experienced significant changes in stratospheric dust only over a period of days, not minutes. I feel the assumption of horizontal homogeneity for the stratospheric aerosol is, therefore, a good one under most atmospheric conditions.

Shettle: He showed a graph of the transmissions. If anything, the transmissions were smaller than the ones that were being measured by ground base measurements because you are dealing with long-slant paths that are showing transmissions in a 50 to 80 percent range.

Drayson: I've got a comment on this idea of measuring the ozone and then going backwards. It seems that here you are really trying to have your cake and eat it too. On the one hand, you are assuming that you know everything about the atmospheric aerosols sufficiently well to find the ozone profile. And then you are turning around and saying, I can put the ozone in and get another parameter about

INVERSION OF STRATOSPHERIC AEROSOL

the atmospheric aerosols. If you don't know the parameters of the atmospheric aerosols well enough in the first place, you can't do the ozone experiment. We have done similar types of calculations for the infrared region of the spectrum and with lots of potential sources of error. The geometry is the same and much of the ideas are the same in here, and we find that the accuracy you get is not nearly as good as the results in today's paper.

Herman: I think all this has been looked at. There are four wavelengths in this experiment. And, they can be solved simultaneously for four pieces of information. You can get the transmission at two wavelengths; you can get the ozone amount and you can get the Rayleigh scatter or any other combination. Anything beyond that, you have to make assumptions.

Drayson: No, I don't think so. If you have done that you haven't got the ozone amount. You've got a parameter, but it isn't the ozone amount, unless you can prove definitely that those other two parameters completely describe the aerosols and the ozone amount comes straight out of it without any further information about the aerosols.

Herman: There are four simultaneous equations and one can solve for four pieces of information.

Drayson: If the atmosphere is described by four pieces of information, then you've got it. But, if it is really described by five pieces of information, then you haven't got your ozone profile but some combination of aerosol and ozone.

Herman: My answer to your statement was that you can get four pieces of data out of this experiment. Any four pieces you want. If you need any more than that, then you have to make assumptions.

Drayson: The question is, do you need more? Can you describe the atmosphere in these four parameters? If not, what you have presented as ozone data is not really ozone data, but is a combination of that plus aerosol parameters.

Herman: You can get ozone data if you make proper assumptions-- enough assumptions to get the aerosol data out of it and vice-versa.

Drayson: But are they correct assumptions?

INVERSION OF SOLAR EXTINCTION DATA FROM THE

APOLLO-SOYUZ TEST PROJECT STRATOSPHERIC

AEROSOL MEASUREMENT (ASTP/SAM)

EXPERIMENT

Theodore J. Pepin
University of Wyoming

The ASTP/SAM Experiment was flown to demonstrate that direct solar occultation measurements by photometers and from photographs can be used for defining stratospheric aerosols. This paper contains a description of the inversion methods that have been used to determine the vertical profile of the extinction coefficient due to the stratospheric aerosols from the data measured during the ASTP/SAM solar occultation experiment. Inversion methods include the "onion skin peel technique" and methods of solving the Fredholm equation for the problem subject to smoothing constraints. The latter of these approaches involves a double inversion scheme that has been employed.

Comparisons are made between the inverted results from the SAM experiment and near simultaneous measurements made by lidar and balloon born dustsonde. The results are used to demonstrate the assumptions required to perform the inversions for aerosols.

I. INTRODUCTION

This experiment was designed to demonstrate the feasibility of remotely sensing aerosols in the stratosphere from a low-orbiting manned spacecraft. Increasing interest in the stratosphere has led to the investigation of methods for remote sensing from Earth-orbiting satellites (Ref. 1). Information gained from the Stratospheric Aerosol Measurement Experiment performed during the Apollo-Soyuz Test Project mission will be used in the design of remote-sensing equipment for future satellite missions (Ref. 2).

The instrument used for making these stratospheric aerosol measurements consisted of a photometer and associated electronics that provided a signal to the command module telemetry system. A Hasselblad data camera (HDC) equipped with special infrared (IR) film and filter was used to photograph the sunset and sunrise events. The experiment technique involved directly measuring solar intensity (photometer) and Sun shape (photographs) in the spectral region centered at approximately 0.84 µm. Immediately before the satellite night, as the spacecraft approached the shadow of the Earth, the line of sight to the Sun passed first through the upper layers of the atmosphere and then steadily down into the lower layers of the troposphere. During the 1.5 minutes (approximately) required for the instrument line of sight to pass through the lower 150 km of the atmosphere, the solar intensity was recorded by the photometer, and solar disk shape changes were photographed. The same set of measurements was made at satellite dawn as the spacecraft emerged from the shadow of the Earth.

The total extinction coefficient was obtained from the variation of the solar intensity as a function of total airmass distributed along the line of sight. At the effective wavelength of the photometer and the photographic system, the extinction was principally produced by the atmospheric aerosols; the measurements are being used to determine the aerosol concentration.

To verify the performance of the SAM Experiment, ground-truth data were acquired by dustsonde (a balloon-borne aerosol optical counter) and a lidar system (ground-based laser radar). The dustsonde was flown from the Richards-Gebaur Air Force base (latitude (lat.) $38.8°N$, longitude (long.) $94.7°W$) near Kansas City, Missouri, at the same time and place of the ASTP second sunset SAM. The lidar measurements were also made from Richards-Gebaur Air Force Base on Earth nights bracketing and during the second sunset SAM.

II. DESCRIPTION OF INSTRUMENTATION

A. The SAM Instrument

The photometer (Fig. 1) used for the SAM Experiment utilizes a pin diode detector having a 10° field of view, and looks at the Sun through the command module window at a wavelength centered at approximately 0.84 µm. The detector was used in the photovoltaic mode to detect the solar signal, and samples were taken at a rate of 10 samples/second using a 12-bit analog-to-digital converter. The signal output of the converter was recorded and transmitted by the command module data system.

Fig. 1. The photometer used for making stratospheric aerosol measurements.

B. Camera System

The photographic portion of the experiment consisted of a series of timed IR photographs of spacecraft sunrises and sunsets taken through the command module window with a HDC. The 250-mm-focal-length (12.5° by 12.5° field of view (FOV)), f/5.6 lens was fitted with a quality glass IR filter capable of obtaining at least three orders of magnitude blocking in the visible and the ultraviolet. Two 50-frame Kodak multispectral IR aerial films (ESTAR-AII base) SO-289 were used in photography. An intervalometer automated the advance of each frame every 2.5 seconds during the experiment.

C. Balloonborne "Dustsonde"

The University of Wyoming balloonborne dustsonde was used in this program for the ground-truth *in situ* aerosol measurement that was made from the Richards-Gebaur Air Force Base (Ref. 3). Air was sampled during balloon ascent and parachute descent with a 2.5 liter light-scattering sizing counter, into the test volume of which a well-defined stream of air is pumped at approximately 0.75 liter/minute. Individual particles scatter light into the microscopes. The light pulses are amplified and by using pulse-height discrimination, the integral concentration of aerosol particles larger than 0.3 and 0.5 μm in diameter can be determined.

The background noise for the system, mainly due to Rayleigh scattering from the air molecules in the chamber at low altitudes and from cosmic-ray scintillation in the photomultiplier glass at high altitudes, was measured approximately every 15 minutes during the flight by passing filtered air through the chamber. The background was negligible above a 10-km altitude; and it was corrected for below this altitude. The dustsonde was also equipped with a rawinsonde temperature element for recording the atmospheric temperature profile.

INVERSION OF SOLAR EXTINCTION DATA

D. Lidar

The lidar measurements for the SAM experiment were provided by the NASA Langley Research Center (LaRC) 122-cm (48-in.) laser radar system, which consists of two temperature-controlled lasers (ruby and neodymium-doped glass) mounted on either side of an f/10 Cassegrainian-configured telescope composed of a 122-cm (48 in.) diamter, f/2, all-metal primary and a 25.4-cm (10 in.) diameter secondary (Ref. 4). The output from the detector package is recorded by a high-speed data acquisition system. Analog signals are amplified and bandwidth limited, digitized at a 5- or 10-MHz rate with 8-bit accuracy, and recorded on magnetic tape. A 16-bit-word storage computer is used to control the data acquisition system and to process the data. An X-band microwave radar, coincident with the laser system axis, is used to ensure safe operation in the atmosphere. A rotating shutter reduces laser fluorescence after Q switching.

III. RESULTS OF OBSERVATIONS

A. SAM Photometry

The SAM Experiment was designed to observe four events on July 22, 1975. During Apollo revolution 95, measurements were made of a sunset observation (00:07:04 Universal Time (UT)) off the coast of New Jersey (lat. $39°10'$ N, long. $72°45'$ W) and of a sunrise observation (00:45:52 UT) over the Indian Ocean off the coast of Australia (lat. 43' S, long. $99°55'$ E). During Apollo revolution 96, a sunset observation (01:37:52 UT) was taken near Kansas City, Missouri (lat. $38°57'$ N, long. $95°06'$ W), followed by a sunrise observation (02:14:39 UT) over the Indian Ocean (lat. $42°55'$ S, long. $77°39'$ E). The SAM photometer was used to obtain radiometric measurements during each of these four events, and the measured photometric intensities have been inverted by using the inversion procedures outlined below to obtain total extinction as a function of altitude. Figure 2 shows the sub satellite tracks for the 95th and 96th revolutions and the locations of the tangent points of the 0-km altitude grazing ray of the Sun for the four events.

B. Photography

Photographs were made of the first sunrise and the second sunset with the HDC. Figure 3 is a composite of frames AST-28-2400 to AST-28-2406 taken during the first sunrise. These photographs have been printed with high contrast to show the observed refracted images of the solar disk and have been superimposed on a grid showing the horizon and the tangent altitudes. Figure 4 shows frame AST-28-2402, an "isodensity tracing" of this frame, and an outline of the theoretical computed shape of the Sun as expected for the orbital and atmospheric conditions present. The distribution of the observed isophotes is due to limb darkening, extinction by atmospheric constituents, and the effects of refraction. These photographs confirm, at spacecraft altitude, that effects due to refraction must be considered in the design of solar

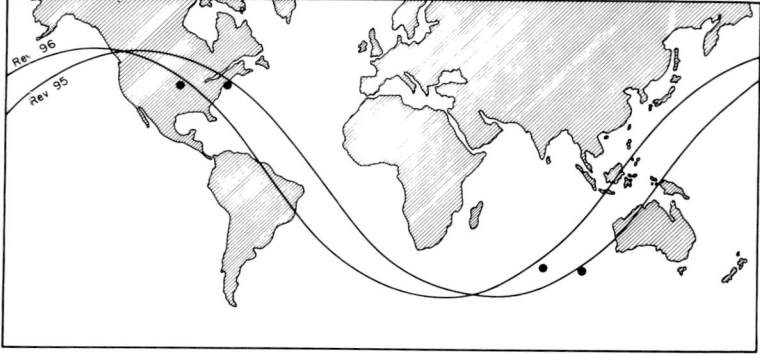

Fig. 2. Sub-satellite tracks for SAM observations.

INVERSION OF SOLAR EXTINCTION DATA 535

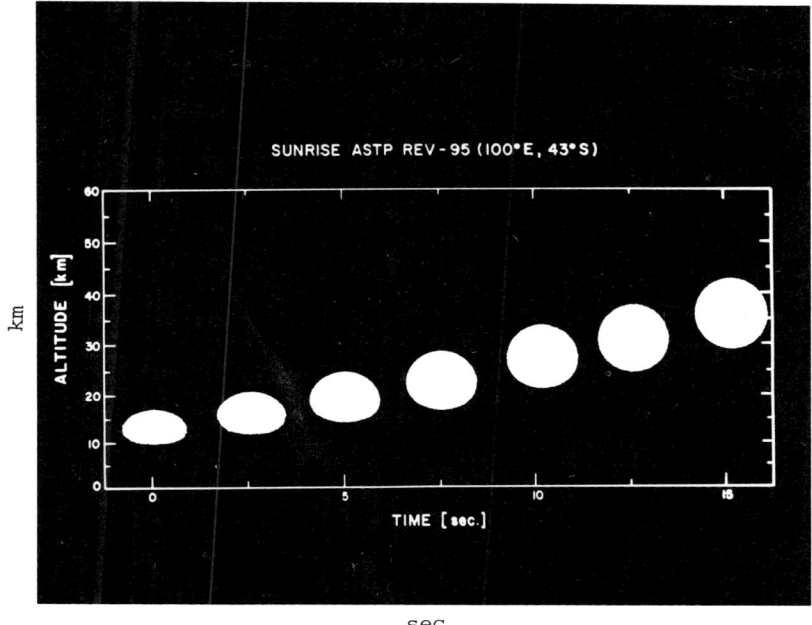

Fig. 3. Composite of photographs taken during first sunrise event (AST-28-2400 to 2406).

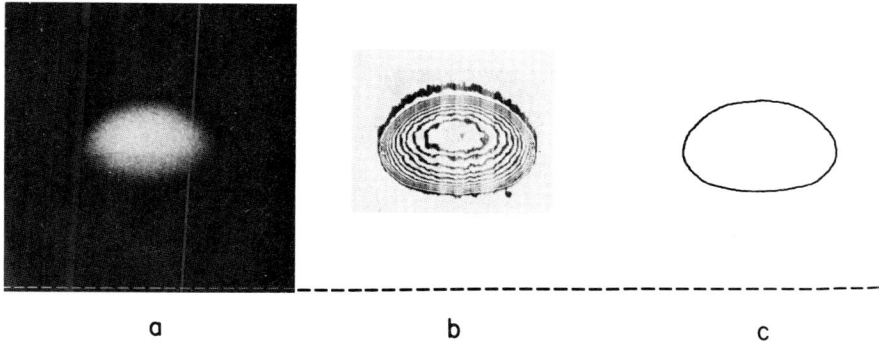

Fig. 4. Sun-shape comparison. (a) High-contrast sun photograph (AST-28-2402); (b) Isodensity tracing; and (c) Expected refracted sun shape.

occultation experiments. The photographs taken during the SAM events are currently being analyzed to substantiate theoretical models that are under development for use in future occultation experiments.

C. Ground-Truth Data

Data from the dustsonde balloon flight launched on July 21, 1975, at 23:58 UT are shown in Figs. 5 to 7. Figure 5 contains the measured temperature profile, Fig. 6 illustrates the measured aerosol concentration as a function of altitude for particles larger than 0.3 μm in diameter, and Fig. 7 shows the aerosol count ratio (ratio of particles greater than 0.3 μm in diameter to particles greater than 0.5 μm in diameter) as a function of altitude.

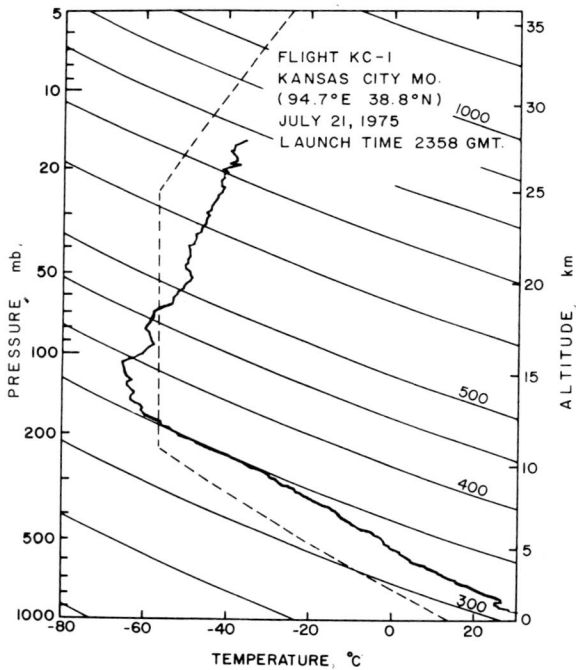

Fig. 5. Measured temperature profile above Kansas City, Missouri.

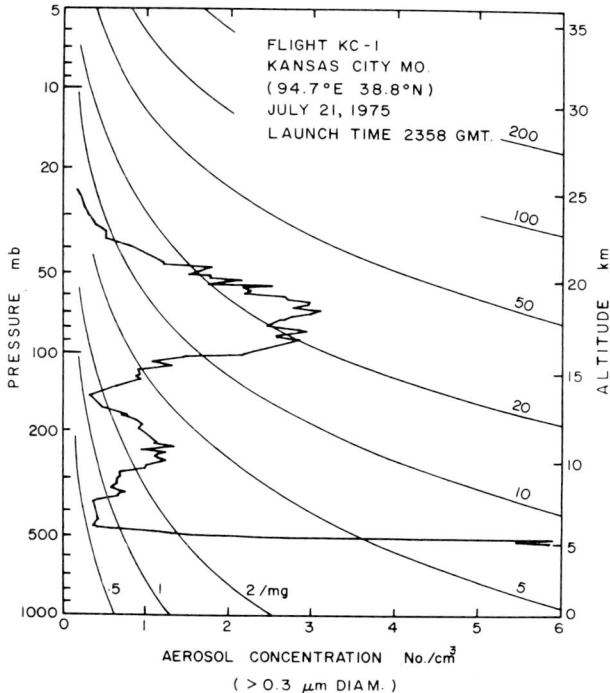

Fig. 6. Measured dustsonde aerosol concentration (>0.3 μm diameter) above Kansas City, Missouri. Curved lines indicate mixing ratio.

The LaRC lidar system was used during the nights of July 22 and July 23 to obtain laser backscatter measurements of the stratospheric aerosols. Figure 8 shows the backscatter ratios obtained during the measurements made on July 23 at 07:30 UT, and Fig. 9 shows the measured backscatter ratios made on July 22 at 04:51 UT. (Backscatter ratio is the ratio of total observed backscatter to molecular backscatter.)

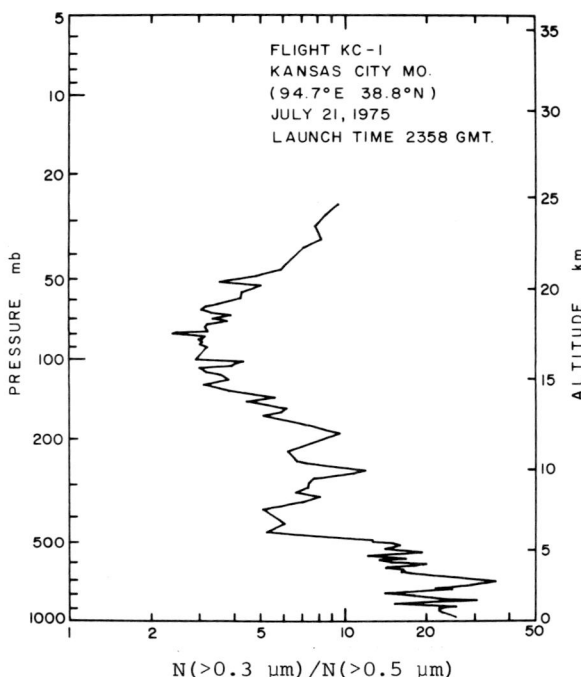

Fig. 7. Ratio of measured aerosol count for particles >0.3 μm diameter to particles >0.5 μm diameter above Kansas City, Missouri.

IV. METHODS OF INVERSION

Two methods have been employed for the inversion of the SAM photometric measurements in order to obtain stratospheric aerosol extinction. Both of the methods assume a model of the stratosphere that contains horizontal homogeneity. The first method has been referred to as the onion skin peel inversion and allows for the analytical inversion of the results. The second method involves a technique of double inversion.

A. Onion Skin Peel Inversion

For this method of inversion, the atmosphere is divided into a number of homogeneous, spherical layers that are considered to

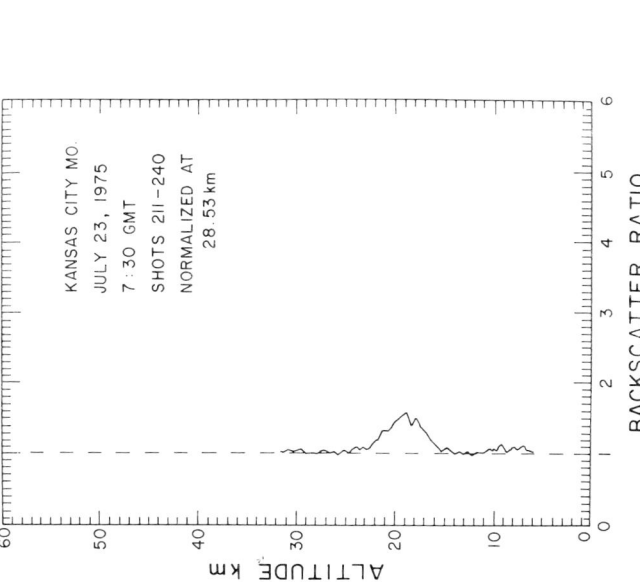

Fig. 8. Profile of lidar aerosol backscatter ratio taken at Kansas City, Missouri, on July 25, 1975, as 07:30 UT (normalized at 28.53 km).

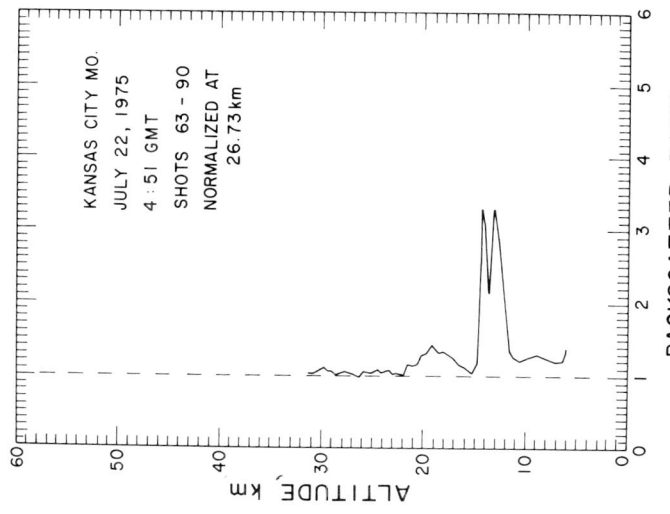

Fig. 9. Profile of lidar aerosol backscatter ratio taken at Kansas City, Missouri, on July 22, 1975, at 04:51 UT.

have uniform aerosol concentrations. Figure 10 shows a cross section of the measurement geometry and illustrates the path lengths of solar rays at different atmospheric layers. Figure 11 shows the inversion geometry and illustrates the solar intensity contribution transmitted through the different layers as seen from the ASTP instrument during a sunset event.

As the bottom limb of the Sun becomes tangent to each of the layers of the onion skin atmospheric model, the intensity can be computed from limb darkening and refraction by summing the intensities over the layers above. For example, following the notation for the layers identified in Figs. 10 and 11, one finds that when the lower limb is tangent to the first and second layers, the observed solar fluxes are given by

$$\left. \begin{array}{l} F_1 = I_{10}^{\Theta} + I_{11} \\ \\ F_2 = I_{20}^{\Theta} + I_{21} + I_{22} \end{array} \right\} \quad (1)$$

where, in general, F_1 and F_2 are the observed fluxes and I_{10}^{Θ} and I_{20}^{Θ} are the intensities above the upper layer of the inversion model (above the point at which extinction is observed). The intensities I_{11}, I_{21}, and I_{22} are the transmitted intensities to be determined by inversion. By using the Lambert-Beer law, Eq. (1) can be written in the form

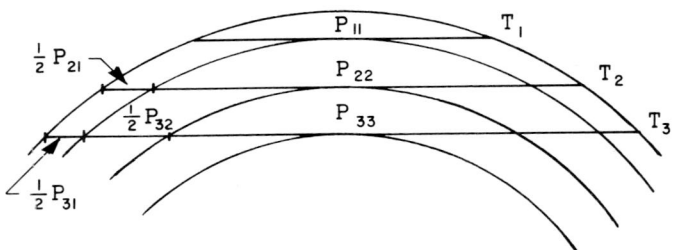

Fig. 10. Cross section of the onion skin atmospheric model showing the ray path lengths, where P_{ij} is the path length in the jth layer for the solar ray tangent to the ith layer.

INVERSION OF SOLAR EXTINCTION DATA

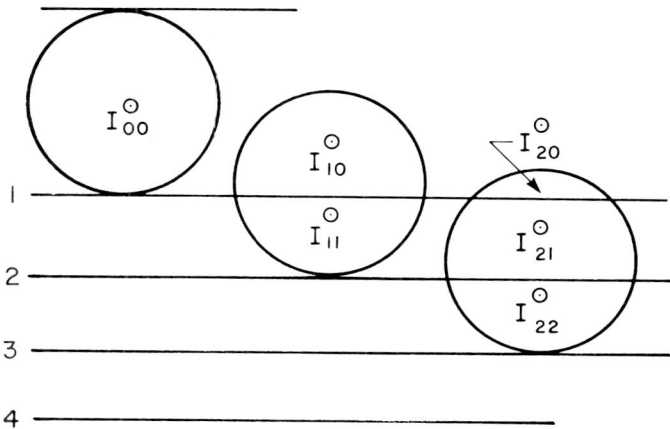

Fig. 11. Inversion geometry showing the contributions to the total transmitted Sun intensity, where I_n is the observed intensity for the nth layer, I_{11}, I_{21}, and I_{22} are the transmitted intensities to be determined by inversion, and I_{ni}^{Θ} is the portion of the unattenuated solar intensity that contributes to I_n.

$$\left. \begin{array}{l} F_1 = I_{10}^{\Theta} + I_{11}^{\Theta} \exp(-\beta_1 P_{11}) \\ \\ F_2 = I_{20}^{\Theta} + I_{21}^{\Theta} \exp(-\beta_1 P_{11}) + I_{22}^{\Theta} \exp(-\beta_1 P_{21} - \beta_2 P_{22}) \end{array} \right\} \quad (2)$$

where I_{ni}^{Θ} is the portion of the unattenuated solar intensity through the ith layer that contributes to the observed flux F_n; P_{ij} is the path length in the jth layer for the solar ray tangent to the ith layer; and β_i is the total extinction at the wavelength of the SAM system in the ith layer. Equation (2) yields

$$\left. \begin{array}{l} \beta_1 = \dfrac{-1}{P_{11}} \ln \left(\dfrac{I_1 - I_{10}^{\Theta}}{I_{11}^{\Theta}} \right) \\ \\ \beta_2 = \dfrac{-1}{P_{22}} \ln \left(\dfrac{I_2 - I_{20}^{\Theta} - I_{21}^{\Theta} \exp(-\beta_1 P_{11})}{I_{22}^{\Theta}} \right) - \beta_1 \dfrac{P_{21}}{P_{22}} \end{array} \right\} \quad (3)$$

This process can be continued as the sunset event occurs; thus, allowing for the determination of the vertical profile of the total extinction. Figures 12 and 13 show the results of the onion skin inversions for the ASTP sunrise and sunset events.

B. Technique of Double Inversion

Using the measurements of the solar flux from the full solar disk, during a sunrise or sunset event, the vertical profile of the aerosol volume extinction coefficient can be determined by a series of solutions of two integral equations. Each of the integral equations can be transformed into integral equations of the first kind and, therefore, lend themselves to solution by standard inversion methods.

The first of these equations is

$$F(t) = \int\int_W I^\Theta(t, x, y) T(x) \cos \phi \, dw \qquad (4)$$

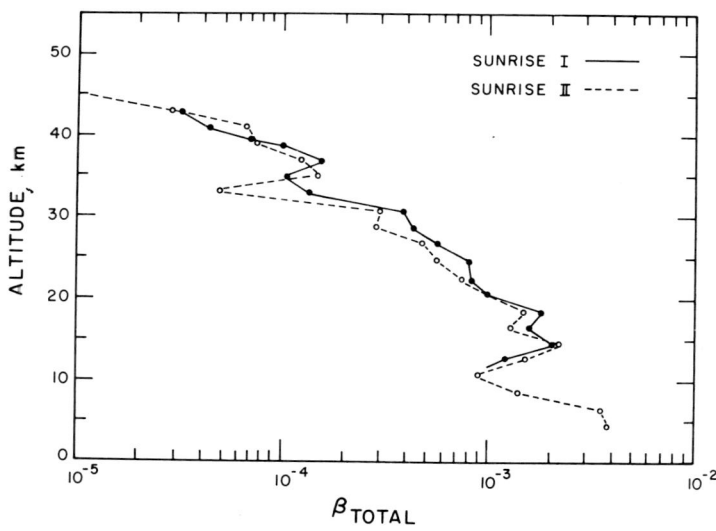

Fig. 12. Results of onion skin inversions for β_{TOTAL} for sunrise events.

INVERSION OF SOLAR EXTINCTION DATA

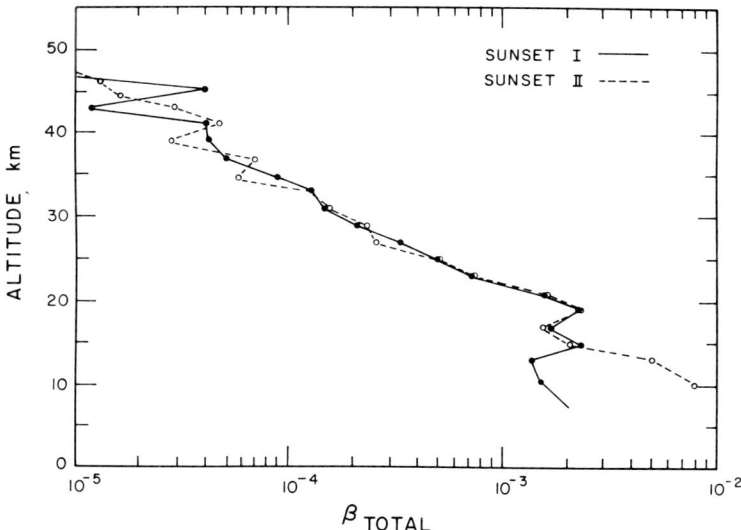

Fig. 13. Results of onion skin inversions for β_{TOTAL} for sunset events.

where F(t) is the time-resolved measured flux, w is the detector field of view and x and y are orthogonal coordinates in the field of view. The y axis has been taken tangential to the Earth's surface. The intensity distribution of the solar disk is given by $I^{\Theta}(t, x, y)$, transmission through the Earth's atmosphere by $T(x)$, and the angle between $I^{\Theta}(t, x, y)$ and the detectors normal by ϕ.

Equation (4) can be reduced to

$$F(t) = \int_{X_o}^{X_m} I^{\Theta}(t, x) T(x) dx \quad (5)$$

by integrating $I^{\Theta}(t, x, y)$ over y to find the function $I^{\Theta}(t, x)$.

Equation (5) is recognized as an integral equation of the first kind and can be written in quadrature form as

$$g(t) = A(t, x) f(x) \quad (6)$$

where

$$\left.\begin{aligned} g(t) &= F(t) \\ A(t, x) &= W(x) \, I^0(t, x) \\ f(x) &= T(x) \end{aligned}\right\} \qquad (7)$$

Equation (6) can be solved by the well-known constrained linear inversion technique developed by Phillips (Ref. 5) and Twomey (Ref. 6)

$$f = (\underline{A}^T \underline{A} + \gamma \, \underline{DH})^{-1} \underline{A}^T g \qquad (8)$$

The second integral equation involves the volume extinction coefficient $\beta(Z)$ for the altitude Z and can be found by considering the transmission along the path P which is given by Beer's law

$$T(x) = e^{-\int c \, \beta(Z) \, dP} \qquad (9)$$

It involves the evaluation of the line integral and can be put in the form of an integral equation of the first kind as

$$\ln\left(\frac{1}{T(x)}\right) = \int_0^{Z_m} \beta(Z) \frac{dP}{dZ}(t, Z) \, dZ \qquad (10)$$

Equation (10) can be put in the form of Eq. (6) by making the substitution

$$\left.\begin{aligned} g(t) &= \ln\left(\frac{1}{T(x)}\right) \\ A(t, Z) &= w(Z) P(t, Z) \\ f(Z) &= \beta(Z) \end{aligned}\right\} \qquad (11)$$

and can be solved for $\beta(Z)$ by the same method as the first of the equations of the inversion.

The linear inversion employed to obtain the solutions for both the transmission and the extinction coefficient made use of smoothing constraints ($\underline{H}$ being the smoothing matrix). That is,

INVERSION OF SOLAR EXTINCTION DATA

the solution was constrained so that the second derivitive of the function was minimized. In addition to this constraint, the diagonal matrix $\underset{\sim}{D}$ allowed for smoothing to be applied to the solution as a function of altitude. The elements of the $\underset{\sim}{D}$ matrix were calculated from the measured fluxes for the first inversion and from the transmissions for the second inversion. They reflect the percent error in each measurement. The larger the measured fluxes the smaller the corresponding extinction coefficients. The elements of the $\underset{\sim}{D}$ matrix apply smoothing differentially with altitude.

Figure 14 shows the inversions for $\beta_{AEROSOL}$ obtained from two sunset events. The molecular extinction coefficient $\beta_m(Z)$ was calculated from Rayleigh scattering theory for the effective wavelength of the SAM system. For the Sunset II observation, the balloon measured temperature profile was used to determine $\beta_m(Z)$.

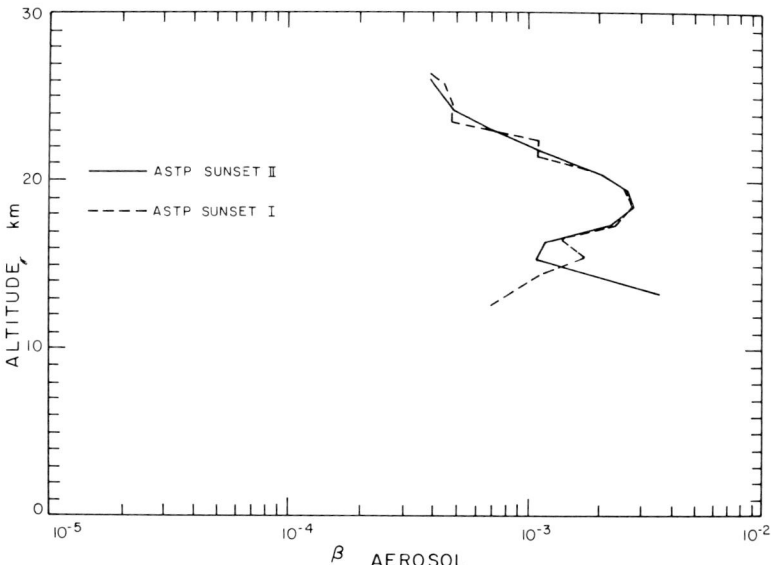

Fig. 14. Results of double inversion for $\beta_{AEROSOL}$ for sunset events.

For the sunset I observation, a standard atmosphere was used. $\beta_{Aerosol}$ was then determined by subtracting the molecular contribution from the inverted results.

$$\beta_{Aerosol}(Z) = \beta(Z) - \beta_m(Z)$$

V. COMPARISON OF RESULTS

The balloon dustsonde measurements of number density of particles greater than 0.3 μm diameter and the aerosol count ratio have been used as a function of altitude to fit various size distributions. These size distributions have been used to predict the SAM photometer measurements of the aerosol extinction observed during the second sunset event and the lidar observations of the backscatter ratio. Figure 15 shows a comparison between a plot of the SAM inversion results obtained for the second sunset event using the onion skin peel inversion method and the Mie-scattering computations performed using the log-normal aerosol size distribution developed by Pinnick et al. (Ref. 7). Figure 16 compares the results of the double inversion technique with the Mie-scattering computations. This distribution was adjusted as a function of altitude so that the total number and mode radius fit the dustsonde observations. For these calculations, the index of refraction for the stratospheric aerosol was taken to be $1.43 - i(0.0)$, $1.50 - i(0.0)$, and $1.60 - i(0.0)$ (where $i = \sqrt{-1}$).

Figure 17 shows the comparison between the lidar backscatter ratios observed on the night of July 22 and the computed backscatter ratios for index of refraction $1.43 - i(0.0)$, $1.50 - i(0.0)$, and $1.60 - i(0.0)$. These backscatter ratios were computed by using the same size distribution fits to the dustsonde observations that were used for the previously mentioned SAM extinction dustsonde comparisons. The large enhancement at approximately 13 km in the lidar data is due to cirrus clouds.

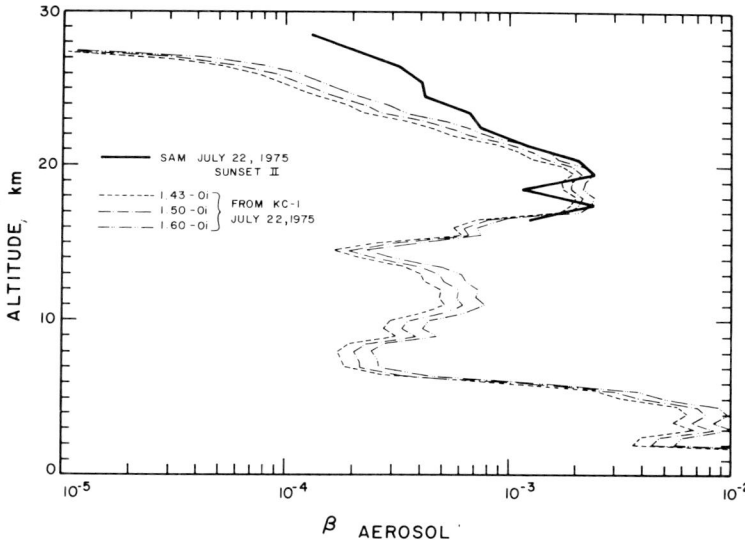

Fig. 15. Comparison of SAM results from onion skin inversion on July 22, 1975 (second sunset event), and Mie computations of dustsonde results on July 22, 1975, for different optical properties of the aerosols ($\beta_{AEROSOL}$ = aerosol extinction).

Results of the three different techniques agree in the placement of the peak altitude of aerosol concentration, and they are consistent with a single size distribution and particle refractive index. The author would like to point out that the log-normal distribution used to fit these observations is not necessarily unique. Other distributions might fit as well.

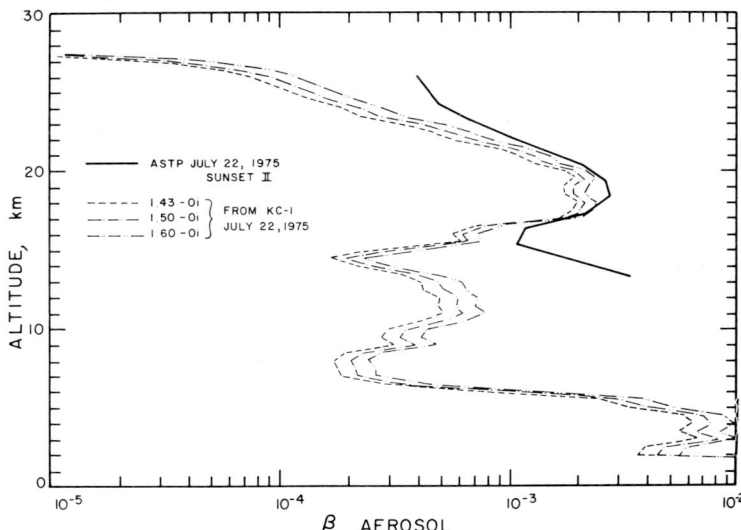

Fig. 16. Comparison of SAM results from double inversion on July 22, 1975 (second sunset event), and Mie computations of dust-sonde results on July 22, 1975, for different optical properties of the aerosols ($\beta_{AEROSOL}$ = aerosol extinction).

INVERSION OF SOLAR EXTINCTION DATA 549

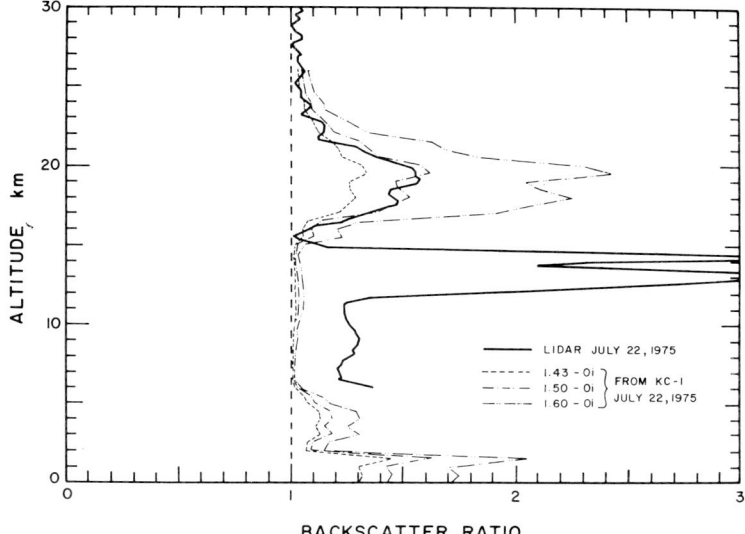

Fig. 17. Comparison of lidar results of July 22, 1975, and Mie computations of dustsonde results on July 22, 1975, for different optical properties of the aerosols.

SYMBOLS

$\underset{\sim}{D}$	diagonal matrix
F_n	flux in nth layer
$\underset{\sim}{H}$	smoothing matrix
I_n	intensity in nth layer
$I_{ni}^{\odot}$	portion of the unattenuated solar intensity through the ith layer that contributes to the observed flux F_n
n	layer number
P	path length
P_{ij}	path length in the jth layer for ray tangent to the ith layer
t	time resolved measured flux
T_i	transmittance in the ith layer

w	detector field of view
x	orthogonal coordinate in field of view
y	orthogonal coordinate in field of view
Z	altitude
$\beta(Z)$	total volume extinction coefficient km^{-1} at altitude Z
$\beta_{Aerosol}(Z)$	aerosol volume extinction coefficient km^{-1} at altitude Z
$\beta_m(Z)$	molecular volume extinction coefficient km^{-1} at altitude Z
β_n	total extinction at wavelength of SAM system in nth layer
ϕ	intensity distribution of solar disk
θ	angle between I (t, x, y) and detectors normal
$\odot$	superscript denoting Sun

ACKNOWLEDGMENT

This work was supported under contract NAS-13213 from NASA Langley Research Center. The author would also like to acknowledge the help of Mrs. F. Simon and Mr. T. Cerni for their help in the inversion of the ASTP/SAM results contained in this paper.

REFERENCES

1. T. J. Pepin, Remote-sensing the Stratospheric Aerosols, *Proc. of the 2d Joint Conference on Sensing Environmental Pollutants Instrument Society of America, JSP 6715*, 1973, p. 333.

2. T. J. Pepin, M. P. McCormick, W. P. Chu, F. Simon, T. J. Swissler, R. R. Adams, W. H. Fuller, Jr., and K. H. Crumbly, "Stratospheric Aerosol Measurement," in *ASTP Summary Science Report, Vol. 1, NASA SP 412*, 1977.

3. James Rosen, Simultaneous dust and ozone soundings over North and Central America, *J. Geophys. Res. 73*, 479 (1968).

4. M. P. McCormick and W. H. Fuller, Jr., Lidar Applications to Pollution Studies, *Paper presented at Joint Conference on Sensing of Environmental Pollutants, Palo Alto, California,* November 8-10, 1971 (AIAA Paper 71-1056).

5. D. L. Phillips, A technique for the numerical solution of certain integral equations of the first kind, *J. Assoc. Comp.*, March 9, 1962, p. 84.

6. S. Twomey, On the numerical solution of Fredholm integral equations of the first kind by the inversion of the linear system produced by quadrature, *J. Assoc. Comp.*, March 10, 1963, p. 97.

7. R. G. Pinnick, J. M. Rosen, and D. J. Hofmann, Stratospheric aerosol measurements III: Optical model calculations, *J. Atmos. Sci. 33,* 304 (1976).

DISCUSSIONS

Shettle: On your deviation at the upper altitudes, one possibility I think there could be some additional large particles. Some of the curves that Franklin Harris showed indicated size distributions that are suggestive of a kink with a somewhat less deep slope. You are measuring particles .15 and .25 µm there and when you get out to one micrometer, the number densities are sufficiently low that they don't show up at all in your data.

Pepin: As you will notice,[1] the size ratio is getting larger at high altitudes. The size ratio is the number of small particles to the number of large particles. And, consequently, the balloon data indicates that the distribution is getting steeper rather than flatter at the high altitudes.

Shettle: Yes. I am suggesting a second component which mainly contributes out in one micrometer region.

Pepin: That would be very interesting if there is a second size distribution at high altitudes.

Shettle: It is suggestive of meteoric dust coming down. I know I talked about it earlier this week. Another possibility that the size of the solar disk is getting larger that you see. What about possible uncertainty in the intensity across the solar disk?

Pepin: The brightness theorem has to be satisfied and the refraction effects are getting very small at those altitudes. Consequently, the errors that one would make in making an estimate are going to zero very quickly at those altitudes. It is low in the atmosphere where the refractive effects are large.

Park: How much do your inversion results depend on the pressure profile you adopt in the retrieval?

Pepin: It is independent of pressure. All I have to do is to know roughly where the ray goes through the atmosphere.

Park: But you have to know the pressure profile.

Pepin: To first order, then I can determine the entire atmosphere independent of the geometrical properties of the atmosphere, because the spacecraft scans with a uniform rate across the atmosphere. It is only the second order coupling with the refracting geometry that produces the errors in the experiment.

---

[1] See Fig. 7 in this paper.

INVERSION OF SOLAR EXTINCTION DATA

Fraser: In your first figure, I think you showed the refraction effects of the sunlight. It looked to me as though when the sun got closer to the horizon, the horizontal dimensions decreased significantly. Why was that?

Pepin: (Showing graph[2]). I think maybe that was an optical illusion, because you can very carefully compare the horizontal distances across each of these images.

Fraser: Are they the same?

Pepin: Yes, they are the same.

Green: Aren't stratospheric aerosols predominantly smaller than tropospheric aerosols?

Pepin: Yes, I think they are very often. The peak seems to be at small sizes for stratospheric aerosols. Whereas Ben showed with his inversions very often there is a second peak for the tropospheric aerosols--move out to something of the order of half to one micron in size.

Green: So might you not be in a regime where the smaller particles would best be described by $1/\lambda^4$, like Rayleigh particles, so that perhaps a better parameterization would be something which interpolates $1/\lambda$ to $1/\lambda^4$?

Twitty: I think we have seen a lot of evidence in the last hour that indicates that the stratospheric aerosols variation of its scattering coefficient with eavelength is more like $1/\lambda$, not $1/\lambda^4$.

Green: If you have lots of small particles, you would be analogous to the situation which Ben was indicating. Perhaps a better two-parameter description would be--not a single power law--but one that interpolates between two powers. You could perhaps still find a two-parameter description that would do that, but. . .

Twitty: Well, you could parameterize it that way. The argument earlier was that you cannot possibly get any information about the two parameters out. The question of what you are putting into your model that you think is the closest representation is open to your own decision. I think you talked about nonpower laws in your simulated analysis; you are using log normal laws.

---

[2] See Fig. 3 in this paper.

Pepin: Yes, in fact, I think you can just look at this measurement that was taken in Kansas City. It is typical of aerosols in the stratosphere. You will notice, in the stratosphere, that between the 0.3 and 0.5-µm diameter size particles, there is a factor of like 4 or 5 in terms of the number of particles larger than that size. If you go to 0.01, the number is ten. So what this is saying is that the distribution has peaked around the 0.1 to 0.2 micrometer region.

EFFECTIVE AEROSOL OPTICAL PARAMETERS FROM

POLARIMETER MEASUREMENTS

Jacob G. Kuriyan
University of California

In this paper, the theory underlying the interpretation of polarimeter measurements is described. The assumptions of the model are carefully stated so that the results obtained from the ground-based experiment can be understood without ambiguity. The meteorological significance of the parameters is also deduced in the paper. With a satellite-borne polarimeter that monitors the upwelling radiation field, the effect of the ground must be taken into account in order to obtain the aerosol parameters. Two methods that hold promise are described.

I. INTRODUCTION

A great deal of attention has centered on the impact of artificial and natural pollutants on man. Of particular interest to meteorologists is their effect on the heat budget of the atmosphere so as to estimate the modifications they induce on weather and climate. This paper shall address itself to the determination of the optical properties of atmospheric particulates that are of significance to meteorologists.

The atmosphere is often likened to an engine driven by the solar energy that is absorbed by the Earth-atmosphere system. In numerical studies of the circulation of the atmosphere, the heat budget of the atmosphere is a significant input to the problem. As the general circulation models have become more sophisticated, it has become necessary to take into account the radiative effects

of aerosols, which can be viewed as a perturbative correction to
the other terms in the heat budget.

The aerosol effects are but a minor (yet significant, in some
cases) correction to the total heat budget calculation, and so,
unless the scheme that is devised to incorporate these effects is
simple, meteorologists will find it easy to continue to ignore
their effects. Theoretical techniques are available for calculation of radiative properties of aerosols of pathological shapes
and distributions, but we must resist the temptation of using these
models unless irrefutable evidence is presented that simpler models
are insufficient. The needs of the user (in the present or foreseeable future) dictates that simple models be used and then
empirical formulae (involving the aerosol model parameters) for the
radiative effects be derived so as to be used in the circulation
models. We may compare this to a method sometimes employed in
French cuisine: a piece of pheasant meat is cooked between two
slices of veal, which are then discarded (Ref. 1). The experimental determination of the model parameters will then assist in
the heat budget calculation.

Most of the radiative transfer programs model aerosols as a
collection of spherical (Mie) scatterers, with a uniform refractive index and their size distribution described by a function $n(r)$.
The parameters of the model are then the refractive index m (in
general complex) and parameters in the size distribution function.
In reality, atmospheric aerosol particles are not all spherical
nor do they have a uniform refractive index. It is often argued
that aerosol particles, no matter what their shape, acquire a water
coating and become spherical. This heuristic argument, reasonable
though it may be, should not be confused as a substitute for proof
of our hypothesis. Rigorously speaking, the intent of the approach
adopted here is to describe the effects of atmospheric aerosols in
terms of a simple model. There is no prior assurance that it is
possible to incorporate all the observed aerosol effects in such

AEROSOL OPTICAL PARAMETERS

a simple model. Thus, initially, the experiments must be used to justify the model. That is to say, not only must the model parameters be determined, but the experimental proof of the self-consistency of the method must be displayed. If the results of the simple model cannot account for the measurements then, and only then, is the consideration of more complicated aerosol models justified.

To digress for a moment, in the literature, the laboratory measurements of scattering off irregular particles floating in a nitrogen jet are compared with the scattering patterns of spherical particles. The lack of agreement is then used as a justification for considering horrendously complicated aerosol models that can only be handled by statistical methods. But it will be found that the ground-based measurements of the scattered radiation field in the atmosphere (that have been obtained at UCLA for the last two years) can, indeed, be explained in terms of the simple model described. Perhaps it is safe to conclude that irregular particles floating in a nitrogen jet do not simulate the behavior of the atmosphere.

The experiments conducted by the late Professor Z. Sekera at UCLA established that the presence of particulates in the atmosphere alters the polarization characteristic of skylight. How the polarization measurements can be used to infer the aerosol model parameters, and the self-consistency checks that have been devised to verify the validity of the assumptions of the aerosol model are described.

The method of approach is to construct the radiation field for all possible values of the aerosol parameters. Admittedly, this is primitive, and the viability of the method hinges crucially on the table being complete and on the development of an efficient algorithm that can help match measurements and calculations. Redundancy in the model can increase the size of the tables and frustrate the search for a fit.

II. CONSTRUCTION OF THE TABLES AND CHOICE OF A SIZE DISTRIBUTION

The compilation of a complete table of radiation field in a turbid atmosphere necessarily involves the identification of a model that is of sufficient generality that it can accommodate the commonly occurring radiation field patterns. Consistent with our philosophy of selecting the simplest such model, we considered the power law distribution due to Junge, $n(r) = a/r^{\nu + 1}$, a deceptively simple functional representation. It is often assumed that there are only two parameters in this representation, a (proportional to the concentration) and ν (that determines the distribution in sizes). Actually, however, the integrals involving n(r) require a cut-off at r_{min} and/or r_{max}, and the results can be sensitive to the value of the cut-off point, i.e., the cut-off point becomes a parameter in the model. This deficiency of the power law distribution is not well recognized in the literature and so it would be worth while to illustrate this with an example. The optical depth τ in our model is given by $\tau = \pi \int r^2 n(r) Q_{ext}(x) \, dr$, where Q_{ext} is the van de Hulst extinction efficiency factor. Let us examine the integral at the lower limit of integration. For small values of r, when m is complex, from van de Hulst's book we find that $Q_{ext}(x) \approx x$, where $x = kr$. Thus, the integrand in the expression for τ behaves like $1/r^{\nu - 2}$. When $\nu > 3$, a condition that occurs quite often, the lower limit of the integral will behave worse than $\ell n \, r_{min}$. Thus, when $\nu > 3$, r_{min} can become a significant parameter in the calculation of τ. This is an undesirable and unphysical restriction on this model.

The modified gamma distribution used by Deirmendjian is well behaved at both limits of r and his analysis of commonly occurring aerosols led him to classify them into three types--haze $H(a_H r^2 \exp(-b_H r))$, haze $L(a_L r^2 \exp(-b_L \sqrt{r}))$ and haze $M(a_M r \exp(-b_M \sqrt{r}))$, with a_H, a_L, a_M, b_H, b_L, and b_M being assigned fixed values. If we accept this classification of Deirmendjian, then the most general aerosol distribution would be an admixture of these three types and we would be led to a tri-modal distribution with the modes determined by the values of the various b's. The admixture ratio would have to be height dependent and, thus, the problem will be complicated even further. This goes against the grain of our philosophy, which is to start with the simplest possible model at the outset, and introduce complications only when the simpler models are found to be deficient in accounting for the measurements.

Using analytical approximations, we investigated (Ref. 2) the redundancy of the description of aerosols and we found that the gamma (or the haze H *type*) distribution $n(r) = a\,r^2 \exp(-b\,r)$, with a and b considered as arbitrary parameters could simulate the results due to haze M or haze L. This feature was found to persist even when the exact (Mie) theory was used. That is to say, the haze H *type* distribution could produce the same phase matrix element as the three haze distributions used by Deirmendjian. Thus, for a start, we are led to use the haze H *type* distribution, with two parameters, a (proportional to concentration) and b (inversely related to the modal radius), and our table entries would correspond to various values of these and other parameters (such as refractive index, wavelength, ground albedo, etc.). We are at liberty to expand the table by the inclusion of other independent models, if experiments warrant it.

It is important to appreciate the implications of the redundancy that is described in the preceding paragraph. Many different size distributions give rise to the same phase matrix element and, hence, the same radiation field. Thus, from a measurement of the radiation field, we will not be led to a unique size distribution but to a class of distributions, all equivalent to one another in that they correspond to the same radiation field (and, hence, fluxes). This observation is perfectly in consonance with our attempt to model the actual atmospheric aerosols in terms of an idealized collection of spherical scatterers, and what we are observing is that there is still another degeneracy allowed in these models, in that even the spherical scatterers can have different size distributions and yet yield the same phase matrix element. Needless to say, we will expect that the data can be fitted by other distributions, such as log-normal or multimodal, but in all cases the fluxes derived must be the same. (This statement on fluxes has been proved only for the Deirmendjian distributions. It is anticipated that the result will hold for other distributions as well.)

Let us then enumerate the assumptions inherent in the equivalent aerosol model and in the calculation of the tables of radiation field.

1. Aerosols are spherical scatterers and so Mie theory can be used.

2. Aerosols have a uniform refractive index.

3. Size distribution of aerosols is given by $n(r) = a\,r^2 \exp(-br)$ where the concentration (related to a) varies with height.

4. The atmosphere is assumed plane parallel and so there is a horizontal homogeneity of aerosols.

5. Multiple scattering code (that due to J.V. Dave, developed for NASA) is used to calculate the radiation field.

6. The ground reflection is assumed to be Lambertian, an assumption that is not serious in the study of the downwelling radiation but quite unacceptable for the upwelling radiation studies (as in a satellite borne experiment). We will comment on this later in the paper.

The multiple scattering calculations are performed iteratively. When the size of the angular integration is small, the error decreases but the cost of computation increases. Thus, there is an optimal angular integration step that is used in the calculation of the entries of the table and, corresponding to this, there is an error in the calculated radiation field. Any fit with experiment need be only consistent within these error bounds.

So as to develop the algorithm, we studied the sensitivity of the radiation field to the variation of the aerosol model parameters (Refs. 3, 4). For our analysis, we considered a conical scan of the sky, with a fixed zenith angle of observation and variation of the azimuth angle ϕ (with respect to the Sun plane) as in Fig. 1. We did not include the almucantar sweep since our polarimeter was not designed to view the direct beam of the Sun. In this type of a scan the intensity curve (I vs. ϕ) exhibits a bell-shaped structure, and it was found that the half-width of the intensity curve was determined mainly by b. [This is a physically reasonable condition since the width of the intensity curve is the forward peak of scattering and that is sensitive to the size of the particles; b is a parameter that describes (the inverse of) the average (modal) radius of the distribution.] Thus, our algorithm consists of choosing a reasonable value (but fixed for a location) for the ground albedo, then determining b by examining the width of the curve of I against ϕ, and then determining τ and m by examining the curve of P against ϕ, and the tail of the curve of I against ϕ. This is actually a graphical algorithm more difficult to explain than to use. Numerical simulations were used to check the algorithm that are more difficult to explain than to use. Numerical simulations were used to check the algorithm. These were found satisfactory and, thus, we were ready to examine the field of measurements.

III. THE FIELD MEASUREMENT PROGRAM AT UCLA

A ground-based polarimeter was used to measure the intensity and polarization of skylight. The scan of the sky was the conical scan described in Fig. 1. The measurements were made at 2 wavelengths and at several (conical scans) zenith angles of observation. From a measurement of the I and P in *one* conical scan (at various azimuth angles), we infer the aerosol parameters using the algorithm that was developed. These inferred aerosol parameters can be inserted in the radiative transfer code and the radiation

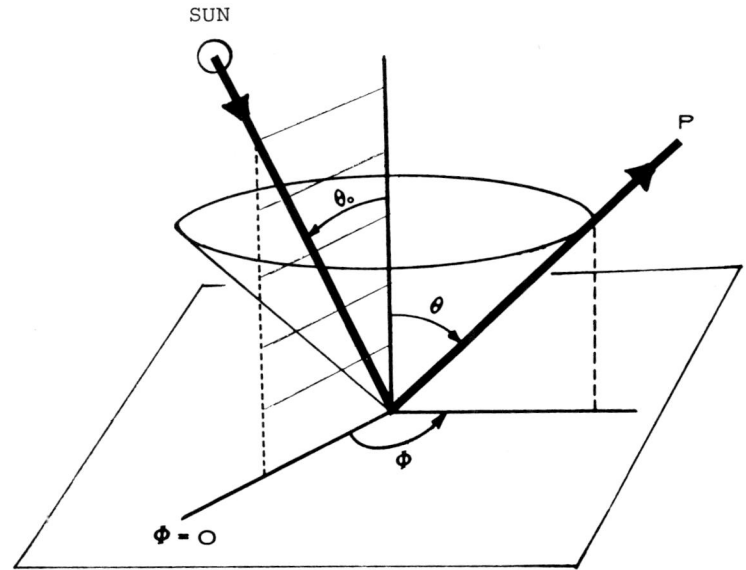

Fig. 1. The conical scan mode of the ground-based polarimeter (P). The azimuth ϕ is defined with respect to the Sun vertical plane (hatched). θ and θ_o are the observation and solar zenith angles, respectively.

field (I and P) can be calculated at the other wavelength and the other zenith angles of observation. If these calculations compare favorably with the measurements at the other wavelength and other zenith angles of observation, then we have a consistency check on the assumption of the model. If they do not match, then we will have to analyze the assumptions in the model and see if any of the assumptions have to be relaxed to fit the data. For instance, if there are horizontal inhomogeneities, as when one air mass is replacing the other, we would find some disparity in the multi-angular scans.

These measurements were carried out at various locations so as to determine if the experimentally inferred effective parameters had any meteorological significance (Ref. 5). For instance, if the aerosols near the coastal area are monitored, there would be a preponderance of marine aerosol particles, and the refractive index would be close to that of water. So also the desert aerosol must yield indices of refraction closer to that of silicates (near 1.55). Santa Ana conditions (dry dusty winds) that bring clear days at Los Angeles should have particles of small size and so on.

A large number of measurements were made over these chosen sites at various times of the year with the idea of arriving at average values of these effective parameters for various geographical and meteorological situations.

IV. RESULTS OF THE GROUND-BASED MEASUREMENT PROGRAM

Measurements, indeed, validated the assumptions of the model. Figures 2 and 3 were made at two different wavelengths, one after the other, and we find that the parameters, indeed, are the same. Note that the optical depth at $\lambda = 0.575$ μm is greater than that at $\lambda = 0.7$ μm, as it ought to be. Figures 3 and 4 show the multiple zenith angle correlation. Again, the parameters derived are the same. We have found this type of a correlation in numerous measurements. Further, we find that the effective parameters do have a meteorological significance, the refractive index approaching 1.34 when there is marine-type aerosols and approaching 1.54 when the dry desert aerosols are present. The extensive measurements that we have carried out also confirms the fact that it is possible to characterize aerosols occurring during certain meteorological conditions with particular average values. This is summarized in the table. It is now possible to assign representative values for these aerosol parameters based on the geographical location and use these in the heat budget calculations. We are, of course, extending the local measurements made in Los Angeles basin on a global basis, a procedure that is reasonable only because the Los Angeles basin includes within it representative regions of the coastal, urban and desert air masses.

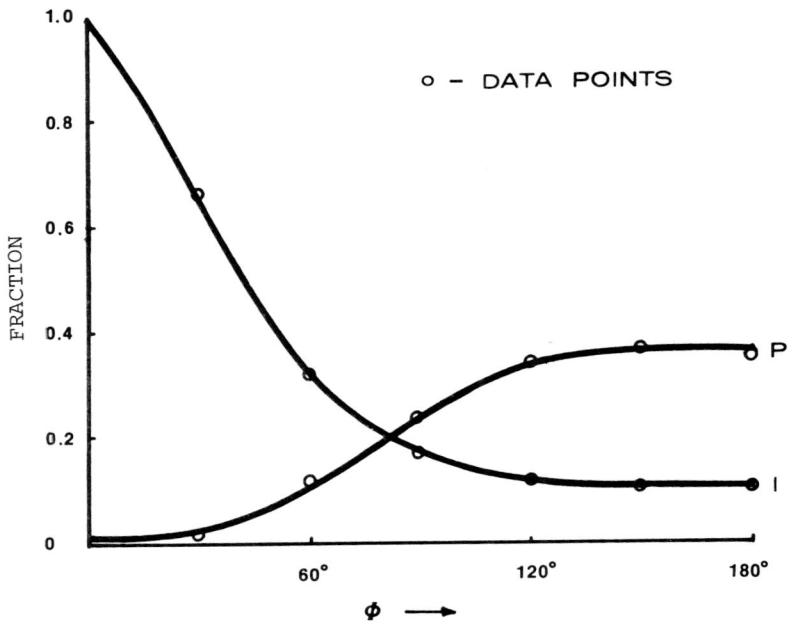

Fig. 2. May 22, 1976 at UCLA. Solar zenith angle = $40°$, observation zenith angle = $60°$, wavelength $(\lambda) = 0.7$ µ. Inferred aerosol parameters (assuming ground albedo (A) = 0.2) are: $m = 1.5$, $b = 20$, $\tau = 0.15$.

The parameters that are determined have the following error bounds: $\delta b = 2$, $\delta\tau = 0.02$, $\delta m = 0.02$. It is possible to sharpen these error bounds by having a more detailed catalog of tables. But it is not clear that this is needed for the use that we envision for the inferred parameters. In some cases, we have found that by choosing a Haze M representation we will be able to obtain better agreement with the experimental measurements, i.e., the parameters can be determined with better accuracy. In all cases, however, we were able to fit the data within the limits of the errors given previously. This, of course, means that if we can live with that amount of error in the parameters, then our model is quite satisfactory.

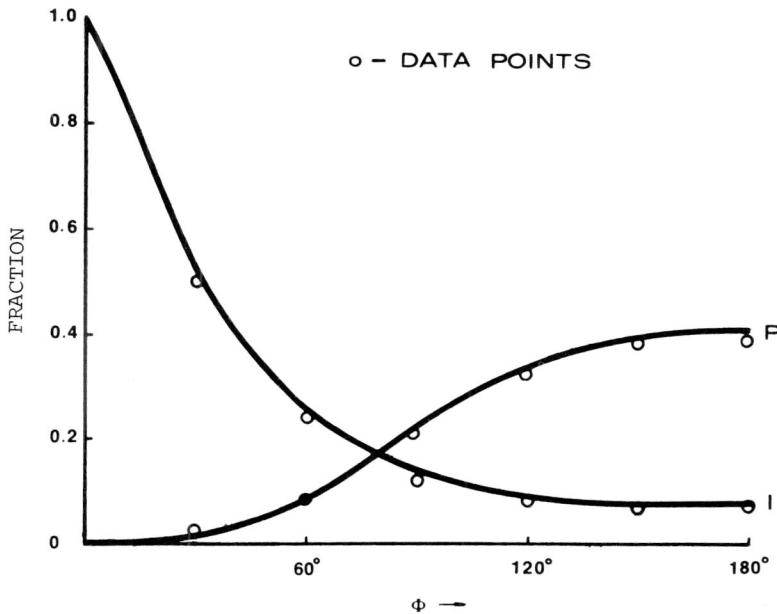

Fig. 3. May 22, 1976 at UCLA. Solar zenith angle = $40°$, observation zenith angle = $50°$, wavelength (λ) = 0.575 μ. Inferred aerosol parameters (assuming ground albedo (A) = 0.2) are: $m = 1.5$, $b = 18$, $\tau = 0.20$.

V. STUDY OF UPWELLING RADIATION--SATELLITE BORNE POLARIMETER

The radiative transfer equation is an integral equation. Therefore, in order to determine the radiation field in any one direction, it is necessary to know it in every other direction. Thus, the check of the assumptions involved in the model has ensured us the validity of the calculations of the upwelling radiation field. However, the experimental check of this is complicated by the presence of ground reflection. The model that we use for the ground, as a Lambertian source, while satisfactory for the study of the downwelling radiation (since the ground signal is only an insignificant part of the downwelling radiation), is

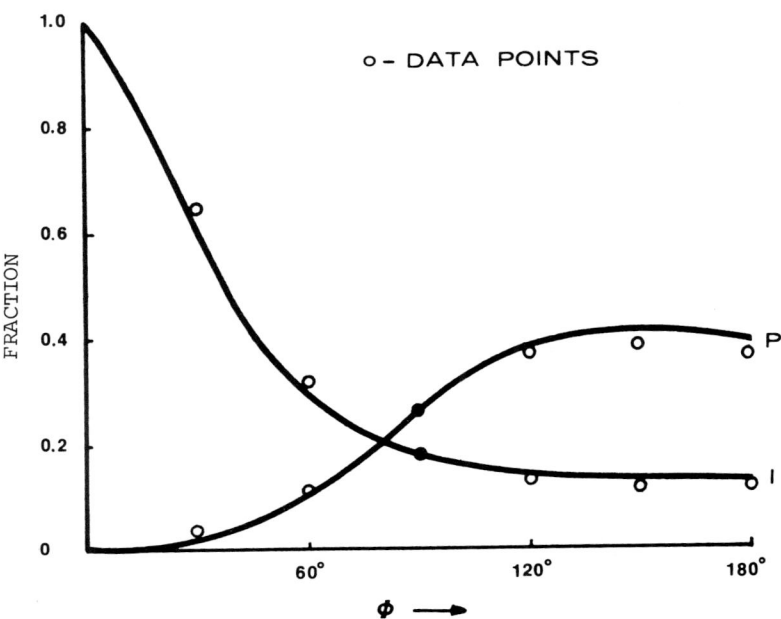

Fig. 4. May 22, 1976 at UCLA. Solar zenith angle = $40°$, observation zenith angle = $60°$, wavelength $(\lambda) = 0.575$ μ. Inferred aerosol parameters (assuming ground albedo $(A) = 0.2$) are: $m = 1.5$, $b = 20$, $\tau = 0.15$.

totally unsatisfactory for the study of upwelling radiation. In fact, in the long wavelength region (near 0.7 μm), the signal from the ground completely dominates the aerosol and the molecular part and can be used as a monitor of the ground reflection property. This can be introduced in the short wavelength radiative transfer calculation as the ground source. The preliminary calculations we have done seem to justify this approach. We will, however, consider the simpler case of the radiation field over the ocean, off from the specular region, where it is not unreasonable to expect a case of zero ground reflection. In actuality, however, the specular component due to the water must also be modeled and included with some gross features of sea-surface roughness.

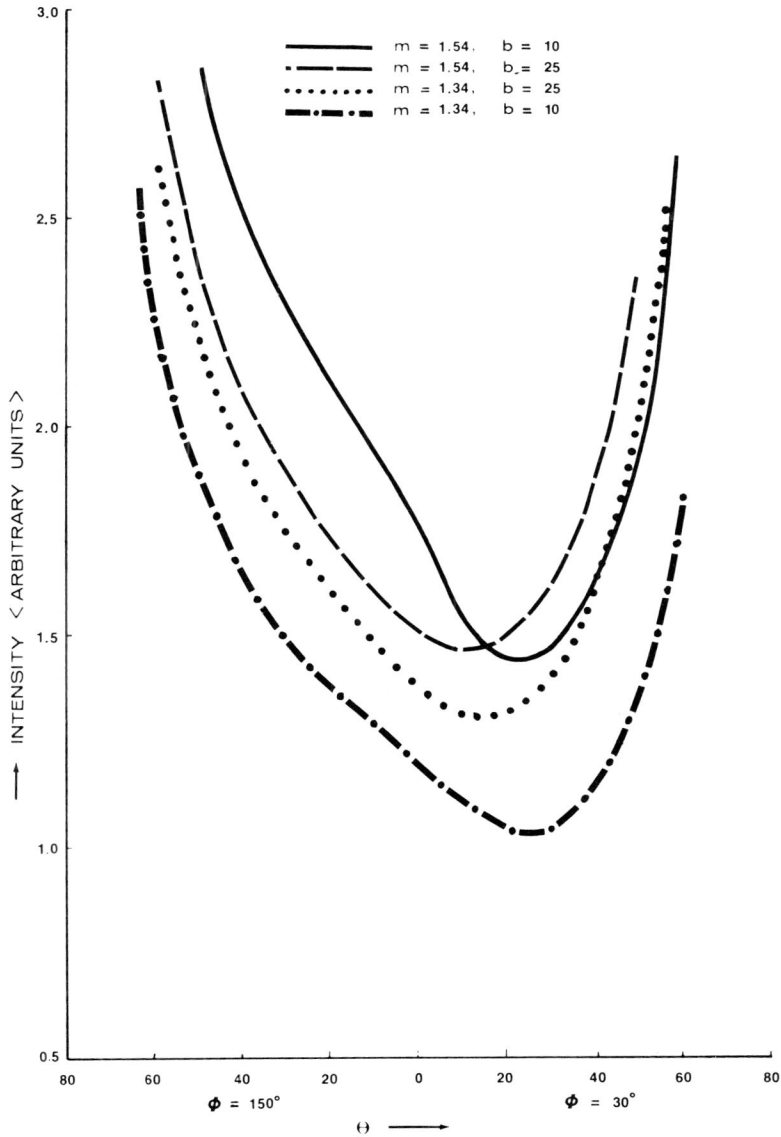

Fig. 5. Comparison of intensities of upwelling radiation due to different aerosol size parameters and refractive indices for, solar zenith angle = $30°$, ground albedo (A) = 0.0, wavelength (λ) = 0.7 µ and optical depth (τ) = 0.25. (Vertical plane scan. Angle between scan-plane and solar vertical plane = $30°$.)

As a preliminary effort we have ignored these effects and assumed that the sea surface is a near-zero reflection surface and analyzed the upwelling radiation in order to determine the algorithm for inference of the aerosol parameters. The analysis is made more difficult here by the fact that the observation geometry must also be consistent with the satellite path. We have considered the mode of observation where the target is situated on the earth in the satellite's path, and the polarimeter will lock itself onto the target, viewing it (in the satellites orbit plane) as the satellite continues in its orbit. The observation angles will be restricted such that the specular component of the reflected light is not encountered by the instrument. The rules that have been developed for the algorithm are quite different, but we do find a marked difference in the intensity and the polarization (Fig. 5) patterns as the aerosol parameters are varied. One other difference in this mode of analysis is that we find it is necessary to determine the actual intensity values and not just the normalized values. In the absence of measurements, we have checked the algorithm that has been developed against numerical simulations. We find that the algorithm is quite satisfactory. Details of these analyses are expected to be published at a later date.

VI. CONCLUDING REMARKS

Theoretical studies of aerosol radiative effects suggest that the heating and cooling due to aerosols can be as much as $1°$ per day, and in order to incorporate their effects in the heat budget of the atmosphere, we must know the values of the parameters of the aerosol model (that was used to estimate their heating and cooling effect). Thus, local measurements can yield the actual values of the aerosol parameters but they are not the values that we use in flux calculations. Radiative calculations involve bulk properties of atmospheric aerosols and it is these that must be experimentally determined. It is possible to use the exact calculations to derive empirical formulas for the fluxes in terms of the aerosol model parameters and, thus, the incorporation of the flux calculation in the circulation model will be rather inexpensive and simple.

Rather than consider very complicated models for aerosols, we decided to use the haze H type distribution. In the literature, it is quite usual to find exotic multi-modal and other distributions, but for our purposes, to obtain a rough estimate of all the parameters, the simpler description will suffice. There was, however, a need to consider the consistency of the whole method. We have devised such a scheme, where multiangular and multispectral measurements are used to verify the internal consistency of the method. The large number of experiments performed at UCLA with a ground-based polarimeter have verified these hypotheses. (The experiments off the coast of Dakar, as part of the GARP's[1] Atlantic Tropical Experiment (GATE) project, yielded aerosol parameters that seem to be of the same magnitude as those determined by using other methods by Quenzel, Prospero and others.)

After settling the question of the internal consistency of the method, our next task was to infer the aerosol parameters of representative air masses. Los Angeles basin has within its environs urban, marine, and desert type (geographic) locations. It seemed sensible to monitor the air masses in these regions and deduce some average values of the relevant parameters so as to obtain an aerosol climatology of sorts. This measurement program yielded the parameters summarized in the table. We also found that typical meteorological conditions resulted in predictable changes in the aerosol parameters. For instance, it was found that the marine air masses invariably resulted in lowering of the refractive index to near that of water and Santa Ana winds yielded smaller particles with higher refractive index.

---

[1] Global Atmospheric Research Program

The imaginary part of the index of refraction was less than 0.02. To determine this any better, we would have to obtain a catalog of radiation field that is finer than what we have. This is not a very serious limitation, because we can easily state the error that this imprecision in the imaginary part introduces in the flux calculations. There is a feeling, in certain sections of the scientific community, that the imaginary part of the aerosol refractive index is the sole parameter of interest to weather and climate. This is an erroneous statement, for a nonabsorbing aerosol (with no imaginary part) can reflect the solar radiation very effectively, lowering the available solar energy, and cause a serious modification of the heat budget. Perhaps the imaginary part is of consequence in the statements of local heating and cooling that can be of importance in the smaller scale processes. Unfortunately, the type of measurements we propose, where the bulk properties are monitored, is probably of very limited value for these studies. Thus, the imprecision in the determination of the imaginary part of the index of refraction is not a serious drawback for this type of an experiment.

The ground-based experiments have validated the theoretical scheme that parameterizes the aerosol effects in the atmosphere. The radiative transfer equation that was solved to determine this scheme was an integral equation, and these have the property that in order to obtain the value of the radiation field in the downward direction (this is what is monitored using the ground-based polarimeter), the radiation field in all other directions must be known. That is, a solution results in the prediction of the radiation field in all other directions. Thus, we can proceed with a great deal of assurance to use this scheme to study the upwelling radiation field, a mode of measurement that would be used by a satellite-borne polarimeter. In this measurement mode, the ground reflection no longer plays an innocuous role and, therefore, the use of Lambert reflection for the ground is not appropriate.

Our sensitivity studies show that at the long wavelength region the signal received by a satellite is for all intents and purposes the ground source. Thus, a long wavelength measurement of the intensity and polarization would be approximately a measurement of the ground reflection intensity and polarization. If we then make the assumption that these characteristics of the ground do not change much in the short wavelength region (an assumption that is quite reasonable or if one prefers the spectral variation can be modeled on the basis of experimental results of ground reflection properties), we will be able to calculate the radiation field in the short wavelength region for a realistic ground reflection matrix. Now, if an algorithm can be developed for this type of a ground, then we will be able to infer the aerosol parameters from satellite measurements. This is the general scheme that we are using but before attempting the full problem, we addressed ourselves to the simpler question, "Can algorithms be devised to infer aerosol particle characteristics from satellite measurements, if there is no ground reflection problem?"

Physically speaking, sea surface, if it is smooth, can approximate a zero ground reflection surface, if the specular reflecting part is excluded. Since the oceans form a large part of the scene viewed from a satellite, this is also a physically relevant study. Our analysis shows that aerosol parameters can, indeed, be inferred, even though the algorithm in this case is more complicated. Over 80 sets of synthetic data were analyzed using the algorithm and we had no trouble arriving at the bulk properties of the aerosol parameters. The viewing geometry used was also chosen to be reasonably close to that of a realistic satellite orbit. It is, of course, possible to extend the analysis to other viewing geometries so as to arrive at the optimum measurement configuration.

The case of zero ground reflection assured us that a polarimeter measurement from a satellite over the oceans can, indeed, be used to infer the aerosol characteristics. We have proceeded to consider the case of non-zero ground reflection. This will involve a multi-spectral analysis. Preliminary work that has been completed shows that when a non-Lambert ground is introduced in the short wavelength radiative transfer code, the algorithm can still be used. We will, during the course of the year, extend this work to various types of non-Lambertian ground surfaces and carry out checks on the algorithm.

SYMBOLS

a	constant of proportionality in Junge's power law distribution
a_H, a_L, a_M	parameters in size distribution for haze H, L, and M, respectively
A	Lambert ground albedo
b_H, b_L, b_M	parameters in size distribution for haze H, L, and M, respectively
I	specific intensity of scattered radiation field
k	$= 2\pi/\lambda$
m	refractive index of scatterer
$n(r)$	size distribution function
P	degree of polarization of scattered radiation field
Q_{ext}	extinction efficiency factor
r	particle radius
r_{min}, r_{max}	minimum and maximum value of r
θ	observation zenith angle
θ_o	Sun zenith angle
λ	wavelength of light
τ	optical depth due to aerosol particles
ϕ	observation azimuth angle
ν	power exponent on r in Junge's power law distribution

ACKNOWLEDGMENT

This research was supported by NASA Grants: NGR 05 007 328 and NSG 1270.

REFERENCES

1. M. Gell-Mann, The symmetry group of vector and axial vector currents, *Physics, 1,* 63 (1964).

2. J. G. Kuriyan and Z. Sekera, Forward scattering in a liquid haze-analytical approximations, *Q. J. R. Meteorol. Soc. 100,* 67 (1974).

3. J. G. Kuriyan, "Particulate Sizes from Polarization Measurements," (J. G. Kuriyan, Ed.). Western Periodicals Co., Hollywood, California, 1974. [Proceedings of the UCLA Conference on Radiation and Remote Probing of the Atmosphere.]

4. J. G. Kuriyan, D. H. Phillips and R. C. Willson, Determination of optical parameters of atmospheric particulates from ground based polarimeter measurements, *Q. J. R. Meteorol. Soc. 100,* 665 (1974).

5. J. G. Kuriyan, Remote measurements of aerosol particle characteristics and their significance to meteorology, *Opt. Eng. 14,* 332 (1975).

DISCUSSIONS

Chahine: Have you compared the results from the polarimeter to those from the multi-spectral measurements?

Kuriyan: Yes. About two or three years ago, Dr. Chahine, a student and I wrote a paper on multi-spectral extinction measurements. We found that if we have four precise measurements, then it is possible to infer the effective parameters. But unfortunately, the precision that was required was not available in the UCLA instrument. So we have used the UCLA instrument strictly to measure the optical depth and make sure that at least one parameter is fixed. What I would like to correct is the feeling, at least in certain circles, that it is the imaginary part of the refractive index that is significant for weather and climate. There is nothing farther from the truth, because a very highly reflecting aerosol (with zero imaginary part) can change the albedo sufficiently to affect weather and climate. It is, therefore, very important to determine the total refractive index--both the real and imaginary parts. Our method, however, will not allow us to determine imaginary parts better than 0.02.

Barkstrom: I think on the downward looking experiment one ought to use considerable care regarding the model of the lower boundary condition. It is possible to do some modeling of that and things seem to fall into two categories. But, there will be very strong limb brightening for a number of surfaces, including things like ocean and vegetation.

Kuriyan: There are two approaches that we adopt. One is to look at zero ground reflection areas (sea surfaces) and to do what Dr. Quenzel talked about, which is to model the roughness for the sea and so on. But the other approach is based on our sensitivity studies. Any measurement at about 0.7 or 0.8 microns seems to be entirely due to the signal from the ground, because the aerosol and molecular effects are so small and what we get is really a good measurement of the reflection property of the ground, the polarization of the ground and so on. Nor, if we make the assumption that this property does not vary when you go to the short wavelength region, then you can build this ground reflection property into the radiative transfer code. So, in the next stage of the development when we consider land-based targets, we include a reflection matrix which is measured at the long wavelength frequencies.

Barkstrom: Yes, if you want to model that with a doubling code you may want to adopt a low single scattering albedo model and a high single scattering albedo model.

AEROSOL OPTICAL PARAMETERS

Kuriyan: Yes, I am sure we will have to consider the full range.

Remsberg: Have you run into any situations where your catalog did not give you a proper answer or an answer?

Kuriyan: Some of the data that we gathered did not correspond exactly to the tables. These occurred during some meteorological activity in the area (such as moving air masses) and so it is not clear if this is an evidence of the deficiency of the model. In other words, during stable conditions, the data that we gathered could be fit with the Deirmendjian H-type distribution.

Remsberg: The point I was getting at was: suppose you had distributions that really couldn't be handled by the modified Deirmendjian case. As I understand it, your catalog is really set up on that type of a distribution where you vary b as your parameter. If you had some multi-modal distributions, that might not hold.

Kuriyan: Well, I didn't discuss this point. The first paper that I wrote on the subject had to do with the fact that Deirmendjian's haze H, L and M were equivalent in that they gave rise to the same phase matrix. I can then take any linear combination of the three models and produce the same phase matrix and I will have a tri-modal distribution. Thus, I am not worried about multi-modal distributions. They don't interest me, because I am interested in the radiative fluxes and both types of distributions yield the same fluxes.

EXPERIENCE WITH THE INVERSION OF NIMBUS 4

BUV MEASUREMENTS TO RETRIEVE THE

OZONE PROFILE

Carlton L. Mateer
Atmospheric Environment Service

The relative merits of pressure increment and partial derivative formulations of the ozone inversion problem are discussed briefly. The height range of validity of the retrieved ozone profile and the effects of adding wavelengths to or of dropping wavelengths from the inversion system are indicated. Illustrative results are presented for profiles retrieved from BUV data using Backus-Gilbert, minimum information (Twomey), and quasi-optimum procedures.

I. INTRODUCTION

The Backscatter Ultraviolet (BUV) experiment on Nimbus 4 measures nadir direction BUV radiances in 12 wavelength channels from 0.2555 µm to 0.3398 µm. Four long wavelength channels (0.3125, 0.3175, 0.3312, 0.3398 µm) are used to infer total atmospheric ozone content, and the eight shorter wavelength channels (0.2555, 0.2735, 0.2830, 0.2876, 0.2922, 0.2975, 0.3019, 0.3058 µm) are used to infer the ozone profile above the main ozone density maximum. In this paper, we are concerned only with the latter problem.

The normalized contribution or weighting functions for the experiment are illustrated in Fig. 1. At the shortest wavelengths, absorption is so strong that penetration of the photons into the atmosphere is severely impeded and a single scattering model is very accurate for calculating the backscattered radiance.

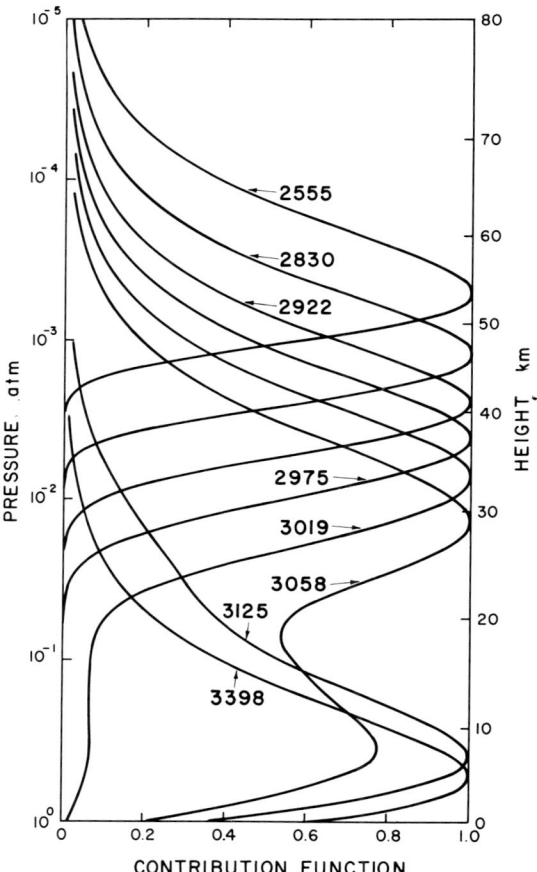

Fig. 1. Contributions to the nadir direction radiance by backscattering at various levels in the atmosphere. (1 atm = 101.3 kPa.)

However, as we progress to longer wavelengths, the absorption coefficient decreases, and the photons penetrate more deeply into the ozone layer. Eventually, the absorption is weak enough to ensure that a sufficient number of photons pass through the ozone layer and multiple scattering in the troposphere has to be considered for the accurate calculation of the backscattered radiance (wavelengths 0.3058 μm and longer in Fig. 1).

In estimating the ozone profile, we have restricted ourselves to a consideration of those wavelengths which are relatively unaffected by multiple scattering--namely, those for which at most secondary scattering is required and the small correction to primary scattering may be parameterized. There are two reasons for this restriction. First, Yarger (Ref. 1) has shown that there is little retrievable information about the ozone profile below the ozone density maximum in measurements at those wavelengths which are strongly affected by multiple scattering. Second, with many ozone profiles to be retrieved, the necessary multiple scattering calculations would be very expensive. As we shall see, this effectively limits the validity range of our retrieved profiles to levels above the ozone density maximum.

II. FORMULATION OF THE INVERSION PROBLEM

For a single scattering model, the basic integral equation may be written as

$$Q(\lambda, \theta_o) = \int_0^{p_o} \exp[-(1 + \sec\theta_o)(\alpha_\lambda X_p + \beta_\lambda p)] \, dp \qquad (1)$$

where $Q(\lambda, \theta_o)$ is the BUV measurement at wavelength λ and solar zenith angle θ_o. For convenience, Q incorporates the extraterrestrial solar flux as well as certain constants and geometric factors;

α_λ is the ozone absorption coefficient; X_p is the amount of ozone above pressure level p; β_λ is the Rayleigh scattering coefficient; and p_o is the surface pressure.

For the four shortest wavelengths, $\alpha_\lambda X_p \gg \beta_\lambda p$. If we set $k = \alpha_\lambda (1 + \sec \theta_o)$ and use the exponential ozone profile assumption at high levels, namely,

$$X_p = Cp^{1/\sigma} \qquad (2)$$

where C and σ are the constants specifying the profile, then it is easy to show that

$$Q(k) = (kC)^{-\sigma} \Gamma(\sigma + 1) \qquad (3)$$

where $\Gamma(\sigma + 1)$ is the gamma function for parameter $\sigma + 1$. In short, a linear regression of ln Q against ln k for the four shortest wavelengths will yield an estimate of the profile constants C and σ. We shall refer to this profile as the upper first guess. Next, from the estimate of the total atmospheric ozone content (see Ref. 2), a lower first guess is constructed from average balloon profiles. The lower and upper first guesses are objectively combined to provide the complete first guess or *a priori* constraint profile from which the final estimated ozone profile is obtained.

We have used two convenient formulations for this final important step. The first of these is the partial derivative (P.D.), one in which

$$\sum_i \left(\frac{\partial \ln Q_\lambda}{\partial \ln x_i} \right) \delta \ln x_i = \delta \ln Q_i \qquad (4)$$

where x_i is the amount of ozone in layer i for the first guess. Typical derivatives are shown as the dashed curves in Fig. 2. Use of the logarithmic derivative has the advantage that the peak values for each wavelength are about the same magnitude. The second formulation is the pressure increment (P.I.) one, which we have used in a normalized form, as follows, from the quadrature form of Eq. (1):

$$\sum_i \{(p_i/Q_\lambda^*)\exp[-(1+\sec\theta_o)(\alpha_{\lambda_j} \sum_{j=1} \beta_\lambda x_j + \beta_\lambda p_i)]\} \frac{\Delta \hat{p}_i - \Delta p_i}{p_i} = \frac{Q_\lambda^* - Q_\lambda}{Q_\lambda^*} \quad (5)$$

where the asterisk refers to the observation and the circumflex ("hat") symbol refers to the solution profile, p_i is the mean pressure in layer i and Δp_i is the pressure change across the layer. The quantity in the curled braces is the kernel for which typical values are shown by the continuous curves in Fig. 2. The normalization (multiplication of each exponential by p_i/Q_λ^*) has the effect of making the peaks of the kernel functions approximately the same magnitude.

In the P.D. method, the unknowns are the logarithmic ozone content changes for each layer. Total atmospheric pressure is conserved exactly, but total ozone will generally differ somewhat from the first guess value. In the P.I. method, the unknowns are the fractional changes in the scattering mass in each layer and the ozone against pressure relationship is recovered by integrating from the top down. Total ozone is conserved exactly, but total atmospheric pressure will differ somewhat from the first guess value. These are trivial considerations for profile retrievals where the lower limit of validity of the final profile is above the ozone density maximum and, consequently, well above the Earth's surface.

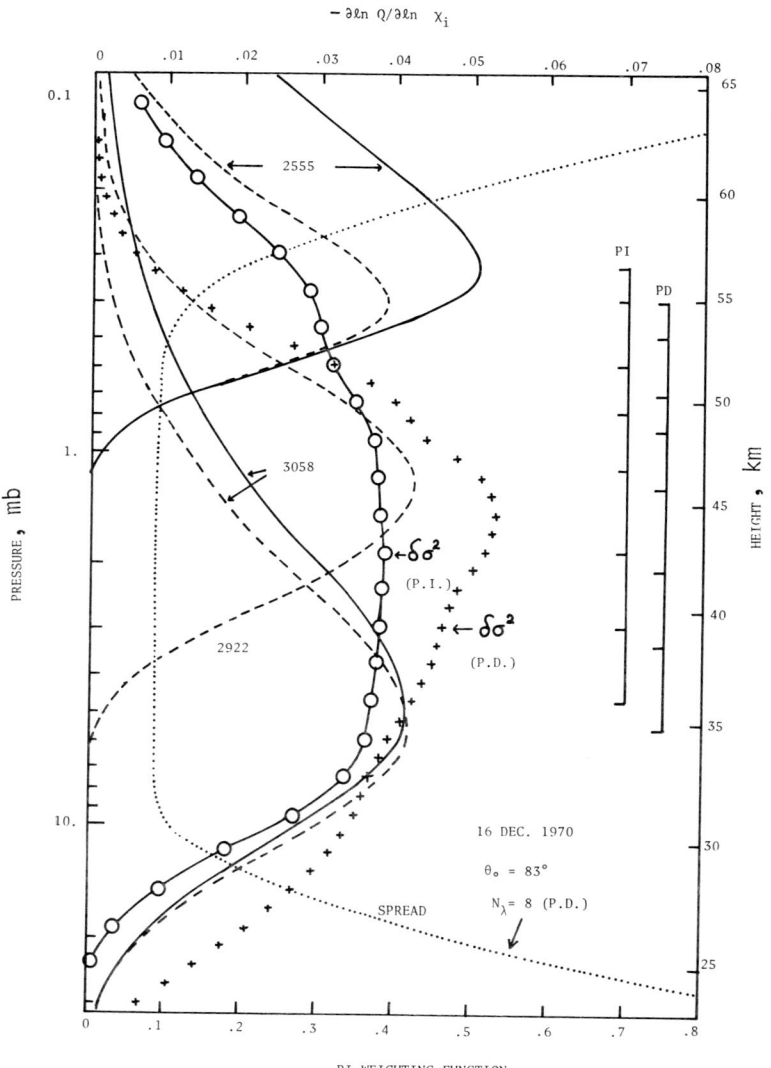

Fig. 2. P.I. weighting functions (solid curves); partial derivatives (dashed curves); Backus-Gilbert spread (dotted curve); profile variance reduction by quasi-optimum (P.I.) method (small circles; profile variance reduction by quasi-optimum (P.D.) method (plus signs Vertical scales at right show location of maximum for all wavelengths for P.I. and P.D. methods, respectively. (1 bar = 100 kPa.)

III. INVERSION PROCEDURES

Results from three inversion procedures are presented here. These are the Backus-Gilbert, the minimum information (Twomey), and the quasi-optimum procedures. The first two of these have been amply described by other authors at this Workshop. The quasi-optimum procedure, as used herein, differs from the optimum one in that the covariance matrices for the constraint profile (first guess profile) and for the observational errors are purely diagonal, having nonzero elements only on their diagonals. The Backus-Gilbert method is used here primarily in a diagnostic sense to indicate the pressure (height) range of validity of the derived profiles and to illustrate the effects of adding wavelengths to or of dropping wavelengths from the inversion system as dictated by the multiple scattering limitations stated earlier.

IV. RELATIVE MERITS OF THE PARTIAL DERIVATIVE AND PRESSURE INCREMENT FORMULATIONS

If there are relatively few data sets to be inverted and/or computer time is of no serious consequence, the partial derivative method offers the advantage of simplicity in interpretation--it gives the change in ozone in each layer directly--and we are, after all, seeking an ozone profile. In an optimum or quasi-optimum procedure, the profile covariance matrix may be interpreted directly in terms of profile errors at different levels. However, unless the first guess or constraint profile is extremely close to the final solution, the P.D. system is nonlinear--a minimum of two iterations is required before the system is quasi-linear. If there are many data sets to be inverted and computer time is a consideration, the computation (and re-computation with each iteration) of the partial derivatives is time consuming.

On the other hand, the pressure increment formulation avoids this time-consuming partial derivative calculation and is very nearly linear on the first iteration so that further iterations are not required. With the optimum or quasi-optimum procedure, the profile covariance matrix can probably be interpreted directly in terms of ozone profile errors at the different levels. The P.I. method does require an integration to recover the pressure-ozone relationship, followed by an interpolation back to the standard layers used in the atmospheric model, a relatively minor disadvantage.

V. PRESSURE RANGE OF SOLUTION VALIDITY

Figures 3 and 4 contain illustrative results for the Backus-Gilbert method using the P.D. formulation. The solid curves labeled "SPREAD" show the Backus-Gilbert spread (in pressure scale-heights) for a constant (profile) error of 5% as a function of atmospheric pressure. The measurement error was assumed to have the form

$$\sigma_\lambda^2 = 0.0001 + \frac{0.00005^2}{I_\lambda} \quad (6)$$

where σ_λ is the measurement error of the backscattered radiance (I_λ) at wavelength λ. This provides for a basic measurement error of ± 1% (1st term) plus a noise term which is about 10% for the lowest backscattered radiances measured at 0.2555 μm. Curves for a constant profile error of 10% (for a more realistic basic measurement error of 2%) are very nearly coincident with the

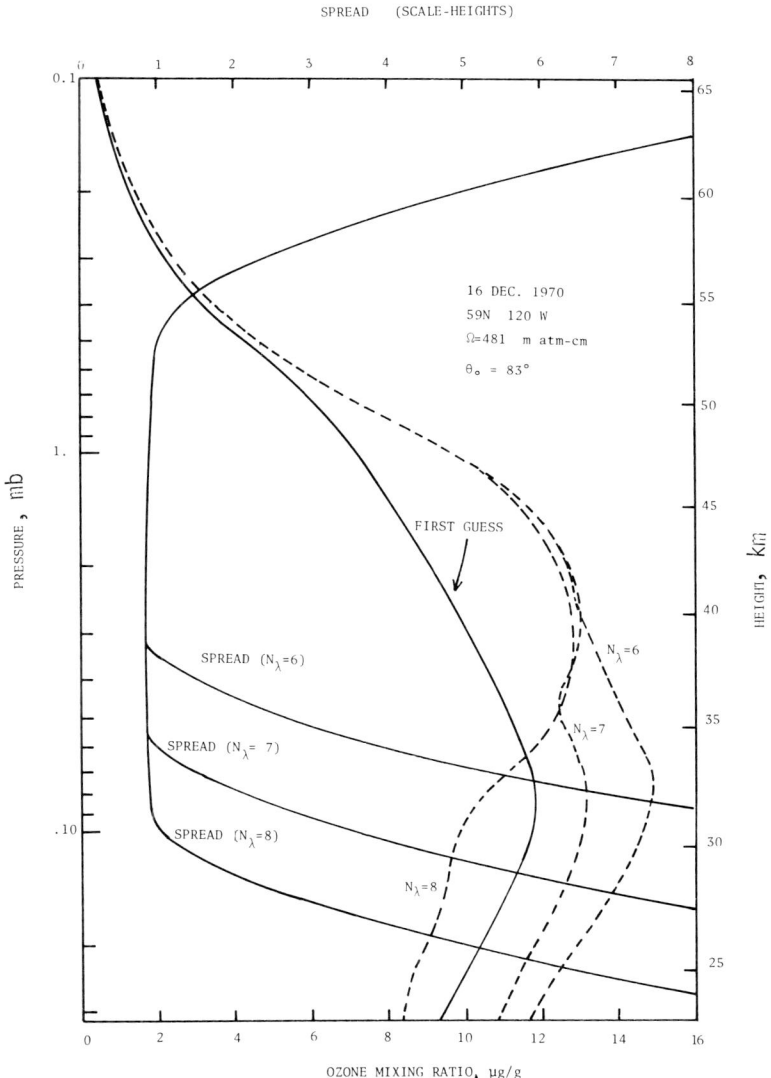

Fig. 3. Spread, first-guess, and solution profiles for N_λ = 6, 7, and 8 for Backus-Gilbert (P.D.) method. (1 bar = 100 kPa; 1 atm = 101.3 kPa.)

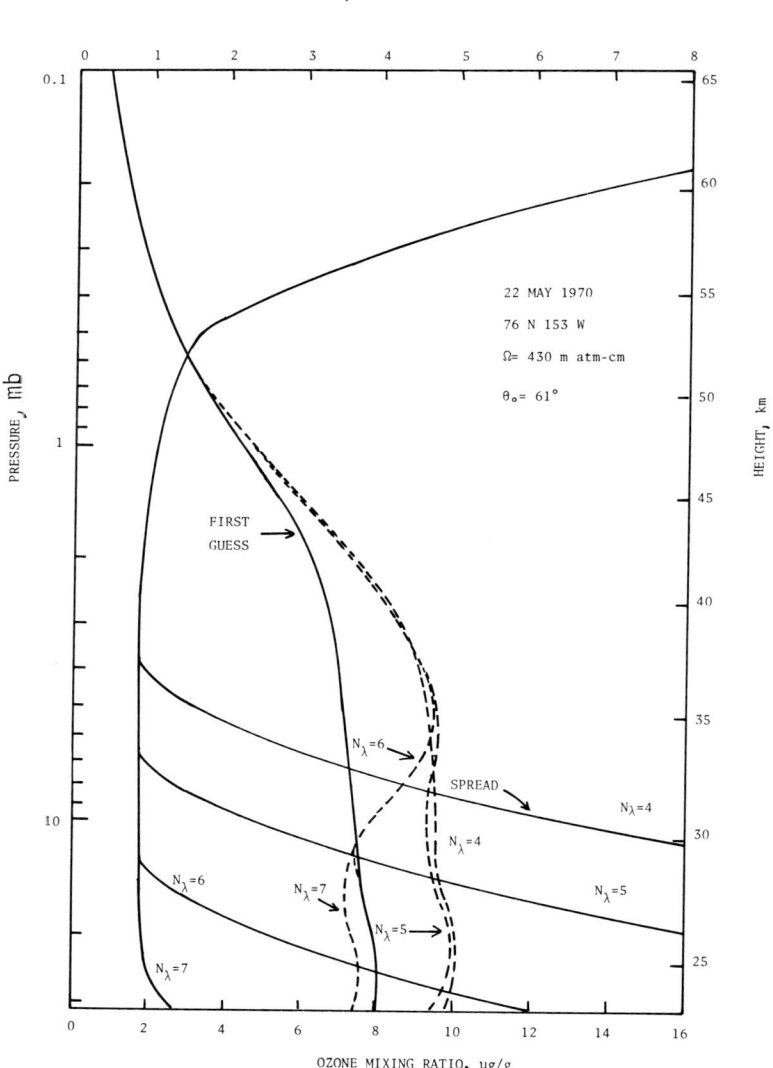

Fig. 4. Spread, first-guess and solution profiles for N_λ = 4,5,6, and 7 for Backus-Gilbert (P.D.) method. (1 bar = 100 kPa; 1 atm = 101.3 kPa.)

INVERSION OF NIMBUS 4 BUV MEASUREMENTS 587

plotted curves. Figure 3 is a high-latitude northern hemisphere case for December 1970. The "regular" solution, involving small secondary scattering corrections at 0.3019 and 0.3058 µm, would utilize the measurements at 8 wavelengths ($N_\lambda = 8$). The solution profiles are the dashed curves. The solution profile for $N_\lambda = 6$ diverges appreciably from the $N_\lambda = 8$ solution profile at about 3 mb and the $N_\lambda = 7$ solution profile diverges appreciably at about 6 mb (1 bar = 100 kPa). These are the levels at which the spread begins to increase rapidly and one is tempted to infer that these levels represent the respective lower validity limits for the $N_\lambda = 6$ and $N_\lambda = 7$ solution profiles. If this is correct, then the lower validity limit for the 8 wavelength solution profile is near 10 mb, and the upper validity limit is near 0.4 to 0.5 mb. Figure 4 is a high-latitude case for the northern hemisphere in May 1970, for a solar zenith angle of $61°$. The "regular" solution would utilize 7 wavelengths and would involve small secondary scattering corrections at 0.2975 and 0.3019 µm. In this case, the picture is not so clear. The $N_\lambda = 4$ and $N_\lambda = 5$ solution profiles remain close together down to 30 mb, but they diverge appreciably from the $N_\lambda = 6$ and $N_\lambda = 7$ solution profiles below about 7 mb, which is also the level at which the spread curve for $N_\lambda = 5$ begins to increase rapidly. The $N_\lambda = 6$ and $N_\lambda = 7$ solution profiles are virtually coincident down to about 16 mb and thereafter are separated by about 5%. The 16 mb level is also the point where the spread curve for $N_\lambda = 6$ begins to increase rapidly. It appears that the lower validity limit for the $N_\lambda = 7$ solution profile should be about 25 to 30 mb. In this case, in common with others, the upper validity limit is not so clear-cut as in Fig. 3, but would appear to be at about 0.5 mb.

The dotted curve in Fig. 2 is the spread for $N_\lambda = 8$. If we wanted to estimate the upper and lower validity limits of derived profiles without calculating the spread, it appears that the upper validity limit is approximately at the level where the

0.2555 μm partial derivative is a maximum, and the lower validity limit is at the level, below the maximum of the partial derivative for the longest wavelength used, where the partial derivative is about 70% of the maximum value.

Information about the probable range of solution validity may also be estimated from the quasi-optimum method. This has been done for both the P.I. and P.D. formulations by using a constant value 0.4 for the variance of the first guess or constraint profile at all levels and noting the variance reduction in obtaining the solution profile. Curves of this variance reduction are labeled $\delta\sigma^2$ (P.I.) and $\delta\sigma^2$ (P.D.) on Fig. 2. These curves do not delineate validity limits as well as the Backus-Gilbert spread. They suggest a possible upward extension of the upper limit for the P.I. formulations and a possible downward extension of the lower limit for the P.D. formulation, as compared to Backus-Gilbert. In all cases, it is clear that the lower validity limit is above the main ozone density maximum, which is generally within the 15 to 25 km range, depending on latitude and season.

VI. SOME ILLUSTRATIVE RESULTS

Figures 5 to 8 inclusive show final solutions derived by various methods for four BUV wavelength scans. In each case, the first guess, or constraint, profile is a solid curve; and the remaining profiles may be identified as follows. *Case 1*--Profile curve of short dashes: This is the one-iteration Backus-Gilbert solution for the P.D. formulation; secondary scattering corrections were not applied; profile error is a constant 5% or 10% for ± 1 or ± 2% basic measurement error, respectively (as in Eq. (6)). *Case 2*--Profile curve of long dashes: This is the quasi-optimum solution for the P.D. formulation; constraint profile variance is a constant 0.4; basic measurement error is 2%. *Case 3*--Profile curve of square symbols: This is the quasi-optimum solution for the P.I. formulation; constraint profile variance is a constant

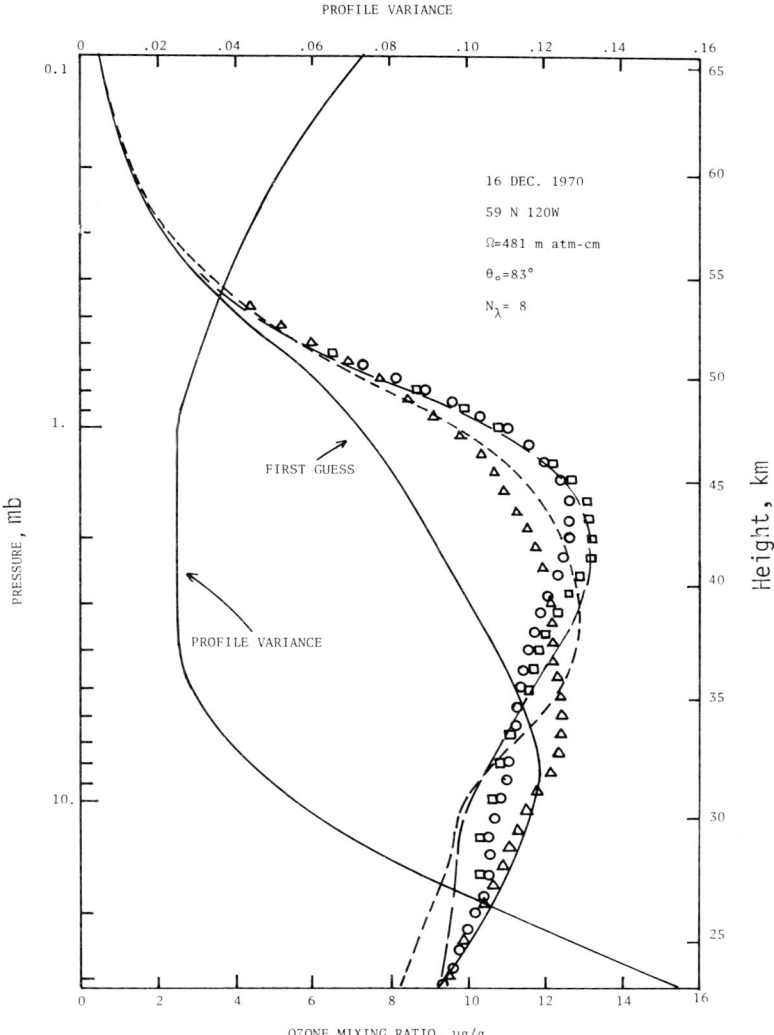

Fig. 5. First-guess, assumed first-guess profile variance and various solution profiles. See text for explanation of symbols. (1 bar = 100 kPa.)

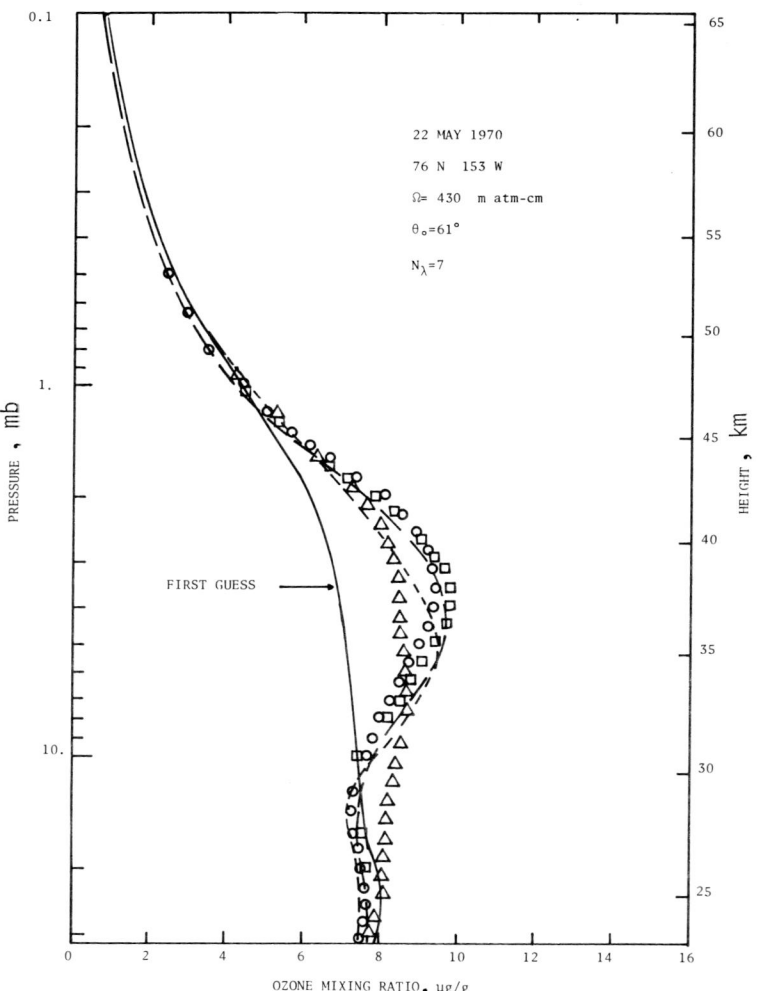

Fig. 6. First-guess and various solution profiles. See text for explanation of symbols. (1 bar = 100 kPa; 1 atm = 101.3 kPa.)

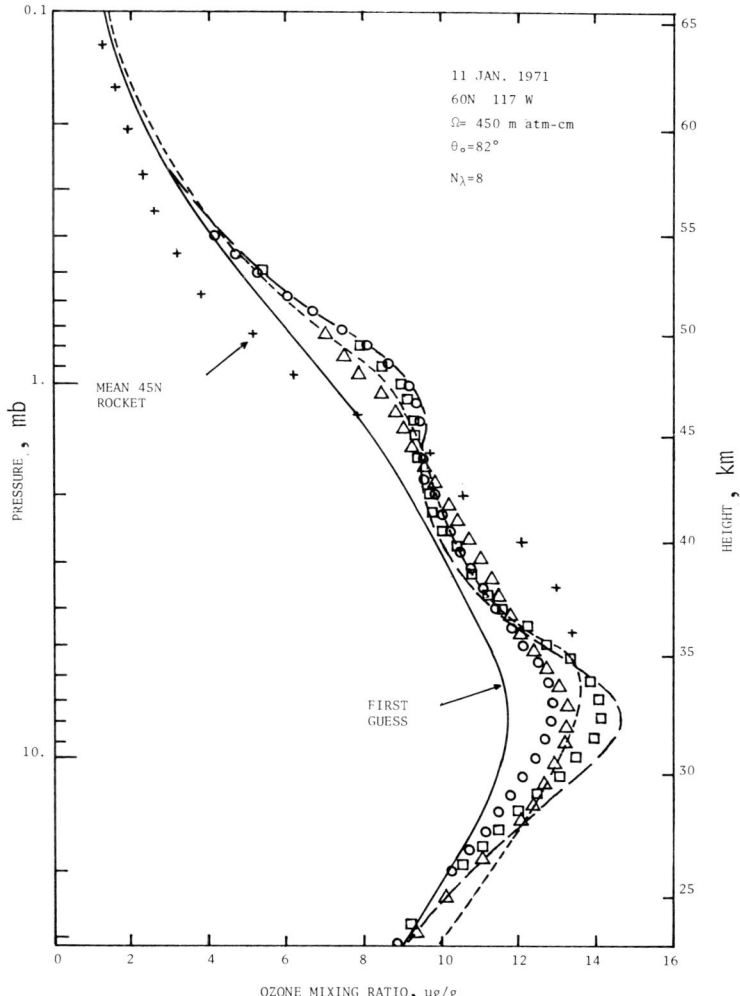

Fig. 7. First-guess and various solution profiles. See text for explanation of symbols. Plus signs show average 45°N rocket profile down to 36 km. (1 bar = 100 kPa; 1 atm = 101.3 kPa.)

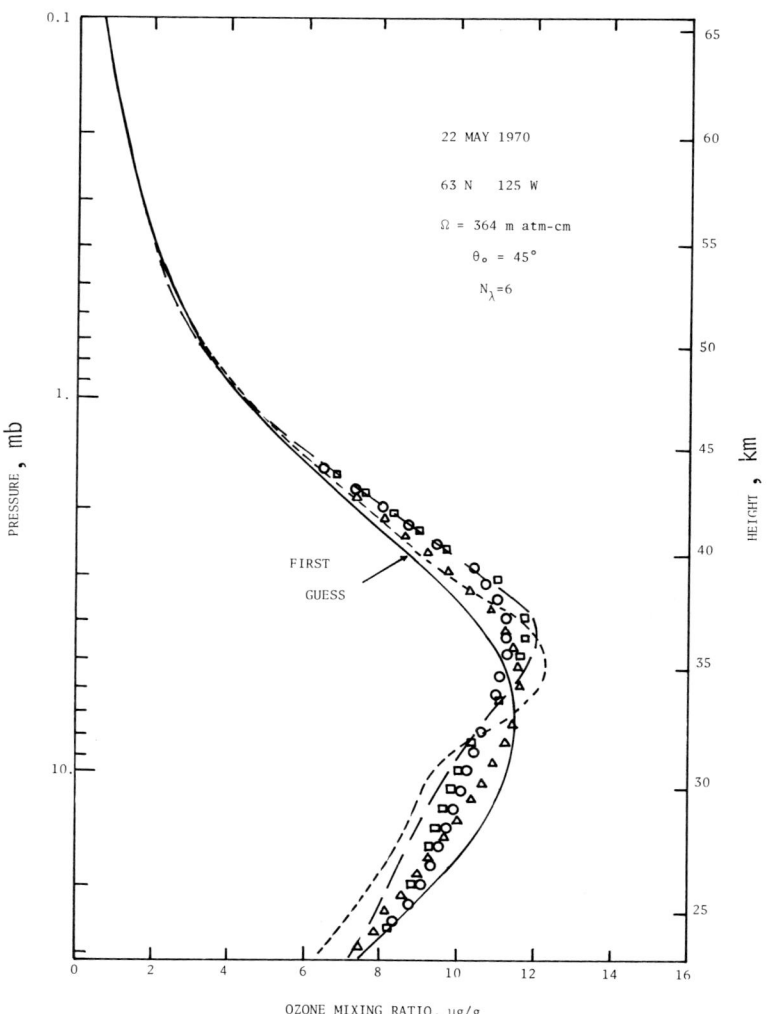

Fig. 8. First-guess and various solution profiles. See text for explanation of symbols. (1 bar = 100 kPa; 1 atm = 101.3 kPa.)

0.02; basic measurement error is ± 2%; for this combination. The overall constraint is about the same as in Case 2. *Case 4*-- Profile curve of triangular symbols: This is the quasi-optimum solution for the P.D. formulation; constraint profile variance is given by the curve so identified in Fig. 5; basic measurement error is ± 2 %. This is the strongest constraint of the five cases presented and the solution profile, therefore, deviates least overall from the constraint profile. *Case 5*--Profile curve of small circles: This is the minimum information (Twomey) solution for the P.I. formulation; constraint profile variance is the identity matrix; the measurement variance is invariant with wavelength and is 0.02 times the trace of the kernel "cross-product" matrix; secondary scattering corrections were not applied.

For comparison, the average $45°N$ rocket profile derived by Krueger and Minzner (Ref. 3) is shown in Fig. 7. At most levels, we find solution profiles which are both higher and lower than the mean rocket profile for the four cases presented. For each individual BUV scan, the difference between the highest and lowest solution profile at any level does not exceed about 20%. Although total ozone is nearly the same for the BUV scans profiled in Figs. 5 and 7, the inferred high-level ozone profiles are very different. These differences illustrate changes that appear to occur between profiles before (Fig. 5) and after (Fig. 7) the well-known stratospheric warming phenomenon of high-latitude winters.

VII. CONCLUDING REMARKS

From the illustrative solutions themselves, there appears to be little to choose between the P.I. and P.D. formulations (compare curves of long dashes and of squares). However, if one has a million or more BUV scans to evaluate, the P.I. formulation should lead to significant savings in computer time.

Since the solution profiles depend significantly on the constraint or first guess profile, and on the variances assigned to this profile and to the instrument observations, considerable attention should be paid to these factors in establishing a BUV evaluation system. The merits of using full covariance matrices as opposed to the diagonal, or variance matrices used here, is not clear. One might be tempted to conclude that the use of the complete profile covariance matrix could lead to spin-off "correlation" information about the ozone profile below the main density maximum. I doubt this very much; more such information is likely available from total ozone-profile correlations and should be incorporated in the first guess profile. In any event, the information necessary to establish the *full* covariance matrix is not yet available. A partial covariance matrix can, of course, be generated from the BUV data themselves by operating "along the track" or on a regional-seasonal basis. Such an approach may be adopted for the evaluation of the data from the Scanning BUV experiment on Nimbus G.

SYMBOLS

C, σ	constants in equations for exponential ozone profile
I_λ	backscattered radiance at wavelength λ
k	$\sigma_\lambda (1 + \sec \theta_o)$
N_λ	number of wavelengths used in an inversion
p	atmospheric pressure
p_o	atmosphere pressure at the Earth's surface
p_i	average atmospheric pressure in layer i
Δp_i	atmospheric pressure change across layer i
$\Delta \hat{p}_i$	estimate of Δp_i for solution
P.D.	partial derivative inversion formulation
P.I.	pressure increment inversion formulation
Q_λ	quantity defined by Eq. (2)
Q^*_λ	observed value of Q

$Q(\lambda_1, \theta_o)$ integral function of wavelength and zenith angle which is proportional to backscattered radiance

Q_i value of Q in layer i

x_i, x_j amount of ozone in lyaers i and j, respectively

x_p amount of ozone above pressure level p

α_λ ozone absorption coefficient at wavelength

β_λ molecular scattering coefficient at wavelength

$\Gamma(\sigma)$ gamma function of argument

$\delta\sigma^2$ reduction in error variance for an ozone profile achieved in an inversion

θ_o solar zenith angle

λ wavelength

X_i amount of ozone in layer i

Ω total amount of ozone

REFERENCES

1. D. N. Yarger, An evaluation of some methods of estimating the vertical atmospheric ozone distribution from the inversion of spectral ultraviolet radiation, *J. Appl. Meterol. 9*, 921 (1970).

2. C. L. Mateer, D. F. Heath and A. J. Krueger, Estimation of total ozone from satellite measurements of backscattered ultraviolet earth radiance, *J. Atmos. Sci. 28*, 1307 (1971).

3. A. J. Krueger and R. A. Minzner, A proposed mid-latitude ozone model for the U.S. standard atmosphere, *Goddard Space Flight Center, Greenbelt, Maryland* [NASA Report X-651-73-22], 1973.

DISCUSSION

Cerni: Were any rocket flights taken simultaneously as the satellite passed overhead which could serve as ground truth measurements?

Mateer: We do not have very many. There are two for which I have the results, although I understand that there are more. Here is one taken on February 24, 1971, at Barking Sands, Hawaii, which is not too bad. (Shows Fig. D-1). This was done with the pressure increment formulation and the Twomey minimum information method with a rather small value for γ; I believe it was 0.005 times the trace, which is a little too small.

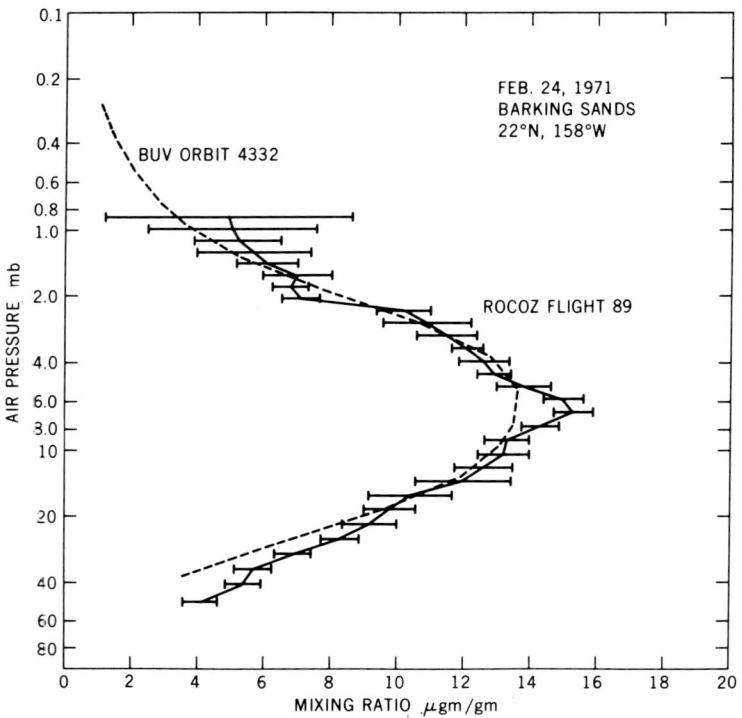

Fig. D-1. *Mixing ratio comparison for Barking Sands data (1 bar = 100 kPa).*

Cerni: What is the difference between the dashed and solid line there?

Mateer: The solid line is the rocket result with the error bars on it and the dashed line is the BUV inversion. There is another one here (shows Fig. D-2). I guess it is not too bad. This is for Point Mugu on June 18, 1970. Again, the dashed line is the BUV result.

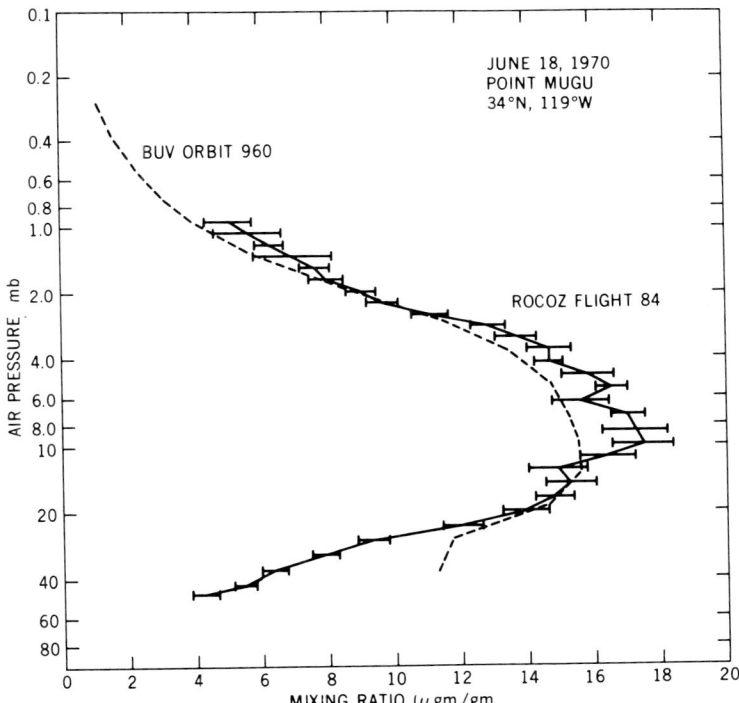

Fig. D-2. Mixing ratio comparison for Point Mugu data (1 bar = 100 kPa).

TEMPERATURE SENSING: THE DIRECT ROAD TO INFORMATION

Lewis D. Kaplan
The University of Chicago

The retrievability of detailed temperature soundings from remote measurements of emission spectra depends not so much on how the data are treated as on what the data are. It is shown that the shape of the weighting functions depends on the nature of the pressure and temperature dependence of the transmittance, which differ from one part of the spectrum to another as well as with spectral resolution.

Included are results of a study by Kaplan, Chahine, Susskind, and Searl, which show that careful selection of channels can result in much narrower weighting functions than those corresponding to channels that have actually been used. More detailed and more accurate retrievals can be obtained, therefore, if the instrumental design is more carefully planned on the basis of sound theoretical principles.

If there is any common thread of this Workshop, it is the question of information content. In particular, much of the discussion has been devoted to ways of obtaining more information out of the measurements that are being made. Dr. Rodgers, for example, suggested introducing other *a priori* information, which he calls "virtual measurements." The difficulty, however, is that we are using the measurements for a purpose, namely, the use of the retrievals in numerical weather prediction and the fact is that numerical circulation models are already using more layers than we can possibly get out of the soundings.

The principal problem, as is evident from Dr. Twomey's paper, is that the weighting functions are too broad. I would like to try to resolve this problem along the lines already touched on by Dr. Chahine. The proper question is not how to get adequate soundings from inadequate data. It is, rather, is there more information in total atmospheric spectra than we are measuring? In particular, are the weighting functions broad because we are not looking at the spectra carefully enough? These are the questions I am going to address. I will restrict myself to the temperature sounding in order to treat the subject properly within the time constraints. I will utilize, in this talk, the treatment and results given in a paper by Drs. Chahine, Susskind, Searl and myself that appeared in February 1977 issue of Applied Optics (Ref. 1) and in which further details may be found.

What we are measuring is a flux of radiation, which is given by this familiar integral:

$$I_\nu = \int B_\nu \frac{d\tau_\nu(0,p)}{d\ln p} d\ln p + B_{\nu G} \tau_G , \qquad (1)$$

where the derivative of the transmittance with respect to the logarithm of the pressure, $d\tau_\nu/d\ln p$, is the weighting function of the black body radiance $B_\nu(\nu, T)$. Our information from each element $d\ln p$ is $B_\nu d\tau_\nu/d\ln p$. This is the quantity we will examine later on, since its shape tells us what sort of resolution we have for getting information out of an individual layer.

The transmittance $\tau_{\nu j}$ in a band pass $\Delta\nu_j$ from a pressure level p to satellite height can be approximated in terms of the equivalent widths w_{ji} of random lines in $\Delta\nu_j$. If it is further assumed that the emitting gas is uniformly mixed and if the temperature dependence of the line intensities and widths are, for the moment, neglected, the transmittance can be shown to be

$$\tau_{\nu j} = \exp(-\Sigma w_{ji}/\Delta\nu_j) = \exp(-a_j p) \qquad (2)$$

The weighting function then assumes the simple form:

$$-d\tau_{\nu j}/d\ell n p = a_j p \exp(-a_j p) \qquad (3)$$

This weighting function peaks at $a_j p = 1$, and the ratios of the pressures at which it has half its peak value is 11-1/2, corresponding to 2-1/2 scale heights. Now this is very broad, and we want to narrow it.

One obvious way of narrowing the weighting functions that we have heard about from Dr. Gille is by limb scanning; but that is not of practical use in the troposphere. But there is another way of doing it, and that is by making measurements in the line wings. This was first suggested by John Houghton and is the basic principle of the selective chopper. The selective chopper, however, also tends to lose its usefulness in the troposphere, where the central portions of weak lines have increasingly substantial influence on the selectively chopped radiances. This is unfortunate because, again for constant composition and temperature-independent line intensities and widths, the transmittance from the top of the atmosphere to pressure p, in the wings of Lorentz lines, is related to the pressure by

$$\tau_{\nu j} = \exp(-b_j p^2), \qquad (4)$$

for which the form of the weighting function is

$$-d\tau_{\nu j}/d\ell n p = 2b_j p^2 \exp(-b_j p^2) \qquad (5)$$

This peaks at $b_j p^2 = 1$ where it has the value 2/e, twice the value at the peak of a weighting function of the form of Eq. (3); and it is twice as narrow. In fact, if the transmittance is given by

$$\tau_{\nu j} = \exp(-c_j p^n), \tag{6}$$

the weighting function has the form

$$-d\tau_{\nu j}/d\ln p = nc_j p^n \exp(-c_j p^n) \tag{7}$$

which we may rewrite as

$$-d\tau_{\nu j}/d\ln p = np^n \exp(-p^n) \tag{7a}$$

if we take p in units of $p_{max} = 1/c_j$; and the ratio of the weighting function at p to its value at p_{max} is given by

$$R = p^n \exp(1 - p^n) = f(p^n) = R(n\ln p) \tag{8}$$

Thus, the entire shape is compressed by a factor of n on a $\ln p$ scale relative to that of Eq. (3).

It turns out that it is not so easy to look directly in the wings of lines, particularly in the 15 μm band of CO_2, which is strewn with very many isotope and Fermi resonance bands. In fact, the only part of the CO_2 atmospheric emission spectrum in which the line wings are clean is the R branch of the 4.3 μm band. This region of the spectrum has the added advantage of strong temperature dependence of the black body function, which further narrows the tropospheric part of the information function $Bd\tau/d\ln p$. There is a broadening effect in the stratospheric part, but it is more than compensated for by the strong exponential decay in a weighting function of the form of Eq. (5).

This is illustrated by the curves of Fig. 1, taken from the aforementioned Applied Optics paper. The left-hand side contains weighting functions for very narrow band passes centered in the wings between some of the closely spaced lines of the 4.3 μm band R-branch. They are indeed very narrow and become even narrower in the troposphere when multiplied by the black body intensity B,

TEMPERATURE SENSING

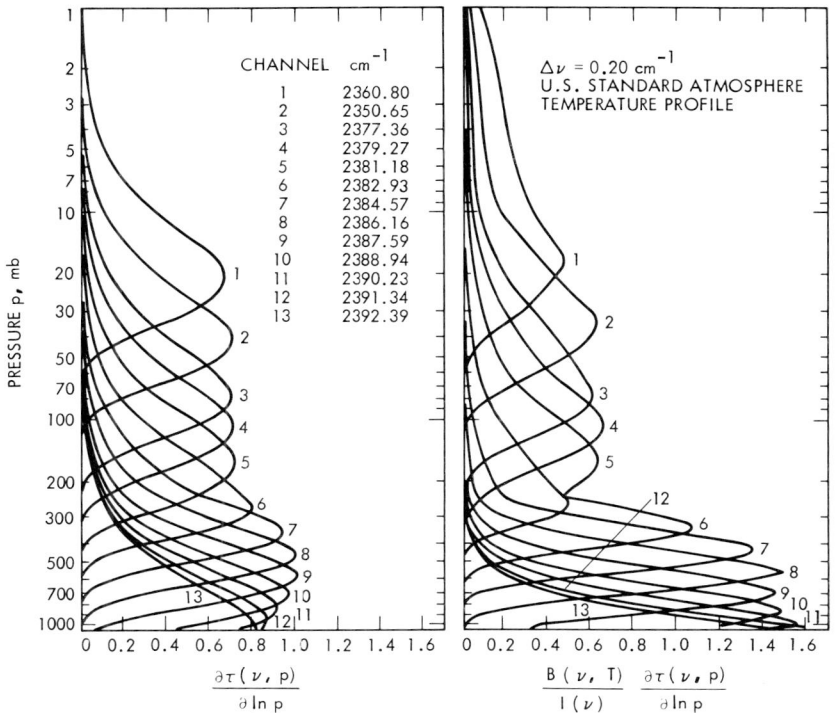

Fig. 1. 4.2 μm weighting functions and normalized integrand for $\Delta\nu = 0.2$ cm^{-1}. (1 bar = 100 kPa.) (From Ref. 1.)

as shown on the right-hand side, to obtain the source distribution of the emerging photons. In the stratosphere, multiplication by B has a broadening effect, however, which illustrates the point made earlier by Dr. Chahine that less information comes from the regions of minimum temperature, because the signal is mixed with effective noise on either side.

But, in the troposphere, both sets of curves are very narrow indeed, and we have many more channels than have been usable for sounding up to this time. The ratio of the pressures at the half-peak points of the weighting functions, which should be about 3-1/2 according to our previous discussion, turns out, in fact, to be about 2-1/2! The reason for this is that there is a further weighting besides the p^2 factor in the wings of the lines, namely, that due to the Boltzmann factor, which enters linearly into the wing absorption coefficients. We had assumed in arriving at Eqs. (4) and (5) that the line intensities were independent of temperature. They are not, however, and the Boltzmann factor has the effect of further narrowing the weighting functions of the channels near the high-J lines, which peak in the middle and lower troposphere. The effective power n of p for the middle troposphere channels, which appears in Eqs. (6), (7) and (8), is about three. For the lowest weighting functions, it is closer to two since the temperature independent induced N_2 absorption and far wings of the strongest CO_2 lines dominate.

These weighting functions were calculated at Goddard Institute for Space Studies by using the Susskind line-by-line transmission program with an assumed resolution of 0.2 cm^{-1}, which is very difficult to accomplish in practice. We, therefore, decided to see how much narrowing of the weighting functions would result, at not so high a resolution, from the temperature effect combined with the p^2 effect of the large amount of background emission from the collision broadened N_2 fundamental.

Figure 2 shows the results of calculations for an optimized set of 2 cm^{-1} channels. At this spectral resolution, each channel includes three to five lines. The vertical resolvability of the temperature profile is definitely not as good, but it is not really bad either. The ratios of the pressures at half-peak levels of the weighting functions are generally of the order 3 in the troposphere,[1] and the effect of the black body factor results in a restriction of the information content to a very narrow region for each of the tropospheric channels.

---

[1] The sharp decay of the lower parts of the weighting functions results in part from the regular spacing of the lines, which maximizes the rate of decrease of transmittance with depth.

TEMPERATURE SENSING

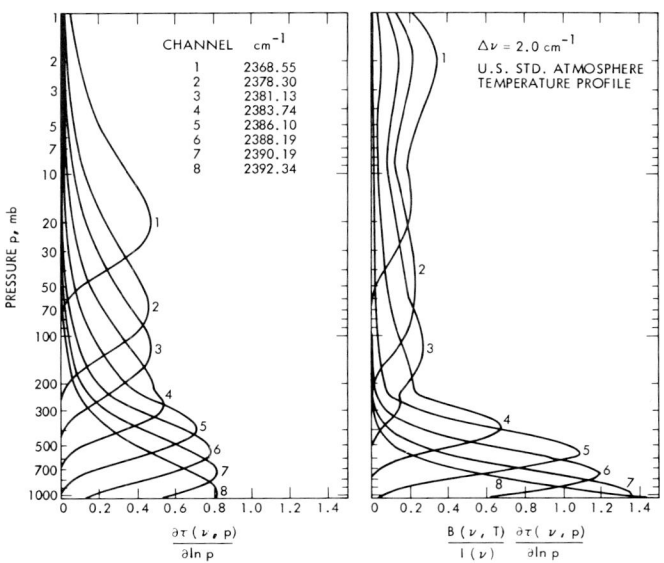

Fig. 2. 4.2 μm weighting functions and normalized integrand for $\Delta \nu = 2$ cm^{-1}. (1 bar = 100 kPa.) (From Ref. 1.)

Another effect of the B factor at this decreased spectral resolution, however, is an unacceptable smearing of the stratospheric information content curves on the right-hand side of the figure. This can be seen to be the effect of the B factor by comparison with Fig. 3, which presents similar curves for 0.5 cm^{-1} channels in the 15-μm band of CO_2. The 15-μm channels are definitely better for sounding the stratosphere.

An obvious solution to the sounding problem is to graft together, like Dr. van de Hulst's apples, the 4.2-μm channels for the troposphere, where they give better vertical resolution, with the 15-μm channels for the stratosphere, where B-smearing is minimal. It is not too difficult to make an instrument like this and Dr. Chahine is, in fact, starting a design study for such a

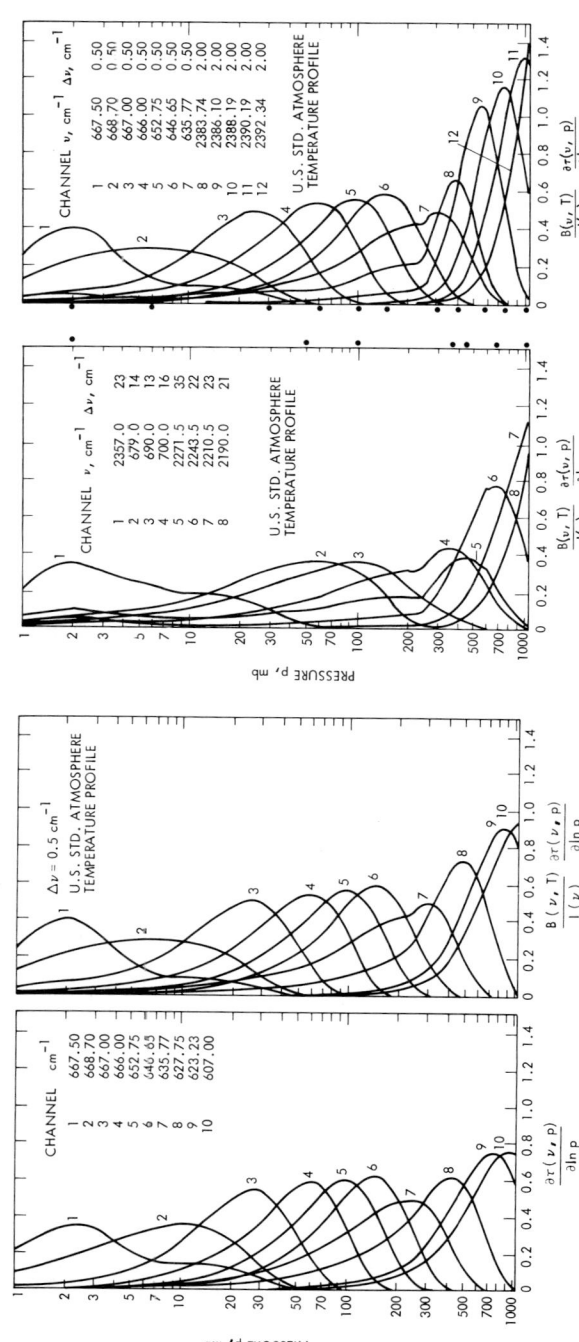

Fig. 3. 15-μm weighting fucntions and normalized integrand for $\Delta\nu = 0.5$ cm^{-1}. (1 bar = 100 kPa.) (From Ref. 1.)

Fig. 4. Normalized integrand for NIMBUS 6 HIRS channels on the left, and combined channels from Figs. 2 and 3 on the right. (1 bar = 100 kPa.) (From Ref. 1.)

sounder. Figure 4 contains curves of $Bd\tau/d\ln p$ for such a grafted sounder on the right. It is compared with similar curves on the left for the channels of the NIMBUS 6 High Resolution Infrared Radiation Sounder (HIRS) that Dr. Susskind believes to give non-redundant information, and which he uses for his retrievals.

This comparison shows the improvement in discretization of the information content of sampled emission spectra that can be obtained by careful selection of channels. It illustrates that proper planning of the measurements themselves may resolve some of the problems that we have devoted so much time to during this Workshop.

SYMBOLS

a_j	a coefficient equal to $\ln \tau_{\nu j}/p$
b_j	a coefficient equal to $\ln \tau_{\nu j}/p^2$
B	blackbody radiance
B_ν	blackbody radiance at ν and T
$B_{\nu G}$	value of B_ν at underlying surface
c_j	a coefficient equal to $\ln \tau_{\nu j}/p^n$
I_ν	radiation flux at ν
p	pressure
p_{max}	$1/c_j$
R	ratio of weighting function at p to its value at p_{max}
T	temperature
w_{ji}	equivalent widths of random lines in $\Delta\nu_j$
ν	radiation frequency
$\Delta\nu$	bandpass width
$\Delta\nu_j$	bandpass width at frequency j
τ	transmittance
τ_G	value of τ_ν from underlying surface to satellite
τ_ν	transmittance at ν
$\tau_{\nu j}$	transmittance in a bandpass $\Delta\nu_j$ from pressure level p to satellite height

ACKNOWLEDGMENT

The work reported in this paper was supported, in part, by NSF Grant ATM 72-01381.

REFERENCE

1. L. D. Kaplan, M. T. Chahine, J. Susskind, and J. E. Searl, Spectral band passes for a high precision satellite sounder, *Appl. Opt. 16,* 322 (1977).

DISCUSSION

Pepin: I would like to ask about the atmospheric contaminants that are in the four micron region. Have you looked at the effects of some of the contaminants?

Kaplan: In this part of the spectrum there aren't any. There are a few very weak water lines all the way down on the tails of the bands. There may be a few nitrous oxide lines but they are very weak transitions; they would be among the isotope bands. The main contaminant is actually a help. It is the pressure broadened nitrogen fundamental. It isn't even contaminated by carbon dioxide lines in this region. The extra lines--the hot bands and the isotope bands--are almost all in the P-branch.

Rodgers: I know it is very attractive to think that the number of photons is proportional to the amount of information you've got, but it isn't true. It's the rate of change of the number of photons with respect to the thing you are interested in.

Kaplan: Yes, okay.

Rodgers: So you should have dB/dT in there.

Kaplan: No, you shouldn't have dB/dT in there. What usually is done is to write $(dB/dT)(dT/d\ell np)$ for what you really want, which is $(d/dT) \int B(d\tau/d\ell np) d\ell np$. And that is all right as long as the transmission is independent of temperature. What we are talking about here are very strong temperature dependent transmissions. So one should actually do the derivatives. Dr. Susskind has done the derivatives and they turn out to be very much the same as what I have shown.

Rodgers: And the other point is that there is no reason why you shouldn't use selective absorption to improve the shape of the weighting functions. This is almost as good as going to very high resolution.

Kaplan: Well, one of the problems with selective absorption is that when you are making measurements over a broad band you are taking into account not only the wings of the strong lines that you are after, but the central portions of weaker lines that fall within this region. If you compare the SCR (Selective Chopper Radiometer) weighting functions with the weighting functions that we have gotten, you will see this effect coming in because the SCR doesn't really get what one would expect from the p^2 effect. I attribute it, at least in part, to the weak lines.

Rodgers: Have you tried computing SCR-type weighting functions for your case to, say, five wave number intervals?

Kaplan: Yes, we did.

Wark: There happens to be another little factor here called aerosols and these are the main contaminants that you have. There are clouds. And for one thing one must first solve for the amount or the effects of clouds, because these weighting functions are grossly distorted by the influence of clouds. Furthermore, in this particular spectral region those very clouds also reflect sunlight very severely. Would you please comment on these two things?

Kaplan: Yes, I am very glad you brought up the subject. First of all, I really promised in the original abstract to talk about the combination of the bands for the various things. One of the advantages of making measurements using an array of detectors as was used on the SIRS instrument, as you know, is that one can make measurements in different orders. It turns out that if you look at the 4 micrometer band in fourth order, you are looking at a 15 micrometer band in first order, and that is why I have a half a wave number rather than two wave numbers for that 15 μm band. You are looking in the window region in second order and 6 micrometer water band in third order. Probably the most effective way, I think, of trying to eliminate the effects of clouds--we usually have partial cloud cover--is to make temperature sounding measurements, not necessarily complete soundings, throughout the whole region in both the 4 micrometer and the 15 μm band as has been outlined by Dr. Chahine. Also to take advantage of the changes of the black body function and their differences in the two bands, and it gives us much more precise information about the clouds. And getting this high resolution in the 4 micrometer band means that you will be able to specify it even more. I think that being able to sound in the presence of clouds is to us, in a way, the most important reason for trying to do the analysis in this direction. With regard to the scattering, it is a problem. But it is possible again to make measurements of sunlight outside of the band. This is being planned. But the other thing that should be remembered is that when you see the scattered sunlight it is going through two traversals. You are usually looking straight down, so you usually have an air mass that is somewhat larger than two-- relative to the vertical soundings. And this does help a great deal to eliminate the effect; however, it probably still remains there, particularly for high clouds. It is necessary to look at reflected radiation at several wavelengths to try to at least get the outline of the sunlight and hope that when you extrapolate it into the R-branch of the four micron band that it keeps the same properties.

TEMPERATURE SENSING

Barkstrom: It seems to me that there is, perhaps, a general question that you have raised which is related to the question of how the measurements are to be used. The question you seem to be asking is, what does the meteorological community need in terms of accuracy on temperature structure and what is the resolution that is needed in order to derive a model? And it seems that some of the things Dr. Rodgers mentioned earlier regarding the uncertainty in a statistical sense could, perhaps, be tied into the things that you are saying here. And then the question is raised, what do we need to know from the meteorological community in order to do the job? Let me make this a little more quantitative. If I remember Dr. Rodgers' presentation, it seems that there is a relation between the uncertainty in the measurement and the error in the satellite measurements. The question would be whether we could get estimates of the uncertainty that is needed from the meteorological community which could then be inverted to estimates of the error and the resolution needed in the measurements?

Kaplan: Yes, first of all I am a meteorologist and that's why I am doing this. I feel that I need this data. The uncertainties themselves really are related to the resolution. For example, in the discussion of Chahine's relaxation method--when it is used in connection with the HIRS data or other data, it is an *a priori* method in principle. But it turns out not to be an *a priori* method in practice when the levels you are sounding are widely removed because you do have to make assumptions about how the atmosphere is behaving all along the way. The atmospheric temperature does not change *very* much with height and if you can make a measurement, say, every kilometer, then it would be very difficult to obtain a temperature distribution which, in connection with numerical weather prediction, can turn out to be very much different from what you want. It also turns out, contrary to what the dynamic meteorologists had believed, that in certain parts of the atmosphere, particularly in the tropics, small changes in static stability of the upper troposphere makes a difference in the tropical circulation. And these small changes are particularly what we are looking for. If you constrain by introducing statistics, you are throwing out information, you are keeping yourself from gaining information that is the most interesting part of the meteorological information you want. Also, if we are looking for climatic change, and we feel that there has been a change, due to the introduction of aerosols, and the changes in ozone distribution and photochemistry, we cannot use past data in order to interpret. And if we try to make models based upon temperature soundings in order to use the statistics we would not have been able to reproduce, for example, the last month and a half's soundings because the temperature distribution during that period would not have ever really existed in the 90-year history of radiosonde soundings. It was at least 94 years since we had weather conditions east of the Rockies as we have now. These anomalous conditions are what we are mostly after.

TEMPERATURE SENSING

OPEN DISCUSSIONS II

van de Hulst: We have gradually entered into the open discussion and I think Dr. Barkstrom put the very appropriate question for this entire conference. I have some comment on my own on that, but they are not very expert. Does somebody else want to comment?

Susskind: I want to make a comment that is not very expert at all. It is just a feeling that I have about real measurements and virtual measurements and something like this. I never thought about virtual measurements before, but what we are looking for in soundings... The point I want to make is I feel there is certainly information content in the observations, the real measurements, but the information content is not sufficient to give you everything you are looking for. As a matter of fact, if you are looking for a detail temperature profile, it is not sufficient to give that at all. And to do this you have to make some kind of assumptions, but to start saying that assumptions that are based on statistics or some feeling of what you think is happening is a real measurement, I think that one might lose sight that you were not really measuring something. Let us say you think there is a tropopause at 100 millibars and you force this into your solution when you have not got it in your vertical resolution, you might come out feeling convinced that I have found the tropopause at 100 millibars when, in fact, that maybe it is not there. And I think you have to bear in mind limitations that this really is not there and you should not convey the impression to someone else, say, who actually does not know exactly what you are doing, that you are actually determining something that really is not there. It is just a feeling I have about this. I would like someone to make a comment about whether they think they are really determining something that is there or that maybe is likely to be there or what?

Strand: With regard to these virtual measurements, you put them in with the correct expression of how much faith you have in them. If you put them in with grossly incorrect--for instance, there is something that deserves a large covariance matrix and you put in a small one, then you can expect trouble. But normally on these things we put in covariances that are commensurate with the degree of our knowledge of whatever we put in for a virtual measurement. I am inclined to agree with Dr. Rodgers that these are every bit as valid as measurements made with some kind of machinery. Just because some kind of machine measures it does not mean it is any more or less sacred than something that you put in from your statistical knowledge.

Rodgers: Well, I was going to say that I agree with Dr. Susskind. It is exactly the sort of thing I meant when I said you have got to make quite sure any virtual measurements you put in are right.

Chahine: How?

Rodgers: That is entirely up to you.

Staelin: Let me attempt to format the question a litle differently. I would say that one of the biggest unsolved problems in the statistical approach, which I think is the proper approach, is how best to select the *a priori* statistics. If the atmospheric statistics were stationary, then it would be an easy matter. An appropriate set of statistics would simply involve a sufficiently large ensemble to make the errors arbitrarily small. The problem arises because the atmospheric statistics are nonstationary in space and in time. The means and standard deviations are both very slowly varying functions. It seems incumbent on people who wish to exploit these techniques further to devise procedures other than the *ad hoc* ones that we are presently using. For example, for the microwave data, we now simply pick an *a priori* ensemble that is overly large. We use the entire globe. We would prefer to take small regions but we once erred by overspecializing the statistics. We broke the globe into seven climatic regions and different seasons. Although we incorporated enough years to get adequate data, even then the statistics were too narrow; they were overconstrained and introduced too much noise. Now somewhere between those two extremes--being overly general or too restrictive-- there must be some optimum region of operation. If one could close the definition of the problem, one ought to be able to arrive in principle at what is more nearly an optimum solution.

King: I think we must be careful not to subvert the language. I think it is incorrect to speak of using virtual measurements as putting information into the inversion. I think that one should say instead that virtual measurements put constraint into the inversion. The only information you have is the radiances.

Chahine: What I want to say is the following: virtual data are not and cannot be a substitute for real physical measurements. We have to do our best to get the most accurate measurements for the best information content to solve the problem. If the data are incomplete, then we have to do something to improve the solution. But first we have to look for the best available measurements, best physics, best real information.

Twomey: I would like to make a point. The distinction is not always as clear as some of the statements just made might make it appear. I think the tropopause is a good example. We have problems, obviously, getting by any straightforward inversion technique; we have trouble getting the tropopause because we have to put in a smoothing operation somewhere. Nevertheless, we know almost always or very often that there is a tropopause there. In principle, one could say let us solve the whole sequence of

TEMPERATURE SENSING

inversion problems in which we define the height and temperature of the tropopause and then get the best solution we can, some constrained solution. This would take a long time. I am just pointing out that it is possible in principle. Now whether this procedure is reasonable or not depends on the error of the measurements and how small we can make our residuals for each combination. In other words we would have to do a search through all possible tropopause levels and temperatures, but this would remove the conflict between the smoothing constraint, which we require for stability, and the existence of the tropopause.

van de Hulst: I came to this meeting with a very firm prejudice and the prejudice is that instead of inversion one should always do modeling. The reason why I thought so is that the inversion problem is similar to buying a suit. You go to a store and, at least in Europe, they sometimes have different floors. On one floor is a tailor. He takes about 40 different measurements. From those measurements finally after a week or two, he comes up with a suit that does not quite fit. And so he takes another couple of measurements and then finally it fits perfectly; it is also expensive. The alternative is that you address someone on the other floor and say: my size is 54, and they look at your belly to see if it is extra big or not and those two parameters together more or less define where on the rack you will find something which suits you. Now, if, in fact, you try to pursue this analogy, in which both the number of parameters and the cost is somewhat representative, then you find that the trade-off is not as simple as you might think. Both methods are wasteful. If you go along the rack, which has been called a catalog or a library in these different talks, then there is waste because extra material is used to manufacture also suits that nobody will eventually buy. The other method is wasteful because you take a certain number of measurements which are redundant. Finally, the customer orientation, i.e., the needs of the public and the responsible bodies in government who have to put up the money will determine where the actual trade-off is. I think in that respect this particular meeting has been quite helpful in helping all of us to get a feel of where the optimum point in this trade-off might be.

INDEX

A

Aerosol(s)
　altitude distribution of, 283, 545-547, 515-520
　radiative effects of, 568
　refractive index of, *see* Refractive index
　size distribution of, *see* Size distribution
　stratospheric, 294, 505, 553
　tropospheric, 553, 563
　volume extinction coefficient of, 244, 544, *see also* Coefficient
　volume scattering function of, 271
Accuracy of solution, *see* Solution
Albedo, 4, 10
　ground, 313, 559
　single scattering, 14, 15, 22, 313, 574
Atmosphere
　homogeneous, 26
　inhomogeneous, 33
　Jovian planets, 177-180
　Martian, 175-176
　plane parallel, 21
　realistic, 33
　semi-infinite, 4
　spherical, 34, 40, 274, 433, 509
Atmospheric model, onion skin, 540
Atmospheric optics, 297-299
Atmospheric state, 218, 436
　inverted, 233
　optimal estimate of, 222, 232
　prior estimate of, 222, 232
　true, 238
Attenuation
　cloud, 400
　oxygen, 398
　rain, 400
　water vapor, 397

B

Bias, effects of, 46
Brightness temperatures, 364, 398, 426
　bias in, 426
　clear, 402
　cloudy, 402
　fluctuations in, 405-406
Backus-Gilbert, *see* Inversion methods and solutions

C

Clouds
　amount of, 102
　effects of, 419-420, 610
　finite, 21
　height of, 102
　multiple, 103
　multiple layers of black, 106
　remote sounding in the presence of, 96-107
　single layer of black, 100
　with spectrally unknown characteristics, 97
Codon(s), 140-146
　H-function, 140
　transfer theory of, 141-146
Coefficient
　absorption, 22, 580
　backscatter, 295
　emissivity, 214
　extinction, 22, 33, 240, 270, 295, 507, 544
　scattering, 22, 270

spectral extinction, 443
volume extinction coefficient, 544
volume scattering, 270
Composition, atmospheric
remote sounding of, 377-383
Convergence, 8, 9, 230
factor, 352
properties, 337
Covariance matrix, 122-124, 221
cross, 52
error, 124, 158
full, 594
partial, 594
profile, 347
statistical, 125
Cost function, 221

D

Depolarization factor, 270
gaseous, 436
Direct-to-diffuse ratio, 323
sensitivity of, 323
spectral irradiance of, 308
Duality, 390
principle, 326-327
virtual, 326
true, 356
Dual(s)
closeness of, 340, 342
convergence for iterative, 352
equations, 344-346
iterative, 325
virtual iterative, 331

E

Eigenvalue(s)
closeness of, 340-342
decomposition of, 332-334
virtual equality of, 334-337
Equation(s)
Dirac, 148, 150
Fredholm first kind, 326, 446-450
Fredholm second kind, 450
linear transfer, 67
nonlinear transfer, 67
pseudoscalar transfer, 139
radiative transfer 21, 33, 67, 432, 434
remote sensing, 431-437
Schrödinger, 150

Error(s)
expected value of, 487
linear, 122
magnification of, 52
nonlinear, 126
norm of, 339
norm of radiance, 351
profile of, 183
quadrature, 86
random, 86, 115
Error analysis
linear, 122-126
nonlinear, 126-127
sensitivity of, 514
statistics of, 261
systematic, 91
Existence of solution, *see* Solution
Extinction law
Lambert-Beer, 509
Extinction ratios
insensitivity of spectral, 457
method of spectral, 453

F

Filters
Kalman, 132, 223
Kalman-Bucy, 376
Flame model, 324
Function(s)
cost, 221
delta, 28
empirical orthogonal, 39, 45, 337, 347
H-, 4, 25
kernel, 42
Planck, 168
reflection, 24
sought, 42
source, 10, 22, 24
transmission, 24
volume phase, 271, 277, 303, 435
volume scattering, 270-271, 277
Wood-Saxon, 310
X-, 3, 25
Y-, 3, 25

G

Gain, 2
Kalman, 223
matrix, 223
Gauss-Markov theorem, 221

INDEX

I

Information, *a priori*, 77, 78, 119, 614
Invariance, principles of, 25
Inversion method(s), 429
  assessment of, 56-60
  Backus-Gilbert, 49-50, 155-162, 583, 585-586
  Cholesky decomposition, 450
  classification of nonlinear, 450-451
  commonality of, 429
  constrained linear inversion, 46-47, 49, 505, 510
  direct, 203-207
  differences of, 429
  disadvantages of, 60
  double, 538
  Gautier-Revah, 156, 181-182
  invariant imbedding, 450
  interpolation, 77-78
  inverse matrix methods, 325, 346-347
  iterative, 195, 201-202, 325, 505, 511
  iterative Crone, 347
  iterative Twomey H-matrix, 346-347
  Landweber, 347
  least-squares, 450
  linear, 325-351
  linearized Chahine, 347, 356
  linearized statistical, 208
  Marquardt algorithm, 450, 512
  minimum information, 583, 596
  merits of, 60
  Moore-Penrose pseudo-inverse, 327
  modified Gram-Schmidt orthogonalization, 450
  modified Twomey, 469
  Newtonian iteration, 450
  nonlinear, 195, 228-231, 450-451
  nonlinear iterative, 53-56
  nonlinear matrix, 139
  onion skin peel, 214, 226, 538
  partial derivative, 580, 583
  pressure increment, 581, 583
  quasi-optimum, 582-583, 588
  regularized linear, 345
  relaxation, 67, 450
  resolution of, 58
  search, 450
  sequential estimation in, 127-132
  smoothing constraints in, 357
  spectral expansion, 45, 450
  stepwise regression, 450
  statistical, 51-53, 117-121, 207-211, 402
  synthesis, 47-50
  Twomey-Tichenov, 124
  truncated singular value decomposition, 450
Inversion problem
  grossly nonlinear, 122
  limb scan, 196, 220
  linear, 121
  moderately nonlinear, 122
  nearly linear, 122
  well-conditioned, 359
Inversion systems, square, 347
Inversion theory, statistical principles of, 117, 395
Irradiance, solar, 23, 308
Iteration, 224

K

Kernel(s), 24, 157
  averaging, 157, 183
  Elsasser band, 44
  exponential absorption, 44
  Fourier transform, 43
  Mie scattering, 43-44
  overlapping, 115
  property of, 67-68
  smooth, band-limited, 45

L

Laplace transform, 14
Lapse-rate, super adiabatic, 41
Legendre polynomials, 28
Limb-darkening, solar, 513
Limb emission, 195
Limb scans, 196, 222, 371

M

Markov
  processes, 376
  theorem of Gauss, 221
Marquardt algorithm, 450, 512
Matrix(ices), *see also* Covariance matrix
  gain, 223
  ill-conditioned, 328, 357
  phase, 5, 22, 435
  radiance, 22
  smoothing, 544
Measurements
  stratospheric aerosol, 529
  BUV, 155, 310, 577
  Dobson, 320

Diffuse-to direct ratio, 307-308
GATE project, 569
lidar, 469
limb radiance, 199-201
limb scan, 217, see also Limb scan
multispectral radiometer, 406, 410
polarimeter, 561-568
selective absorption, 609
solar aureole, 265, 300
solar extinction, 469, 505-520, 529-548
Umkehr, 321
virtual, 120, 599
VTPR, 114
Measurement technique/system
  BUV, 155, 310
  "Dustsonde" balloon borne, 532
  HIRS, 6, 607
  lidar, 471-486, 533
  limb scanning systems, 371
  LRIR, 199
  passive radiometer, 195
  photographic, 272-274, 301
  polarimeter, 555, 565
  pressure sounder, 394
  SAM camera, 534
  SAM photometer, 529, 533
  SAPE, 300
  SCAMS radiometer, 366, 413
  selective chopper radiometer, 129, 601
  solar aureole, 313
  solar radiometer, 486, 490
  Umkehr, 320
  VTPR, 330
Mie
  code, 227
  efficiency factor, 443
Milne problem, 26
Model
  analytic, 297
  atmospheric, 540
  flame, 324
  nuclear shell, 299
  onion skin, 540
Multiple scattering, 1, 293-294, 319, 579, see also Radiative transfer methods, multiple scattering
  code for, 560
  solar aureole, 295-305

N

Noise, 234-255
  instrumental, 326

Nonlinearity
  degree of, 121-126
  sources of, 440

O

Onion skin peel method, 214, 226, 538
Optical depth, 4-5, 22, 26, 269-270
  error bounds, 564
Orthogonality relations, 6
Ozone, see also Profile
  amount of, 527
  low altitude, 308-314
  mixing ratio, 210, 215, 597-598
  profile retrieval for, 163-167, 519, 577-593
  solution, 163, 165, 238
  tropospheric, 308
Optics
  atmospheric, 297-299
  nuclear, 297-300

P

Phase function, 22, 24, 227, 271, 303, 435
Phase matrix, 5, 22, 435
Photon path distribution, 14
Photon path length, 15
Polarization, 5, 12, 14, 23, 268, 293, 433, 472, 560
  degree of, 13
Pressure, atmospheric, 383-385
Profile(s)
  absorbing gases mixing ratio, 92, 96, 204
  aerosol, see Remote sounding, aerosols
  CO, 364
  CO_2, 204, 324
  constraint, 594
  gaseous concentration, 217
  humidity, 410-420, see also Profile, water vapor
  ozone, 163-167, 210, 215, 308-314, 519, 527, 577-593
  pressure, 406
  probability density function of, 118
  Rayleigh scatterers, 219, 553
  refractive index, 293
  scattered radiance horizon, 217
  trace constituent, 195, 198
  temperature, 78-92, 155, 162-183, 402-410, 419, 437-441, 451-452, 599-607
  water vapor, 97, 364, 379, 395, 410-420

INDEX

R

Radiation
  middle UV, 309
  upwelling, 565
Radiative transfer methods, *see also* Multiple scattering
  adding, 26
  analytical, 25
  approximation, 30
  asymptotic, 29
  asymptotic fitting, 7
  Case, 25
  DART, 26, 221
  Delta-Eddington, 31-32
  discrete ordinates, 26-33
  double Delta-Eddington, 31-32
  doubling method, 9, 26
  Eddington method, 31
  exponential kernel, 31
  Gauss-Seidel, 26
  Hartel, 39
  invariant imbedding, 6, 26
  matrix operator, 26-27
  modified two-stream, 31
  Monte-Carlo, 26, 34, 227
  singular eigenfunction, 6, 25
  spherical harmonics, 26, 33
  standard two-stream, 31
  successive scattering, 7, 9, 26
  two-stream, 31-32
  Weiner-Hopf, 25
Reflection
  diffuse, 4
  functions, 24
  ground, 23, 565, 571
  Lambert, 565, 570
  surface, 364
Refractive index
  aerosol, 268, 453-458, 559, 564
  error bounds on, 564
  imaginary part of, 232, 570
  real part of, 232, 570
Relaxation method, *see* Inversion methods
Remote sounding
  aerosol, 217-257, 265-287, 297-314, 441-446, 451-460, 505-520, 529-548, 555-571
  aerosol size distribution, *see* Remote sounding, aerosols, and Size distribution
  aerosol refractive index, *see* Remote sounding, aerosols, and Refractive index
  composition, atmospheric, 377
  horizon scattered radiance, 217
  humidity profile, 410-420, *see also* Profiles, water vapor
  limb scanning, 195, 220, 371
  microwave, 361-385, 395-421
  microwave, ground-based passive, 395, 421
  microwave, satellite passive, 361-385
  pressure, atmospheric, 383-385
  radar pressure, 394
  sunlit limb, 260
  techniques, *see* Measurement techniques
  temperature, atmospheric, 45, 78-92, 167-183, 366-376, 402-410, 437-441, 451, 599-607
Residuals
  properties of, 75-76, 92-95, 115
  root mean square, 93
  variance of, 124-125, 326, 403
Resolution, 50
  degree of, 189
  inversion method, 58
  length of, 162, 176
  optimum, 189
  vertical, 166, 183-184, 372
Risk function, 127
  quadratic, 126
Recursion, 222-224

S

Scale height, tropospheric, 320
Scanning function, 48, 59
  effective, 59
  width of, 50
Scattering
  aerosol, *see* Aerosol
  amplitude function of, 443
  anisotropic, 5
  corrections due to secondary, 587
  higher order, 12
  isotropic, 5, 8, 14
  linearly isotropic, 6
  microwave, from clouds, 363
  multiple, 1, 293-294, 305, 319, 560-579, *see also* Radiative transfer methods
  near-forward, 265, 445
  nth order, 14
  Rayleigh, 25
  Rayleigh-Cabannes, 436
  resonantly fluorescent line, 437
  secondary (2nd order), 579, 587
  single scattering (1st order), 5, 7, 268-306
  UV, 310, 579

INDEX

Size distribution(s)
  aerosols, 217-257, 266, 278, 302, 441-446, 451-460, 469-490, 558-571
  altitude, 266
  columnar, 486
  cumulative, 311
  Deirmendjian, models of, 29, 466, 558
  error bounds on, 564
  Junge, 219
  mean, 469
  multi-modal, 575
  oversize, 302
  two-parameter, 304
  unique, 559
Smoothing parameter, 330
Solar aureole, 265, 299
  almucantar, 266-267, 275
  isophotes, 267, 285
  photographic methods, 266-267, 313
  UV, 319
Solution, inversion
  acceptability of, 47
  accuracy of, 68, 83, 117, 420
  ambiguity of, 42
  Backus-Gilbert, 48-49, 56
  comparison of, 56-58
  construction of, 429, 431, 447, 449
  convergence of, 74
  constrained linear inversion, 56
  dual, 340-344
  existence of, 69, 429, 431, 447
  inverse matrix, 345, 352
  iterative, 73-74, 82, 327
  linear regularized, 351
  least-squares, 329-330
  minimum information, 330
  oscillations in, 510
  quality of, 404
  sensitivity of, 458
  stability of, 45, 51, 68, 74, 86, 429, 447, 431
  Twomey-Phillips, 330-332, 351
  uniqueness of, 69, 84, 429, 431, 447
  validity of, 584, 588
Stability, *see* Solution

Stokes parameters, 22, 433, 472-473
Surface emissivity effects, 393

T

Temperature
  blackbody function for, 602
  brightness, 364, 398, 402, 405, 426, *see also* Brightness temperatures
  Kinetic, 364
  receiver noise, 362
  source, 362
  surface, 393-394
Temperature profiles, inversion for, 78-92, 167-183, 366-376, 402-410, 437-441, 599-607
Thermal emission
  microwave, downweling, 395
  spectrum of atmosphere, 402
Transformation
  mapping, 68, 70-72
  relaxation, 70-72
Turbidity
  factor, 436
  spectral, 304

U

Uniqueness of solution, *see* Solution

W

Water vapor, *see* Profiles
Weighting function(s)
  CO_2, 336
  Fourier spectrum of, 449
  partial derivative, 582
  pressure increment, 582
  properties of, 439
  SCR, 609
  selective chopper radiometer, temperature, 405
  universal, 448
  water vapor, 379